MATTER & INTERACTIONS II
ELECTRIC & MAGNETIC INTERACTIONS

MATTER & INTERACTIONS II
ELECTRIC & MAGNETIC INTERACTIONS

RUTH W. CHABAY

BRUCE A. SHERWOOD

Carnegie Mellon University

JOHN WILEY & SONS, INC.

New York • Chichester • Brisbane • Toronto • Singapore • Weinheim

ACQUISITIONS EDITOR   Stuart Johnson

MARKETING MANAGER   Robert Smith

PRODUCTION EDITOR   Rebecca Rothaug

SENIOR DESIGNER   Maddy Lesure

This book was set by the authors, and printed and bound by Courier Westford.
The cover was printed by Lehigh Press.

To order books or for customer service please call 1 (800) 225-5945.
For instructor supplements visit http://www.wiley.com/college/chabay.

ISBN 0-471-44255-0

Printed in the United States of America

10 9 8 7 6 5 4 3 2 1

Preface to Volume II

Matter & Interactions II: Electric & Magnetic Interactions emphasizes the concepts of electric and magnetic fields and extends the study of the atomic structure of matter to include the role of electrons. This second volume continues to emphasize the fact that there are only a small number of fundamental principles that underlie the behavior of matter, and that using these powerful principles it is possible to construct models that can explain and predict a wide variety of physical phenomena.

Prerequisites

This volume is intended for introductory college physics courses taken by science and engineering students. It is assumed that you have had a college course which included classical mechanics (Newton's laws of motion) and that you can use vectors and vector components. The math needed is a basic knowledge of derivatives and integrals.

Desktop experiments

In order to use this book you need an experiment kit. Integrated with discussions of theory are critical "desktop" experiments to be carried out using simple equipment in a regular classroom, a lab, or even at home. It is important that you apply the same care and attention to carrying out the desktop experiments that you apply to learning the theory, since the theory and the experiments reinforce each other. You will find that by using very simple equipment you can gain insight into rather deep scientific issues.

? Stop and Think

As you read the text, you will frequently come to a paragraph that asks you to stop and think by making a prediction, carrying out a step in a derivation or analysis, or applying a principle. Usually these questions are answered in the following paragraphs, but it is important that you make a serious effort to answer the questions on your own before reading further. Be honest in comparing your answers to those in the text; paying attention to surprising or counterintuitive results can be a useful learning strategy.

Exercises

Small exercises that require you to apply new concepts are found at the end of many sections of the text. These may involve qualitative reasoning or simple calculations. You should work these exercises when you come to them, to consolidate your understanding of the material you have just read. Answers to the exercises are at the end of each chapter.

Worked-out example problems

Following the summary page of each chapter are one or more worked-out example problems. The purpose is to show you how to apply the fundamental principles in complex situations.

Conventions used in diagrams and equations

The conventions most commonly used to represent vectors and scalars in diagrams in this volume are shown in the adjacent figure. In equations and text, a vector will be written with an arrow above it:

$$\vec{p}$$

The magnitude of $\vec{p}$ is written as $|\vec{p}|$ or simply as p.
The x component of $\vec{p}$ is written as p_x.

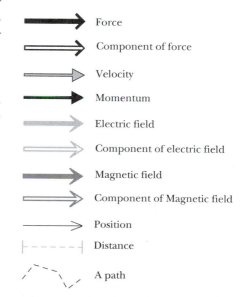

Force

Component of force

Velocity

Momentum

Electric field

Component of electric field

Magnetic field

Component of Magnetic field

Position

Distance

A path

A vector may be written in terms of components or in terms of unit vectors along the x, y, and z axes:

$$\vec{p} = <p_x, p_y, p_z> = p_x\hat{\imath} + p_y\hat{\jmath} + p_z\hat{k}$$

To the Instructor

The approach to electricity and magnetism taken in this volume differs significantly from that in most introductory physics texts. (If you used an earlier version of this E&M textbook, see the section "Earlier editions" below.) Key emphases of our approach to E&M are these:

- Starting from fundamental principles rather than secondary formulas
- Modeling the real world through idealizations and approximations
- Atomic-level description and analysis
- Qualitative reasoning that complements the quantitative reasoning
- Unification of electrostatics and circuits
- Tight integration of theory and experiment

The preceding one-semester introductory course on classical mechanics and statistical mechanics is addressed by *Matter & Interactions I: Modern Mechanics*, Ruth Chabay & Bruce Sherwood, Wiley 2002.

Textbook web site

To obtain useful supplements to our textbook, go to the Wiley web site:

http://www.wiley.com/college/chabay

There you will find a daily log of what we do in our own classrooms, sample assignment sheets, problem solutions, sample exams, etc. You will also find a link to the authors' web site, which includes additional information of a general character, including useful educational software.

Desktop experiment kit

Students need desktop experiment kits, which may be purchased directly from PASCO Scientific, www.pasco.com, 1-800-772-7200 (USA only) or 916-786-3800. This kit makes it possible for students to make key observations of electrostatic, circuit, and magnetic phenomena in the classroom, tightly integrated with the theory. The daily log available at Wiley's web site provides information on how to procure the components if you choose to assemble kits yourself.

At some institutions students buy kits at the bookstore. Another arrangement is for the physics department to order kits, distribute the kits to students in class, and charge a lab fee for the cost of the kit. Yet another option is for the department to loan kits to students and charge a breakage fee. Whatever scheme is used, the kits are an important part of the course.

This does not preclude having other, more complex laboratory experiences associated with the course. For historical reasons we have not had a regular laboratory component in our own course, but we do have the students sign up for two experiments that go beyond what can be done with the kit (writeups are available on the Wiley web site). One experiment adds modern multimeters to the kit, and the other deals with Faraday's law and requires signal generators, large coils, and oscilloscopes. You may have lab experiments already in place that will go well with our textbook.

Computer homework problems

Some homework problems are designed for the student to write a computer program. Students who have done the computer modeling in *Matter & Interactions I: Modern Mechanics* will probably enjoy the computer visualization and calculation problems in this volume, and will find that 3D computer visualization is very helpful in thinking about electric and magnetic fields in three dimensions. We strongly recommend VPython as the best tool for student use. An adequate subset of the underlying Python programming language can be taught in an hour or two, even to students who have never written a program before. Real-time 3D animations are generated as a side effect of student computations, and these animations provide powerfully motivating and instructive visualizations of fields and motions. VPython supports true vector computations, which encourages students to begin thinking about vectors as powerful tools rather than barriers to understanding. VPython can be obtained at no cost for Windows, Macintosh, and Linux through the authors' web site (accessible from the Wiley web site).

Homework problems

Given the emphases of this textbook, it is appropriate for homework to consist of a small number of rather large problems, rather than a large number of small, routine problems. In addition to the large homework problems provided at the end of a chapter, there are many small-scale problems in the form of exercises distributed throughout the chapter, and small review questions at the end of the chapter. You may wish to ask students to write out some exercises or review questions as part of the homework.

In-line problems

Some of the homework problems are central to the course and cannot be omitted without compromising the enterprise; an example is the in-line homework problem in Chapter 14 on the amount of charge on a tape. In-line homework problems should normally be assigned unless the associated topic is omitted.

Earlier editions

There were earlier editions of this volume on E&M, in 1995 (in a workbook format) and in 1999 (a preliminary second edition). If you taught from one of these earlier versions, we should alert you to a major rearrangement of topics. Magnetic field is introduced in the fifth chapter of this volume, in order to address problems we observed with its usual position late in the course. We found that during the many weeks of working with electric field before encountering magnetic field, students often did not fully appreciate the field concept, because they could mentally think in terms of the Coulomb force law for much of their reasoning. It was only when they began their study of magnetic field that the necessity of a two-part analysis drove home to them the true nature of the field concept (current makes field, field affects current).

Introducing magnetic field early also gives many more weeks of practice with magnetic field than is usually the case. It made it possible to discuss the relationship between electric field and magnetic field, a topic in which students expressed keen interest.

Finally, magnetic field is a somewhat difficult topic, in part because the applications are inherently three-dimensional. We found that introducing this topic late in the semester was problematic for students who were beginning to droop a bit.

Choosing topics

In our own teaching, with daily 50-minute periods (3 lectures and 2 small-group workshops every week), we are able to complete most of this E&M volume in a 15-week semester. You can read the organizational details in the daily log available at the Wiley web site.

What can be omitted if there is not enough time to do everything? Let us say that the one thing we feel should *not* be omitted is electromagnetic radiation and its effects on matter (Chapter 23). This is the climax of the whole E&M enterprise. One way to decide what can be omitted is to be guided by what needs to be done to ensure dealing adequately with radiation.

Any starred section (*) can safely be omitted. Material in these sections is not referenced in later work. Next we offer chapter by chapter discussions of what other topics might be omitted, and what the ramifications are of such omissions.

Chapter 13 (Electric Field): The field concept is the backbone of the course, so none of this brief chapter should be omitted. The first volume of our textbook (*Matter & Interactions I: Modern Mechanics*) gives students experience with Coulomb's law and with the atomic nature of matter. You may have to spend extra time reviewing these topics if your E&M students did not use our Volume I.

Chapter 14 (Matter and Electric Fields): The case study on sparks in air can be omitted, because nothing later depends critically on this topic. The one aspect that should be retained is the experimental fact that air becomes a conductor if a field of more than about 3×10^6 N/C is applied, as this gives a useful measure of what constitutes a "large" electric field. The disadvantage to omitting the spark topic is that it is an introductory-level example of a phenomenon where an intuitively appealing model fails utterly, while a different model predicts several key features of the phenomenon. The latter model is also a good example of a long chain of reasoning, each step of which is simple.

Chapter 15 (Electric Field of Distributed Charges): You might choose to downplay the derivations and concentrate entirely on the pattern of electric fields in space made by the various charge distributions, and the distance dependence of the fields. You need to decide how important it is that your students learn to be competent in setting up physical integrals.

We delay Gauss's law until Chapter 21, and have structured the textbook so that Gauss's law is not needed in the earlier chapters. If you prefer, you can do the Chapter 21 sections on Gauss's law immediately after Chapter 15 (however, see the notes on Chapter 21). In Chapter 15 the students obtain experience with patterns of field around several standard charge distributions, and this experience is necessary background for Gauss's law. Most textbooks attempt to teach Gauss's law during the first week of the E&M course, and this attempt fails for most students, who in the first week don't yet have enough experience with the relationship between charge and field, and with patterns of field in space, to grasp the subtle topological nature of Gauss's law.

Chapter 16 (Electric Potential): You can safely omit the section on dielectric constant, as long as you do not assign homework problems that use this concept. Our volume I gives students experience with $1/r$ potential energy, both electric and gravitational. If your E&M students did not use our Volume I, you may have to spend additional time on potential energy before introducing potential.

Chapter 17 (Magnetic Field): In the sections on the atomic structure of magnets, you might choose to discuss only the first part, in which we find that the magnetic moment of the bar magnet is consistent with an atomic model. Omitting the remaining sections on spin and domains will not cause significant difficulties later.

If you wish, you can discuss the Ampere's law sections of Chapter 21 immediately after Chapter 17.

Chapter 18 (A Microscopic View of Electric Circuits): It is best to do everything, except that you might choose not to do all of the final applications.

Chapter 19 (Capacitors, Resistors, and Batteries): We recommend doing everything, though you might skimp a bit on meters. Note that multiloop resistive circuits are relegated to a starred section. We have come to see this as mere bookkeeping with little real physics content, and we ourselves omit this topic (and do not assign homework problems of this kind). Physics and engineering students who need to analyze complex multiloop circuits will take specialized courses on the topic, and it seems inappropriate to spend time on this in the introductory course, where the emphasis should be on giving all students a good grounding in the fundamental mechanisms underlying circuit behavior.

In contrast, the capacitor circuits in Chapter 19 provide an excellent review of many aspects of electric interactions, and we strongly recommend that this topic not be omitted.

Chapter 20 (Magnetic Force): We recommend discussing Jack and Jill and Einstein, but it is safe to omit the sections on relativistic field transformations. However, students often express high interest in the relationship between electric fields and magnetic fields, and here is an opportunity to satisfy some aspects of their curiosity. While the sections on motors and generators deal with interesting and practically important applications, these sections are starred in keeping with the emphasis on fundamentals.

Chapter 21 (Patterns of Field in Space): We recommend not omitting anything (other than the starred sections, of course). Gauss's law and Ampere's law are important foundations for the following chapters on Faraday's law and electromagnetic radiation. It is possible to do Gauss's law just after Chapter 15, and Ampere's law just after Chapter 17. However, we have found that studying Gauss's law just before Faraday's law makes the concept of flux fresh and unproblematic in Chapter 22.

Chapter 22 (Faraday's Law): Though it can safely be omitted, we recommend retaining the section on superconductors, because students are curious about this topic. Note that we have downgraded Lenz's rule and instead emphasize the pattern of non-Coulomb electric field in space. We found that students tend to try to use Lenz's rule quantitatively rather than qualitatively, in the incorrect form that "a magnetic field is made which cancels the change," a version that is valid only for superconductors. The starred sections on RL and LC circuits are not essential for what comes later, though you may wish to include this material.

Chapter 23 (Electromagnetic Radiation): This is the climax toward which much of the course has been heading. We strongly recommend omitting nothing (other than the starred section).

Chapter 24 (Waves and Particles): This chapter contrasts the classical wave model of light developed in Chapter 23 with the particle model that was used in Volume I, and attempts to show the relationship between these two very different models. It also develops key concepts in physical optics. In our own teaching we rarely have time for the sections on standing waves.

Chapter 25 (Semiconductor Devices): This whole chapter is optional. The topic is intrinsically interesting and provides a nice review of a number of important topics, such as Gauss's law.

Acknowledgments for Volume II

We owe much to the unusual working environment provided by the Center for Innovation in Learning and the Department of Physics at Carnegie Mellon, which made it possible during the 1990's to carry out the research and development leading to our two-volume textbook. We are grateful for the open-minded attitude of our colleagues in the Carnegie Mellon physics department toward curriculum innovations.

We thank Hermann Haertel for opening our eyes to the fundamental mechanisms of electric circuits. Robert Morse, Priscilla Laws, and Mel Steinberg stimulated our thinking about desktop experiments. Bat-Sheva Eylon offered important guidance at an early stage. Ray Sorensen provided deep analytical critiques that influenced our thinking in several important areas. Randall Feenstra taught us about semiconductor junctions. Thomas Moore showed us how to think about divergence and curl. Fred Reif helped us devise an assessment of student learning of basic E&M concepts. Uri Ganiel suggested the high-voltage circuit used to demonstrate the reality of surface charge. The unusual light bulb circuits at the end of Chapter 22 are based on an article by P. C. Peters, "The role of induced emf's in simple circuits," *American Journal of Physics* **52**, 1984, 208 - 211.

We thank David Andersen and David Scherer for the development of tools that enabled us and our students to write associated software. We also thank Thomas Ferguson, Thomas Foster, Robert Hilborn, Leonardo Hsu, Barry Luokkala, Matthew Kohlmyer, Sara Majetich, Robert Nichol, Robert Swendsen, and Hugh Young.

This project was supported, in part, by the National Science Foundation (grants MDR-8953367, USE-9156105, DUE-9554843, and DUE-9972420). Opinions expressed are those of the authors, and not necessarily those of the Foundation.

Ruth W. Chabay
Bruce A. Sherwood
Center for Innovation in Learning and Department of Physics
Carnegie Mellon University
May 2001

Image credits

Barry Luokkala (Figure 24.18). Jeremy Henriksen (Figure 24.20).

All other figures were created by the authors, using Adobe Illustrator, Adobe Photoshop, VPython, POV-Ray, and Specular Infini-D.

The text was produced by the authors as camera-ready copy, using Adobe FrameMaker. The font is 10-point New Baskerville.

Volume I: Modern Mechanics

Chapter 13

Electric Field

Chapter 13

Electric Field

13.1 What is new in this volume?

There are two important new ideas that will form the core of our study of electric and magnetic interactions. The first is the concept of electric and magnetic fields. This concept is more abstract than the concept of force, which we used extensively in our study of modern mechanics. The reason we want to incorporate the idea of "field" into our models of the world is that this concept turns out to be a very powerful one, which allows us to explain and predict important phenomena which would otherwise be inaccessible to us.

The second important idea is a more sophisticated and complex model of matter. In our previous study of mechanics and thermal physics it was usually adequate to model a solid as an array of electrically neutral microscopic masses (atoms) connected by springs (chemical bonds). As we consider electric and magnetic interactions in more depth, we will find that we need to consider the individual charged particles—electrons and nuclei—that make up ordinary matter.

The material in this chapter is not difficult, but it is very important that you do the in-line exercises! It is important to practice using basic concepts before getting to complex and challenging problems.

13.2 Review

In this section we review briefly concepts familiar to you from your previous studies.

13.2.1 Point charges

There are two kinds of electric charge, which are called positive and negative. Particles with like charges repel each other (two positive or two negative particles); particles with unlike charges (positive and negative) attract each other. By "point particle," we mean an object whose radius is very small compared to the distance between it and all other objects of interest, so we can treat the object as if all its charge and mass were concentrated at a single mathematical point. Small particles such as protons and electrons can almost always be considered to be point particles.

13.2.2 The Coulomb force law for point particles

The force law called Coulomb's law describes the electric force between two point-like electrically charged particles:

$$\left| \vec{F} \right| = F = \frac{1}{4\pi\varepsilon_0} \frac{|Q_1 Q_2|}{r^2}$$

where Q_1 and Q_2 are the magnitudes of the electric charge of objects 1 and 2, and r is the distance between the objects. The electric force acts along a line between two point-like objects. Like charges repel; unlike charges attract. As is clear from the mathematical expression above, two charged objects interact even if they are some distance apart.

13.2.3 Units and constants

The SI unit of electric charge is the coulomb, abbreviated "C." The charge of one proton is 1.6×10^{-19} C in SI units. The constant e is often used to represent this amount of positive charge: $e = +1.6 \times 10^{-19}$ C. An electron has a charge of $-e$.

The constant $1/4\pi\varepsilon_0$ has the value $9 \times 10^9 \text{N} \cdot \text{m}^2/\text{C}^2$. We write this constant in this way, instead of using the letter k, for two reasons. First, since the letter k represents the Boltzmann constant, we avoid confusion. Second, the constant ε_0 will appear by itself in important situations. $\varepsilon_0 = 9 \times 10^{-12} \text{C}^2/\text{N} \cdot \text{m}^2$.

13.2.4 Charged particles

There are many microscopic particles, some of which are electrically charged, and hence interact with each other through the electric interaction. The characteristics of some charged particles are shown in the table below.

Particle	Mass	Charge	Radius
electron	9×10^{-31} kg	$-e$ (-1.6×10^{-19} C)	? (too small to measure)
positron	9×10^{-31} kg	$+e$ ($+1.6 \times 10^{-19}$ C)	?
proton	1.7×10^{-27} kg	$+e$	$\sim 10^{-15}$ m
antiproton	1.7×10^{-27} kg	$-e$	$\sim 10^{-15}$ m
muon	1.88×10^{-28} kg	$+e$ (μ^+) or $-e$ (μ^-)	?
pion (π^+ or π^-)	2.48×10^{-28} kg	$+e$ (π^+) or $-e$ (π^-)	$\sim 10^{-15}$ m

This is not an exhaustive list; you may also have learned about other charged particles, such as the W and Δ particles, and other particles which are short-lived and not commonly encountered in everyday circumstances.

Ordinary matter is composed of protons, electrons, and neutrons (which have about the same size and mass as protons, but are uncharged). However, some other charged particles do play a role in everyday processes, as we will see, for example, in our study of sparks in Chapter 14.

13.2.5 Size and structure of atoms

As you have learned in chemistry courses or from other sources, matter is made of tiny atoms. In a solid metal, atoms are arranged in a regular three-dimensional array, called a lattice (Figure 13.1). A cubic centimeter of solid metal in which atoms are packed right next to each other contains around 10^{24} atoms, which is an astronomically large number.

A neutral atom has equal numbers of protons and electrons. The protons and neutrons are all found in the nucleus at the center of the atom. The electrons are spread out in a cloud surrounding the nucleus.

Each atom consists of a cloud of electrons continually in motion around a central "nucleus" made of protons and neutrons. If we imagine taking a "snapshot" of an iron atom, with a nucleus of 26 protons and 30 neutrons surrounded by 26 moving electrons, it might look something like Figure 13.2. In this figure the nucleus is hardly visible, because it is much smaller than the electron cloud, whose radius is on the order of 10^{-10} m.

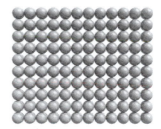

Figure 13.1 A cross-sectional "slice" through a metal lattice. The radius of a single atom is on the order of 10^{-10} m, and in a cube that is 1 cm on a side, there are on the order of 10^{24} atoms!

Figure 13.2 The electron cloud of an iron atom. The radius of the electron cloud is approximately 10^{-10} meter. On this scale the tiny nucleus, located at the center of the cloud, is hardly visible.

Figure 13.3 The nucleus of a iron atom contains 26 protons and 30 neutrons. Its radius is approximately 4×10^{-15} meter.

Figure 13.4 Acceleration of a proton.

Figure 13.5 Acceleration of another proton at a later time.

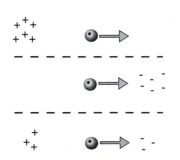

Figure 13.6 Three possible arrangements of charged particles that might be responsible for the acceleration of a proton.

Figure 13.7 An electron.

The nucleus of the iron atom, depicted in Figure 13.3, has a radius of roughly 4×10^{-15} meter, about 25000 times smaller than the tiny electron cloud. If an iron atom were the size of a football field, the nucleus would have a radius of only 4 millimeters! Yet almost all of the mass of an atom is in the nucleus, because the mass of a proton or neutron is about 2000 times larger than the mass of an electron.

13.3 The concept of "electric field"

Consider the following thought experiment. Having evacuated the air from the room (to avoid collisions with air molecules), you hold a proton in front of you and release it. There are no other objects near by. You observe that the proton begins to move downward, picking up speed at the rate of 9.8 m/s each second (Figure 13.4).

? What do you think is responsible for this change in the velocity of the proton?

You probably inferred that the gravitational interaction of the Earth and the proton caused the downward acceleration of the proton. This is a reasonable explanation for this observation.

Now suppose that at a later time, you release another proton in the same location. This time you observe that the proton begins to move to the right, picking up speed at a rate of 10^{11}m/s^2 (Figure 13.5).

? What might be responsible for this change in the velocity of the proton?

This acceleration cannot be due to a gravitational interaction of the proton with the Earth, since the magnitude of the effect is too large, and the direction is not appropriate. Could it be due to a gravitational interaction with a nearby black hole? No, because if there were a black hole very near by we would not be here to observe anything! Could it be an interaction via the strong (nuclear) force? No, because the strong force is a very short range force, and there are no other objects near enough.

It is, however, plausible that the interaction causing the acceleration of the proton could be an electric interaction, since electric interactions can have large effects, and can occur over rather large distances.

? What charged objects might be responsible for this interaction, and where might they be?

There are many possible configurations of charged particles that might produce the observed effect. As indicated in Figure 13.6, there might be positively charged objects to your left. Alternatively, there could be negative charges to your right. Perhaps there are both—you can't draw a definite conclusion from a single observation. There are many possible arrangements of charges in space that could produce the observed effect.

(If, however, you made several observations of the proton over some time period, you might note a change in the proton's acceleration. For example, if you noticed that the acceleration of the proton increased as it moved to the right, you might suspect that there was a negative charge to your right, which would have an increasingly large effect on the proton as it moved closer to the charge.)

? On the basis of your observations so far, can you predict what you would observe if you released an electron instead of a proton at the same location (Figure 13.7)?

An electron would accelerate to the left rather than to the right, since it has a negative charge. Its acceleration would be greater than 10^{11}m/s^2, because the mass of the electron is much less than the mass of the proton.

? What would happen if you placed an alpha particle (a helium nucleus, with charge $+2e$) at the observation location (Figure 13.8)?

Figure 13.8 An alpha particle.

Since the alpha particle has a positive charge, it would accelerate to the right as the proton did. However, because the charge of the alpha particle is twice the charge of the proton, and its mass is about four times that of a proton, the magnitude of its acceleration would be one-half of 10^{11}m/s^2.

? If you released a neutron at the observation location, what would you expect to observe (Figure 13.9)?

Figure 13.9 A neutron.

Since the neutron has no electric charge, it would experience no electric force, and should simply accelerate toward the Earth at 9.8 m/s^2.

? Finally, suppose we do not put any particle at the observation location. Is there anything there?

Since we know that if we were to put a charged particle at that location, it would experience a force, it seems that in a certain sense there is something there, waiting for a charged particle to interact with. This "virtual force" is called "electric field." The electric field created by a charge is present throughout space at all times, whether or not there is another charge around to feel its effects. The electric field created by a charge penetrates through matter. The field permeates the neighboring space, lying in wait to affect anything brought into its web of interaction.

We define the electric field at a location by the following equation:

DEFINITION OF ELECTRIC FIELD

$$\vec{F}_2 = q_2 \vec{E}_1$$

This equation says that the force on particle 2 is determined by the charge of particle 2 and by the electric field $\vec{E}_1$ made by all other charged particles in the vicinity, as shown in Figure 13.10.

Figure 13.10 A particle with charge q_2 experiences a force $\vec{F}_2$ because of its interaction with the electric field $\vec{E}_1$ produced by all other charged particles in the vicin-

How might we measure the electric field at a given location in space? We can't see it, but we can measure it indirectly. If we place a charge at that location in space, we can measure the force on the charge due to its interaction with the electric field at that location. We can determine the magnitude and direction of $\vec{E}$ by measuring a force on a known charge q:

$$\vec{E} = \frac{\vec{F}}{q}$$

Electric field therefore has units of newtons per coulomb. Note that it doesn't matter whether the affected charge q is positive or negative. If $\vec{E}$ points north, a positive charge will experience a force to the north, while a negative charge would experience a force to the south (Figure 13.11); dividing this force by a negative q would still yield a vector $\vec{E}$ pointing north. Though we will usually write $\vec{E}$ for the value of the electric field at a particular location and time, this is really shorthand for $\vec{E}(x, y, z, t)$, since the electric field has a value at all locations and times.

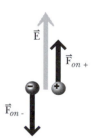

Figure 13.11 A positive charge experiences a force in the direction of the electric field at its location, while a negative charge experiences a force in the opposite direction.

Drawing electric field vectors

In diagrams we will use arrows to represent electric field vectors. By convention, such a vector is always drawn with its tail at the location where the field is measured.

Ex. 13.1 An electron in a region where there is an electric field experiences a force of magnitude $3.2{\times}10^{-16}$ N. What is the magnitude of the electric field at the location of the electron?

Ex. 13.2 A proton is observed to have an instantaneous acceleration of $10^{11} \mathrm{m/s^2}$. What is the magnitude E of the electric field at the proton's location?

13.3.1 The electric field of a point charge

We can now write an algebraic expression for the electric field created by a point charge at a location in space. If there is a particle with charge q_1 at some location in space, we can show that the electric field produced by that point charge q_1 at a location a distance r away has magnitude

$$E_1 = \left(\frac{1}{4\pi\varepsilon_0} \frac{q_1}{r^2} \right)$$

because a particle with charge q_2 at that location will experience a force

$$F_2 = q_2 E_1 = q_2 \left(\frac{1}{4\pi\varepsilon_0} \frac{q_1}{r^2} \right)$$

? How do we include the direction of the electric field in the expression for electric field?

Figure 13.12 Electric field of a positive point charge, in a plane containing the charge.

As indicated by Figure 13.12, the electric field of a positive point charge points away from the charged particle, no matter where it is measured. In vector terms, we can say that it points in the direction of a vector $\vec{r}$ pointing from the charged particle to the observation location, as shown in Figure 13.13. To add this direction information to the expression above, we can multiply the magnitude of the field by a unit vector $\hat{r}$ pointing from the source charge to the observation location (in the direction of $\vec{r}$)

$$\vec{E}_1 = \left(\frac{1}{4\pi\varepsilon_0} \frac{q_1}{r^2} \right)\hat{r}$$

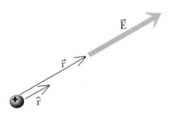

Figure 13.13 The vector $\vec{r}$ points from the point charge to the observation location. $\hat{r}$ is a unit vector with the same direction as the vector $\vec{r}$.

Example: A particle with charge +2 nC (a nanocoulomb is $1{\times}10^{-9}$ C) is located at the origin. What is the electric field due to this particle at a location $\langle -0.2, -0.2, -0.2 \rangle$ m?

Solution: The vector $\vec{r}$ points from source to observation location:

$$\vec{r} = \langle \text{observation location} \rangle - \langle \text{source location} \rangle$$
$$= \langle -0.2, -0.2, -0.2 \rangle - \langle 0, 0, 0 \rangle$$
$$\vec{r} = \langle -0.2, -0.2, -0.2 \rangle$$

The magnitude of $\vec{r}$ is

$$|\vec{r}| = \sqrt{(-0.2)^2 + (-0.2)^2 + (-0.2)^2} = 0.35$$

The direction of the electric field is given by $\hat{r}$:

$$\hat{r} = \frac{\vec{r}}{|\vec{r}|} = \frac{\langle -0.2, -0.2, -0.2 \rangle}{0.35} = \langle -0.57, -0.57, -0.57 \rangle$$

The magnitude of the electric field is:

$$E = \frac{1}{4\pi\varepsilon_0}\frac{q}{r^2} = \left(9\times10^9\frac{\text{Nm}^2}{\text{C}^2}\right)\left(\frac{2\times10^{-9}\text{C}}{0.35^2\text{m}^2}\right) = 147\frac{\text{N}}{\text{C}}$$

So the electric field is:

$$\vec{E} = E\hat{r}$$

$$= \left(147\frac{\text{N}}{\text{C}}\right)\langle-0.57, -0.57, -0.57\rangle$$

$$= \langle-84, -84, -84\rangle\frac{\text{N}}{\text{C}}$$

The electric field points away from the positively charged particle, as shown in Figure 13.14.

Ex. 13.3 A particle with charge +1 nC (a nanocoulomb is 1×10^{-9} C) is located at the origin. What is the electric field due to this particle at a location $\langle 0.1, 0, 0\rangle$ m?

Ex. 13.4 What is the electric field at a location $\vec{b} = \langle-0.1, -0.1, 0\rangle$ m, due to a particle with charge +3 nC located at the origin?

Ex. 13.5 A charged particle located at the origin creates an electric field of $\langle-1.4\times10^3, 0, 0\rangle$ N/C at a location $\langle 0.15, 0, 0\rangle$ m. What is the particle's charge?

Ex. 13.6 In a hydrogen atom in its ground state, the electron is on average a distance of about 10^{-10} m from the proton. What is the magnitude of the electric field due to the proton a distance of 10^{-10} m away?

Figure 13.14 The electric field calculated in the example.

13.3.2 The physical concept of "field"

The word "field" has a special meaning in mathematical physics. A field is a physical quantity that has a value at every location in space. Its value at every location can be a scalar or a vector.

For example, the temperature in a room is a scalar field. At every location in the room, the temperature has a value, which we could write as $T(x,y,z)$, or as $T(x,y,z,t)$ if it were changing with time. The air flow in the room is a vector field. At every location in the room, air flows in a particular direction with a particular speed. Electric field is a vector field; at every location in space surrounding a charge the electric field has a magnitude and a direction.

? Think of another example of a quantity that is a field.

The field concept is also used with gravitation. Instead of saying that the earth exerts a force on a falling object, we can say that the mass of the earth creates a "gravitational field" surrounding the earth, and any object near the earth is acted upon by the gravitational field at that location (Figure 13.15). Gravitational field has units "newtons per kilogram." At a location near the earth's surface we can say that there is a gravitational field $\vec{g}$ pointing downward (that is, toward the center of the earth) of magnitude $g = 9.8$ newtons per kilogram.

A falling object of mass $m = 10$ kg experiences a gravitational force of magnitude $mg = (10\text{ kg})(9.8\text{ N/kg}) = 98$ newtons. If the object is moving at a speed slow compared to the speed of light, we can calculate its acceleration $a = F/m = mg/m = 9.8$ m/s^2.

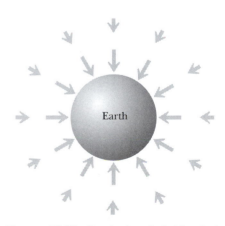

Figure 13.15 Gravitational field of the Earth, in a plane cutting through the center of the Earth.

Of course, farther away from the surface of the Earth, the gravitational field of the Earth is smaller in magnitude, just as the electric field of a charge decreases in magnitude as the distance from the charge increases.

The field concept also applies to magnetism. As we will see in Chapter 17, electric currents, including electric currents inside the Earth, create "magnetic fields" that affect compass needles.

? Find an example of a quantity that is *not* a field.

In contrast, many things are not fields. The position and velocity at one instant of a car going around a race track are not fields. The distance between the Earth and the Sun is not a field. The rate of change of the momentum of the Moon is not a field.

13.3.3 Patterns of electric field near point charges

It is helpful to see what the electric field vectors look like at representative locations near a positive point charge and near a negative point charge. Figure 13.16 shows electric field vectors at a few selected locations in a two-dimensional slice of space containing a positive point charge. The tail of each vector is placed at the location where the electric field was measured. Notice that the vectors representing the electric field all point radially out from the point charge, and that the magnitude of the field decreases rapidly with distance from the charge.

The electric field near a negative point charge points radially inward toward the charge, so a positively charged particle would experience a force toward the negative charge (Figure 13.17). The magnitude of the field decreases rapidly with distance from the charge. We will frequently use two-dimensional diagrams because they are easy to draw, but it is important to remember the 3-D character of electric fields in space.

13.3.4 Predicting forces due to a known electric field

One reason the concept of electric field is useful is that if we know the value of the electric field at some location, we can easily predict the force that will be experienced by any charged particle placed at that location.

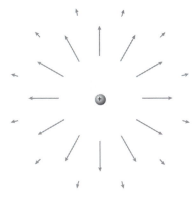

Figure 13.16 Electric field of a positive point charge, in a plane containing the charge.

Figure 13.17 Electric field of a negative point charge, in a plane containing the charge.

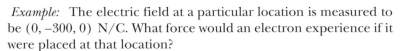

Example: The electric field at a particular location is measured to be $(0, -300, 0)$ N/C. What force would an electron experience if it were placed at that location?

Solution: To find the force on the electron, we simply multiply the electric field by the charge on the electron:

$$\vec{F} = (-e)\vec{E}_1$$

$$\vec{F} = -1.6 \times 10^{-19} \text{C} \langle 0, -300, 0 \rangle \text{N/C} = \langle 0, 4.8 \times 10^{-17}, 0 \rangle \text{N}$$

Since the electron's charge is negative, the force on the electron is in a direction opposite to the direction of the electric field. Note that we did not need to know the source of the electric field in order to calculate this force.

Ex. 13.7 What force would a proton experience if placed at the same location (see preceding example)?

Ex. 13.8 What force would an alpha particle at the same location experience?

Ex. 13.9 What force would a neutron at the same location experience?

13.3.5 No "self-force"

It is important to note that a point charge is not affected by its *own* electric field. A point charge does not exert a force on itself! Mathematically this is reassuring, since at the location of a point charge its own electric field would be infinite ($1/0^2$). Physically, this makes sense too, since after all, the charge can't start itself moving, nor is there any way to decide in what direction it should go.

When we use the electric field concept with point charges we always talk about a charge q_1 making a field $\vec{E}_1$, and a different charge q_2 in a different place being affected by that field with a force $\vec{F}_{\text{on } q_2} = q_2\vec{E}_1$.

13.4 Superposition of electric fields

An important property of electric field is a consequence of the superposition principle for electric forces: the net electric field due to two or more charges is the vector sum of each field due to each individual charge. Superposition as a vector sum holds true for forces in general, so it also holds true for fields, because field is force per unit charge.

We restate the superposition principle here in terms of electric field, in a form intended to emphasize an important aspect of the principle:

THE SUPERPOSITION PRINCIPLE

> The net electric field at a location in space is the vector sum of the individual electric fields contributed by all charged particles located elsewhere.

> The electric field contributed by a charged particle is unaffected by the presence of other charged particles.

This may seem obvious, but it actually is a bit subtle. The presence of other particles does not affect the fundamental electric interaction between each pair of particles! The interaction doesn't get "used up;" the interaction between one particle and another is unaffected by the presence of a third particle.

A significant consequence of the superposition principle is that matter cannot "block" electric fields; electric fields "penetrate" through matter.

Applying the superposition principle

Consider three charges q_1, q_2, and q_3 that interact with each other, as shown in Figure 13.18. We could calculate the net force that q_1 and q_2 exert on q_3 by using Coulomb's law and vector addition.

However, let's use the field concept to find the force on q_3 due to the other two charges. We pretend that q_3 isn't there, and we find the electric field $\vec{E}_{\text{net}}$ due to q_1 and q_2, at the location where q_3 would be. We can then calculate the force on q_3 as $q_3\vec{E}_{\text{net}}$. In Figure 13.19 the scheme is shown graphically. $\vec{E}_{\text{net}}$ is the vector sum of $\vec{E}_1$ (due to q_1) and $\vec{E}_2$ (due to q_2), at the location of q_3.

Advantage of electric field in calculations

We could just as easily have calculated the force on q_3 by using Coulomb's law directly, without using the more abstract concept of electric field. What advantage is there to using electric field? One of the advantages is that once

Figure 13.18 Three interacting charged particles.

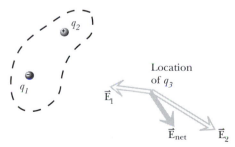

Figure 13.19 Electric field at the location of q_3.

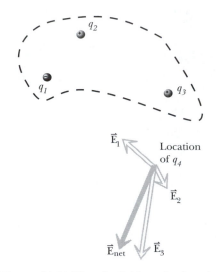

Figure 13.20 Electric field at the location of q_4.

Figure 13.21 An electric dipole.

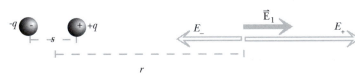

Figure 13.22 Electric field of a dipole at a location on the x axis (which is the axis of the dipole).

you have calculated the field $\vec{E}_{net}$ due to q_1 and q_2, you can quickly calculate the force $Q\vec{E}_{net}$ on *any* amount of charge Q placed at that location, not just the force on q_3. The "force per unit charge" idea simplifies these kinds of calculations.

If we are interested in the effect that all three charges would have on a fourth charge q_4 placed at some location, we would include all three charges as the source of the electric field at that location, but exclude q_4, as shown in Figure 13.20.

Knowing the electric field $\vec{E}_{net}$ at a particular location, we know what force would act on a charge q_4 placed at that location: it would simply be $q_4\vec{E}_{net}$.

13.4.1 The electric field of a dipole

Neutral matter contains both positive and negative charges (protons and electrons). The simplest piece of neutral matter that we can analyze in detail is an "electric dipole" consisting of two equally but oppositely charged point-like objects, separated by a distance s, as shown in Figure 13.21. Dipoles occur frequently in nature, so this analysis has practical applications. For example, a single molecule of HCl is an electric dipole: the H end is somewhat positive and the Cl end is somewhat negative. Moreover, the analysis of a single dipole provides a basis for analyzing more complicated forms of matter.

By applying the superposition principle we can calculate the electric field due to a dipole at any location in space. We will be particularly interested in the electric field in two important locations: at a location along the axis of the dipole, and at a location along the perpendicular bisector of the axis.

Along the *x* axis

Figure 13.22 shows a horizontal dipole with charge $+q$ and $-q$, and separation s. Consider the electric field $\vec{E}_1$ at a location $(r,0,0)$ along the x axis, measured from the center of the dipole.

The net field $\vec{E}_1$ is in the $+x$ direction, because at this location the field $\vec{E}_+$ of the closer $+q$ charge is larger than the field $\vec{E}_-$ of the more distant $-q$ charge. At locations along the x axis only the x component of E is nonzero, so we do not need to worry about the y and z components:

$$E_{1,x} = E_{+,x} + E_{-,x} = \frac{1}{4\pi\varepsilon_0}\frac{q}{\left(r-\dfrac{s}{2}\right)^2} + \frac{-1}{4\pi\varepsilon_0}\frac{q}{\left(r+\dfrac{s}{2}\right)^2} \qquad \textit{superposition principle}$$

$$= \frac{1}{4\pi\varepsilon_0}\left[\frac{2\,qsr}{\left(r-\dfrac{s}{2}\right)^2\left(r+\dfrac{s}{2}\right)^2}\right] \qquad \begin{array}{l}\textit{common denominator}\\ \textit{expand numerator}\\ \textit{simplify numerator}\end{array}$$

Approximation: Far from the dipole

If we are far from the dipole, as is normally the case when we interact with molecular dipoles, the magnitude of the field can be further simplified.

$$\text{If } r \gg s, \text{ then } \left(r-\frac{s}{2}\right)^2 \approx \left(r+\frac{s}{2}\right)^2 \approx r^2$$

Using this approximation, we can simplify the expression to

$$E_{1,\,x} \approx \frac{1}{4\pi\varepsilon_0}\frac{2qs}{r^3}$$

A vector expression for the electric field at this location would be:

$$\vec{E}_1 \approx \langle \frac{1}{4\pi\varepsilon_0}\frac{2qs}{r^3}, 0, 0 \rangle$$

Is this result reasonable?

? Should we have expected the magnitude of the electric field of the dipole to be proportional to $1/r^2$?

The electric field of a single point charge is proportional to $1/r^2$, but when we add the electric fields contributed by more than one charge, the result may have quite a different distance dependence.

? Does our result have the correct units?

The units of electric field come out to N/C, as they should.

$$\left(\frac{\mathrm{N}\cdot\mathrm{m}^2}{\mathrm{C}^2}\right)\frac{(\mathrm{C})(\mathrm{m})}{(\mathrm{m}^3)} = \frac{\mathrm{N}}{\mathrm{C}}$$

Along the y axis

Next consider the electric field $\vec{E}_2$ at a location $<0,y,0>$ along the y axis, perpendicular to the axis of the dipole (Figure 13.23). First we need to find $\vec{r}_+$ and $\vec{r}_-$, the vectors from each source charge to the observation location.

$$\vec{r}_+ = \langle 0, y, 0 \rangle - \langle \frac{s}{2}, 0, 0 \rangle = \langle -\frac{s}{2}, y, 0 \rangle$$

$$\vec{r}_- = \langle 0, y, 0 \rangle - \langle -\frac{s}{2}, 0, 0 \rangle = \langle \frac{s}{2}, y, 0 \rangle$$

The magnitudes of $\vec{r}_+$ and $\vec{r}_-$ are equal, because each charge is the same distance from the observation location.

$$|\vec{r}_+| = \sqrt{\left(\frac{s}{2}\right)^2 + y^2 + 0^2} = \sqrt{\left(\frac{s}{2}\right)^2 + y^2}$$

$$|\vec{r}_-| = \sqrt{\left(-\frac{s}{2}\right)^2 + y^2}$$

The net field $\vec{E}_2$ is the vector sum of $\vec{E}_+$ and $\vec{E}_-$.

$$\vec{E}_+ = \frac{1}{4\pi\varepsilon_0}\frac{q}{\left(\frac{s}{2}\right)^2 + y^2}\frac{\langle -\frac{s}{2}, y, 0\rangle}{\sqrt{\left(\frac{s}{2}\right)^2 + y^2}}\frac{\mathrm{N}}{\mathrm{C}}$$

$$\vec{E}_- = \frac{1}{4\pi\varepsilon_0}\frac{-q}{\left(\frac{s}{2}\right)^2 + y^2}\frac{\langle \frac{s}{2}, y, 0\rangle}{\sqrt{\left(\frac{s}{2}\right)^2 + y^2}}\frac{\mathrm{N}}{\mathrm{C}}$$

$$\vec{E}_2 = \vec{E}_+ + \vec{E}_- = \frac{1}{4\pi\varepsilon_0}\frac{qs}{\left[\left(\frac{s}{2}\right)^2 + y^2\right]^{3/2}}\langle -1, 0, 0 \rangle\frac{\mathrm{N}}{\mathrm{C}}$$

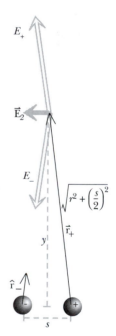

Figure 13.23 Electric field of a dipole at a location on the y axis. Shown on the diagram are $\vec{r}_+$ and $\hat{r}_-$.

The y components of the two fields cancel, and the net field $\vec{E}_2$ is horizontal to the *left*, which may come as a bit of a surprise.

If we are far from the dipole, as is normally the case when we interact with molecular dipoles, the magnitude of the field can be further simplified. You should be able to show that in this important case ($r \gg s$), by making an appropriate approximation, you obtain:

$$E_{2,\,x} \approx \frac{-1}{4\pi\varepsilon_0} \frac{qs}{r^3}$$

Since the dipole may be oriented in any direction, we have replaced y with r in our expressions. The full vector expression for $\vec{E}_2$ is:

$$\vec{E}_2 \approx \langle \frac{-1}{4\pi\varepsilon_0} \frac{qs}{r^3}, 0, 0 \rangle \text{ at a location } <0,r,0>.$$

Note that the magnitude of the electric field at a location along the y axis is half of the magnitude of the electric field at a location the same distance away on the x axis.

Along the z axis

? What would the electric field of a dipole be at a location $(0,0,r)$ on the z-axis?

By looking at the symmetry of the system, we can conclude by inspection that the electric field along the z axis should have exactly the same distance dependence as the electric field along the y axis, as shown in Figure 13.24.

$$\vec{E}_2 \approx \langle \frac{-1}{4\pi\varepsilon_0} \frac{qs}{r^3}, 0, 0 \rangle \text{ at a location } <0,0,r>.$$

Other locations

By applying the superposition principle we can calculate the electric field of a dipole at any location, though we may not get a simple algebraic expression. Figure 13.25 shows the electric field of a dipole at locations in a plane containing the dipole, calculated numerically. Problems 13.7 and 13.9 are computer exercises that allow you to visualize the electric field of a dipole at any location.

The utility of approximations

By considering the common situation where $r \gg s$, we were able to come up with a simple algebraic expression for the electric field of a dipole at locations along the x and y axes which gives us a clear insight into the distance dependence of the dipole field. If we had not made this simplification, we would have been left with a messy expression, and it would not have been easy to see that the electric field of a dipole falls off like $1/r^3$. (This turns out to be true for locations not on the axes as well, but the algebra required to demonstrate this is more complicated.)

? How would we calculate the electric field of a dipole at a location very near the dipole, where $r \approx s$?

Since the approximation we made in the calculations above is no longer valid, we would have to go back to adding the fields of the two charges exactly, as we did in the first step of our work above.

Figure 13.24 Electric field of a dipole at locations on the x, y, and z axes.

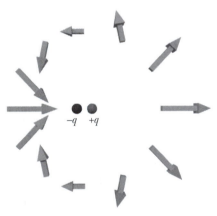

Figure 13.25 Electric field of a dipole, at locations in the xy plane.

Ex. 13.10 A dipole is located at the origin, and is composed of charged particles with charge $+e$ and $-e$, separated by a distance

2×10^{-10} m along the *x* axis. Calculate the magnitude of the electric field due to this dipole at a location $\langle 0, 2 \times 10^{-8}, 0 \rangle$ m.

Ex. 13.11 Calculate the net electric field due to the two negative charges shown in Figure 13.26, at the location indicated by letter *A* on the diagram.

13.4.2 Interaction of a point charge and a dipole

Since we have analytical expressions for the electric field of a dipole at locations along the *x, y* or *z* axes, we can calculate the force exerted on a point charge that interacts with the field of the dipole. Figure 13.27 shows a dipole acting on a point charge +*Q* that is a distance $d \gg s$ from the center of the dipole.

Figure 13.26 Two negative charges (Exercise 13.11).

> *Example:* What are the direction and the magnitude of the force exerted by the dipole on the point charge shown in Figure 13.27?
>
> *Solution:* The electric field due to the dipole at the location of the point charge points in the −*x* direction (Figure 13.28) The dipole exerts a force on *Q* toward the dipole:

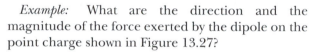

d >> s

Figure 13.27 A dipole and a point charge.

$$\vec{F} = Q\vec{E}_{\text{dipole}} = Q\langle \frac{-1}{4\pi\varepsilon_0} \frac{2qs}{d^3}, 0, 0 \rangle$$

? Does the direction of this electric force make sense?

Yes. The negative end of the dipole is closer to the location of interest, so its contribution to the net electric field is larger than the contribution from the positive end of the dipole.

d >> s

Figure 13.28 The electric field of the dipole at the location of the point charge, and the force on the point charge due to this field.

? What are the magnitude and direction of the force exerted on the dipole by the point charge?

By the principle of reciprocity of electric forces (Newton's third law), the force exerted by *Q* on the dipole must be equal in magnitude and opposite in direction to the force on the point charge by the dipole.

? But how is this possible? Since the particles making up the dipole have equal and opposite charges, why isn't the net force on the dipole zero?

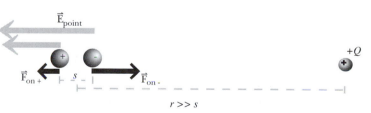

r >> s

The key lies in the distance dependence of the electric field of the point charge. Although the two ends of the dipole are very close together, the positive end is very slightly farther away from the point charge. Hence, the magnitude of the point charge's electric field is slightly less at this location, and the resulting repulsive force on the positive end of the dipole is slightly less than the attractive force on the negative end of the dipole. The net force is small, but attractive.

Figure 13.29 The electric field of the point charge is slightly larger in magnitude at the location of the negative end of the dipole. Hence, the net force on the dipole is to the right. The arrows representing field and force vectors are offset slightly from the location of the charges for clarity.

Ex. 13.12 If we double the distance *r*, by what factor is the force on the point charge due to the dipole in Figure 13.28 reduced?

Ex. 13.13 How would the magnitude of the force change if the point charge had a charge of $+3Q$?

Ex. 13.14 If the charge of the point charge were $-2Q$, how would the force change?

Ex. 13.15 Refer to your answer to Exercise 13.10. Calculate the magnitude of the force on a proton placed at the location where you calculated the electric field.

13.4.3 Dipole moment

The electric field of a dipole is proportional to the product qs, called the "dipole moment" and denoted by p $(= qs)$. Far away from a dipole, the same electric field could be due to a small q and a large s, or a large q and a small s. The only thing that matters is the "dipole moment" qs. Moreover, this is the quantity that is measurable for molecules such as HCl and H_2O that are permanent dipoles, not the individual values of q (the amount of charge on an end) and s (the separation of the charges).

We can rewrite our expressions for the electric field of a dipole in terms of the dipole moment p:

$$\vec{E}_1 \approx \langle \frac{1}{4\pi\varepsilon_0} \frac{2p}{r^3}, 0, 0 \rangle \text{ at a location } <r,0,0> \text{ on the } x \text{ axis}$$

$$\vec{E}_2 \approx \langle \frac{-1}{4\pi\varepsilon_0} \frac{p}{r^3}, 0, 0 \rangle \text{ at a location } <0,r,0> \text{ on the } y \text{ axis}$$

Dipole moment as a vector

The dipole moment can be defined as a vector $\vec{p}$ which points from the negative charge to the positive charge, with a magnitude $p = qs$ (Figure 13.30). Note that the electric field along the axis of the dipole points in the same direction as the dipole moment $\vec{p}$, which is one of the useful properties of the vector $\vec{p}$.

In Chapter 17 we will see a similar pattern of magnetic field around a magnetic dipole characterized by a magnetic dipole moment that is a vector.

13.4.4 A dipole in a uniform electric field

We will see in Chapter 15 that it is possible to arrange a collection of point charges in a configuration which produces an electric field that is nearly uniform in magnitude and direction within a particular region. What would be the force on a dipole in such a uniform electric field?

The net force on the dipole would be zero, since the forces on the two ends would be equal in magnitude and opposite in direction, as shown in Figure 13.31. However, there would be a torque on the dipole about its center of mass, and the dipole would begin to rotate.

Ex. 13.16 What would be the equilibrium position of the dipole shown in Figure 13.31?

13.4.5 The electric field of a uniformly charged sphere

In Chapter 15 we will apply the superposition principle to calculate the electric field of macroscopic objects with charge spread out over their surfaces.

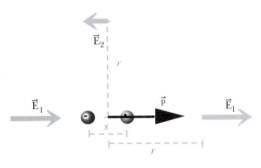

Figure 13.30 Electric field of a dipole on the x and y axes, and the dipole moment vector.

Figure 13.31 A dipole in a uniform electric field would experience no net force. It would experience a torque about its center of mass (in this example $\vec{\tau}_{CM}$ is out of the page, and the dipole would begin to rotate counter-clockwise).

One of the results of such a calculation is sufficiently useful that we present it now, without going through the calculations involved.

As shown in Figure 13.32, a spherical object of radius R with charge Q uniformly spread out over its surface produces an electric field with the following distance dependence, if $\hat{r}$ is a vector from the center of the sphere to the observation location:

$$\vec{E}_{sphere} = \frac{1}{4\pi\varepsilon_0}\frac{Q}{r^2}\hat{r} \text{ for } r > R \text{ (outside the sphere)}$$

$$\vec{E}_{sphere} = 0 \text{ for } r < R \text{ (inside the sphere)}$$

In other words, at locations outside the charged sphere, the electric field due to the sphere is the same as if all its charge were located at its center. Inside the sphere, the electric field due to all the charge on the surface of the sphere adds up to zero!

Informally, we may say that a uniformly charged sphere "acts like" a point charge, at locations outside the sphere. This is not only because the electric field due to the sphere is the same as the electric field of a point charge located at the center of the sphere, but because the charged sphere responds to applied electric fields (due to other charges) in the same way as a point charge located at its center would.

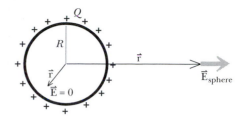

Figure 13.32 The electric field of a uniformly charged sphere, at locations outside and inside the sphere.

Ex. 13.17 A sphere with radius 1 cm has a charge of 5×10^{-9} C spread out uniformly over its surface. What is the magnitude of the electric field due to the sphere at a location 4 cm from the center of the sphere?

Ex. 13.18 A sphere with radius 2 cm is placed at a location near a point charge. The sphere has a charge of -8×10^{-10} C spread uniformly over its surface. The electric field due to the point charge has a magnitude of 500 N/C at the center of the sphere. What is the magnitude of the force on the sphere due to the point charge?

13.5 Choice of system

In our previous study of classical mechanics, when we applied principles such as conservation of energy or conservation of momentum to multiparticle systems we had to decide what objects to include in our "system" and what objects to consider as external to the system. Similarly, when we use the field concept to calculate a force rather than using Coulomb's law directly, we split the Universe into two parts:
- the charges that are the sources of the field, and
- the charge that is affected by that field.

This is particularly useful when the sources of the field are numerous and fixed in position. We calculate the electric field due to such sources throughout a given region of space, and then we can predict what would happen to a charged particle that enters that region. (It is of course true that the moving charged particle exerts forces on all the other charges, but often we're not interested in those forces, or the other charges are not free to move.)

For example, in an oscilloscope, charges on metal plates are the sources of an electric field that affects the trajectory of single electrons (Figure 13.33). The electric field has nearly the same magnitude and direction everywhere within a box (this can be achieved with a suitable configuration of positive charges below and negative charges above). The trajectory of an

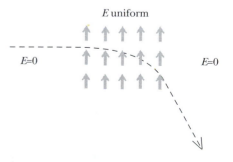

Figure 13.33 Path of an electron through a region of uniform electric field.

electron outside this box, where the field is nearly zero, is a straight line. Upon entering the box (through a small hole) the electron experiences a downward force (it has negative charge, hence the force on it is opposite to the field) and has a trajectory as shown.

Ex. 13.19 If this upward-pointing electric field has a magnitude of 5000 N/C, what is the magnitude of the force on the electron while it is in the box?

Ex. 13.20 If a particle experiences a force of 1.6×10^{-15} N when passing through the same region, what is the charge of the particle?

13.6 A fundamental rationale for electric field

Reflecting on what we have done so far, we can give two reasons for why we introduced the concept of electric field:

1) Once we know the electric field $\vec{E}$ at some location, we know the electric force $\vec{F} = q\vec{E}$ acting on *any* charge q that we place there.

2) In terms of electric field, we can describe the electric properties of matter, independent of how the applied electric field is produced. Later we'll see that if the electric field in air exceeds about 3×10^6 N/C, air becomes a conductor, no matter how this electric field is produced. This is an example of the value of the electric field concept for parameterizing the behavior of matter. It would be very awkward to describe the properties of matter if we didn't split the problem into two parts using the field concept, in which we talk about the field made by other charges, and the effect that a field has on a particular kind of matter.

Even with these advantages of the field concept, it may seem that it doesn't matter whether we think of electric field as merely a calculational convenience, or as something real. However, Einstein's special theory of relativity, which has been experimentally verified in a wide variety of situations, rules out instantaneous action at a distance and implies that the field concept is necessary, not merely convenient, as we will show next.

13.6.1 Retardation

Special relativity predicts that nothing can move faster than the speed of light, not even information, and no one has ever observed a violation of this prediction. The speed of light is $c = 3 \times 10^8$ m/s, or about one foot (30 cm) per nanosecond, where a nanosecond is 10^{-9} seconds.

Consider a charge that is 30 feet away (9 meters), as shown in Figure 13.34. It takes light 30 nanoseconds (30 ns) to travel from the charge to you.

Suppose the charge is suddenly moved to a new location, as shown in Figure 13.35. You cannot observe change in the electric field until 30 nanoseconds have elapsed! If the field could change instantaneously, that would provide a mechanism for sending signals faster than the speed of light.

For a while (30 nanoseconds), the electric field has some kind of reality and existence independent of the original source of the field. Bizarre! If you place a positive charge at your observation location during this 30 ns interval, you will see it accelerate to the right, in the direction of the old electric field that is still valid.

At the end of the 30-nanosecond delay, the field suddenly changes to correspond to the place to which the charge was moved 30 nanoseconds ago, as shown in Figure 13.36. You *see* the new position of the charge at the same instant that you notice the change in the electric field, because of course light itself travels at the speed of light.

H----------------------------I
30 ft (9 meters)

Figure 13.34 Electric field of a point charge.

Figure 13.35 At $t = 0$ charge is moved, but the electric field at the observation location does not change for 30 ns.

Figure 13.36 At $t = 30$ ns, the field at the observation location finally changes.

An even more dramatic example concerns the electric field made by a remote electric dipole. Suppose an electron and a positron are separated by a small distance *s*, making a dipole, and creating an electric field throughout space. The electron and positron may suddenly come together and react, annihilating each other and releasing a large amount of energy in the form of high-energy photons (gamma rays):

$$e^- + e^+ \rightarrow \gamma + \gamma$$

At a time shortly after the annihilation has occurred, there are no longer any charged particles in the vicinity. However, if you monitor the electric field in this region, you will keep detecting the (former) dipole's field for a while even though the dipole no longer exists! If you are a distance *r* away from the original location of the dipole, you will still detect the dipole's electric field for a time *r/c*, which is the same as the time it would take light to travel from that location to your position.

This strange behavior, called "retardation," means that Coulomb's law is not completely correct, since the formula

$$\vec{F} = \frac{1}{4\pi\varepsilon_0} \frac{q_1 q_2}{r^2} \hat{r}$$

doesn't contain time *t* or speed of light *c*. Coulomb's law is an approximation that is quite accurate as long as charges are moving slowly. But when charges move with speeds that are a significant fraction of the speed of light, Coulomb's law is not adequate. Similarly, the formula for the electric field of a point charge,

$$\vec{E} = \frac{1}{4\pi\varepsilon_0} \frac{q}{r^2} \hat{r}$$

is only an approximation, valid if the charge is moving at a speed small compared with the speed of light.

Later we will consider the fields of moving charges, which have some intriguing properties. We will also consider the fields of *accelerated* charges, which produce a very special kind of electric field—a component of the electromagnetic field that makes up radio and television waves, and light.

Evidently the electric field isn't just a calculational device. Space can be altered by the presence of an electric field, even when the source charges that made the field are gone.

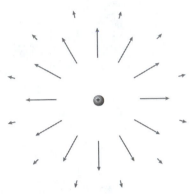

Figure 13.37 Electric field of a positive point charge, at locations in a plane containing the charge.

13.7 Summary

Fundamental Principles

Electric field

A point charge q_1 at a location x_1, y_1, z_1 makes an electric field throughout space. A different charge q_2 at some other location x_2, y_2, z_2 experiences a force due to that field.

$$\vec{F}_2 = q_2\vec{E}_1$$

A point charge is not affected by its own electric field.

The superposition principle

The net electric field at a particular location due to two or more charges is the vector sum of each field due to each individual charge. Each individual contribution is unaffected by the presence of the other charges. Thus, matter cannot "block" electric fields.

Retardation

Changes in electric fields (due to changes in source charge distribution) propagate through space at the speed of light. Therefore the electric field at a distant point does not change instantaneously when the source charge distribution changes.

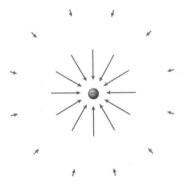

Figure 13.38 Electric field of a negative point charge, in a plane containing the charge.

New Concepts

Electric field of a point charge (Figure 13.37 and Figure 13.38).

$$\vec{E} = \frac{1}{4\pi\varepsilon_0}\frac{q}{r^2}\hat{r}$$

Dipole moment (Figure 13.39)

The dipole moment of an electric dipole consisting of point charges $+q$ and $-q$ separated by a distance s is defined as

$$p \equiv qs$$

The vector $\vec{p}$ points from the negative charge to the positive charge.

Results

Electric field of a dipole (Figure 13.39)

At a location $<r,0,0>$ on the x axis, where $r \gg s$:

$$E_x \approx \frac{1}{4\pi\varepsilon_0}\frac{2qs}{r^3} \text{ and } E_y = 0 \text{ and } E_z = 0$$

At a location $<0,r,0>$ on the y axis, where $r \gg s$:

$$E_x \approx -\frac{1}{4\pi\varepsilon_0}\frac{qs}{r^3} \text{ and } E_y = 0 \text{ and } E_z = 0$$

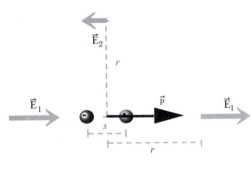

Figure 13.39 Electric field of a dipole on the x and y axes, and the dipole moment vector.

13.8 Example problem: Dipole and charged ball

A dipole consisting of point charges $+q$ and $-q$ separated by a distance s is located a distance b to the left of a hollow plastic ball with radius R and charge $-Q$ distributed uniformly over its surface, as shown in Figure 13.40. What is the net electric field at location C, a distance a directly below the dipole? The distances a and b are both much larger than the dipole separation s (the diagram is not drawn to scale).

Solution

$$\vec{E}_{net} = \vec{E}_{dipole} + \vec{E}_{ball}$$

Let's put the origin at location C. Then the location of the center of the ball is $\langle b, a, 0 \rangle$, and the location of the center of the dipole is $\langle 0, a, 0 \rangle$.

Because location C is far from the dipole, we can approximate the magnitude of the dipole field as:

$$\left| \vec{E} \right| = \frac{1}{4\pi\varepsilon_0} \frac{2qs}{a^3} \quad \text{(along axis of dipole)}$$

The direction of the dipole field is <0,1,0> (along the positive y axis, because the negative end is closer to location C). So the electric field at location C due to the dipole is:

$$\vec{E}_{dipole} = \left\langle 0, \frac{1}{4\pi\varepsilon_0} \frac{2qs}{a^3}, 0 \right\rangle$$

The electric field due to the ball is the same as if the ball were a point charge located at its center. The vector $\vec{r}$ from the center of ball to location C is:

$$\vec{r} = \langle 0, 0, 0 \rangle - \langle b, a, 0 \rangle = \langle -b, -a, 0 \rangle$$

and the magnitude of $\vec{r}$ is:

$$\left| \vec{r} \right| = \sqrt{b^2 + a^2 + 0}$$

so the unit vector $\hat{r}$ is:

$$\hat{r} = \frac{\vec{r}}{\left| \vec{r} \right|} = \frac{\langle -b, -a, 0 \rangle}{\sqrt{b^2 + a^2}}$$

The electric field due to the charged ball is then:

$$\vec{E}_{ball} = \frac{1}{4\pi\varepsilon_0} (-Q) \frac{\langle -b, -a, 0 \rangle}{\sqrt{b^2 + a^2}} = \frac{1}{4\pi\varepsilon_0} Q \frac{\langle b, a, 0 \rangle}{\sqrt{b^2 + a^2}}$$

As shown in Figure 13.41, both the x and y components of $\vec{E}_{ball}$ are positive, which is correct, since the ball is negatively charged.

The net electric field is:

$$\begin{aligned}
\vec{E}_{net} &= \vec{E}_{dipole} + \vec{E}_{ball} \\
&= \left\langle 0, \frac{1}{4\pi\varepsilon_0} \frac{2qs}{a^3}, 0 \right\rangle + \frac{1}{4\pi\varepsilon_0} Q \frac{\langle b, a, 0 \rangle}{\sqrt{b^2 + a^2}} \\
&= \frac{1}{4\pi\varepsilon_0} \left\langle \frac{Qb}{\sqrt{b^2 + a^2}}, \frac{Qa}{\sqrt{b^2 + a^2}} + \frac{2qs}{a^3}, 0 \right\rangle
\end{aligned}$$

The direction of the net electric field is shown in Figure 13.41. (The diagram indicates that the contribution of the charged ball is larger in magnitude than the contribution of the dipole; of course we can't be sure of this without knowing numerical values for all the parameters in the problem.)

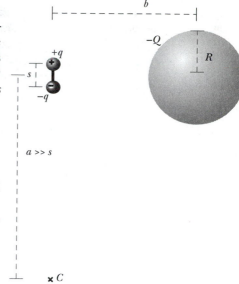

Figure 13.40 A dipole and a charged ball. Not to scale ($s \ll a$).

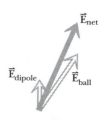

Figure 13.41 Contributions of dipole and ball, and net electric field at location C.

13.9 Review questions

Terminology and units

RQ 13.1 What is the relationship between the terms "field" and "force"? What are their units?

Electric field

RQ 13.2 At a particular location in the room there is an electric field $\vec{E} = (1000, 0, 0)$ N/C. Figure out where to place a single positive point particle, and how much charge it should have, in order to produce this electric field (there are many possible answers!). Do the same for a single negatively charged point particle. Be sure to draw diagrams to explain the geometry of the situation.

RQ 13.3 What is the magnitude and direction of the electric field $\vec{E}$ at location <20,0,0> cm if there is a negative point charge of 1 nC (1×10^{-9} C) at location <40,0,0> cm? Include units.

RQ 13.4 Where must an electron be to create an electric field of <0,160,0> N/C at a location in space? Calculate its displacement from the observation location and show its location on a diagram.

RQ 13.5 An electron is observed to accelerate in the +z direction with an acceleration of 1.6×10^{16} m/s^2. Explain how to use the definition of electric field to determine the electric field at this location, and give the direction and magnitude of the field.

RQ 13.6 The electric field at a location C points north, and the magnitude is 10^6 N/C. Give numerical answers to the following questions:
- Where relative to C should you place a single proton to produce this field?
- Where relative to C should you place a single electron to produce this field?
- Where should you place a proton and an electron, at equal distances from C, to produce this field?

Superposition

RQ 13.7 Where could you place one positive charge and one negative charge to produce the pattern of electric field shown in Figure 13.42? (As usual, each electric field vector is drawn with its tail at the location where the electric field was measured.) Briefly explain your choices.

RQ 13.8 Draw a diagram showing two separated point charges placed in such a way that the electric field is zero somewhere, and indicate that position. Explain your reasons.

RQ 13.9 Criticize the following statement: "A proton can never be at rest, because it makes a very large electric field near itself which accelerates it."

Dipoles

RQ 13.10 We found that the force exerted on a distant charged object by a dipole is given by

$$F_{\text{on Q by dipole}} \approx Q\left(\frac{1}{4\pi\varepsilon_0}\frac{2qs}{r^3}\right)$$

(a) In this formula, what is the meaning of the symbols "q", "Q", "s", and "r"?

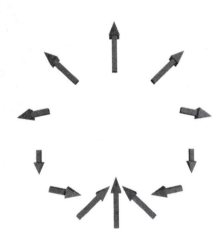

Figure 13.42 Electric field (RQ 13.7).

(b) At a given instant in time, three charged objects are located near each other, as shown in Figure 13.43. Explain why the formula above cannot be used to calculate the electric force on the ball of charge $+Q$.

RQ 13.11 Draw a diagram like the one in Figure 13.44. On your diagram, draw vectors showing
  (a) the electric field of the dipole at the location of the negatively charged ball
  (b) the net force on the ball due to the dipole
  (c) the electric field of the ball at the center of the dipole
  (d) the net force on the dipole due to the ball

RQ 13.12 If the distance between the ball and the dipole in RQ 13.11 were doubled, what change would there be in the force on the ball due to the dipole?

RQ 13.13 The dipole moment of the HF (hydrogen fluoride) molecule has been measured to be 6.3×10^{-30} C·m. If we model the dipole as having charges of $+e$ and $-e$ separated by a distance s, what is s? Is this plausible?

Relativity

RQ 13.14 You make repeated measurements of the electric field $\vec{E}$ due to a distant charge, and you find it is constant in magnitude and direction. At time t = 0 your partner moves the charge. The electric field doesn't change for a while, but at time t = 45 nanoseconds you observe a sudden change. How far away was the charge originally?

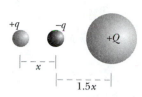

Figure 13.43 Three charged particles are located near each other (RQ 13.10).

Figure 13.44 The ball is far from the other charges (RQ 13.11).

13.10 Homework problems

Problem 13.1 Lithium nucleus affected by an electric field
A lithium nucleus consisting of 3 protons and 4 neutrons accelerates to the right due to electric forces, and the initial magnitude of the acceleration is 3×10^{13} meters per second per second.
  (a) What is the direction of the electric field that acts on the lithium nucleus?
  (b) What is the magnitude of the electric field that acts on the lithium nucleus? Be quantitative (that is, give a number).
  (c) If this acceleration is due solely to a single helium nucleus (2 protons and 2 neutrons), where is the helium nucleus initially located? Be quantitative (that is, give a number).

Problem 13.2 Electric field of two ions
An Fe^{3+} ion is located 400 nm (400×10^{-9} m, about 40 atomic diameters) from a Cl^- ion.
  (a) Determine the magnitude and direction of the electric field $\vec{E}_A$ at location A, 100 nm to the left of the Cl^- ion.
  (b) Determine the magnitude and direction of the electric field $\vec{E}_B$ at location B, 100 nm to the right of the Cl^- ion.
  (c) If an electron is placed at location A, what are the magnitude and direction of the force on the electron?

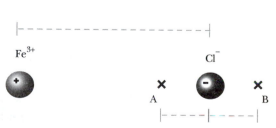

Figure 13.45 (Problem 13.2).

Problem 13.3 Making a zero electric field
  (a) On a clear and carefully drawn diagram, place a helium nucleus (consisting of 2 protons and 2 neutrons) and a proton in such a way that the electric field due to these charges is zero at a location marked ×, a distance 10^{-10} m from the helium nucleus. (A helium nucleus contains two protons and two neutrons.) Explain briefly but carefully, and use diagrams to help in the explanation. Be quantitative about the relative distances.

Continued on next page

(b) On a clear and carefully drawn diagram, place a helium nucleus and an electron in such a way that the electric field due to these charges is zero at a location marked ×. Explain briefly but carefully, and use diagrams to help in the explanation. Be quantitative about the relative distances.

Problem 13.4 Field and force with three charges
At a particular moment, one negative and two positive charges are located as shown in Figure 13.46. Your answers to each part of this problem should be vectors.
(a) Find the electric field at the location of Q_1, due to Q_2 and Q_3.
(b) Use the electric field you calculated in part (a) to find the force on Q_1.
(c) Find the electric field at location A, due to all three charges.
(d) An alpha particle (He^{2+}, containing two protons and two neutrons,) is released from rest at location *A*. Use your answer from part (c) to determine the initial acceleration of the alpha particle.

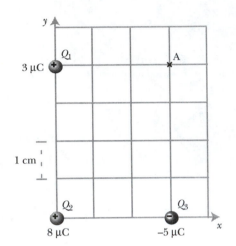

Figure 13.46 (Problem 13.4).

Problem 13.5 Measuring electric field
You are the captain of a spaceship. You need to measure the electric field at a specified location P in space outside your ship. You send a crew member outside with a meter stick, a stopwatch, and a small ball of known mass *M* and net charge +*Q* (held by insulating strings while being carried).
(a) Write down the instructions you will give to the crew member, explaining what observations to make.
(b) Explain how you will analyze the data that the crew member brings you to determine the magnitude and direction of the electric field at location P.

Problem 13.6 Electric field of two objects
A hollow ball with radius 2 cm has a charge of –3 nC spread uniformly over its surface (Figure 13.47). The center of the ball is at $\langle -3, 0, 0 \rangle$ cm. A point charge of 5 nC is located at $\langle 4, 0, 0 \rangle$ cm.
(a) What is the net electric field at location $\langle 0, 6, 0 \rangle$ cm?
(b) Draw an arrow representing the net electric field at that location. Make sure that the arrow you drew makes sense.

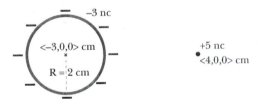

Figure 13.47 (Problem 13.6).

Problem 13.7 Electric field of a dipole (computer program)
(a) Consider a dipole comprising two singly charged ions (with charge +*e* and –*e* respectively) that are 0.2 nm apart (Figure 13.48). Write a computer program to find the magnitude and direction of the electric field due to this dipole at eight locations arranged as shown in Figure 13.48, and eight other locations not in the *xy* plane (if you are working in a 3D environment). Display a vector of appropriate length and direction indicating the field at each location. Make sure the display is comprehensible by scaling the vectors so they are easily visible but do not overlap.
(b) Place a proton at location <0, 0.3,0> nm and release it. Before writing the program to do this, predict qualitatively what you expect its motion to be. Using your expression for electric field from part (a), compute and animate the trajectory of the proton, leaving a trail to show the proton's path (that is, use the field concept in your calculations.) You may wish to start with $dt = 1 \times 10^{-17}$ s.
(c) Simultaneously compute and plot a graph showing the potential energy *U*, kinetic energy *K*, and (*K*+*U*) *vs.* time for the entire system (dipole + proton). Your graph will be more useful as a computational diagnostic tool if you do not include the potential energy associated with the interaction of the pair of charges making up the dipole, which does not change.
(d) Explain the shape of the *K* and *U* graphs.

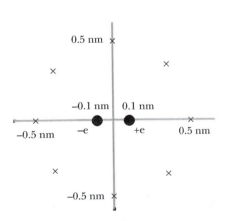

Figure 13.48 (Problem 13.7).

Problem 13.8 Exact vs. approximate calculations (computer program)
Write a computer program to calculate and plot a graph of the magnitude of the electric field of the dipole of Problem 13.7 at locations on the x axis as a function of distance from the center of the dipole. Vary x from 0.2 nm to 0.5 nm from the center of the dipole. Do the calculation two different ways and put both plots (in different colors) on the same axes:

(a) Calculate the electric field exactly as the superposition of the fields due to the individual charges.

(b) Calculate the electric field using the approximate formula for the dipole field derived in section 3.3.1.

Rescale your graph if necessary so you can see differences in the two values.

(c) Comment on the validity of the approximate formula. How close to the dipole (compared to s, the dipole separation) do you have to get before the approximate formula no longer gives good results? What is your criterion for "good results"?

Problem 13.9 Electric field of a dipole using the program *EM Field*
In this chapter we calculated the electric field of a dipole at two special places, marked 1 and 2 on Figure 13.49, which was produced using the computer program *EM Field* by David Trowbridge and Bruce Sherwood (published by Physics Academic Software, Raleigh NC). In this problem you will use a computer program to find the electric field at other locations near a dipole.

(a) Show that the directions and relative magnitudes of the electric field at locations 1 and 2 are in agreement with your previous results for a dipole.

(b) Draw the electric field vectors at positions 3 and 4. Explain your choices of direction and relative magnitude.

(c) Finding the electric field direction and magnitude at the locations 5, 6, 7, and 8 involves quite a bit of messy algebra and trigonometry. Draw your *guesses* for the electric field at these locations.

(d) Use the computer program *EM Field* to check your results for positions 3, 4, 5, 6, 7, and 8. You will be given a handout explaining the details of how to start up the program and how to make screen prints. You have a choice among point charges, long charged rods, and currents (which relate to magnetic fields to be studied later). On the "Sources" menu choose "3D point charges."

You choose a location (by clicking or dragging the mouse), and the program patiently does the messy algebra and trigonometry to add up the vector contributions of each charge, and displays the net electric field at that location.

To make the diagram above, five +8 charges were piled on top of each other (total charge of +40), and five –8 charges were piled on top of each other (total charge of –40), in order to make the electric field vectors for the dipole be large enough to see easily despite the partial cancellation due to the oppositely signed charges. The option "Constrain to grid" was used to make it easier to place the charges exactly on top of each other.

(e) Explore the electric field direction and magnitude at other locations, to get some feel for the complete pattern of electric field surrounding a di-

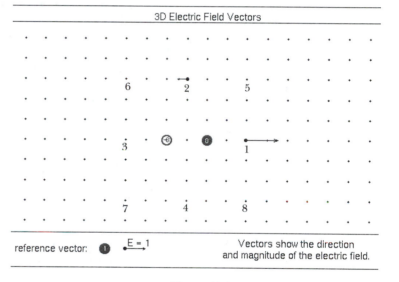

Figure 13.49 (Problem 13.9).

pole. Make a screen print of your full results and comment briefly on the pattern you find.

Problem 13.10 Motion of charges without friction

Electric Field Hockey by Ruth Chabay (published by Physics Academic Software, Raleigh NC) is a computer game that makes it possible for you to experiment with the motion of charged objects under the influence of electric interactions, without friction. Instructions are included in the program. We want you to play the game enough to get a good physical feel for the sometimes surprising ways in which the $1/r^2$ character of the electric interaction determines the motion of charged objects.

(a) Play the game, going through levels 1–5. Make a screen print of your solution to level 5.

(b) Now, using the same level 5 setup, add one more charge in the upper or lower right hand corner of the screen—as far away from your starting position as possible, but so that it makes a big difference in the ball's trajectory. If no effect is seen, move this extra charge to some other far-away location where it *does* make a big difference. Make a screen print of the result.

(c) Look at the trajectory of the ball shown in your first level 5 print (part a). At what locations along the ball's trajectory was the rate of change of momentum large? How can you tell from looking at the trail left by the ball? (The dots are at equal time intervals.)

(d) At what locations along the ball's trajectory was the magnitude of the rate of change of momentum small or zero? How can you tell from looking at the trail left by the ball?

(e) How do your answers to (c) and (d) illustrate the very rapid falloff of the Coulomb force with distance?

(f) Find a point on the trajectory where the net force on the ball was in a different direction from the ball's velocity. Draw and label vectors showing the force on the ball and its velocity at this point.

(g) Since the Coulomb force falls off very rapidly with distance, one would expect that a far away charge would have a negligible effect. However, in part (b) above you saw a large effect. Compare your screen prints carefully, especially along the first segment of the trajectory. Explain why the far-away charge had an effect.

13.11 Answers to exercises

13.1 (page 438)   2000 N/C

13.2 (page 438)   1060 N/C

13.3 (page 439) $\vec{E} = \langle 900, 0, 0 \rangle$ N/C

13.4 (page 439) $\vec{E} = \langle -955, -955, 0 \rangle$ N/C

13.5 (page 439) -3.5×10^{-9} C

13.6 (page 439) 1.44×10^{11} N/C

13.7 (page 440) $\vec{F} = \langle 0, -4.8 \times 10^{-17}, 0 \rangle$ N

13.8 (page 440) $\vec{F} = \langle 0, -9.6 \times 10^{-17}, 0 \rangle$ N

13.9 (page 441)   0 N

13.10 (page 444) 3.6×10^{4} N/C

13.15 (page 446) 1.15×10^{-14} N

13.11 (page 445) $\vec{E} = \left\langle 0, \dfrac{-1}{4\pi\varepsilon_0} \dfrac{2qr}{\left[r^2 + \left(\frac{s}{2} \right)^2 \right]^{3/2}}, 0 \right\rangle$

13.12 (page 445)   The force would be 1/8 as large.

13.13 (page 446)   The force would be 3 times larger.

13.14 (page 446)   The force would be 2 times larger, and would be in the opposite direction.

13.16 (page 446)   The dipole moment of the dipole aligns with the applied electric field:

13.17 (page 447) 2.8×10^{4} N/C

13.18 (page 447) 4.0×10^{-7} N

13.19 (page 448) 8×10^{-16} N

13.20 (page 448) 3.2×10^{-19} C $(=2e)$

Chapter 14

Matter and Electric Fields

Chapter 14

Matter and Electric Fields

14.1 Charged particles

Since ordinary matter is composed of charged particles, electric fields can affect matter. In order to understand the effect of electric fields on matter, in this chapter we will extend our microscopic model of matter to include the fact that matter contains charged particles: protons and electrons.

14.1.1 Net charge

Elementary particles such as protons and electrons are electrically charged. If a proton and an electron combine to form a hydrogen atom, however, the hydrogen atom is electrically "neutral"—its net charge is the sum of the charges of its constituent particles, which in this case is zero:

$$(+e) + (-e) = 0$$

A sodium atom has 11 protons in its nucleus, and 11 electrons surrounding the nucleus, so it has a net charge of zero, and is electrically neutral. However, a sodium atom can lose an electron, becoming a sodium ion, Na^+.

? What is the net charge of a sodium ion, Na^+ (Figure 14.1)?

A sodium ion has 11 protons and 10 electrons, so its net charge is

$$(+11e) + (-10e) = +e = +1.6 \times 10^{-19} \text{ C}$$

Ordinary matter is electrically neutral. However, it is possible to remove or add charged particles, giving an object a nonzero net charge.

14.1.2 Conservation of charge

In an extremely wide variety of experiments, no one has ever observed electric charge to be created or destroyed. These results are summarized by the fundamental principle called "conservation of charge": if the net charge of a system changes, the net charge of the surroundings must change by the opposite amount. For example, if your comb acquires negative charge, your hair acquires an equal amount of positive charge.

CONSERVATION OF CHARGE

The net charge of a system and its surroundings cannot change.

Consider the annihilation reaction between an electron and a positron:

$$e^- + e^+ \rightarrow \gamma + \gamma$$

? In this reaction an electron and a positron are destroyed, creating two high energy photons. Does this reaction violate the principle of conservation of charge?

No. The net charge of the system (electron plus positron) was initially zero; the charge of the two photons is also zero. Even though charged particles were destroyed, the net charge of the system did not change.

Sodium ion contains 11 protons and 10 electrons

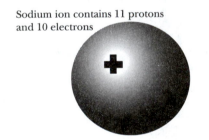

Lost an electron

Figure 14.1 A sodium ion Na^+ consists of a sodium atom that has lost an electron.

14.2 Observing electric interactions

Many of the interactions we observe in our everyday lives are electric in nature. By being systematic in observing the behavior of simple systems, and by thinking carefully in our analysis of this behavior, we can uncover some deep questions about the interaction of ordinary matter with electric fields. In the following sections we will do some simple experiments, and will use the electric field concept to reason about the results.

14.2.1 Is invisible tape electrically charged?

When you pull a long piece of invisible tape, such as Scotch® brand Magic™ Tape (or a generic brand), off a roll, it often curls up or sticks to your hand. We will do some simple experiments with invisible tape. Our first task is to determine whether or not a piece of invisible tape might be electrically charged.

? How can we decide if a piece of invisible tape is electrically charged?

If an object has a net electric charge, it should create an electric field in the surrounding space. Another charged object placed nearby should therefore experience an electric force. If we observe a change in an object's momentum we can conclude that a force acts on the object.

We know that the electric field of a point charge has these characteristics:
- the magnitude of $\vec{E}$ is proportional to the amount of charge
- the magnitude of $\vec{E}$ decreases with distance from the charge
- the direction of $\vec{E}$ is directly away from or toward the source charge

Therefore, since $\vec{F}_2 = q_2\vec{E}_1$, the electric force on object 2 should have the same properties. In addition, we should be able to observe both attraction and repulsion, since charges of different sign will be affected differently by a particular field.

We will observe the interactions of two pieces of invisible tape, and see if they meet the criteria listed above.

Preparing a "U" tape

Use a strip of tape about 20 cm long (about 8 inches, about as long as this paper is wide). Shorter pieces are not flexible enough, and longer pieces are difficult to handle. Fold under one end of the strip to make a non-sticky handle, as shown in Figure 14.2.

Figure 14.2 Fold under one end of a strip of tape to make a non-sticky handle.

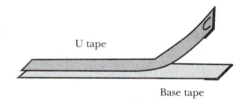

Figure 14.3 The U tape lies on top of the base tape.

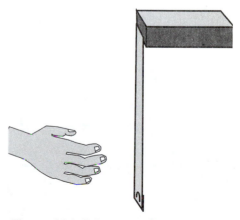

Figure 14.4 Bring your hand near a hanging U tape, and observe what happens.

HOW TO PREPARE A U TAPE
- Stick a strip of tape with a handle down onto a smooth flat surface such as a desk. This is a "base" tape.
- Smooth this base tape down with your thumb or fingertips. This base tape provides a standard surface to work from. (Without this base tape, you get different effects on different kinds of surfaces.)
- Stick another tape with a handle down on top of the base tape, as shown in Figure 14.3.
- Smooth the upper tape down well with your thumb or fingertips.
- Write U (for Upper) on the handle of the upper tape.
- With a quick motion, pull the U tape up and off the base tape, leaving the base tape stuck to the desk.
- Hang the U tape vertically from the edge of the desk, and bring your hand near the hanging tape, as shown in Figure 14.4. If the tape is in good condition and the room is not too humid, you should find that there is an attraction between the hanging strip of tape and your hand when you get close to the tape. If there is no attraction, remake the U tape.

A note on experimental technique: try to handle the tapes only by their ends while you are doing an experiment.

Experiment 14.1 Interaction of two U tapes

(a) If U tapes are electrically charged, how would you expect two U tapes to interact with each other? Would you expect them to repel each other, attract each other, or not to interact at all? Make a prediction, and briefly state a reason.

(b) Make two U ("upper") tapes by following the procedure detailed above. Make sure that both tapes interact with your hand. Hang one on the edge of a desk. Bring the second U tape near the hanging U tape. Since the hanging tape is attracted to your hands, try to keep your hands out of the way. For example, you might approach the vertically hanging tape with the other tape oriented horizontally, held by two hands at its ends. What happens?

You should have seen the two U tapes repel each other. If you did not observe repulsion, try remaking the U tapes (or making new ones, both from the same roll of invisible tape). It is important to see this effect before continuing further.

Making a tape not interact

You may have already discovered that if you handle a U tape too much, it no longer repels another U tape. Next we will learn a systematic way for making this happen.

- Make sure that you have an active U tape, which is attracted to your hand.
- Holding onto the bottom of the U tape, slowly rub your fingers or thumb back and forth along the *slick* side of the tape (Figure 14.5).
- You should find that the U tape no longer interacts with your hand. If it still does, repeat the process.

This is a little odd; if the U tape was originally electrically charged the charges would presumably have been on its sticky side. However, by running a finger along the other side (the slick side) we have apparently "neutralized" it—it now appears uncharged. It will be a while before we can explain this peculiar effect, but now we have a useful way to neutralize a U tape.

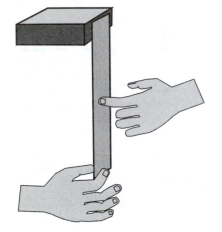

Figure 14.5 Neutralizing a tape by rubbing the slick side.

Experiment 14.2 Is this an electric interaction?

To decide whether the interaction between two U tapes (Figure 14.6) is or is not an electric interaction, we will see if it obeys the criteria for an electric interaction. (As is done throughout the scientific community, it is important to compare your results to the results of other experimenters.)

(a) Does the force act along a line connecting the two tapes? Think of a way to determine whether or not the force between two tapes acts along a line drawn from one object to the other, and do the experiment. What did you find? (What would you see if this were not the case?)

(b) Does the force decrease rapidly as the distance between the tapes increases? How can you determine this?

(c) Is the force proportional to the amounts of both charges? Design and carry out an experiment to test this. One way to vary the amount of charge on a tape is to neutralize part of one of the tapes. by running your finger along the length of the slick side of the tape, being careful that your finger touches only a portion of the width of the tape. What do you observe?

Figure 14.6 Two U tapes repel each other.

The real world is messy! You may have noted several difficulties in making your measurements. For example, the tapes are both attracted to your hand, as well as repelling each other. If you tried to use a ruler to measure the distance between the tapes, you may have found that the tapes are attracted to the ruler, too.

Unlike charges

So far we have observed that two U tapes repel each other, that the force acts along a line between the tapes, that the strength of the repulsion decreases as the tapes get farther away from each other, and that the strength of the interaction depends on the amount of charge on the tape. These observations are consistent with the hypothesis that the U tapes are electrically charged, and that all U tapes have like electric charge.

? How could you prepare a tape that might have an electric charge unlike the charge of a U tape? Think of a plan before reading further.

Perhaps you reasoned along these lines: We don't know how the U tape became charged, but if the tapes started out neutral, maybe the U tape pulled some charged particles off of the lower tape (or vice versa). So now the lower tape should have an equal amount of charge, of the opposite sign.

Making an "L" tape

Here is a reproducible procedure for making an "L" tape, whose charge is unlike the charge of a U tape:

HOW TO PREPARE AN L TAPE
- Stick a strip of tape with a handle down onto a base tape, smooth this tape down thoroughly with your thumb or fingertips, and write L (for Lower) on the handle of this tape.
- Stick another tape with a handle down on top of the L tape, and write U (for Upper) on the handle of this tape. Smooth the upper tape down well with your thumb or fingertips.
- You now have three layers of tape on the desk: a base tape, an L tape, and a U tape (Figure 14.7).
- Slowly lift the L tape off the base tape, bringing the U tape along with it (and leaving the bottom base tape stuck to the desk). Hang the double layer of tape vertically from the edge of the desk and see whether there is attraction between it and your hand (Figure 14.8). If so, get rid of these interactions (hold the bottom of the tape and slowly rub the slick side with your fingers or thumb).
- Check that the tape pair is no longer attracted to your hand. This is important!
- Hold onto the bottom tab of the L tape and quickly pull the U tape up and off (Figure 14.9). Hang the U tape vertically from the edge of the desk, not too close to the L tape!

Repeating exactly the same procedure, make another pair of tapes so that you have at least two U tapes and two L tapes. Before separating the tapes from each other, always remember to make sure that the tapes are not attracted to your hand.

? An important step in preparing an L tape is to neutralize the L/U tape pair before separating the two tapes. Considering the principle of charge conservation, why is this step important? What can go wrong if this step is omitted?

The principle of conservation of charge states that if the pair has a total charge of zero before separation, the two tapes will have a total charge of zero after separation: one tape will have a charge of $+q$ and the other a

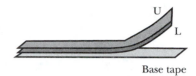

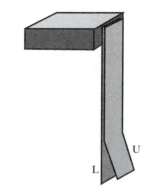

Figure 14.7 Preparing an L tape: First, smooth down three layers of tape—a base tape, an L tape, and a U tape.

Figure 14.8 Preparing an L tape: Second, lift the upper two layers (the L and U tapes) and hang them from the desk. Make sure that they are not attracted to your hand! (Neutralize if necessary.)

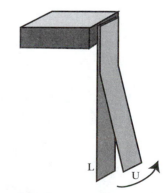

Figure 14.9 Preparing an L tape: Third, quickly pull the U tape off the L tape.

charge of $-q$. But if the total charge before separation is nonzero and positive (say), the separated tapes could both have positive charge, as long as their individual charges add up to the original amount.

Experiment 14.3 Observations of L and U tapes

You should now have two L tapes and two U tapes. Make sure that both the U tapes and the L tapes are active (attracted to your hand).

(a) If an L tape is indeed electrically charged, and its charge is unlike the charge on a U tape, what interaction would you *predict* between an L tape and a U tape?

(b) What interaction do you *observe* between an L tape and a U tape?

(c) What interaction would you *predict* between two L tapes?

(d) What interaction do you *observe* between two L tapes?

(e) Is the pattern of interactions consistent with the statement: "Like charges repel; unlike charges attract"?

A U tape and an L tape: Distance dependence of attraction

If U and L tapes are electrically charged, then we would expect the strength of the attractive interaction to decrease as the distance between the tapes increases. Make the same sort of observations you made with two U tapes.

Experiment 14.4 Distance dependence of force between U and L

Move a U tape very slowly toward a hanging L tape. Observe the deflections of the tapes from the vertical, at several distances (for example, the distance at which you first see repulsion, half that distance, etc.) The deflections of the tapes away from the vertical is a measure of the strength of the interaction.

(a) Does the force decrease rapidly as the distance between the tapes increases?

(b) Why is this measurement more difficult with a U and an L tape than with two U tapes?

Summary and conclusions: U and L tapes

Let's summarize the observations and try to conclude, at least tentatively, whether U and L tapes are electrically charged. Presumably you have observed the following:

- There are two kinds of charge, called "+" and "−".
- Like charges repel, unlike charges attract.
- The electric force
  - acts along a line between the charges,
  - decreases rapidly as the distance between the charges increases, and
  - is proportional to the amounts of both charges.

Our observations of U and L tapes seem to be consistent with a description of the electric interactions between charged objects. We tentatively conclude that U and L tapes are electrically charged, and have unlike charges.

14.2.2 How a plastic comb or pen becomes charged

Charged objects, such as invisible tape, are negatively charged if they have more electrons than protons, and positively charged if they have fewer electrons than protons. Are U tapes positively or negatively charged? How can we tell? Charging an object in a standard manner gives us a "litmus test."

It is known that if you rub a glass rod with silk, the glass rod becomes positively charged and the silk negatively charged. Likewise, if you rub a clear plastic object such as a pen through your hair (or with fur, wool, or even cotton), the plastic ends up having a negative charge and so repels electrons. A similar process occurs when you separate one tape from another.

There are a variety of possible explanations for this phenomenon. Electrons could be removed from one object and transferred to the other object. Large organic molecules in the plastic or your hair may break at their weakest bond in such a way that negative ions (negatively charged fragments) are deposited on the plastic and/or positive ions (positively charged fragments) are deposited on your hair. It may be significant that almost the only materials that can be charged easily by rubbing are those that contain large organic molecules, which can be broken fairly easily. It is typically more difficult to pull single electrons out of atoms or molecules, although we cannot rule out the possibility of stripping a single electron out of a molecule.

Glass (silicon dioxide) is one of the few common inorganic materials that can be charged easily by rubbing. It may be that positive ions break off the large organic molecules in the silk and are deposited on the glass, or that silk strips single electrons off of glass.

Molecular breakage or electron transfer provides an explanation of our puzzle as to why tapes and combs get charged, but such details as to why the plastic rather than your hair becomes negative are the subject of continuing research by physicists, chemists, and materials scientists. Part of the complexity of these phenomena is due to the fact that they are surface phenomena. The special nature of intermolecular interactions at the surface of a solid are generally less well understood than those in the interior, and there is a great deal of current research on the properties of surfaces. Moreover, unless one takes extraordinary precautions, real surfaces are always "dirty" with various kinds of (possibly charged) contaminants, which further complicates any prediction about the effect of rubbing, which may remove or deposit charged contaminants.

It is known that rubbing is not essential to transferring charge from one object to another. Mere contact is sufficient. However, rubbing produces many points of contact, which facilitates transfer.

You can't transfer bare nuclei or protons

One thing is certain: you cannot by rubbing remove bare nuclei from inside the surface atoms nor remove protons from inside the nuclei of the surface atoms. The amount of energy required to do this would be enormous. Removing protons would amount to transmuting one element into another!

The nucleus is buried deep inside the atom, and the protons are bound tightly in the nucleus. So the only charged objects that can be transferred by rubbing are positive or negative ions, or electrons.

Experiment 14.5 Determining the charge on U and L tapes
Prepare a U tape and an L tape, and hang them from your desk. Test them with your hand to make sure they are both charged. Rub a plastic pen or comb on your hair (clear plastic seems to charge best), or on a piece of cotton or wool, and bring it close to each tape. You should observe that one of the tapes is repelled by the pen, and one is attracted to it.

Knowing that the plastic is negatively charged, what can you conclude about the sign of the electric charge on U tapes? On L tapes?

These results may be reversed if you try a different brand of "invisible tape." Be sure to compare your results with those of other students. Make sure you all agree on the assignment of "+" and "–" labels to your tapes. (If other groups are using different brands of tape, you may disagree on wheth-er U tapes or L tapes are positive, but the electric interactions between your "+" tapes and their "+" tapes should be repulsive!)

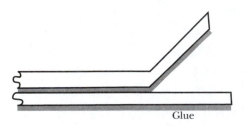

Figure 14.10 On a diagram like this, indi-cate the locations of + and – charges.

Ex. 14.1 Figure 14.10 is a side view of an upper tape (U) being pulled up off a lower tape (L). On a similar diagram, place –'s wherever a surface has gained negatively charged particles (electrons or negative ions) or lost positive ions and therefore has become negatively charged. Place +'s wherever a surface has gained positive ions or lost negatively charged particles (electrons or negative ions) and therefore has become positively charged. Place –'s and +'s only on the appropriate upper or lower surfaces where they actually are for your brand of tape! Be sure to check your completed diagram against the solution at the end of this chapter.

Ex. 14.2 In any of your experiments, did you find any objects, other than tapes or a charged comb or pen, that repelled a U or L tape? If so, those objects must have been charged. List these objects and whether the charge was "+" or "–".

14.2.3 Observing interactions with dipoles

You can also use charged tapes to observe the behavior of a dipole.

? In Figure 14.11, consider the forces that the positive charge Q exerts on the charges making up the dipole and describe the main features of the resulting motion of the dipole.

Figure 14.11 A positive charge Q interacts with a dipole.

There is a twist (torque) that tends to align the dipole along the line con-necting the charge Q and the center of the dipole, with the negative end of the dipole closer to the single positive charge. The dipole has a nonzero net force acting on it which makes it move toward the positive charge.

Experiment 14.6 An electric "compass"
Make a tall dipole and observe the motion (Figure 14.12). Take a + tape and a – tape and stick them together, overlapping them only enough to hold them together. Avoid discharging the tapes with too much handling. Hang the combination from a thread or a hair.

Now approach the tapes with a charged object and admire how sensitive a charge detector you have made! Move the charged object all around the dipole and observe how the dipole tracks the object.

If you draw an appropriately labeled arrow on the tape, you have an elec-tric "compass" that points in the direction of electric field.

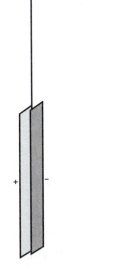

Figure 14.12 A dipole made from U and L tapes.

14.3 Amount of charge on a tape

We have concluded that U tapes and L tapes are electrically charged, but we have no idea how much charge is on one of the charged tapes—we don't even know an approximate order of magnitude for this quantity. Even a rough measurement of the amount of charge on a tape would be useful, be-cause it would give us a feel for the amount of charge there might be on an ordinary object which is observed to interact electrically with other objects. Therefore the following problem is an important one.

Problem 14.1 Amount of excess charge on a tape
In this problem you will design and carry out an experiment to determine the approximate number of excess electron charges on the surface of a negatively charged tape.

Initial estimates

Since we do not know what order of magnitude to expect for our answer, it is important to put upper and lower bounds on reasonable answers.

(a) What is the smallest amount of excess charge that a tape could possibly have?

(b) What is the largest amount of excess charge a tape could possibly have?

Design and perform an experiment

A centimeter ruler is printed on the inside back cover of this textbook. A piece of half-inch-wide (1.2 cm) invisible tape, 20 cm long (8 inches), has a mass of about 0.16 grams.

(c) Make a clear and understandable diagram of your experimental set-up, indicating each quantity you measured. Report all measurements you made.

Analyze the results

(d) Clearly present your physical analysis of your data. Make an appropriate diagram, labeling all vector quantities. Reason from fundamental physics principles. Explicitly report any simplifying assumptions or approximations you have made in your analysis. Report two quantities:

• the amount of charge on a tape, in coulombs
• the number of excess electrons to which this charge corresponds

Present your analysis clearly. Your reasoning must be clear to a reader.

(e) Estimate whether the true amount of excess charge is larger or smaller than the value you calculated from your experimental data. Explain your reasoning briefly.

Is this a lot of charge?

(f) What fraction of the molecules on the surface of the tape have gained an extra electronic charge? To estimate this, you may assume that molecules in the tape are arranged in a cubic lattice, as indicated in the accompanying figure, and that the diameter of a molecule in the tape is about 3×10^{-10} m. Does your answer suggest that it is a common event or a rare event for a molecule to gain an extra electron?

(g) If the electric field at a location in air exceeds 3×10^{6} N/C, the air will become ionized and a spark will be triggered. In Chapter 16 we will see that the electric field in a region very close to a uniformly charged disk or plate depends approximately only on the charge Q per unit area A:

$$E = \frac{1}{2\varepsilon_0}\left(\frac{Q}{A}\right)$$

Use this model (or make a different but justifiable simplifying assumption) to calculate the magnitude of the electric field at a location in the air very close to your tape (less than 1 mm from the surface of the tape). How does it compare to the electric field needed to trigger a spark in the air?

3×10^{-10}m

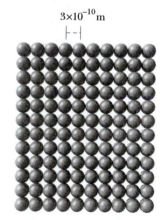

Array of molecules on one surface of a piece of invisible tape.

14.4 Interaction of charges and neutral matter

We have focused on the interactions of U and L tapes with other U and L tapes. Let's look more broadly at the interactions of charged tapes with other objects.

Experiment 14.7 Interactions of U and L tapes with other objects
Which other objects (paper, metal, plastic, etc.) have an attractive interaction with a hanging U or L tape, and which objects have a repulsive interaction? Which objects have no interaction? Record the objects you try, and the interactions observed.

The attraction of both U and L tapes to your hand, and to many other objects, is deeply mysterious. The net charge of a neutral object is 0, so your neutral hand should not make an electric field that could act on a charged tape; nor should your neutral hand experience a force due to the electric field made by a charged tape. Nothing in our statement of the properties of electric interactions allows us to explain this attraction!

14.4.1 The structure of an atom

An external charge can cause a shift in the position of the charges that make up a neutral atom or molecule. To see this clearly we need to look in more detail at the structure of atoms. We'll consider a hydrogen atom because it is the simplest atom, but the effects we discuss occur with other atoms as well.

In Figure 14.13 we show a special kind of picture of a hydrogen atom, based on quantum mechanics, the theory that describes the detailed structure of atoms. A hydrogen atom consists of an electron and a nucleus normally consisting of one proton (and no neutrons). The light-weight electron doesn't follow a well-defined orbit around the heavy nucleus the way the Earth does around the Sun. Rather, there is only a probability for finding the electron in any particular place.

Figure 14.13 shows this probability graphically. You can think of the picture as a multiple exposure. For each exposure, the position of the electron at that time is shown as a dot. Because the electron is most likely to be found near the nucleus, that part of the multiple exposure is so dark you can't see the individual dots. The electron is seldom found a long ways from the nucleus, so as you get farther and farther from the nucleus the density of dots gets less and less.

We call this probability distribution the "electron cloud." In hydrogen the cloud consists of just one electron, but in other atoms the electron cloud is made up of many electrons. The average location of the electron is in the center, at the same location as the nucleus. You're just as likely to find the electron to the right of the nucleus as to the left of the nucleus.

It is impossible to show the nucleus accurately on this scale. Although the mass of a proton is 2000 times the mass of an electron, the radius of the proton, about 10^{-15} m, is only about 1/100000 as big as the radius of the electron cloud, which is itself only about 10^{-10} meters! We used an oversize white dot to mark the position of the tiny nucleus in Figure 14.13.

In the following exercise, remember that in the previous chapter we pointed out that the electric field produced by a uniformly distributed sphere of charge, outside the sphere, is the same as though all the charge were located at the center of the sphere (this will be discussed in more detail in the next chapter).

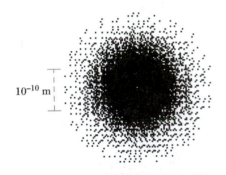

10^{-10} m

Figure 14.13 A quantum-mechanical view of a hydrogen atom. The picture is a two-dimensional slice through a three-dimensional spherical distribution. Each dot represents the location of the electron at the time of a multiple-exposure photo. The tiny nucleus is shown as a white dot at the center of the electron cloud (actual nuclear radius is only about 10^{-15} m).

Ex. 14.3 A student asked, "Since the positive nucleus of the atom is hidden inside a negative electron cloud, why doesn't all matter appear to be negatively charged?" Explain to the student the flaw in this reasoning.

14.4.2 Polarization of atoms

If the electron cloud in an atom could be considered to be spherically uniform and always centered on the nucleus, a neutral atom would have no interaction with an external charge. If the electron cloud is centered on the nucleus, the electric field produced by the *N* electrons would exactly cancel the field produced by the *N* protons. However, the electron cloud doesn't always stay centered, as we'll see next.

In an atom the electron cloud is not rigidly connected to the nucleus. The electron cloud and the nucleus can move relative to each other. If an external charge is nearby, it creates an electric field, which exerts forces on the electron cloud and on the nucleus. Under the influence of this "applied" electric field the electron cloud and the nucleus shift position relative to each other.

Figure 14.14 shows the probability distribution or electron cloud for hydrogen when there is an external positive charge located somewhere to the left of the hydrogen atom. You can see that the cloud has been distorted, because the positive charge attracts the electron to the left and repels the nucleus to the right.

Average location of the electron

The average location of the electron is now not at the center where the nucleus is located, but is displaced somewhat to the left of the nucleus. That is, each time that you take a snapshot, you're more likely to find the electron to the left of the nucleus than to the right of the nucleus.

The hydrogen atom isn't immediately torn apart, because the attraction between the nucleus and the electron is stronger than the forces exerted by the distant external charge. However, if the external charge gets *very* close the hydrogen atom may break up, or react with the external charge. If the external charge were a proton, it could combine with the hydrogen atom to form ionized molecular hydrogen (H_2^+).

You can see in Figure 14.14 that the outer regions of the cloud are affected the most by the external charge. This is because in the outer regions the electron is farther from the nucleus and can be influenced more by the external positive charge. In an atom containing several electrons, the outer electrons are affected the most. The picture is deliberately exaggerated to show the effect: unless the polarization is caused by charges only a few atomic diameters away, the shift in the electron cloud is normally too small to represent accurately in a drawing.

An atom is said to be "polarized" when its electron cloud has been shifted by the influence of an external charge so that the electron cloud is not centered on the nucleus.

Diagrams of polarized atoms or molecules

For most purposes we can approximate the charge distribution of the polarized atom as consisting of an approximately spherical negative cloud whose center is displaced from the positive nucleus (Figure 14.15). A uniform spherical charge distribution acts as if it were a point charge located at the center of the sphere, both in the sense that it makes an electric field outside the sphere identical to the electric field of a point charge, and that it responds to applied fields as though it were a point charge. It is therefore reasonable to model a polarized atom as a dipole, consisting of two opposite point charges separated by a small distance.

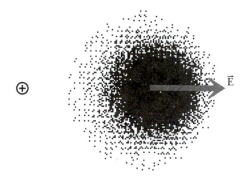

Figure 14.14 A positive charge makes an electric field that shifts the electron cloud of the hydrogen atom to the left (and shifts the hydrogen nucleus to the right). It is now more probable that the electron will be found to the left of the nucleus than to the right.

Figure 14.15 We can approximate a polarized atom as a roughly spherical electron cloud whose center is displaced from the positive nucleus.

Figure 14.16 A simplified representation of a polarized atom or molecule.

To simplify drawing a polarized atom or molecule, and to emphasize its most important aspects, we will often represent it as an exaggeratedly elongated blob, with + and – at the ends (Figure 14.16).

14.4.3 Induced dipoles are created by applied electric fields

Figure 14.15 shows quite clearly that a polarized atom or molecule is a dipole, since there are two opposite charges separated by a distance. However, the polarized atom or molecule is not a permanent dipole. If the applied electric field is removed (for example, by removing the charges making that field), the electron cloud will shift back to its original position, and there will no longer be any charge separation. We call the polarized atom or molecule an "induced" dipole, because the dipole was induced (caused) to form by the presence of an applied electric field.

An "induced dipole" is created when a neutral object is polarized by an applied electric field. The induced dipole will vanish if the applied field is removed.

A "permanent dipole" consists of two opposite charges separated by a fixed distance, such as HCl or H_2O molecules, or the dipole you constructed out of + and – tapes.

14.4.4 Polarizability

It has been found experimentally that for almost all materials, the amount of polarization induced (that is, the dipole moment of the polarized atoms or molecules) is directly proportional to the magnitude of the applied electric field. This result can be written like this:

$$\vec{p} = \alpha \vec{E}$$

The constant α is called the "polarizability" of a particular material. The polarizability of many materials has been measured experimentally, and these experimental values may be found in reference volumes.

Ex. 14.4 In an induced dipole, is the distance between the charges fixed, or can it vary? Explain.

Ex. 14.5 A typical atomic polarizability is 10^{-40} C·m/(N/C). If the q in $p = qs$ is equal to the proton charge e, what charge separation s could you produce in a typical atom by applying a large field of 3×10^6 N/C, which is large enough to cause a spark in air?

14.4.5 A neutral atom and a point charge

In the previous chapter we found that since the electric field of a dipole was proportional to $1/r^3$, the force exerted by a dipole on a point charge was also proportional to $1/r^3$. Because of the reciprocity of the electric force, the force on the dipole by the point charge was therefore also proportional to $1/r^3$. Let us extend this analysis by considering the case of a point charge q_1 and a neutral atom.

Even though the entire process happens very quickly, it is instructive to analyze it as if it occurred in several steps. (Of course, the process is not instantaneous, since information about changes in electric field takes a finite time to propagate to distant locations.)

Figure 14.17 At the location of the atom there is an electric field $\vec{E}_1$ due to the point charge.

(Step 1) At the location of the atom there is an electric field $\vec{E}_1$ due to the point charge (Figure 14.17). This electric field affects both the nucleus and the electron cloud, both of which, due to their spherical symmetry, can be modeled as point charges. The force on the electron cloud and the force on the nucleus are in opposite directions. Since the electron cloud and the nucleus can move relative to each other, they shift in opposite directions, until a new equilibrium position is reached.

The atom is now polarized, with dipole moment $\vec{p}_2 = \alpha\vec{E}_1$ proportional to the applied electric field $\vec{E}_1$.

(Step 2) The polarized atom now has a dipole moment $p = q_2 s$. The atom, which is now an induced dipole, makes an electric field $\vec{E}_2$ at the location of the point charge (Figure 14.18). We can write an expression for the magnitude of $\vec{E}_2$:

$$\left|\vec{E}_2\right| = E_2 = \frac{1}{4\pi\varepsilon_0}\frac{2p}{r^3} = \frac{1}{4\pi\varepsilon_0}\frac{2\alpha E_1}{r^3}$$

Figure 14.18 The polarized atom makes an electric field $\vec{E}_2$ at the location of the point charge.

Since we know $\vec{E}_1$, the electric field of the point charge at the location of the dipole, we can put that into our equation:

$$E_2 = \frac{1}{4\pi\varepsilon_0}\frac{2\alpha}{r^3}E_1 = \left(\frac{1}{4\pi\varepsilon_0}\frac{2\alpha}{r^3}\right)\left(\frac{1}{4\pi\varepsilon_0}\frac{q_1}{r^2}\right)$$

$$= \left(\frac{1}{4\pi\varepsilon_0}\right)^2\left(\frac{2\alpha q_1}{r^5}\right)$$

? (Step 3) What is the force exerted on the point charge by the induced dipole (Figure 14.19)?

$$\vec{F}_1 = q_1\vec{E}_2 = q_1\left(\frac{1}{4\pi\varepsilon_0}\right)^2\left(\frac{2\alpha q_1}{r^5}\right)\hat{r} = \left(\frac{1}{4\pi\varepsilon_0}\right)^2\left(\frac{2\alpha q_1^2}{r^5}\right)\hat{r}$$

We find that the force on the point charge by the polarized atom is proportional to $1/r^5$.

? What is the force on the neutral atom by the point charge?

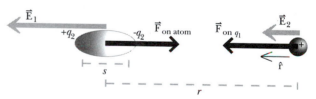

Because of the reciprocity of the electric interaction (Newton's third law), the force on the neutral atom by the point charge is equal in magnitude and opposite in direction to the force on the point charge by the neutral atom:

$$\vec{F}_2 = -\vec{F}_1 = -\left(\frac{1}{4\pi\varepsilon_0}\right)^2\left(\frac{2\alpha q_1^2}{r^5}\right)\hat{r}$$

Figure 14.19 The force on the point charge due to the electric field of the polarized atom is equal in magnitude to the force on the polarized atom due to the electric field of the point charge.

So the force on a (polarized) neutral atom by a point charge also is proportional to $1/r^5$.

Ex. 14.6 Atom A is easier to polarize than atom B. Which atom, A or B, would experience a greater attraction to a point charge a distance r away?

Ex. 14.7 If the distance between a neutral atom and a point charge is doubled, by what factor does the force on the atom by the point charge change?

14.4.6 Interaction of charged tapes and neutral matter

We are now in a position to explain why both positively and negatively charged tapes (U and L tapes) are strongly attracted to neutral matter.

? Try to explain in detail what happens when a positively charged tape is brought near your hand. This is a complex process; consider all the interactions involved.

In considering the interactions of fields and matter, it is useful to follow the following scheme: (1) Identify any sources of electric fields, (2) Identify any charges at other locations which can be affected by these fields, (3) Redistribution of the affected charges may create an electric field at the location of the original source charges: are they affected?

The positively charged tape makes an electric field, which points away from the tape. This electric field is present inside your hand, and affects atoms, molecules, and ions inside your hand. Figure 14.20 shows the polarization caused inside your finger by the electric field of the tape. You should be able to construct a diagram like the one in Figure 14.20 illustrating what happens when a negatively charged tape interacts with your finger.

You may have noticed that the attraction between your neutral hand and a hanging tape changes much more rapidly with distance ($1/r^5$) than does the interaction between two charged tapes ($1/r^2$).

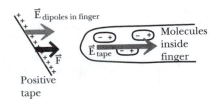

Figure 14.20 The electric field of the tape polarizes your finger. The induced dipoles in your finger create an electric field at the location of the tape, which attracts the tape.

Experiment 14.8 Observing attraction of like-charged objects (!)

Because of the very rapid $1/r^5$ increase in the attraction to neutral matter at short distances, it is sometimes the case that at short distances the attractive effect can actually overcome the repulsion between like-charged objects. Specifically, it could be that

$$N\left(\frac{1}{4\pi\varepsilon_0}\right)^2\frac{2\alpha q_1^2}{r^5} > \frac{1}{4\pi\varepsilon_0}\frac{q_1 q_2}{r^2}$$

where q_1 and q_2 are the excess charges on the surfaces of the two tapes, and N is the total number of neutral atoms in tape 2 (since each neutral atom participates in the attraction.) As can be seen by dividing the inequality by q_1, you can enhance the effect by making q_1 be significantly larger than q_2, so you might wish to partially discharge one of the tapes.

Try it! Hold one of the U tapes horizontally, with its slick side facing away from you and toward the slick side of the hanging tape. Move toward the hanging tape and check that the hanging tape is repelled as you approach. Then move close enough so that the tapes touch each other (a partner may have to hold the bottom of the hanging tape in order to be able to get very close).

You may be able to detect some slight attraction when the tapes are very close together or touch each other, despite the fact that the tapes repel at longer distances. Do you see such an effect? The effect is quite easy to see in the interaction between a highly-charged Van de Graaff generator and a charged tape.

14.4.7 Determining the charge of an object

Suppose you have a negatively charged tape hanging from the desk, and you rub a wooden pencil on a wool sweater and bring it near the tape.

? If the tape swings toward the pencil, does this show that the pencil had been charged positively by rubbing it on the wool?

Not necessarily. Even if the pencil is uncharged, the charged tape will polarize the pencil and be attracted by the induced dipoles.

? Can a charged object repel a neutral object? Why or why not? Draw diagrams to help you make your point.

Polarization always brings the unlike-sign charge closer, yielding a net attraction. Repulsion of an induced dipole can't happen. Therefore repulsion is a good test of whether an object is charged.

14.4.8 Electric field through intervening matter

The superposition principle states that the presence of matter does not affect the electric field produced by a charged object. Intervening matter does not "screen" or "shield" the electric field, just as your desk does not "screen" or "shield" your book from the gravitational field of the Earth.

Interaction through a U tape

You have already observed one case of electric field passing through intervening matter. You saw the same interaction with your hand, or another tape, when approaching either side of a hanging U tape, despite the charges being on the sticky side of the tape (the charges are initially on the sticky side, and we will soon show that the charges can't move through the tape to the slick side).

Experiment 14.9 Interaction through a piece of paper

Have a partner hold a piece of paper close to, but not touching, a hanging U tape. Bring another U tape toward the hanging tape from the other side of the paper, holding both ends of this tape so that it can't swing (Figure 14.21). Can you observe repulsion occurring right through the intervening paper?

This is difficult, because the paper attracts the hanging tape, which masks the repulsion due to the other tape. You can heighten the sensitivity of the experiment by moving the tape rhythmically toward and away from the hanging tape, as though you were pushing a swing. This lets you build up a sizable swing in the hanging tape even though the repulsive force is quite small, because you are adding up lots of small interactions. Using rhythmic movements, are you able to observe repulsion through the intervening paper?

The effect is especially hard to observe if you have weak repulsion due to high humidity. The farther away you can detect repulsion, the better, because the competing attraction falls off rapidly with distance. Under good conditions of low humidity, when tapes remain strongly charged and repulsion is observable with the tapes quite far apart from each other, it is possible to see repulsion with the paper in place, showing that electric field does go right through intervening matter. You have seen evidence of this when you observed attraction between a tape and your hand even when you approached the slick side of the tape.

Figure 14.21 Attempting to observe an electric interaction (repulsion) through a piece of paper.

Intervening matter and superposition

The fact that electric field acts through intervening matter is another example of the superposition principle. It is true that the repulsion is weaker when the paper is in the way, but when viewed in terms of the superposition principle this reduction is not due to the paper partially "blocking" the field of the other tape. Rather, we say that the net field is due to the superposition of two fields: the *same* field that you would have had without the paper intervening, plus another field due to the induced dipoles in the paper.

At this time we can't prove that this view is correct and that there is no "blocking" of electric field. However, during this course we will find repeatedly that the superposition principle makes the right predictions for a broad range of phenomena and offers a simpler explanation than any kind of hypothetical "blocking" effect.

14.5 Conductors and insulators

All materials are made of atoms that contain electrons and protons, but different materials respond in different ways to electric fields. These differences can lead to very different electric phenomena.

14.5.1 Definition of "conductor" and "insulator"

Many materials are made up of molecules which do not easily break apart, and whose electrons are tightly bound to the molecules. These materials are called insulators because such materials can electrically "insulate" one charged object from another, since charged particles cannot flow through the material.

Other materials contain charged particles, such as ions or electrons, that are free to move through the material. These materials are called conductors.

DEFINITION OF "CONDUCTOR" AND "INSULATOR"

A conductor contains mobile charges that can move through the material.

An insulator has no mobile charges.

Experiment 14.10 Is tape a conductor or an insulator?
Prepare a hanging tape that has the top half charged and the bottom half uncharged. After a second or two, check to see whether the bottom half of the tape has become charged.

Based on this observation, is tape a conductor or an insulator? That is, are the charges free or bound? Explain fully and rigorously how your observations justify your conclusion. (Hint: draw a diagram showing what effect the charges on the upper half of the tape have on each other and on charges inside the tape, and reason through what will happen if any of these charges are free to move.)

14.5.2 Polarization of insulators

In insulators, all of the electrons are firmly bound to the atoms or molecules making up the material. We have seen that an individual atom or molecule can be polarized by an applied electric field, producing an induced dipole of atomic or molecular dimensions. The electrons in an atom or molecule of an insulator shift position slightly, but remain bound to the molecule—no charged particles can move more than about one atomic diameter, or 10^{-10} m (most move significantly less than this distance).

In Figure 14.22 we show a solid block of insulating material, each of whose molecules has been polarized by an applied electric field (that is, an electric field made by external charges—in this case a single positive charge). The molecules are of course not shown to scale! This is an example of "induced polarization"—the electric field has induced the normally unpolarized insulator to become polarized. In each molecule the electrons have moved a very short distance, and the molecules themselves are not free to move. However, the net effect can be very large because there are many

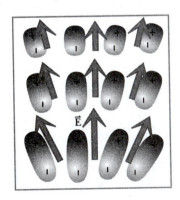

Figure 14.22 A block of insulating material (plastic, glass, etc.) polarized by an applied electric field. The molecules are not shown to scale!

molecules in the insulator to be affected. Note that the polarized molecules align with the electric field that is polarizing them, and that the stronger the electric field the larger the "stretch" of the induced dipole.

Diagrams showing polarization of insulators

In diagrams of insulators we show polarized molecules exaggerated in size, to indicate that individual molecules in an insulator polarize, but the electrons remain bound to the molecule. We show the extent of polarization by the degree to which the molecule is "stretched." Keep this diagrammatic convention in mind, and compare it to diagrams of polarized conductors in the following sections.

Charge on or in an insulator

Since there are no mobile charged particles in an insulator, excess charges stay where they are. Excess charge can be located in the interior of an insulator, or can be bound to a particular spot on the surface without spreading out along that surface (Figure 14.23).

14.5.3 Low-density approximation

When an electric field $E_{applied}$ is applied to a dense material (a solid or a liquid), the induced dipole moment of one of the atoms or molecules in the material isn't simply $p = \alpha E_{applied}$, but is really $p = \alpha |\vec{E}_{applied} + \vec{E}_{dipoles}|$, where $E_{dipoles}$ is the additional electric field at the location of one of the molecules, due to all the other induced dipoles in the material. In this course we make the simplifying assumption of low density and assume that $E_{dipoles}$ is small compared to $E_{applied}$. This is good enough for our purposes, but accurate measurements of polarizability must take this effect into account.

14.5.4 Polarization of conductors

As we stated earlier, a conductor has some kind of charged particles that can move freely throughout the material. In contrast to an insulator, where electrons and nuclei can move only very small distances (around 10^{-10} m), the charged particles in a conductor are free to move large distances.

14.5.5 Ionic solutions

Ionic solutions are conductors, such as a solution of sodium chloride (table salt) in water. In salt water, the mobile charged particles are Na^+ ions and Cl^- ions (Figure 14.24; there are also very small concentrations of H^+ and OH^- ions which are not shown).

? What happens when an electric field is created in the region of the beaker?

When an electric field is applied to a conductor, the mobile charged particles begin to move in the direction of the force exerted on them by the field. However, as the charges move, they begin to pile up in one location, creating a concentration of charge which itself creates an electric field in the region occupied by the remaining mobile charges. The net electric field in the region is the superposition of the applied (external) field and the electric field created by the relocated charges in the material. Figure 14.25 is a diagram of the polarization that occurs in the salt water. The ions (charged atoms or molecules) are in constant motion, so the actual situation isn't simple. Moreover, the interior of the liquid is full of positive and negative ions; there's just a slight excess concentration of ions near the sides of the beaker.

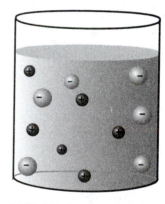

Figure 14.23 In an insulator, charge can occur in patches on the surface, and there can be excess charge inside.

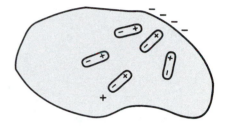

Figure 14.24 A beaker containing an ionic solution (salt water).

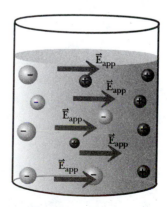

Figure 14.25 Under the influence of an applied electric field (labeled $\vec{E}_{app}$), the liquid polarizes. There is a slight excess ion concentration at the two sides of the beaker. The electric field due to the redistributed ions is not shown.

The polarization process in an ionic solution

Polarization occurs very rapidly, but it is not instantaneous. Let's "slow down time" so we can talk about the process of polarization; we'll operate on a time scale of attoseconds (10^{-18} seconds!). To simplify our analysis, we'll imagine that we are able temporarily to "freeze" the ions in the salt water, and to release them after we have brought charges nearby to apply an electric field.

Consider the net electric field at a location in the interior of the liquid, at a time attoseconds after polarization has begun, but long before the process has finished. The electric field E_{app} due to external charges is shown in Figure 14.26, and a smaller electric field E_{pol} due to the polarization charges present at this time. The net electric field at a location in the middle of the liquid is now smaller than it was before polarization began.

? Will the polarization of the salt water increase beyond what it is now?

At this instant the net electric field in the solution still has magnitude greater than zero, so ions in the solution will still experience forces in the direction to increase the polarization. More ions will pile up at the sides of the beaker, and the net electric field in the interior will be further weakened.

? How far does this weakening process go? In the final state of static equilibrium (when there is no further increase in polarization), how big is the net electric field in the interior of the liquid?

An example of a "proof by contradiction"

You may have correctly deduced that in the final state the net electric field in the conductor goes to zero at equilibrium. A rigorous way to reason about this is to construct a "proof by contradiction." In a proof by contradiction, we assume the opposite of what we want to prove, then, making valid logical deductions from this assumption, show that we reach a conclusion which is impossible or contradictory. We therefore conclude that the original assumption was wrong, and its opposite must be true.

1) Assume that in static equilibrium the net electric field in the interior of an ionic solution is greater than zero.

2) Since $E_{net} > 0$, mobile ions in the solution will experience a force and will begin to move in the direction of the force.

3) Since there is motion of charges, the system cannot be in static equilibrium, because by definition static equilibrium means no motion. This contradicts our assumption that the ionic solution is in static equilibrium.

4) Hence, the assumption that the net electric field in the solution is nonzero in static equilibrium must be wrong. Thus, the net electric field in an ionic solution in static equilibrium must be zero.

Superposition

Note that the electric field inside the liquid is zero not because of any "blocking" of fields due to external charges, but by the superposition of two effects: the effect of the external charges, and the effect of the polarization charges. This is another example of the superposition principle in action.

It is not true that the net electric field in a solution is zero at *all* times. While the ionic solution is in the process of polarizing, it is not in static equilibrium; there is a nonzero electric field, and hence a nonzero force on an ion in the liquid, as you saw above. If electrodes are placed in the ionic solution and connected to a battery, the battery prevents the system from reaching static equilibrium. In such a case (no static equilibrium), there can be a field continuously acting on ions inside the liquid, resulting in contin-

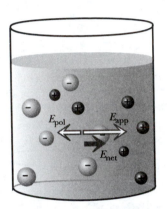

Figure 14.26 An intermediate stage in the polarization process, before polarization is complete.

uous shifting of the ions through the liquid, constituting an electric current.

Since there are a very large number of ions in the solution, none of them has to move very far during the polarization process. Even a tiny shift leads to the buildup of an electric field large enough to cancel out the applied electric field.

Polarization of salt water in the body

Your own body consists mainly of salt water, including the blood and the insides of cells. Look again at the diagrams in which you focused on the way an external charge polarizes individual molecules inside your finger. An additional effect is the polarization of the salt water inside your finger. As shown in Figure 14.25, there will be a shift of Na^+ and Cl^- ions in the blood and tissues. This shift may be a larger effect than the molecular polarization. It is a bit unsettling to realize that a charged tape or comb messes with the insides of your body!

14.5.6 A model of a metal

You probably know that metals are very good electrical conductors. In a metal, the mobile charged particles are electrons.

The mobile electron sea

The atoms in a solid piece of metal are arranged in a regular 3D geometric array, called a "lattice." The inner electrons of each metal atom are bound to the nucleus. Some of the outer electrons participate in chemical bonds between atoms (the "springs" in the ball-and-spring model of a solid). However, some of the outer electrons (usually one electron per atom) join a "sea" of mobile electrons which are free to move throughout the entire macroscopic piece of solid metal (Figure 14.27). In a sense, the entire hunk of metal is like one giant molecule, in which some of the electrons are spread out over the entire crystal. The electrons are not completely free; they are bound to the metal as a whole and are difficult to remove from the metal. (For example, electrons do not drip out when you shake a piece of metal!). Metals are excellent conductors because of the presence of these mobile electrons.

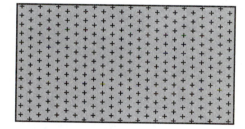

Figure 14.27 An unpolarized metal: uniform mobile-electron sea (gray), positive atomic cores ("+" symbols).

No net interaction between mobile electrons

Although the roaming electrons repel each other strongly, this repulsion between electrons is neutralized on the average by the attractions exerted by the positive atomic cores (a "core" is a neutral atom minus its roaming electron, so it has a charge of $+e$). The effect is that on average, the net electric field inside a piece of metal in equilibrium is zero.

Because of this, in some ways the mobile electrons look like an ideal gas: they move in a region free from electric field, so they appear not to interact with each other or with the atomic cores. In fact, in some simple models of electron motion the mobile electron sea is treated as an ideal gas.

Applying an electric field

When an electric field (due to some external charges) is applied to a metal, the metal polarizes. We can describe the polarization of a metal as shifting the entire mobile-electron sea relative to the fixed positive cores. In Figure 14.28, in response to an applied field, electrons have piled up on the left, creating a very thin negatively charged layer near the surface. There is a corresponding deficiency of electrons on the right, creating a very thin positively charged layer near the surface.

Excess of electrons (excess negative charge)

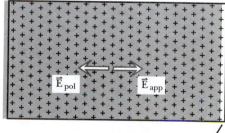

Deficiency of electrons (excess positive charge)

Figure 14.28 Polarized metal: mobile-electron sea shifted left relative to the positive atomic cores, under the influence of an applied electric field. There is an excess of electrons on the left side, and a deficiency of electrons on the right. These charges contribute to the net electric field inside the metal.

Diagrams showing polarization of metals

In Figure 14.29 we show a polarized metal in a simplified way that is both easier to draw and easier to interpret at a glance.

• We show – and + signs outside the surfaces to indicate which surfaces have thin layers of negative charge (electron excess) or positive charge (electron deficiency), as a result of shifts in the mobile electron sea. Note that by convention we draw + and – signs just outside the surface of a metal object to indicate that the excess charge (excess or deficiency of electrons) is on the surface of the object. If the charge is drawn inside the boundary, the diagram is ambiguous—it is not clear whether the charge is inside the object, or on the surface.

• We do not show the positive atomic cores and the mobile electron sea inside the metal, because the interior is all neutral. The diagram is much easier to interpret if we do not clutter up neutral regions with charges that must be counted to see if they balance.

Note that most charge build-up is typically on the ends of the metal, but that there is also a small amount of charge on the sides as well.

Compare these conventions to the convention we used earlier to show that individual atoms or molecules polarize in an insulator.

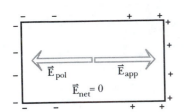

Figure 14.29 A simpler way to depict the polarization of a polarized metal. Excess charges are drawn outside the boundary lines, to indicate that they are on the surface.

Polarized and/or charged

Take care to use technical terms precisely. The metal block shown in Figure 14.29 is *polarized*. It is *not* charged; its net charge is still zero. "Polarized" and "charged" are not synonyms.

14.5.7 Electric field in a metal goes to zero in static equilibrium

The reasoning process which we went through when considering the polarization of ionic solutions applies equally well to metals, or to any conductor. We used proof by contradiction to demonstrate that at equilibrium the net electric field inside a conductor must be zero (because if it were not zero, mobile charged particles would move under the influence of the field, and the system would not be at equilibrium).

It is intriguing that it is possible for mobile charges to rearrange themselves in such a way that the net electric field is zero not just at one single location, but at every location inside the metal. It would be a very difficult problem for us to calculate exactly where to place charged particles to make a net field of zero inside a metal object, but in fact the many mobile charges do rearrange in just such a way as to accomplish this. It can be shown that it is because of the $1/r^2$ distance dependence of the electric field that this is possible—if the exponent were not exactly 2.0 the world would be quite different.

When equilibrium is reached in a metal, things are essentially unchanged in the interior of the metal. There is no excess charge—we still have a uniform sea of electrons filling the space around the positive atomic core. The net electric field inside the metal, which is the sum of the applied field and the field due to the charge buildup on the edges of the metal, is still zero.

$$\vec{E}_{net} = \vec{E}_{app} + \vec{E}_{pol} = 0 \text{ in a conductor at equilibrium}$$

At the surfaces there is some excess charge, so we can represent a polarized metal as having thin layers of charge on its surfaces but being unpolarized in the interior, unlike an insulator.

The shifting of the mobile-electron sea in metals is a much larger effect than occurs in insulators, where the polarization is limited by the fact that all the electrons, including the outermost ones, are bound to the atoms, unlike the situation in metals. One can think of a polarized insulator as a collection of tiny (molecule-sized) dipoles, whereas a polarized metal forms a giant dipole.

E is not always zero inside a metal

Do not over generalize our previous conclusions. It is not true that the net electric field in a metal is zero at all times. While the metal is in the process of polarizing, the metal is not in static equilibrium, and there is a nonzero electric field inside the metal, creating a nonzero force on electrons in the electron sea, as you saw above. In an electric circuit the battery prevents the system from reaching static equilibrium. In such a nonequilibrium situation, there can be an electric field inside the metal, and hence a force continuously acting on electrons in the mobile electron sea, resulting in continuous shifting of the electron sea around the closed circuit, constituting an electric current.

14.5.8 Excess charges on conductors

Another important property of metals (and of the $1/r^2$ property of the electric interaction, as we will see when we study Gauss's law in a later chapter) is that any excess charges on a piece of metal, or any conductor, are always found on an outer or inner surface. This makes intuitive sense, since any excess charges will repel each other and will end up as far apart as possible—on the surface of the conductor. Any multi-atom region in the interior of the conductor has a net charge of zero. Moreover, the mutual repulsion among any excess charges makes the mobile electron sea redistribute itself in such a way that charge appears almost immediately all over the surface (Figure 14.30).

Figure 14.30 In a metal, charge is spread all over the surface (not necessarily uniformly), and there is no excess charge inside.

Summary: Conductors versus insulators

	conductor	insulator
mobile charges	yes	no
polarization	entire sea of mobile charges moves	individual atoms or molecules polarize
static equilibrium	$\vec{E}_{net} = 0$ inside	$\vec{E}_{net}$ nonzero inside
location of excess charge	only on surface	anywhere on or inside material
distribution of excess charge	spread out over entire surface	located in patches

Ex. 14.8 An object can be both charged *and* polarized. On a negatively charged metal ball, the charge is spread uniformly all over the surface (Figure 14.31). If a positive charge is brought near, the charged ball will polarize. Draw the approximate final charge distribution on the ball.

Ex. 14.9 A negatively charged plastic pen is brought near a neutral solid metal cylinder (Figure 14.32). Show the approximate charge distribution for the metal cylinder. Then draw a vector representing the net force exerted by the pen on the metal cylinder, and explain your force vector briefly but completely, including all relevant interactions.

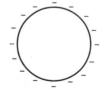

Figure 14.31 *Ex. 14.8.* This is a cross section of the metal ball.

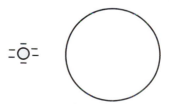

Figure 14.32 *Ex. 14.9* and *Ex. 14.10.* This is a cross section of the cylinder.

Ex. 14.10 A negatively charged plastic pen is brought near a neutral solid plastic cylinder (Figure 14.32). Show the approximate charge distribution for the plastic cylinder. Then draw a vector representing the net force exerted by the pen on the plastic cylinder, and explain your force vector briefly but completely, including all relevant interactions.

14.6 Charging and discharging

We say an object is "charged" when its net charge is nonzero. You charged initially neutral tapes by stripping them off other pieces of tape. You charged an initially neutral pen by rubbing it on your hair. In both these cases, some kind of charged particle was added to or removed from a surface that was originally neutral.

Your own body plays an interesting role in some kinds of charging or discharging phenomena. In the following sections we will see why.

14.6.1 Discharging by contact

If you exercise on a hot day, you sweat, and your body becomes covered with a layer of salt water. Even in a cool place, when you are not moving, there is usually a thin layer of salt water covering your skin. As we saw in section 14.5.5, salt water is a conductor, so you have a conducting film all over the surface of your skin.

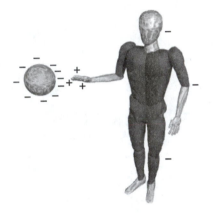

Figure 14.33 The metal is charged, and the person is uncharged but slightly polarized.

When you approach a negatively charged surface, your body polarizes as shown in Figure 14.33. The polarization includes not only atomic or molecular polarization but also polarization of the blood and sweat, which are salt solutions.

When you touch the charged object, the negatively charged object attracts positive Na^+ ions in the film of salt water on the skin. The Na^+ ions pick up an electron, partially neutralizing the excess negative charge of the object (Figure 14.34). The body acquires a net negative charge. (The Na atom can react with the water to form NaOH!) In the case of a small piece of metal, on which charge is free to redistribute itself, this process nearly neutralizes the metal, because the original net amount of charge is now spread out over the much larger area of metal plus human body.

Similarly, a positive metal surface would attract negative Cl^- ions from your skin, which give up an electron to the metal. The body acquires a net positive charge. (Chlorine can be emitted in tiny quantities!)

Grounding

Touching a small charged object is a pretty effective way to discharge the object. An even better way to discharge an object is to "ground" it by making a good connection to the earth or ground (typically through a water pipe that goes into the ground). Earth is a rather good conductor due to the presence of water containing ions. Grounding spreads charge throughout a huge region, neutralizing an object essentially completely.

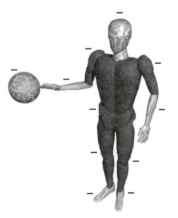

Figure 14.34 The net negative charge is distributed over a much larger area, nearly neutralizing the metal.

14.6.2 How did we neutralize a tape?

We've routinely used what may have seemed a very odd method to discharge a tape, by rubbing the slick side even when it was the sticky side that got charged. Let's try to clear up this puzzle.

You can easily discharge a charged metal foil by briefly touching it anywhere, because it is a conductor. It is more difficult to discharge a charged tape. Since the tape is an insulator, you have to run your fingers all over a surface to neutralize all the charges on that surface. As it happens, rubbing the sticky side of the tape tends to charge the tape, which would compete with your attempt to discharge it. This is why we have had you rub only the slick side of the tape. But the question remains as to why this works.

Ex. 14.11 Figure 14.35 is an edge view of an upper (U) tape whose sticky side is negatively charged (your brand of tape may differ). Imagine running your fingers along the slick side of this tape. Will the tape attract charges from your body? Remember that there are mobile positive and negative ions on your skin (see section 14.6.1). Show and label the charge situation in and on the tape after removing your fingers. How will a tape with these charges interact with other objects, compared to the original charged tape?

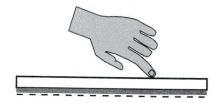

Figure 14.35 You run your finger along the slick side of the tape, and the tape seems to become neutralized.

Experiment 14.11 Discharging a tape

The previous exercise suggests that a key element in neutralizing a tape is the salt solution on the surface of your finger. Design one or more experiments you can do to confirm or reject this explanation of discharging a tape.

14.6.3 Charging by induction

The conductive properties of the body can be used to charge a piece of metal in an efficient way.

Experiment 14.12 Charging by induction

Hang a short piece of aluminum foil (about the width of the tape and half as long as your thumb) from a tape, with another piece of tape added to the bottom of the foil as a handle (Figure 14.36). Now carry out the following operations *exactly* as specified:

① Make sure that the tape and foil are uncharged (touch the foil, and rub the slick side of the tape).

② Have a partner hold onto the bottom tape to keep the foil from moving.

③ Bring a charged plastic pen or comb very close to the foil, *but don't touch the foil with the plastic.*

④ *While holding the plastic near the foil*, tap the *back* of the foil with your finger.

⑤ Move your finger away from the foil, *then* move the plastic away from the foil.

You should find that the metal foil is now strongly charged. This process is called "charging by induction." Now touch the charged aluminum foil with your finger and observe that this discharges the foil, as predicted by the discussion on the previous page.

Figure 14.36 Hang a short piece of aluminum foil from a neutral tape.

Ex. 14.12 Complete the "comic strip" of diagrams illustrating the charging by induction process you carried out (Figure 14.37). Make sure you have the sign of the charges right. In each diagram show charge distributions, polarization, movement of charges, etc. For each frame, explain briefly what happens. Remember that excess charges on the metal foil can only be on the surface.

This process is called charging by induction because the entire piece of foil becomes an induced dipole when it is polarized by the external charge. Charging by induction makes it possible to charge a metal without touching the external charge to the metal.

Cross-sectional side views

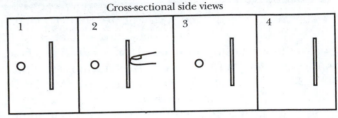

Figure 14.37 Give a step-by-step explanation of charging by induction.

Ex. 14.13 Explain the process of discharging the metal foil by touching it, using the same kind of time-sequence "comic strip" diagrams you used in the previous exercise. Illustrate the important aspects of each step in the process. Include any changes to your body as well as to the foil. Be precise in your use of words.

14.6.4 *Effect of humidity on tapes

Isolated atoms are always symmetrical and unpolarized unless an external charge shifts the electron cloud and makes an induced dipole. But some molecules are permanently polarized even in the absence of an external charge, and this leads to important physical and chemical effects.

For example, water molecules are permanently polarized. The water molecule (H_2O) is not spherically symmetrical but has both hydrogen atoms off to one side of the oxygen atom. In Figure 14.38 the δ^+ and δ^- symbols are used to indicate that slight shifts of the electron clouds to the right leave the right side of the molecule a bit negative, and the left side a bit positive, so the water molecule is a permanent dipole.

Many of water's unusual chemical and physical properties are due to this structure. In particular, the charged ends can bind to ions, which is why many chemicals dissolve well in water.

When water molecules in the air strike a surface they sometimes become attached to the surface, probably because the charged ends bind to the surface. A film of water builds up on all surfaces. Pure water is a very poor conductor but does contain small amounts of mobile H^+ and OH^- ions. More importantly, the water dissolves surface contaminants such as salt, and the impure water provides an effective path for charges to spread onto neighboring objects. After a while a charged surface loses its original charge, so experiments with charged objects work better when the humidity is low.

Figure 14.38 A water molecule is a permanent dipole.

Experiment 14.13 A water film as a conductor
Prepare a hanging tape that has the top half charged and the bottom half uncharged. Let it hang while you do other work, but check every few minutes to see what has happened in the two halves. What do you predict will happen to the state of charge in the two halves? What do you observe over a period of many minutes? (If the room is very dry or very wet you may not be able to see this effect.)

Suppose you were to breathe heavily through your mouth onto the slick and sticky sides of a short section in the middle of a long charged tape. Your breath is very moist. What do you predict you would find immediately afterwards? Try the experiment—what do you observe? (Repeat if you see no effect.)

14.6.5 *Amount of charge transferred by contact

We can determine experimentally how much charge is transferred by contact between two identical objects.

Experiment 14.14 Transferring charge by contact

Make two identical hanging foil arrangements (Figure 14.39) and charge one of the aluminum foils by induction. Discharge the other foil by touching it with your finger. Charge a tape or pen or comb, and note the approximate strength of the interaction between it and the charged foil.

Next make the two foils touch each other, being careful not to touch either foil with your fingers. Note the approximate strength of the interaction that there is now between the plastic and each foil.

Compared with the situation before the two foils touched, what sign of charge, and roughly how much, is there now on each foil? Discuss this fully with your partners and convince yourselves that you understand the process. Make a written explanation, including appropriate diagrams.

What would you expect to happen if one piece of foil were much larger than the other?

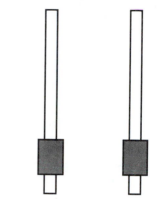

Figure 14.39 Prepare two hanging pieces of aluminum foil.

14.7 When the field concept is less useful

The field concept and the idea of splitting of the universe into two parts are not very useful if the charge affected by the field significantly alters the original distribution of source charges. Consider a negatively charged metal sphere (Figure 14.40). The electric field due to the sphere points radially inward.

If we place a particle with very little charge q near this charged sphere (a single proton, for example), it hardly alters the distribution of charge on the metal sphere. We can reliably calculate the small force on the small charge as $q\vec{E}$, where $\vec{E}$ is the electric field we calculated in the absence of the small additional charge q.

But if we place a particle with a *big* charge Q near the sphere, the sphere polarizes to a significant extent (Figure 14.41). We show the electric field due solely to the new charge distribution on the sphere (we don't show the large additional contribution to the net electric field due to Q). Clearly, the force on Q is not simply Q times the *original* $\vec{E}$, but Q times a significantly larger field.

With these effects in mind, we need to qualify our previous method for measuring electric field, in which we measure the force exerted on a charge q and determine the force per unit charge:

$$\vec{E} = \vec{F} / q$$

This procedure is valid only if q is small enough not to disturb the arrangement of other charges that create $\vec{E}$.

Since no object can have a charge smaller than e (the charge of a proton), sometimes it is not possible to find a charge small enough that it doesn't disturb the arrangement of source charges. In this case, we can't measure the electric field without changing the field!

On the other hand, if we know the locations of the source charges, we can calculate the electric field at a location, by applying the superposition principle and adding up the contributions of all the point charges that are the sources of the field:

$$\vec{E}_{net} = \frac{1}{4\pi\varepsilon_0}\left(\frac{q_1}{r_1^2}\hat{r}_1 + \frac{q_2}{r_2^2}\hat{r}_2 + \frac{q_3}{r_3^2}\hat{r}_3 + \cdots\right)$$

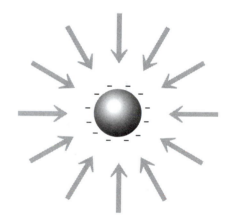

Figure 14.40 Electric field of a uniformly charged metal sphere.

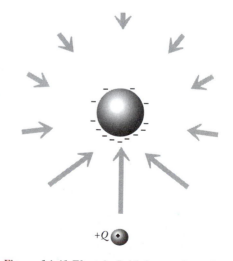

Figure 14.41 Electric field due to the polarized sphere.

If even the smallest possible charge e would disturb this arrangement of source charges, we can't use the calculated field to predict the force that would act on a charge placed at this location. However, we could use the calculated field to predict the polarization of a neutral atom placed at that location, because a neutral atom, even if (slightly) polarized, would disturb the existing arrangement of source charges much less than a charged object would.

14.8 Case Study: Sparks in air

The concept of electric field makes it possible to discuss in detail a very complex electric phenomenon: the occurrence of electric sparks in air. You may occasionally see an electric spark when you flip a switch, unplug a power cord, or when you separate clothing that has stuck together in the dryer or pull a blanket away from a sheet in the winter. During a spark, charge is transferred from one object to another. How can such sparks occur in air?

Like other gases, air is an excellent insulator, consisting mostly of neutral nitrogen and oxygen molecules. (If air were a good conductor, your charged tape would very quickly become discharged.) There are no moveable charged particles—all the electrons are bound to neutral gas molecules. But during a spark, electrons are ripped out of molecules ("ionization"), leaving free electrons and positive N_2^+ and O_2^+ ions, all of which can move and transfer significant amounts of charge from one place to another through the air.

An ionized gas is called a plasma. Plasma physics is a domain of physics concerned with the behavior of plasmas, which are found in stars, fusion reactors, and sparks.

The goal of the following sections is to figure out a physically reasonable mechanism for the creation of electric sparks in air. You will be asked to estimate various physical quantities, in order to judge whether alternative models of sparks are physically reasonable. We will see that sometimes simple calculations make it possible to completely rule out particular models of complex phenomena!

To explain a spark we must explain these aspects of the phenomenon:

> How can electric charge move through air?
>
> Why does a spark last only a short time?
>
> Why is light given off?
>
> And finally, how can the air become ionized?

A long chain of reasoning

There is considerable interest in understanding sparks, since they are one of the most dramatic kinds of electric phenomena. In addition, the kind of explanation we will find for sparks provides a good example of an important kind of scientific reasoning, in which there is a a long chain of reasoning. Each of the steps in the explanation is relatively simple; the complexity lies in linking many simple steps into one long, rigorous, logical chain.

14.8.1 How can electric charge move through air? Two models

In order to construct a model for the motion of electric charge through ionized air, we will first look at a simpler case. Consider two metal balls, one charged positively and the other negatively (Figure 14.42). If the air between them is not "ionized," the air is an insulator, and the charges stay on the balls. The balls merely polarize each other.

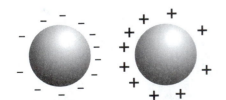

Figure 14.42 Two charged metal spheres.

A simpler case: Charge transfer through a wire

We could transfer charge from one ball to the other by connecting a metal wire between the balls (Figure 14.43), in which case the free electrons in the wire (and balls) shift away from the negative ball and toward the positive ball, making the negative ball less negative and the positive ball less positive. This conduction of charge through the wire (and balls) lasts for a short time, until the system reaches static equilibrium. (Usually we don't draw "+" and "−" charges on diagrams in regions where the net charge is zero, such as inside a wire, but in this chapter we will sometimes show individual charges in order to discuss details at the atomic level.)

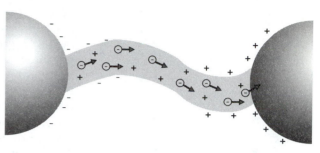

Figure 14.43 Two charged metal spheres connected by a metal wire.

A wire 1 meter long contains around 10^{23} free electrons, but a typical amount of charge on a ball is only about 10^{10} electron charges (about 10^{-9} coulomb; compare with your results for Problem 14.1).

? Estimate how far the free-electron sea must shift inside the wire in order to neutralize the positive ball.

In a fraction $10^{10}/10^{23} = 10^{-13}$ of the 1-meter length of the wire are enough electrons (10^{10} of them) to neutralize the positive ball. So the free-electron sea shifts about 10^{-13} m! This is an extraordinarily short distance; an atom has a radius of about 10^{-10} m, so this is about one thousandth of an atomic radius.

? Since the positive ball becomes neutral, the net effect is as though some electrons on the negative ball had "jumped" to the positive ball. But is that what actually happens? Where exactly did the electrons come from that were added to the positive ball?

The average displacement of a free electron in the wire is only 10^{-13} meter, so electrons on the negative ball do not move far enough to reach the positive ball. The electrons that go onto the positive ball are the ones that were initially very near the ball, not those way at the other end of the wire!

Charge transfer through air

Now consider the case in which the substance between the balls is air instead of a metal wire. How can charge travel through air?

Model 1: Electrons jump between balls

It is tempting to guess that during a spark electrons could simply jump from the negative ball to the positive ball. Let's try to determine whether this is a reasonable model or not. One important issue is how far, on average, an electron could travel through air before colliding with a gas molecule and losing most of its energy. Since molecules in air are relatively far apart, compared to those in a solid or liquid, perhaps an electron could travel all the way from the negative ball to the positive ball without colliding with anything. We can determine this by estimating what is called the "mean free path" of a particle in air. You may recall a similar argument from Chapter 11 on the kinetic theory of gases; we review it here.

Mean free path

The mean free path is the distance that a particle can travel, on average, before colliding with another particle.

? Qualitatively, what factors should the mean free path of a particle in a gas depend on?

You may have guessed that the mean free path d of a particle should be inversely proportional to the density of the gas surrounding it—the higher the gas density, the smaller the mean free path. You may also have guessed that d should be inversely proportional to the size of the gas molecules, or more precisely, since we are interested in collisions, the cross-sectional area of a gas molecule—the larger the molecule, the smaller the mean free path. Both of these factors will indeed prove to be important. If you guessed that the speed of the particle was a factor, think again; because we are not interested in the time between collisions, but only in the distance a particle can travel between collisions, speed does not matter.

It is not difficult to estimate the mean free path d of an electron in air. The calculation depends on a simple geometrical insight. Draw a cylinder along the direction of motion of the electron, with length d and cross-sectional area A equal to the cross-sectional area of an air molecule (Figure 14.44). The geometrical significance of this cylinder is that if the path of the electron comes within one molecular radius of a molecule, the electron will hit the molecule, so any molecule whose center is inside the cylinder will be hit.

By definition, d is the average distance the electron will travel before colliding with a molecule, so the cylinder drawn above should on average contain about one molecule. You can calculate the density of air from the fact that at standard temperature and pressure (STP, one atmosphere at 0° Celsius) one mole of air (6×10^{23} molecules) has a volume of 22.4 liters (22.4×10^{3} cm^3, or 22.4 cubic decimeters). You also know something about the sizes of atoms, so you can estimate the size of an air molecule.

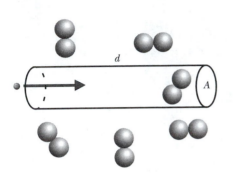

Figure 14.44 A cylindrical volume containing on average one air molecule.

? Using the fact that the cylinder in Figure 14.44 should contain approximately one molecule, calculate approximately the mean free path d of an electron in air.

$$\# \text{ molecules in cylinder} \approx 1$$

$$(\text{number of air molecules per m}^3)(\text{volume of cylinder}) \approx 1$$

$$(\text{number of air molecules per m}^3)(Ad) \approx 1$$

$$\left(\frac{6 \times 10^{23}}{22.4 \times (0.1\,\text{m})^3}\right)(\pi(1.5 \times 10^{-10}\,\text{m})^2)d \approx 1$$

$$d \approx 5 \times 10^{-7}\,\text{m}$$

So we see that a free electron in air can travel approximately 5×10^{-7} meters before colliding with an air molecule or ion.

? During a spark in air, the positive ball becomes less positive and the negative ball becomes less negative. Do electrons jump through the air from one ball to the other?

Apparently not! Since a free electron can travel only about 5×10^{-7} meters before colliding with a gas molecule and losing much of its energy, this is not a possible mechanism. On the basis of this estimate, we can rule out Model 1 for charge transfer through air.

Model 2: Positive ions and electrons move in ionized air

If many of the oxygen and nitrogen molecules in air became ionized, we would have a gas composed of charged particles, all of which are free to move. In other words, the air would now contain both free electrons and positive N_2^+ and O_2^+ ions, and would therefore be a conductor. Leaving aside for the moment the question of how air could become ionized, let's consider the mechanism of charge transfer in ionized air. Just as in the metal wire, there is now a "sea" of mobile charged particles.

? Which direction will the electrons in Figure 14.45 move? Which direction will the positive ions move?

All of the free electrons in the ionized air will move opposite to the electric field, toward the positively charged ball. The positive ion "sea" will be driven by the electric field toward the negatively charged ball.

? How would the negative ball become less negative? How would the positive ball become less positive?

As nearby positive ions come in contact with the negative ball, they can acquire electrons, making neutral molecules once again. The negative ball therefore becomes less negative. Free electrons near the positive ball can come in contact with the ball and join the free electron sea in the metal, making the ball less positive.

? Does any particle jump from one ball to the other?

No, since there are a very large number of charged particles near each ball, the "seas" of free electrons and positive ions must move only a very small distance to supply adequate charge to neutralize both balls.

Model 2 (ionized air) appears to be significantly better than Model 1 (jumping electrons), because in the ionized air model no particle travels farther than around one mean free path. We still need to answer the question of how the air might become ionized, though; we will return to this shortly.

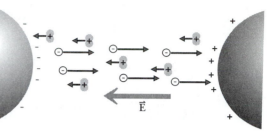

Figure 14.45 Free electrons and positively charged ions (O_2^+ or N_2^+) move through the ionized air between the spheres.

Ex. 14.14 A student said, "During a spark, protons are pulled out of the positive ball and jump across to the negative ball." Give at least two reasons why this analysis cannot be correct.

Ex. 14.15 At high altitudes the air is less dense. If the density of air were half that at sea level, how would the mean free path of an electron in air change?

Ex. 14.16 If the radius of an air molecule were twice as large, how would the mean free path change?

14.8.2 Why does a spark last only a short time?

Assuming that the air can somehow become ionized (we haven't yet explained how), with plenty of mobile electrons and ions, what happens next? Electrons drift toward the positive ball, and the positive ions drift (much more slowly) toward the negative ball. Since the electrons move so much more quickly, the ionized air conducts in a manner that is rather similar to electric conduction in a metal. There is a "sea" of free electrons that drifts through the air. There is a kind of "start-stop" motion of an individual electron, in that it is accelerated until it collides with a molecule and comes nearly to a stop, then accelerates again. This accelerated "start-stop" motion has some average speed, called the "drift speed." We will discuss a similar process inside metals in a later chapter.

? What happens when an electron collides with the positive metal ball? Does the electric field in the air between the balls change?

The electron fills an electron deficiency in the positively charged metal ball, making the ball less positively charged. As a result, the magnitude of the electric field at all locations in the air between the spheres decreases slightly.

? What happens when an ion collides with the negative metal ball? What happens to the ion? Does the electric field in the air between the balls change?

The ion can pick up an excess electron from the negatively charged metal ball, and the ion becomes an ordinary neutral air molecule. The negatively charged ball is now less negatively charged. Again, as a result, the magnitude of the electric field at all locations in the air between the spheres decreases slightly.

We'll see later why it is that as long as the electric field is larger than about 3×10^6 N/C throughout a region of air, the air remains heavily ionized, but if the electric field becomes too small, the ions and electrons recombine with each other, and the gas is no longer a conductor.

? Why does the spark last only a very short time?

The excess charge on both metal balls is quickly reduced to an amount that is not sufficient to make a large enough electric field to keep the air heavily ionized, and the spark goes out. However, if you continually resupply the metal balls with charge by connecting the balls to a battery or power supply, you may be able to maintain a steady spark. A dramatic example is the very bright spark maintained between two carbon electrodes in a searchlight or commercial movie projector (a "carbon arc" light). A gentler example is the continuous glow seen in neon lights, which contain neon at low pressure.

14.8.3 Why is light given off?

What causes the light that we see emitted by a spark (why are photons emitted)? Actually, the emission of light is a sort of side effect of the ionization of the air. Occasionally a free electron comes near enough to a positive ion (not necessarily the original ion) to be attracted and recombine with the ion to form a neutral molecule. When the free electron and the ion recombine (Figure 14.46), there is a transition from the high energy unbound state (free electron and ion) to a low energy bound state (Figure 14.47), with the emission of a photon.

It is important to understand that the emission of light goes on simultaneously with charge conduction through the ionized air. We will see later that as long as the applied electric field is big enough, neutral molecules are continually being ionized, at the same time that some free electrons and ions are recombining to make neutral molecules (with the emission of light). Although the light that we see is the most obvious aspect of a spark, the light can be considered to be essentially a mere side effect of conduction in a gas.

We have now seen the main aspects of sparks in ionized gas. Sparks are often referred to as "gas discharges," and a heavily ionized gas whose net charge is zero is called a "plasma." Plasma physics is an important area of contemporary research.

There remains the big question, how does the gas get ionized in the first place? That is the subject of the remaining sections.

14.8.4 How does the air become ionized? Two models

It is observed experimentally that an electric field of about 3×10^6 N/C is sufficient to ionize air and make it a conductor. We will try to come up with a model of this process. To determine if our model is a good one, we will

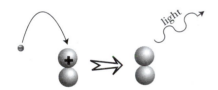

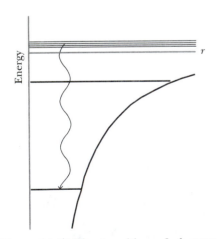

Figure 14.46 A free electron recombines with a positive ion to produce a neutral molecule. The excess energy is given off as visible light.

Figure 14.47 The transition of electron plus ion from an unbound (high energy) state to a bound (lower energy) state is accompanied by emission of a photon.

compare the predictions of the model to experimental observations—in particular, the observation that the critical field value is $E_{crit} = 3 \times 10^6$ N/C.

Since an atom or molecule is polarized by an electric field, we might suppose that an electric field of this magnitude is large enough to pull an electron completely out of an air molecule, creating a positive ion and a free electron. Let's check this model.

Model 1: A strong electric field pulls an electron out of a molecule

? Knowing what you know about atoms, estimate about how big an electric field would be needed to rip an outer electron out of a neutral atom.

If you could not do this calculation, think again for a moment before reading farther. You really do know enough to make this calculation. In particular, remember that the radius of an atom is about 10^{-10} m. For simplicity you may want to think about a hydrogen atom.

You have been told that a spherical charged object acts like a point charge. Think of an outer electron as being attracted by the rest of the atom, which has a charge $+e$ (Figure 14.48). The electric field made by this $+e$ charge that acts on the outer electron is

$$E = \frac{1}{4\pi\varepsilon_0}\frac{e}{r^2} = (9\times10^9 \text{N·m}^2/\text{C}^2)\frac{(1.6\times10^{-19}\text{C})}{(10^{-10}\text{m})^2} = 1.4\times10^{11}\text{N/C}$$

where $r = 10^{-10}$ m is the approximate radius of an atom. We would have to apply at least that big a field in order to pull an outer electron out of an atom.

The enormous difference between the experimentally observed field of 3×10^6 N/C and the value predicted by our model for the electric field necessary to ionize an atom means that we can rule out the possibility of direct atomic ionization by the applied electric field. This is an excellent example of one of the strengths of science: it is often possible to rule out a proposed model for a process, even if we can't figure out a better explanation.

Having ruled out direct atomic ionization by the applied electric field, what other process can we think of?

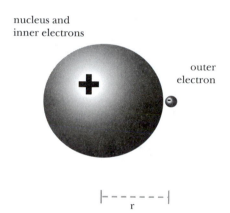

nucleus and
inner electrons

outer
electron

r

Figure 14.48 An outer electron interacts with the nucleus and inner electrons of the atom.

Model 2: Fast moving charged particles knock electrons out of atoms

It takes a dramatic event to ionize air, because it requires a very large force to rip electrons out of air molecules. A fast-moving charged particle that collides with an atom or molecule can knock out an electron, leaving a singly ionized ion behind. Where would such energetic charged particles come from? As it happens, there are fast-moving charged particles passing through your body at this very moment, ionizing some atoms and molecules in your body! Some of these charged particles are "muons" produced by cosmic rays in nuclear reactions at the top of the atmosphere. Others are electrons, positrons, or alpha particles (helium nuclei) emitted by radioactive isotopes present in trace quantities in materials in and around you.

This process is the cause of some potentially dangerous events. A neutral atom in your DNA that suddenly turns into an ion can cause biochemical havoc, and genetic mutations. The passage of fast-moving charged particles through the tiny and very sensitive components in a computer chip can change a bit with potentially disastrous consequences. To guard against such disasters, both DNA and computer circuits have checks built into them to try to compensate for damage caused by the unavoidable passage of high-speed charged particles.

Invaders from outer space

"Cosmic rays," consisting mostly of very high energy protons coming from outside our solar system, strike the nuclei of molecules in the Earth's upper atmosphere. The resulting nuclear reactions produce a spray of other particles. Many of the products of these reactions are particles such as protons, neutrons, and pions that interact very strongly with other nuclei in air molecules. For that reason these particles don't go very far through the upper atmosphere. Only positive and negative "muons" (μ^+ and μ^-) and neutral neutrinos are able to penetrate the atmosphere and reach the surface of the earth.

Electrons, muons, and neutrinos do not interact with nuclei through the strong interaction. Electrons and muons, being charged, do interact electrically with atoms and nuclei. Electrons have so little mass that they undergo large deflections and don't travel very far through the air. Muons are essentially just massive electrons, behaving in almost every way just like electrons except for having a mass about 200 times the electron mass. Because of their large mass, the muons undergo small deflections and can travel long distances through the air.

The neutral neutrinos, lacking electric charge, don't have any electric interactions with matter, and like electrons and muons they have extremely weak nonelectric interactions with matter. The interactions of neutrinos with matter are so unimaginably weak that most of them go right through the entire earth, coming out the other side with no change!

As the rapidly moving charged muons plunge downward through the air, they have enough energy to occasionally knock electrons out of air molecules, producing free electrons and positive ions (Figure 14.49). Each muon leaves a trail of slightly ionized air. Muons produced by cosmic rays are not the only cause of ionization in the air. Materials around you contain trace quantities of radioactive nuclei which can emit high-speed electrons, positrons, or alpha particles (helium nuclei, He^{2+}). These rapidly moving charged particles can also ionize the air.

If the ionization were extensive, air would be a good conductor at all times, but the number of free electrons and ions produced by cosmic-ray muons and natural radioactivity is much too small to make the air be a good conductor. However, in the next section we will see that the small amount of ionization produced in this way can act as a trigger for large-scale ionization.

It is surprising that the muons actually manage to reach the earth's surface, because a positive or negative muon at rest has an average lifetime of about 2 microseconds before splitting up ("decaying") into a positron (if μ^+) or electron (if μ^-), a neutrino, and an antineutrino. Even if moving at nearly the speed of light, a muon would be expected to travel only about $(3{\times}10^8 \text{ m/s})(2{\times}10^{-6} \text{ s}) = 600$ meters before decaying, which is far too short a distance to reach the ground from the top of the atmosphere where the muon was produced.

However, as predicted by Einstein's special theory of relativity, time seems to elapse slowly for a fast-moving object, so that in the reference frame of the muon it takes less than 2 microseconds to reach the ground, and many muons plunge through your room every minute. Another way of thinking about this is that from a fast-moving reference frame, lengths in the direction of motion seem shortened (relativistic "length contraction") and to the fast-moving muon the long distance from the top of the atmosphere to the ground looks quite short.

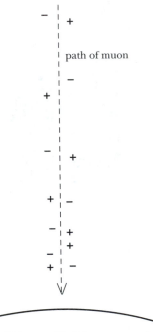

Figure 14.49 Ion trail left by a muon plunging down through the Earth's atmosphere.

A chain reaction

If there happens to be a single free electron in the air (due to muons or natural radioactivity), it can be accelerated by an applied electric field due to nearby charged objects such as the charged metal balls shown earlier (Figure 14.50).

If the electron gets going fast enough before colliding with an air molecule, it can knock another electron out of that molecule, so now there are two free electrons and two ions (Figure 14.51).

These two free electrons can knock out two more electrons, so the number of free electrons grows rapidly: 2, 4, 8, 16, 32, 64, etc. This is an avalanche, or "chain reaction." The air becomes significantly ionized with a large number of mobile charged particles and is now a rather good conductor (Figure 14.52). Note that the air is still neutral overall, just as the interior of a metal wire is neutral overall.

Electric field required to trigger a chain reaction

The few free electrons due to cosmic rays and radioactivity frequently bump into neutral air molecules, but these collisions won't start a chain reaction unless the electrons are traveling fast enough to rip electrons out of the molecules. Let's see if we can predict from fundamental principles and the structure of matter the magnitude of the critical electric field E_{crit} that would be required to accelerate the free electrons to high enough speeds to trigger a chain reaction.

Role of electrons

Consider a region of air that is only very slightly ionized by muons and radioactivity.

? If we apply an electric field E in this region, the applied electric field exerts the same electric force $F = eE$ on free electrons and (singly ionized) ions. Which particles accelerate more, the electrons or the ions? Why?

The magnitude of the electric force qE is the same on an electron (charge $-e$) and on a singly ionized ion (charge $+e$), but the ion is much more massive than the electron. So the electrons accelerate a lot more. As a result, the electrons attain much higher speeds and run into neutral air molecules much more frequently than the ions do. In a short time interval Δt it is the electrons that are mainly responsible for knocking electrons out of neutral air molecules and producing a chain reaction.

Kinetic energy gained by an accelerating electron

A free electron, accelerated by an electric force $F = eE$, travels some average distance d before running into a neutral air molecule. This distance d is called the "mean free path" of the electron (Figure 14.53).

The amount of work done on the electron, $Fd = eEd$, is equal to the increase in kinetic energy ΔK of the electron. If we assume on average that the electron has negligible kinetic energy to start with, we have

$$\Delta K = K - 0 = Fd\cos 0 = eEd$$

If eEd is less than the amount of energy required to ionize the molecule (knock an electron out of the neutral molecule), the incoming electron just bounces off (Figure 14.54), but if eEd is greater than the ionization energy for an air molecule, after the collision there are two electrons where there had been one (Figure 14.55 on the next page).

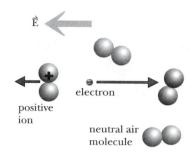

Figure 14.50 A single free electron is accelerated by the electric field, initiating a chain reaction.

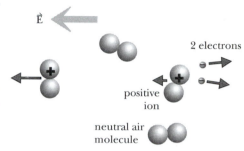

Figure 14.51 The first free electron ionizes an air molecule, producing a second free electron.

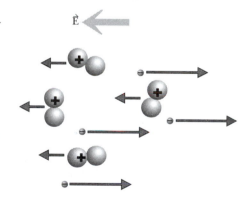

Figure 14.52 As the chain reaction progresses more and more charged particles are created, and the air becomes a conductor.

Figure 14.53 Mean free path d.

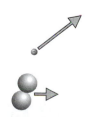

Figure 14.54 An electron with insufficient energy simply bounces off.

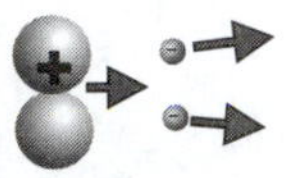

Figure 14.55 An electron with sufficient energy knocks another electron out of an air molecule during a collision.

Both of these electrons in Figure 14.55 can now be accelerated on average through a distance d, and both may ionize molecules, leading to there being 4 free electrons, then 8, 16, 32, 64, etc.—a chain reaction. (After a typical ionizing collision both electrons are moving rather slowly, so each time they are accelerated nearly from rest, and the collision energy is always about equal to eEd.)

The critical condition is that the energy eEd be sufficient to ionize a neutral molecule. This depends on how big an electric field E is applied by the nearby charged objects, but also on the mean free path d of an electron in air, which we estimated in on page 485 to be 5×10^{-7} m.

Estimate of breakdown field for air

? Estimate roughly the amount of energy required to remove one outer electron from an oxygen or nitrogen molecule.

A rough estimate can be made by calculating the electric potential energy for an electron to be about 10^{-10} m from the center of the remainder of the molecule, which is a singly-charged positive ion:

$$U_{el} = \frac{1}{4\pi\varepsilon_0}\frac{(-e)(e)}{(10^{-10}\text{ m})} \approx -2.3 \times 10^{-18}\text{ J}$$

We have to add this much energy to remove the electron. The amount of energy required to singly ionize a nitrogen molecule (that is, knock one outer electron out of the molecule) has been experimentally measured to be 2.4×10^{-18} J , so our estimate happens to be very good.

? Use this measurement, and your approximate value for d, to calculate an approximate value for the magnitude of the critical electric field E_{crit} required to cause massive ionization (breakdown) in air.

$$eE_{crit}d = 2.4 \times 10^{-18}\text{J}$$

$$E_{crit} \approx \frac{2.4 \times 10^{-18}\text{J}}{(1.6 \times 10^{-19}\text{C})(5 \times 10^{-7}\text{m})}$$

$$E_{crit} \approx 30 \times 10^{6}\text{N/C}$$

? Should we consider this good or poor agreement with the experimental value of 3×10^6 N/C?

For the previous model (direct ionization by the applied field), the prediction of the model was off by 5 orders of magnitude (10^5) from the experimental value. It is satisfying that with the chain reaction model we predict the right order of magnitude (we're off by a factor of 10 instead of being off by a factor of 10^5), and we get an intuitive feeling for the issues.

To make an accurate prediction would involve difficult calculations because of the statistical nature of the process. In our simple calculations we assumed that every electron goes one mean free path and picks up an amount of energy eEd. However, occasionally an electron happens to go much farther than d, and a smaller electric field is sufficient to accelerate such an electron to a high enough energy to ionize the molecule that it finally hits. Nor is it necessary that every electron ionize an atom in order to create a chain reaction. That is why our simple calculation has over-estimated the magnitude of electric field required to cause a spark.

14.8.5 Additional tests of the models

It is possible to make additional tests of the two models by comparing their predictions with other experimental observations, such as the effect on the critical field of changing the density of the gas. If a physical model explains a variety of different kinds of experimental results, this increases our confidence in the validity of the model.

? If the density of air were half as large as it is at STP, according to each model how would this affect the electric field required to ionize air (which is normally about 3×10^6 N/C)?

Halving the density of the air would double the mean free path d. According to model 1 (electric field large enough to detach an electron) this would have no effect. According to model 2 (accelerated free electron collides with molecule), since the kinetic energy of the free electron (starting from rest) is $\Delta K = K_{final} - 0 = eEd$, doubling d would decrease $E_{critical}$ by a factor of two. What is observed experimentally is that the required electric field does decrease as air density decreases, supporting model 2.

An ordinary long fluorescent light tube exemplifies this. If the gas density inside the bulb were not very low, it would be difficult to create a sufficiently high electric field throughout the tube to sustain a discharge (a continuous spark). Because the gas density in the tube is low, the electric field required is also low, making such lights practical. On the other end of the scale, gasses at high pressures (hence high densities) are sometimes used as insulators.

Ex. 14.17 If you observed that under certain conditions an electric field three times as great as usual were required to ionize air, what might you conclude about the mean free path under these conditions?

Ex. 14.18 What would happen if an electron with energy less than the ionization energy of a nitrogen molecule collided with a nitrogen molecule?

Ex. 14.19 Approximately how large an electric field would be required to cause a spark in a gas at standard temperature and pressure if the ionization energy for the molecules of this gas is 1.5 times the ionization energy for air molecules, and the cross-sectional area of the molecules is 0.8 times the cross-sectional area of air molecules?

14.8.6 Drift speed of free electrons in a spark

We can now figure out the average "drift speed" of free electrons in a spark. Knowing that the kinetic energy acquired in one mean free path is 2.4×10^{-18} J, we can determine the typical speed v of an electron when it hits a molecule:

$$\frac{1}{2} m v^2 \approx 2.4 \times 10^{-18} \text{J}$$

$$v \approx \left(\frac{2(2.4 \times 10^{-18} \text{J})}{9 \times 10^{-31} \text{kg}} \right)^{1/2} = 2 \times 10^6 \text{m/s}$$

The statistical average speed is difficult to compute, since there is a distribution of free paths. But evidently the drift speed is of the order of 10^6 m/s.

Propagation of ionization

The magnitude of the electric field is not uniform in the region between the two charged metal balls, and it is largest near the balls. (Do you see why?) Suppose the field near the surface of the negative ball is bigger than 3×10^6 N/C, so that the air near the surface of this ball becomes strongly ionized. How can the ionization extend into other regions where E is smaller?

Positive ions are attracted toward the negative ball, and electrons are repelled into regions of the air where the electric field was less than 3×10^6 N/C. These electrons increase the electric field in the un-ionized region, and their large number facilitates a chain reaction, so the spark penetrates into the air farther and farther from the ball. It is possible to form an unbroken column of heavily ionized air leading from one ball to the other, even though initially the field was stronger than 3×10^6 N/C only near the negatively charged ball. Once the air is heavily ionized, an electric field significantly less than 3×10^6 N/C is sufficient to keep the spark going for a while. Recombination of electrons and ions along this column makes the column glow, and you see a spark connecting the two balls.

Something very similar happens in a lightning storm. A column of ionized air propagates from the negatively charged bottom of a cloud downward toward the ground. The speed of propagation from cloud to ground is somewhat slower than the drift speed you calculated (about 10^6 m/s), because the propagation occurs in steps, with pauses between steps due to the statistical nature of triggering a chain reaction in the neighboring air. Once the column of ionized air reaches all the way to the ground, a large current runs through this ionized column. The visible lightning flash is of course due to the recombination of ions and electrons in this column. The large current heats the column explosively, and the rapidly expanding gas makes the noise heard as thunder.

Problem 14.2 Spark from a doorknob

In very dry weather, if you shuffle across the carpet wearing rubber-soled shoes, and then bring your finger near a metal object such as a doorknob, you will probably get a shock and see a spark. How this can occur is puzzling, since rubber is an insulator, so charge can't move through the soles of your shoes.

Explain this process in detail, with appropriate diagrams. Make sure that you answer the following questions:

(a) Carry out an experiment to determine the sign of the charge on the sole of your shoe after rubbing it on carpet, wool, or other cloth. Explain what test you did. (If you are unable to obtain results, choose a sign to use in the rest of the discussion.)

(b) Draw a diagram that includes both the shoe and the rest of your body. Include all relevant charges and fields.

(c) Suppose that a spark occurred when your finger was 1 cm from the doorknob. Draw two diagrams showing your hand, the doorknob, all relevant charges, and all contributions to the electric field at relevant locations:

(i) when your finger is 1.5 cm from the doorknob (no spark yet)
(ii) when your finger is 1 cm from the doorknob (spark starts)

On the basis of these diagrams, explain why the spark starts only when your finger is close enough (1 cm) to the doorknob and not farther away.

(d) At what location do you think the spark starts, and why (why not somewhere else)?

(e) Why does the spark stop?

14.9   Summary

Fundamental Principles

CONSERVATION OF CHARGE

The net charge of a system and its surroundings cannot change.

New Concepts

Polarization of an atom or molecule produces an induced dipole

$$p = qs = \alpha E_{\text{applied}}$$

(where p is dipole moment, and α is atomic polarizability)

DEFINITION OF "CONDUCTOR" AND "INSULATOR"

A conductor contains mobile charges
that can move through the material.

An insulator has no mobile charges.

Model of a metal: mobile electron sea, like an ideal gas.

	conductor	insulator
mobile charges	yes	no
polarization	entire sea of mobile charges moves	individual atoms or molecules polarize
static equilibrium	$\vec{E}_{\text{net}} = 0$ inside	$\vec{E}_{\text{net}}$ nonzero inside
location of excess charge	only on surface	anywhere on or inside material
distribution of excess charge	spread out over entire surface	located in patches

Results

$\vec{E}_{\text{net}} = 0$ inside a conductor in static equilibrium

Excess charges move to the surface of a conductor

Force between point charge and neutral atom is proportional to $1/r^5$

$$F_{\text{on pt}} = q_{\text{pt}} E_{\text{ind. dipole}} = \left(\frac{1}{4\pi\varepsilon_0}\right)^2 \left(\frac{2\alpha q_1^2}{r^5}\right)$$

Sparks: $E_{\text{critical}} = 3\times10^6$ N/C

When the applied electric field is sufficiently large to accelerate free electrons to the energy required to ionize a nitrogen or oxygen molecule, a chain reaction can occur.

14.10 Example problem: A ball and a wire

Here is a rather challenging problem, followed by a full analysis. You may find it useful to attempt your own analysis before studying ours.

The center of a small spherical metal ball of radius R, carrying a negative charge $-Q$, is located a distance r from the center of a short, thin copper wire of length L. The ball and the wire are held in position by threads that are not shown. If $R = 5$ mm, $Q = 10^{-9}$ C, $r = 10$ cm, and $L = 4$ mm, calculate the force that the ball exerts on the wire.

Solution

Start from fundamental principles: What electric field is produced by given charges? What does this field do to matter? If new charges appear due to polarization, what electric field do they produce? If there is metal present, keep in mind that the net electric field inside the metal must be zero in static equilibrium.

It is best to carry out the analysis algebraically and evaluate numerically only at the end, since it makes our work easier to follow.

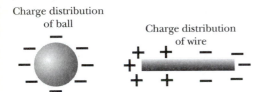

Figure 14.56 The ball polarizes the wire. Neglect effect of wire on ball.

(a) Ball makes a field which polarizes the wire as shown in Figure 14.56. The polarized wire in turn makes a field that polarizes the ball, but assume we can neglect this tiny effect, so we can model the ball as a point charge. The polarized wire will be attracted by the ball.

(b) At any location inside the metal wire, $\vec{E}_{net} = 0$ in static equilibrium (Figure 14.57). Consider a location in the center of the wire, and model the wire as a dipole with $+q$ and $-q$ on the ends, a distance L apart, ignoring the small amount of charge on the rest of the wire:

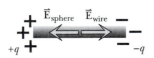

Figure 14.57 Net electric field inside wire must be zero in static equilibrium.

$$2\left(\frac{1}{4\pi\varepsilon_0}\right)\frac{q}{(L/2)^2} - \frac{1}{4\pi\varepsilon_0}\frac{Q}{r^2} = 0 \quad \text{(inside a metal in static equilibrium)}$$

$$q \approx \frac{Q}{8}\left(\frac{L}{r}\right)^2 \quad \text{(check units: coulombs)}$$

(c) In our model the wire is a dipole with dipole moment $= p = qL$, so the force on the ball (which is equal in magnitude to the force on the wire) is this:

$$F \approx Q\left(\frac{1}{4\pi\varepsilon_0}\frac{2qL}{r^3}\right) = \frac{1}{4\pi\varepsilon_0}\frac{Q^2}{4}\left(\frac{L}{r}\right)^2\frac{L}{r^3} = \frac{1}{4\pi\varepsilon_0}\frac{Q^2L^3}{4r^5}$$

Not too surprisingly, we find a force proportional to $1/r^5$.

(d) $q \approx \dfrac{(10^{-9}\ \text{C})}{8}\left(\dfrac{4\times10^{-3}\text{m}}{0.1\ \text{m}}\right)^2 = 2\times10^{-13}$ C, which is a very small charge.

This justifies our assumption that the polarized wire won't polarize the ball to any significant extent.

$$F = \left(9\times10^9\frac{\text{N}\cdot\text{m}^2}{\text{C}^2}\right)\frac{(10^{-9}\ \text{C})^2(4\times10^{-3}\text{m})^3}{4(0.1\ \text{m})^5} = 1.4\times10^{-11}\ \text{N}, \text{ a tiny force.}$$

Note that if we double r, 1/4 as much q, 1/32 as much force.
If we double Q, 2 times as much q, 4 times as much force.

14.11 Review questions

Charges and neutral matter

RQ 14.1 Criticize the following statement: "Since an atom's electron cloud is spherical, the effect of the electrons cancels the effect of the nucleus, so a neutral atom can't interact with a charged object." ("Criticize" means to explain why the given statement is inadequate or incorrect, as well as correcting it.)

RQ 14.2 Criticize the following statement: "A positive charge attracts neutral plastic by polarizing the molecules and then attracting the negative side of the molecules." ("Criticize" means to explain why the given statement is inadequate or incorrect, as well as correcting it.)

RQ 14.3 Carbon tetrachloride (CCl_4) is a liquid whose molecules are symmetrical and so are not permanent dipoles, unlike water molecules. Explain briefly how the effect of an external charge on a beaker of water (H_2O) differs from its effect on a beaker of CCl_4. (Hint: consider the behavior of the permanent dipole you made out of U and L tapes.)

Induced dipoles

RQ 14.4 Is the following statement true or false? If true, what principle makes it true? If false, give a counter-example or say why.

"The electric field E_{point} at the center of an induced dipole, due to the point charge, is equal in magnitude and opposite in direction to the electric field E_{dipole} at the location of the point charge, due to the induced dipole."

E_{dipole} ?

$\oplus$

Point charge

E_{point} ?

Induced dipole

RQ 14.5 Explain briefly why the attraction between a point charge and a dipole has a different distance dependence for induced dipoles ($1/r^5$) than for permanent dipoles ($1/r^3$). (You need not explain either situation in full detail: just explain why there is this difference in their behavior.)

Testing the sign of charge

RQ 14.6 Explain briefly why repulsion is a better test for the sign of a charged object than attraction is.

Insulators and conductors

RQ 14.7 A positive charge is located between a neutral block of plastic and a neutral block of copper (Figure 14.58). Draw the approximate charge distribution for this situation.

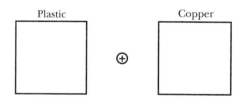

Figure 14.58 Draw the approximate charge distribution.

RQ 14.8 Make a table showing the major differences in the electric properties of plastic, salt water, and copper. Include diagrams showing polarization by an external charge.

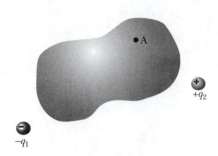

Figure 14.59 A neutral, solid metal object near two point charges (RQ 14.9).

Metals

RQ 14.9 Figure 14.59 shows a neutral, solid piece of metal placed near two point charges. Copy this diagram.

(a) On your diagram, indicate the polarization of the piece of metal.

(b) Then, at location A *inside* the solid piece of metal, carefully draw and label three vectors:

1) $\vec{E}_1$, the electric field due to $-q_1$
2) $\vec{E}_2$, the electric field due to $+q_2$
3) $\vec{E}_3$, the electric field due to all of the charges on the metal

(c) Explain briefly why you drew the vectors the way you did.

Moist skin is a conductor

RQ 14.10 A student said, "When you touch a charged piece of metal, the metal is no longer charged: all the charge on the metal is neutralized." As a practical matter, this is pretty nearly correct, but it isn't exactly right. What's wrong with saying that all the charge on the metal is neutralized?

RQ 14.11 You are wearing shoes with thick rubber soles. You briefly touch a negatively-charged metal sphere. Afterwards, the sphere seems to have little or no charge. Why? Explain in detail.

Discharging a tape

RQ 14.12 Criticize the following statement: "When you rub your finger along the slick side of a U tape, the excess charges flow onto your finger, and this discharges the tape." Draw diagrams illustrating a more plausible explanation.

RQ 14.13 Jill stuck a piece of invisible tape down onto another piece of tape. Then she yanked the upper tape off the lower tape, and she found that this upper tape strongly repelled other upper tapes and was charged positive. Jack ran his thumb along the slick (upper) side of the upper tape, and the tape no longer repelled other upper tapes. Jill and Jack explained this by saying that Jack rubbed some protons out of the molecules in the tape.

Give a critique of their explanation. If Jill's and Jack's explanation is deficient, give a physically possible explanation for why the upper tape no longer repelled other upper tapes. Include explanatory diagrams.

Charging by induction

RQ 14.14 Can you charge a piece of plastic by induction? Explain, using diagrams. Compare with the amount of charging obtained when you charge a piece of metal by induction.

When the field concept is less useful

RQ 14.15 Suppose you try to measure the electric field $\vec{E}$ at a location by placing a charge Q_1 there and observing the force $\vec{F}_1$, so that you measure $E_1 = F_1/Q_1 = 1000$ N/C. Then you remove Q_1 and place a much larger charge $Q_2 = 30Q_1$ at the same location, and observe the force $\vec{F}_2$. This time you measure $E_2 = F_2/Q_2 = 1100$ N/C, though you expected to measure 1000 N/C again. What's going on here? Why didn't you get $E = 1000$ N/C in your second measurement? Sketch a possible situation that would lead to these measurements.

Sparks

RQ 14.16 "When you connect a metal wire between two oppositely charged metal blocks, electrons on the negatively charged block jump to the positively charged block." Explain briefly what is wrong with this statement.

RQ 14.17 Explain briefly why there is a limit to how much charge can be placed on a metal sphere in the classroom. If the radius of the sphere is 15 cm, what is the maximum amount of charge you can place on the sphere? (Remember that a uniform sphere of charge makes an electric field outside the sphere as though all the charge were concentrated at the center of the sphere.)

14.12 Homework problems

14.12.1 In-line problems from this chapter

Problem 14.1 (page 467) Amount of excess charge on a tape
In this problem you will design and carry out an experiment to determine the approximate number of excess electron charges on the surface of a negatively charged tape.
Initial estimates
   Since we do not know what order of magnitude to expect for our answer, it is important to put upper and lower bounds on reasonable answers.
   (a) What is the smallest amount of excess charge that a tape could possibly have?
   (b) What is the largest amount of excess charge a tape could possibly have?
Design and perform an experiment
   A centimeter ruler is printed on the inside back cover of this textbook. A piece of half-inch-wide (1.2 cm) invisible tape, 20 cm long (8 inches), has a mass of about 0.16 grams.
   (c) Make a clear and understandable diagram of your experimental set-up, indicating each quantity you measured. Report all measurements you made.
Analyze the results
   (d) Clearly present your physical analysis of your data. Make an appropriate diagram, labeling all vector quantities. Reason from fundamental physics principles. Explicitly report any simplifying assumptions or approximations you have made in your analysis. Report two quantities:
   • the amount of charge on a tape, in coulombs
   • the number of excess electrons to which this charge corresponds
Present your analysis clearly. Your reasoning must be clear to a reader.
   (e) Estimate whether the true amount of excess charge is larger or smaller than the value you calculated from your experimental data. Explain your reasoning briefly.
Is this a lot of charge?
   (f) What fraction of the molecules on the surface of the tape have gained an extra electronic charge? To estimate this, you may assume that molecules in the tape are arranged in a cubic lattice, as indicated in the accompanying figure, and that the diameter of a molecule in the tape is about 3×10^{-10} m. Does your answer suggest that it is a common event or a rare event for a molecule to gain an extra electron?
   (g) If the electric field at a location in air exceeds 3×10^{6} N/C, the air will become ionized and a spark will be triggered. In Chapter 16 we will see that

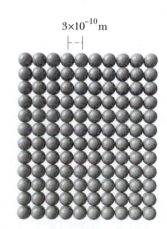

3×10^{-10} m

Array of molecules on one surface of a piece of invisible tape.

the electric field in a region very close to a uniformly charged disk or plate depends approximately only on the charge Q per unit area A:

$$E = \frac{1}{2\varepsilon_0}\left(\frac{Q}{A}\right)$$

Use this model (or make a different but justifiable simplifying assumption) to calculate the magnitude of the electric field at a location in the air very close to your tape (less than 1 mm from the surface of the tape). How does it compare to the electric field needed to trigger a spark in the air?

Problem 14.2 (page 494) Spark from a doorknob

In very dry weather, if you shuffle across the carpet wearing rubber-soled shoes, and then bring your finger near a metal object such as a doorknob, you will probably get a shock and see a spark. How this can occur is puzzling, since rubber is an insulator, so charge can't move through the soles of your shoes.

Explain this process in detail, with appropriate diagrams. Make sure that you answer the following questions:

(a) Carry out an experiment to determine the sign of the charge on the sole of your shoe after rubbing it on carpet, wool, or other cloth. Explain what test you did. (If you are unable to obtain results, choose a sign to use in the rest of the discussion.)

(b) Draw a diagram that includes both the shoe and the rest of your body. Include all relevant charges and fields.

(c) Suppose that a spark occurred when your finger was 1 cm from the doorknob. Draw two diagrams showing your hand, the doorknob, all relevant charges, and all contributions to the electric field at relevant locations:

> (i) when your finger is 1.5 cm from the doorknob (no spark yet)
> (ii) when your finger is 1 cm from the doorknob (spark starts)

On the basis of these diagrams, explain why the spark starts only when your finger is close enough (1 cm) to the doorknob and not farther away.

(d) At what location do you think the spark starts, and why (why not somewhere else)?

(e) Why does the spark stop?

14.12.2 Additional problems

Problem 14.3 Charging with specified sign

You have three metal blocks marked A, B, and C, sitting on insulating stands (Figure 14.60). Block A is charged + but blocks B and C are neutral.

Without using any additional equipment, and without altering the amount of charge on block A, explain how you could make block B be charged + and block C be charged −. Explain your procedure in detail, including diagrams of the charge distributions at each step in the process.

Problem 14.4 Two spheres and a pen

You have two identical metal spheres labeled A and B, mounted on insulating posts, and you have a plastic pen which charges negatively when you rub it on your hair (Figure 14.61).

(a) (+ and −) Explain in detail, including diagrams, what operations you would carry out to make sphere A have some positive charge and to make sphere B have an *equal* amount of negative charge (the spheres are initially uncharged).

(b) (+ and +) Explain in detail, including diagrams, what operations you would carry out to make sphere A have some positive charge and to make sphere B have an *equal* amount of positive charge (the spheres are initially uncharged).

Figure 14.60 Block A is charged +; blocks B and C are initially neutral (Problem 14.3).

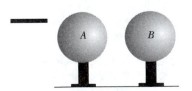

Figure 14.61 Two metal spheres and a plastic pen (Problem 14.4).

Problem 14.5 Charging by induction

Here is a variant of "charging by induction." Place two uncharged metal objects so as to touch each other, one behind the other. Call them *front* object and *back* object. While you hold a charged comb in front of the *front* object, your partner moves away the *back* object (handling it through an insulator so as not to discharge it). Now you move the comb away. Explain this process. Use only labeled diagrams in your explanation (no prose!).

Problem 14.6 Metal spheres and a block of plastic

Two identical metal spheres are suspended from insulating threads. One is charged with excess electrons, and the other is neutral. When the two spheres are brought near each other, they swing toward each other and touch, then swing away from each other.

(a) Explain in detail why both these swings happen. In your explanation, include clear diagrams showing charge distributions, including the final charge distribution.

(b) Next the spheres are moved away from each other. Then a block of plastic is placed between them (Figure 14.62). The original positions of the spheres are indicated, before the plastic is placed between them. Sketch the new positions of the spheres. Explain, including charge distributions on the spheres.

(c) Show the polarization of a molecule inside the plastic at points A, B, C, D, and E. Explain briefly.

Figure 14.62 A plastic block is placed between the two spheres (Problem 14.6).

Problem 14.7 Effect of intervening material

A large positive charge pulls on a distant electron. How does the *net* force on the electron change if a slab of glass is inserted between the large positive charge and the electron? Does the net force get bigger, smaller, or stay the same? Explain, using only labeled diagrams.

(Be sure to show *all* the forces on the electron before determining the net force on the electron, not just the force exerted by the large positive charge. Remember that the part of the net force on the electron contributed by the large positive charge does not change when the glass is inserted: the electric interaction extends through matter.)

Figure 14.63 (Problem 14.8a).

Problem 14.8 Glass and silk

A small glass ball is rubbed all over with a small silk cloth and acquires a charge of +5 nC. The silk cloth and the glass ball are placed 30 cm apart.

(a) On a diagram like that in Figure 14.63, draw the electric field vectors qualitatively at the locations marked ×. Pay careful attention to directions and to relative magnitudes. Use dashed lines to explain your reasoning graphically, and draw the final electric field vectors with solid lines.

(b) Next, a neutral block of copper is placed between the silk and the glass. On a diagram like that in Figure 14.64 carefully show the approximate charge distribution for the copper block and the electric field vectors inside the copper at the locations marked ×.

(c) The copper block is replaced by a neutral block of plastic.Carefully show the approximate molecular polarization of the plastic block at the locations marked × in Figure 14.65.

(d) Even if you have to state your result as an inequality, make as quantitative a statement as you can about the electric field at the location of the glass ball, and the net force on the ball, when the plastic block is in place compared to there being no block. Explain briefly.

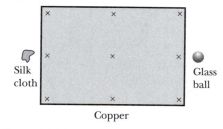

Figure 14.64 (Problem 14.8b).

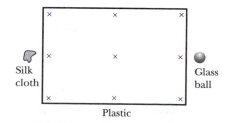

Figure 14.65 (Problem 14.8c).

Figure 14.66 An electroscope (Problem 14.10).

Figure 14.67 A positively charged glass rod is brought near an electroscope (Problem14.10a).

Figure 14.68 Metal sphere A is charged negatively, and B has a net charge of zero (Problem 14.12).

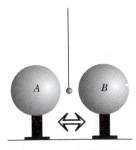

Figure 14.69 A small, initially uncharged metal ball swings back and forth for a while, then comes to a stop (Problem 14.12).

Problem 14.9 Two metal balls

A metal ball with diameter of a half a centimeter and hanging from an insulating thread is charged up with 10^{10} excess electrons. An initially uncharged identical metal ball hanging from an insulating thread is brought in contact with the first ball, then moved away, and they hang so that the distance *between their centers* is 20 cm.

(a) Calculate the electric force one ball exerts on the other, and state whether it is attractive or repulsive. If you have to make any simplifying assumptions, state them explicitly and justify them.

(b) Now the balls are moved so that as they hang, the distance *between their centers* is only 5 cm. Naively one would expect the force that one ball exerts on the other to increase by a factor of $4^2 = 16$, but in real life the increase is a bit less than a factor of 16. Explain why, including a diagram. (Nothing but the distance between centers is changed—the charge on each ball is unchanged, and no other objects are around.)

Problem 14.10 An electroscope

As shown in Figure 14.66, an electroscope consists of a steel ball connected to a steel rod, with very thin gold foil leaves connected to the bottom of the rod (in good electric contact with the rod). The bottom of the electroscope is enclosed in a glass jar, and held in place by a rubber stopper.

(a) The electroscope is brought near to but not touching a positively charged glass rod (Figure 14.67). The foil leaves are observed to spread apart, as shown at right. Explain why, in detail, using as many diagrams as necessary.

(b) The electroscope is moved far away from the glass rod and the steel ball is touched momentarily to a metal block. The foil leaves spread apart and stay spread apart when the electroscope is moved away from the block. As the electroscope is moved close to but not touching the positively charged glass rod, the foil leaves move closer together. Is the metal block positive, negative, or neutral? How do you know? Explain.

Problem 14.11 Interactions among U and L tapes and a plastic pen

You take two invisible tapes of some unknown brand, stick them together, and discharge the pair before pulling them apart and hanging them from the edge of your desk. When you bring an uncharged plastic pen within 10 cm of either the U tape or the L tape you see a slight attraction. Next you rub the pen through your hair, which is known to charge the pen negatively. Now you find that if you bring the charged pen within 8 cm of the L tape you see a slight repulsion, and if you bring the pen within 12 cm of the U tape you see a slight attraction. Briefly explain all of your observations.

Problem 14.12 A bouncing ball

Metal sphere A is charged negatively and then brought near an uncharged metal sphere B (Figure 14.68). Both spheres rest on insulating supports, and the humidity is very low.

(a) Use +'s and −'s to show the approximate distribution of charges on the two spheres. (Hint: think hard about *both* spheres, not just B.)

(b) A small, lightweight hollow metal ball, initially uncharged, is suspended from a string and hung between the two spheres (Figure 14.69). It is observed that the ball swings rapidly back and forth, hitting one sphere and then the other. This goes on for 5 seconds, but then the ball stops swinging and hangs between the two spheres. Explain in detail, step by step, why the ball swings back and forth, and why it finally stops swinging. Your explanation must include good physics diagrams.

Problem 14.13 Plastic shell and metal block

In Figure 14.70 a very thin spherical plastic shell of radius 15 cm carries a uniformly-distributed negative charge of –8 nC (–8×10^{-9} C) on its outer surface (so it makes an electric field as though all the charge were concentrated at the center of the sphere). An uncharged solid metal block is placed nearby. The block is 10 cm thick, and it is 10 cm away from the surface of the sphere.

(a) Sketch the approximate charge distribution of the neutral solid metal block.

(b) Draw the electric field vector at the center of the metal block that is due solely to the charge distribution you sketched (that is, excluding the contributions of the sphere).

(c) Calculate the magnitude of the electric field vector you drew. Explain briefly. If you must make any approximations, state what they are.

Problem 14.14 Plastic and metal rods

(a) A lightweight (conducting) metal ball hangs from a thread, to the right of an (insulating) plastic rod, as in Figure 14.71. Both are initially uncharged. You rub the left end of the plastic rod with wool, depositing charged molecular fragments whose total (negative) charge is that of 10^9 electrons. You observe that the ball moves toward the rod as shown in Figure 14.72. Explain. Show all excess charged particles, polarization, etc. clearly in a diagram. Make it clear whether charged particles that you show are on the surface of an object or inside it.

(b) You perform a similar experiment with a (conducting) metal rod. You touch the left end of the rod with a charged metal object, depositing 10^9 excess electrons on the left end. You then remove the object. As shown in Figure 14.73, you see the ball deflect more than it did with the plastic rod in part (a). Explain. Show all excess charged particles, polarization, etc. clearly in the diagram. Make it clear whether charged particles that you show are on the surface of an object or inside it.

Problem 14.15 Polarizability of a carbon atom

Try rubbing a plastic pen through your hair, and you'll find that you can pick up a tiny scrap of paper when the pen is about one centimeter above the paper. From this simple experiment you can estimate how much an atom in the paper is polarized by the pen! You will need to make several assumptions and approximations. Hints may be found on page 508.

(a) Suppose that the center of the outer electron cloud ($q = -4e$) of a carbon atom shifts a distance s when the atom is polarized by the pen. Calculate s algebraically in terms of the charge Q on the pen.

(b) Assume that the pen carries about as much charge Q as you found on a piece of invisible tape in Problem 14.1. Evaluate s numerically. How does this compare with the size of an atom or a nucleus?

(c) Calculate the polarizability α of a carbon atom. Compare your answer to the measured value of 1.96×10^{-40} C·m/(N/C) (T. M. Miller and B. Bederson, "Atomic and molecular polarizabilities: a review of recent advances," *Advances in Atomic and Molecular Physics*, 13, 1–55, 1977).

(d) Carefully list all assumptions and approximations you made.

Problem 14.16 Two plastic balls and a metal ball

Two plastic balls are charged equally and positively and held in place by insulating threads (Figure 14.74). They repel each other with an electric force of magnitude F. Then an uncharged metal ball is held in place by insulating threads between the balls, closer to the left ball. State what change (if any) there is in the net electric force on the left ball, and on the net electric force on the right ball. Show relevant force vectors. Also show the charge distribution on the metal ball. Explain briefly but completely.

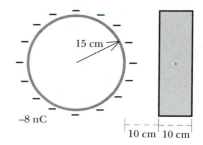

Figure 14.70 A uniformly charged sphere and an uncharged metal block (Problem 14.13).

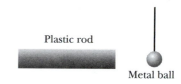

Figure 14.71 Plastic rod and metal ball: initial state (Problem 14.14).

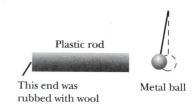

Figure 14.72 Plastic rod and metal ball after end of rod was rubbed with wool (Problem 14.14a).

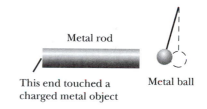

Figure 14.73 Metal rod (Problem 14.14b).

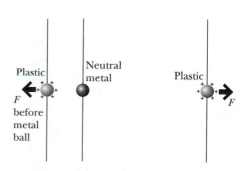

Figure 14.74 Two plastic balls repelled each other. Then a metal ball was introduced.

Problem 14.17 Plastic shell with four point charges

A thin, hollow spherical plastic shell of radius R carries a uniformly distributed negative charge $-Q$. (A slice through the plastic shell is shown in Figure 14.75.) To the left of the spherical shell are four charges packed closely together as shown (the distance "s" is shown greatly enlarged for clarity). The distance from the center of the four charges to the center of the plastic shell is L, which is much larger than s ($L \gg s$). Remember that a uniformly charged sphere makes an electric field as though all the charge were concentrated at the center of the sphere.

(a) Calculate the x and y components of the electric field at location B, a distance b to the right of the outer surface of the plastic shell. Explain briefly, including showing the electric field on a diagram. Your results should not contain any symbols other than the given quantities R, Q, q, s, L, and b (and fundamental constants). You need not simplify the final algebraic results except for taking into account the fact that $L \gg s$.

(b) What simplifying assumption did you have to make in part (a)?

The plastic shell is removed and replaced by an uncharged metal ball (Figure 14.76):

(c) At location A inside the metal ball, a distance b to the left of the outer surface of the ball, accurately draw and label the electric field $\vec{E}_{ball}$ due to the ball charges and the electric field $\vec{E}_4$ of the four charges. Explain briefly.

(d) Show the distribution of ball charges.

(e) Calculate the x and y components of the net electric field at location A.

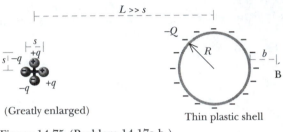

(Greatly enlarged)   Thin plastic shell

Figure 14.75 (Problem 14.17a,b.)

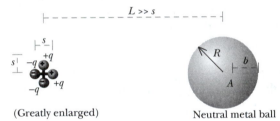

(Greatly enlarged)   Neutral metal ball

Figure 14.76 (Problem 14.17c,d,e.)

Problem 14.18 An electron and a carbon atom

An electron and a neutral carbon atom are initially $d = 10^{-6}$ m apart (about 10000 atomic diameters), and there are no other particles in the vicinity. A careful measurement of the polarizability of a carbon atom gives the value $\alpha = 1.96 \times 10^{-40}$ m^2/(N·m^2/C^2). Carbon has an atomic mass of 12 (6 protons and 6 neutrons in the nucleus).

(a) Calculate the initial magnitude and direction of the acceleration of the electron. Explain your steps clearly. Pay particular attention to clearly defining your algebraic symbols. Don't put numbers into your calculation until the very end.

(b) If the electron and carbon atom were initially twice as far apart, how much smaller would the initial acceleration of the electron be?

Problem 14.19 Sparks with two metal spheres

Metal sphere 1 is charged negatively, and metal sphere 2 is not charged, as shown in Figure 14.77.

(a) Using the conventions for diagrams described and used in this textbook, show the charge distribution for both spheres.

(b) The spheres are moved closer to each other (using insulating supports so as not to change their charge). When they are a certain short distance apart, a spark is seen for a brief instant. Explain qualitatively what determines this distance for the spark to be produced. (Why doesn't the spark occur when the spheres are farther apart? What's special about this distance?)

(c) After the spark stops, show the charge distribution for both spheres, and explain briefly but completely how this new charge distribution came about.

Negative   Neutral

Figure 14.77 (Problem 14.19.)

Problem 14.20 Van de Graaff generator

A Van de Graaff generator, shown in Figure 14.78, pulls electrons out of the Earth and transports them on a conveyor belt onto a nearly spherical metal shell. The diameter of this generator's metal shell is 24 cm.

(a) Suppose that the conveyor belt in the Van de Graaff generator is running so fast that the generator succeeds in building up and maintaining just enough charge on the metal shell to cause the air to steadily glow bluish near the surface of the shell. Under these conditions, how much net charge $|Q|$ is on the metal shell? Calculate a numerical value for $|Q|$ and explain briefly. (Remember that the electric field outside a uniformly charged sphere is like that of a point charge located at the center of the sphere.)

(b) Under these conditions (with the air steadily glowing), the Van de Graaff generator continually delivers additional electrons to the metal shell and yet the net charge Q on the shell does not change. Explain briefly but in detail why the charge on the shell does not change.

(c) Now assume that the Van de Graaff generator is run more slowly, and the buildup of charge on the metal shell is limited by the inability of the motor to force any more electrons onto the negatively-charged shell. The air no longer glows. We have a pear-shaped piece of metal that is attached by a metal wire to the earth ("grounded"). When we bring the grounded piece of metal to a location near the Van de Graaff metal shell, as shown in Figure 14.79, we observe a big spark. Explain why a spark occurs now but doesn't occur without the additional piece of metal nearby.

(d) This spark lasts only a very short time. Why?

(e) If we keep holding the grounded piece of metal 5 cm from the metal shell, we observe that there are big, brief sparks every two seconds. But if we reduce the distance to 2 cm, the sparks that we observe occur more frequently, and they are less intense. Explain briefly.

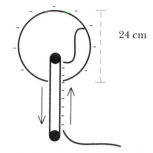

Figure 14.78 Van de Graaff generator (Problem 14.20a,b).

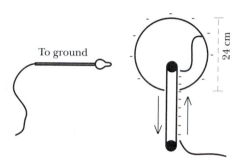

Figure 14.79 Van de Graaff generator and grounding device (Problem 14.20c,d,e).

Problem 14.21 Two permanent dipoles

Two identical permanent dipoles, each consisting of charges $+q$ and $-q$ separated by a distance s, are aligned along the x axis, a distance r from each other, where $r \gg s$ (Figure 14.80).

(a) Draw a diagram like Figure 14.80. Draw vectors showing all individual forces acting on each particle. Draw heavier vectors showing the net force on each dipole.

(b) Show that the magnitude of the net force exerted on one dipole by the other dipole is

$$F \approx \frac{1}{4\pi\varepsilon_0} \frac{6q^2 s^2}{r^4}$$

Show all of the steps in your work, and briefly explain each step.

Figure 14.80 Two permanent dipoles interact with each other (Problem 14.21).

14.13 Answers to exercises

14.1 (page 466)   If your U tape is negatively charged:

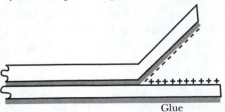

Glue

If your U tape is positively charged, the charges in this diagram should be reversed.

14.2 (page 466)   Did not find any such objects.

14.3 (page 468)   The student has forgotten the superposition principle. Electric interactions go right through matter, so the effect of the positive nucleus is not blocked by the surrounding electron cloud. There are exactly as many protons in the nucleus as there are electrons, and normally the electron cloud is centered on the nucleus, so the net effect is zero.

14.4 (page 470)   Varies; proportional to strength of applied field.

14.5 (page 470) 2×10^{-15} m, about the diameter of the nucleus!

14.6 (page 471)   If the material is more easily polarizable, the separation of charge in a molecule will be greater. Larger dipole moment, larger force.

14.7 (page 471)   1/32

14.8 (page 479)   Note shift of charge distribution; no longer uniform:

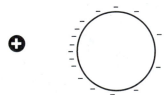

14.9 (page 479)   Negative pen polarizes the neutral metal cylinder by shifting the electron sea; + charges are closer than − charges, so the pen exerts a net attraction on the cylinder.

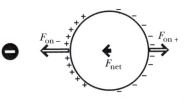

14.10 (page 480)   Negative pen polarizes the neutral plastic cylinder by polarizing the molecules; + charges are closer than − charges, so the pen exerts a net attraction on the cylinder.

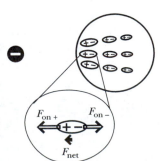

14.11 (page 481) Positive ions from the salt solution on the skin are attracted to the negatively charged tape and are deposited on its slick surface, so the tape becomes neutral (net charge becomes zero). The + charges on the top and the − charges on the bottom make dipoles (and there are induced dipoles inside the tape), but these dipoles exert much weaker forces on other objects than the negatively charged tape did. So the tape acts like ordinary matter.

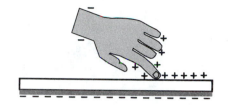

14.12 (page 481) Charging by induction:

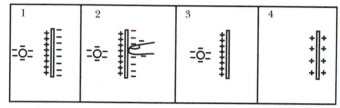

1) Negative pen repels electron sea, polarizing foil (deficiency of electrons on left, excess of electrons on right).

2) Positive ions from finger are attracted by negative charge, and remove electrons from foil. Finger is now negative.

3) Finger removes negative charge. Foil is still polarized by pen.

4) When pen is removed, electron sea shifts toward positive side of foil. Electron deficiency spreads over entire surface of foil.

14.13 (page 482) Discharging by touching the metal foil:

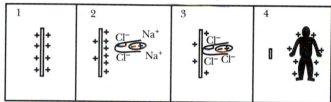

1) Initially foil has positive charge (deficiency of electrons, spread over whole surface).

2) Foil polarizes finger, which in turn polarizes foil.

3) Negative ions from skin are attracted to positive foil, and give electrons to foil.

4) Foil is left nearly neutral when finger is removed. Body is now very slightly positive (excess of positive ions, spread all over body).

14.14 (page 487) First, protons cannot be pulled out of the nuclei of the metal atoms because this would take a huge amount of energy! Second, no single particle can travel the whole distance between the balls in a single jump. Each electron travels only a very small distance between collisions with air molecules or ions. (All of the free electrons in the ionized air move toward the positive ball, just like the shift of the free electron sea in a metal. All of the positive ions move toward the negative ball.)

14.15 (page 487) *d* would double.

14.16 (page 487) *d* would decrease by a factor of 4.

14.17 (page 493) *d* is one third as large as usual.

14.18 (page 493)   It would bounce off without ionizing the molecule (the molecule might gain some kinetic energy).

14.19 (page 493)   About 3.6×10^6 N/C.

Hints

Hints for Problem 14.15 (page 503) Polarizability of a carbon atom

(a1) What must the force on a single carbon atom in the paper be at the moment the paper is lifted by the pen?

(a2) You know how to calculate the force on a point charge due to a dipole. How does this relate to the force on the dipole by the point charge? In this problem, is there something you can model as a dipole and something else you can model as a point charge?

(b) Note that the dipole moment ($p = qs$) of a polarized atom or molecule is directly proportional to the applied electric field (see page 470). In this case the charged pen is generating the applied electric field.

Chapter 15

Electric Field of
Distributed Charges

Chapter 15

Electric Field of Distributed Charges

In previous chapters you calculated the electric field due to two individual charged particles as the superposition of the contributions of each particle. In a similar way, you could calculate the net electric field due to three or four charged particles. But how can we calculate the electric field due to millions of individual charged particles? For example, what is the electric field near a charged tape, which has millions of charged particles distributed all over a rectangular area?

In the process of developing a technique for answering this question, we will begin to become familiar with some common patterns of electric field in space. This familiarity will be important in later chapters.

15.1 Overview

In this chapter we will study mathematical techniques for adding up the contributions to the electric field of large numbers of point charges distributed over large areas. The most general technique is to divide the charge distribution into a large but finite number of pieces and use a computer to add up the contributions ("numerical integration"). One of the goals of this chapter is to show you how to set up such a computation.

Figure 15.1 The electric field at locations near a positively charged rod. Near the middle of the rod the electric field vectors lie in a plane perpendicular to the rod.

In a few special but important cases we can get an analytical solution by using an integral to add up the contributions. We are able to do this for some locations near a charged rod, ring, disk, capacitor, and sphere. A major advantage of the analytical approach is that we see how the field varies with distance from the charge distribution, which is important in many applications. These results turn out to be useful because it is often possible to model ordinary objects as combinations of spheres, rods, rings, and disks, and thereby estimate their electric fields.

15.2 A uniformly charged thin rod

As an example of how to calculate the electric field due to large numbers of charges, we'll consider the electric field of a uniformly charged thin rod. The rod might, for example, be a glass rod that was rubbed all over with silk, giving it a nearly uniform distribution of positive charge on its surface. We'll consider a thin rod of length L and total positive charge Q.

? Before launching into a calculation, think about the pattern of electric field you would expect to observe around a rod. Viewed end-on, what would it look like?

The field of a point charge was spherically symmetric, since a point charge itself has spherical symmetry. Since a rod is shaped like a cylinder, we should expect cylindrical symmetry in the electric field of a uniformly charged rod. Near a positively charged rod, the electric field might look like Figure 15.1. If we took a slice perpendicular to the rod, near the middle of the rod, the pattern of field might look like Figure 15.2.

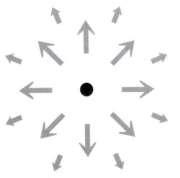

Figure 15.2 The electric field near a charged rod should be cylindrically symmetric. This is a representation of the field in a plane perpendicular to the rod, near the middle of the rod.

Step 1: Divide the distribution into pieces; draw $\Delta\vec{E}$

To apply the superposition principle, we imagine cutting up the thin rod into very short sections each with positive charge ΔQ, as shown in Figure 15.3. The Greek capital letter delta (Δ) denotes a small portion of something, or a change in something. Here ΔQ is a small portion of the total charge Q of the rod, which contributes $\Delta\vec{E}$ to the net field at the observation location. We can treat the piece ΔQ as though it were a point charge, which should be a fairly good approximation as long as its size is small compared to the distance to the observation location.

In Figure 15.3 we have picked a representative piece of the rod that is not at a "special" location (in this case, neither at an end nor at the middle), and drawn its contribution $\Delta\vec{E}$ to the net field at our chosen observation location.

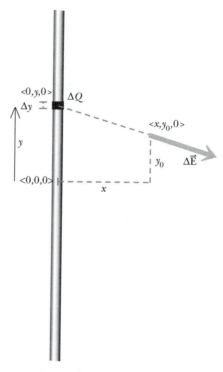

Figure 15.3 $\Delta\vec{E}$ is the contribution to the total field at location $<x, y_0, 0>$ made by a small piece of the rod of length Δy.

Assumptions

We assume that the rod is so thin that we can ignore the thickness of the rod. We choose the piece of charge ΔQ to be small enough that it can be modeled as a point charge.

Step 2: Write an expression for the electric field due to one piece

Origin and axes

In order to write an algebraic expression for the contribution to the electric field of one representative piece of the rod, we need to pick an origin and axes. It is possible to put the origin anywhere, but some choices make the algebra easier than others. Here we put the origin at the center of the rod; the x axis extends to the right, and the y axis extends up.

Location of one piece

The location of one piece of the rod (shown in Figure 15.3) depends on y.

Integration variable

? What variables should remain in our answer?

The coordinates of the observation location should remain, but we will sum (integrate) over all pieces of the rod, so the coordinates of the piece of the rod should not remain. However, in the expression we want to integrate it is okay to have variables representing the location of the rod segment; they are called "integration variables," and will disappear after we do the sum.

A piece of the rod is a distance y from the origin. The length of the piece is Δy, a small increment in the integration variable y, shown in Figure 15.3. Our invented variable y will not appear in the final result.

Note that if there is no Δ-*something* (like Δy) in the expression, we can't do a sum or evaluate an integral.

Vector from piece to observation location

The vector $\vec{r}$ points from the source (the representative little piece of charge ΔQ) to the observation location. We can read the components of $\vec{r}$ from the diagram in Figure 15.4:

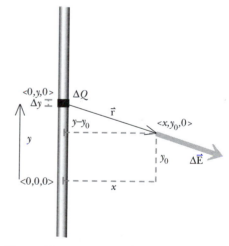

Figure 15.4 The vector $\vec{r}$ points from the source to the observation location.

$$\vec{r} = \langle \text{obs. loc.} \rangle - \langle \text{source} \rangle = \langle x, y_0, 0 \rangle - \langle 0, y, 0 \rangle$$
$$= \langle x, (y_0 - y), 0 \rangle$$

Note that $(y_0 - y)$, the y component of $\vec{r}$, is negative (as it should be).

Magnitude of $\hat{r}$:

The magnitude of $\hat{r}$ is the distance from the piece of charge ΔQ to the observation location:

$$r = [x^2 + (y_0 - y)^2]^{1/2}$$

Unit vector $\hat{r}$:

$$\hat{r} = \frac{\vec{r}}{|\vec{r}|} = \frac{\langle x, (y_0 - y), 0 \rangle}{[x^2 + (y_0 - y)^2]^{1/2}}$$

Again, note that the y component of $\hat{r}$ is negative, as it should be.

Magnitude of $\Delta\vec{E}$

This piece contributes a field of magnitude

$$\Delta E = \frac{1}{4\pi\varepsilon_0} \frac{\Delta Q}{r^2} = \frac{1}{4\pi\varepsilon_0} \frac{\Delta Q}{x^2 + (y_0 - y)^2}$$

Vector $\Delta\vec{E}$:

We can now write the vector $\Delta\vec{E}$ (Figure 15.5):

$$\Delta\vec{E} = (\Delta E)\hat{r} = \frac{1}{4\pi\varepsilon_0} \frac{\Delta Q}{x^2 + (y_0 - y)^2} \frac{\langle x, (y_0 - y), 0 \rangle}{[x^2 + (y_0 - y)^2]^{1/2}}$$

$$\Delta\vec{E} = \frac{1}{4\pi\varepsilon_0} \frac{\Delta Q}{[x^2 + (y_0 - y)^2]^{3/2}} \langle x, (y_0 - y), 0 \rangle$$

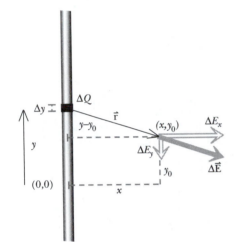

Components of $\Delta\vec{E}$:

The components of $\Delta\vec{E}$ (Figure 15.5) are then:

$$\Delta E_x = \frac{1}{4\pi\varepsilon_0} \frac{x\Delta Q}{[x^2 + (y_0 - y)^2]^{3/2}}$$

$$\Delta E_y = \frac{1}{4\pi\varepsilon_0} \frac{(y_0 - y)\Delta Q}{[x^2 + (y_0 - y)^2]^{3/2}}$$

$$\Delta E_z = 0$$

Figure 15.5 The components of $\Delta\vec{E}$ may be calculated by multiplying $|\Delta\vec{E}|$ by the components of $\hat{r}$.

? Which things are constants and which things are variables in the expressions for ΔE_x and ΔE_y?

You should have identified y as a variable, since it differs for each small piece. We need to rewrite the expression in a form in which it is easy to add up all the ΔE_x's and ΔE_y's due to all the pieces of the rod. This means expressing everything in terms of one integration variable, related to the coordinates of the piece; in this case y is the integration variable.

ΔQ in terms of integration variable

In particular, we need to express the charge ΔQ in terms of the integration variable y. The rod is uniformly charged with a total charge Q, so the amount of charge on a section of length Δy is equal to

$$\Delta Q = \left(\frac{\Delta y}{L}\right)Q$$

since $\Delta y / L$ is the fraction of the whole rod represented by Δy. Alternatively, there is a linear charge density of Q/L in coulombs/meter, so on a length of Δy meters there is an amount of charge $\Delta Q = (Q/L)\Delta y$. The important thing is to realize that ΔQ is a small fraction of the total charge Q, and that you must express ΔQ in terms of Δy.

Components of $\Delta\vec{E}$

Putting it all together, we have an expression for the x component:

$$\Delta E_x = \frac{1}{4\pi\varepsilon_0}\frac{Q}{L}\frac{x}{[x^2+(y_0-y)^2]^{3/2}}\Delta y$$

A similar approach yields an expression for the y component:

$$\Delta E_y = \frac{1}{4\pi\varepsilon_0}\frac{Q}{L}\frac{(y_0-y)}{[x^2+(y_0-y)^2]^{3/2}}\Delta y$$

A note on the symbol Δ

We have used the Greek letter Δ (capital delta) in two ways in setting up this problem. ΔE or ΔQ refer to a small contribution to a total quantity. Δy refers to a change in the integration variable y, which determines the location of the piece of the rod currently under consideration. You will need to use the symbol Δ in both these ways when setting up problems like this one.

Step 3: Add up the contributions of all the pieces

Simplified problem

Before adding up the contributions of all the pieces, let's simplify the problem by deciding to find the electric field at the location $<x, 0, 0>$, as shown in Figure 15.6. That is, we let $y_0 = 0$.

y components

In the case of a uniformly charged rod, if $y = 0$, we see right away that we can simplify the problem. When we add up the contributions of all the little pieces that make up the rod, the y components at a location on the x axis add up to zero, due to the symmetry of the situation. One way to see this is to slice the rod up into pairs of pieces whose y components of electric field are equal and opposite, as shown in Figure 15.6. Had we chosen a location where y_0 was not zero, we would need to add up all the y components explicitly, too.

x components

Each ΔQ contributes ΔE_x to the net field. If we number each piece 1, 2, 3, etc., we have

$$\Delta E_x = \Delta E_{x_1} + \Delta E_{x_2} + \Delta E_{x_3} + \dots$$

It is standard practice to write such sums in a more compact form using a Σ (the Greek capital Sigma stands for "Summation"):

$$E_x = \sum \Delta E_x = \sum \frac{1}{4\pi\varepsilon_0}\frac{Q}{L}\frac{x}{(x^2+y^2)^{3/2}}\Delta y$$

Some of these quantities are the same for every piece and can be taken outside the summation as common factors:

$$E_x = \frac{1}{4\pi\varepsilon_0}\frac{Q}{L}x\sum\frac{1}{(x^2+y^2)^{3/2}}\Delta y$$

At this point we have a problem. Each term in this summation is different. Although we can choose to have all the pieces be of the same length Δy and take Δy out of the sum, y itself is different for every piece along the rod. So how can we add up all the contributions of all the pieces?

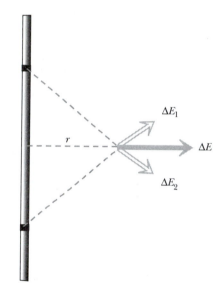

Figure 15.6 Simplified problem: $y_0 = 0$. Along the midplane only the x component of the electric field is nonzero.

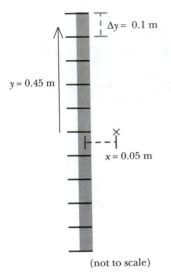

y = 0.45 m

Δy = 0.1 m

x = 0.05 m

(not to scale)

Figure 15.7 Division of the rod into 10 equal slices.

Table 1:

Slice #	y	$\dfrac{\Delta y}{(x^2 + y^2)^{3/2}}$ m^{-2}
1	+0.45	1.08
2	+0.35	2.26
3	+0.25	6.03
4	+0.15	25.30
5	+0.05	282.84
6	−0.05	282.84
7	−0.15	25.30
8	−0.25	6.03
9	−0.35	2.26
10	−0.45	1.08
Sum	All y	635.03

Table 2:

10 slices	$31.75 \dfrac{Q}{4\pi\varepsilon_0}$
20 slices	$39.31 \dfrac{Q}{4\pi\varepsilon_0}$
50 slices	$39.80 \dfrac{Q}{4\pi\varepsilon_0}$
100 slices	$39.80 \dfrac{Q}{4\pi\varepsilon_0}$

Numerical summation

One way to add up all these different contributions would be to divide the rod into 10 slices as shown in Figure 15.7, calculate the contribution of each of those 10 slices (using our formula with $\Delta y = L/10$), and add up these 10 numbers. Of course that would be only an approximate result, because each slice isn't really a point particle, but it might be good enough for many purposes. Let's do this calculation for the case of $L = 1$ meter, $\Delta y = 0.1$ meter, and $x = 0.05$ meter. For each slice we'll take y to be at the center of the slice, and we'll use a calculator to evaluate each varying term in the summation. The results are shown in Table 1, including the sum for all pieces.

If we had thought ahead a bit, we could have avoided half the work, because we can see that the contributions of slices with positive y are exactly the same as the contributions of slices with negative y, due to the y^2 in the formula. It is often the case that we can simplify electric field calculations by taking advantage of such symmetries.

With $x/L = 0.05$, we have

$$E_x = \frac{1}{4\pi\varepsilon_0}\frac{Q}{L}x\sum\frac{1}{(x^2+y^2)^{3/2}}\Delta y = \frac{Q}{4\pi\varepsilon_0}(0.05)(635.03) = (31.75\ m^{-2})\frac{Q}{4\pi\varepsilon_0}$$

If $Q = 1$ nC, then we find that $E_x = 286$ N/C.

Because of the way we organized our calculation, we can quickly determine the electric field for any other value of Q.

It would be possible to get a more accurate value by cutting the rod into more slices, but it would be tedious to do the calculations by hand. We wrote a little computer program to do these calculations for various numbers of slices, and the results of these calculations are shown in Table 2.

Judging from the computations summarized in Table 2, our calculation with just 10 slices was not very accurate (though it might be good enough for some purposes), while taking more than 50 slices makes almost no difference. The accuracy depends on how good an approximation it is to consider one slice as a point charge. With 50 slices, each slice Δy is 1/50th of a meter long (0.02 meters), and the nearest slice is 0.05 meters away from the observation location. Apparently the 0.02-meter slices are adequately approximated by point charges at their centers, since using smaller 0.01-meter slices (100 slices) gives practically the same result.

How do we know that our computer summations are correct? For 10 slices they agree with the calculator results. For larger and larger numbers of slices the computer results approach a constant value, which is expected. Later we will discuss additional ways to check such work.

Summation as an integral

A major disadvantage of adding up the contributions numerically is that we don't get an analytical (algebraic) form for the electric field. This means that we can't easily answer such questions as, "How does the electric field vary with r near a rod?" For a point charge we know that the field goes like $1/r^2$, and far from a permanent dipole it goes like $1/r^3$. Is there a way to do the summation in such a way as to get an algebraic rather than a numerical answer for the rod?

That is what integral calculus was invented to do. The key idea of integral calculus applied to problems like ours is to imagine taking not 50 or 100 slices, but an infinite number of infinitesimal slices. We let the number of slices $N = L/(\Delta y)$ increase without bound, and the corresponding slice length

$\Delta y = L/N$ decreases without bound. We take the limit as Δy gets arbitrarily small:

$$E_x = \lim_{\Delta y \to 0} \frac{1}{4\pi\varepsilon_0} \frac{Q}{L} x \sum \frac{1}{(x^2 + y^2)^{3/2}} \Delta y$$

This limit is called a "definite integral" and is written like this:

$$E_x = \frac{1}{4\pi\varepsilon_0} \frac{Q}{L} x \int_{-\frac{L}{2}}^{+\frac{L}{2}} \frac{1}{(x^2 + y^2)^{3/2}} \, dy$$

The integral sign $\int$ is a distorted "S" standing for "Summation," just as the Greek sigma Σ stands for summation. It is important that in this context you think of an integral as a sum of many contributions.

The integration variable y ranges from $y = -L/2$ (the bottom of the rod) to $y = +L/2$ (the top of the rod), so these are the limits on the definite integral. If you have drawn an appropriately labeled diagram, you should be able to read the correct limits of the integration variable right off the diagram. This is another of the many benefits of drawing a well-labeled physics diagram.

Our small length Δy has now changed into dy, an "infinitesimal" increment in y that *must* appear in the integrand. This is why Δy *must* appear in the algebraic expression for the contribution of one piece of the charge distribution.

Evaluating the integral

Most of the physics in this problem went into setting up the integral. Evaluating the integral is just mathematics. In some cases it is easy to evaluate the integral. In this particular case, the integral is not a very simple one, but if you would like to exercise your skill at integration, give it a try! Otherwise, it can be found in tables of integrals in mathematical handbooks and in some calculus textbooks. Looking up the result in a table of integrals, we get the following:

$$E_x = \frac{1}{4\pi\varepsilon_0} \left(\frac{Q}{L} x \right) \left(\frac{y}{x^2 \sqrt{x^2 + y^2}} \right) \Big|_{-L/2}^{+L/2}$$

Plugging in the end values of z and simplifying, we have a value for E_x:

$$E_x = \frac{1}{4\pi\varepsilon_0} \left[\frac{Q}{x\sqrt{x^2 + (L/2)^2}} \right]$$

Note that as we expected this result does not contain the integration variable y, which was simply a variable referring to the coordinates of one piece of the rod, and which was necessary in setting up the summation.

Replace x with r

Because the rod, and its associated electric field, are cylindrically symmetric, the axis we called the x axis could have been rotated around the rod by any angle, and we would have obtained the same answer. To indicate this, we replace x with r in our result:

ELECTRIC FIELD OF A UNIFORMLY CHARGED THIN ROD

$$E = \frac{1}{4\pi\varepsilon_0} \left[\frac{Q}{r\sqrt{r^2 + (L/2)^2}} \right]$$

at a distance r from the midpoint, perpendicular to the rod (Figure 15.8).

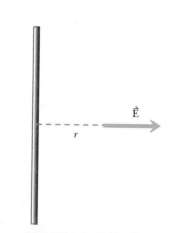

Figure 15.8 Electric field of a uniformly charged thin rod, near the midpoint of the rod.

Written as a vector:

$$\vec{E} = \frac{1}{4\pi\varepsilon_0}\left[\frac{Q}{r\sqrt{r^2 + (L/2)^2}}\right]\hat{r}$$

where $\hat{r}$ is, as usual, a vector of magnitude 1 pointing from the source (the middle of the rod) to the observation location.

Step 4: Check the result

Because there are many opportunities to make mistakes in this procedure, it is extremely important to check the result in as many ways as you can. Different checks provide information about different kinds of possible errors.

Direction

First, is the direction qualitatively correct? We have the electric field pointing straight away from the midpoint of the rod, which is correct, given the symmetry of the situation. The vertical component of the electric field should indeed be zero.

Units

Second, do we have the right units? Comparing with the electric field for a single point particle,

$$\frac{1}{4\pi\varepsilon_0}\frac{Q}{r^2}$$

we easily verify that our answer does have the right units, since

$$\frac{1}{r\sqrt{r^2 + (L/2)^2}} \text{ has the same units as } \frac{1}{r^2}.$$

Special case: $r \gg L$

? Next let's try a special case for which we already know the answer. If r is very much larger than L, the distant rod looks almost like a point charge, so the net field ought to look like the field of a point charge.

Show that if $r \gg L$, $E \rightarrow \dfrac{1}{4\pi\varepsilon_0}\dfrac{Q}{r^2}$, as it should.

Special case: $L \ll r$

For another special case, suppose r is fairly near the rod, but the rod is so short as to look almost like a point charge. In that case $L \ll r$, and we again get $E \rightarrow \dfrac{1}{4\pi\varepsilon_0}\dfrac{Q}{r^2}$, as we should.

Compare to numerical calculation

Earlier we carried out a numerical calculation for the case of $L = 1$ meter and $r = 0.05$ meter, and we found that if we used 50 or more slices the result was $39.80\,Q/4\pi\varepsilon_0$. The analytical solution gives

$$E = \frac{1}{4\pi\varepsilon_0}\frac{Q}{r\sqrt{r^2 + (L/2)^2}} = \frac{Q}{4\pi\varepsilon_0}\left(\frac{1}{(0.05)\sqrt{0.05^2 + 0.5^2}}\right) = 39.80\frac{Q}{4\pi\varepsilon_0}$$

which agrees very well with the numerically calculated value.

Checking a numerical solution

Similar techniques can be used to check a numerical integration done on a computer. For example, set the length of the rod to be very short, or the distance to be very large, and the numerical integration should give a result equal to what you calculate by hand for a point charge.

Ex. 15.1 If the total charge on a rod of length 0.4 m is 2.5 nC, what is the magnitude of the electric field at a location 1 cm from the midpoint of the rod?

Ex. 15.2 Graph the magnitude of E vs. r. Does E fall off monotonically with distance?

Special case: A very long rod

? A very important special case, which we will refer to often, is the case of a rod which is very long (or alternatively a rod very close to the observation location). In either case $L \gg r$. Show that if $L \gg r$, the electric field is approximately

$$\frac{1}{4\pi\varepsilon_0}\frac{2(Q/L)}{r}$$

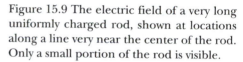

Try to work this out for yourself before reading further. The exact expression we derived for the electric field of a rod is

$$E = \frac{1}{4\pi\varepsilon_0}\left[\frac{Q}{r\sqrt{r^2 + (L/2)^2}}\right]$$

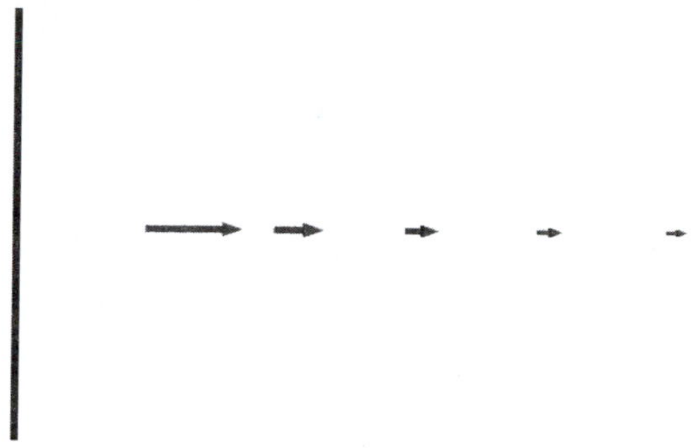

Figure 15.9 The electric field of a very long uniformly charged rod, shown at locations along a line very near the center of the rod. Only a small portion of the rod is visible.

If $L \gg r$, then the denominator is approximately $r\sqrt{(L/2)^2} = r(L/2)$, and we get the result given above, and illustrated in Figure 15.9. This also holds for a short rod if we are very close, so that $r \ll L$.

ELECTRIC FIELD OF A UNIFORMLY CHARGED THIN ROD

At a distance r from the midpoint along a line perpendicular to the rod,

$$E = \frac{1}{4\pi\varepsilon_0}\left[\frac{Q}{r\sqrt{r^2 + (L/2)^2}}\right]$$

For a very long rod ($r \ll L$),

$$E \approx \frac{1}{4\pi\varepsilon_0}\frac{2(Q/L)}{r}$$

or, written as a vector:

$$\vec{E} \approx \frac{1}{4\pi\varepsilon_0}\frac{2(Q/L)}{r}\hat{r}$$

Our calculations apply to the electric field near the midpoint of a charged rod (Figure 15.9). As illustrated in Figure 15.10, near the ends of the rod the electric field is not perpendicular to the rod. We would have to do a numerical calculation to determine the field near the ends of the rod (the calculations to generate Figure 15.10 were done numerically).

Figure 15.10 Electric field of a positively charged rod, calculated numerically.

15.3 A general procedure for calculating electric field

We summarize the procedure we use to calculate electric field:

Step 1: Cut up the charge distribution into pieces and draw $\Delta \vec{E}$

- Divide the charge distribution into pieces whose field is known. In particular, very small pieces can be approximated by point particles.

- Pick a representative piece, and at the location of interest draw a vector $\Delta \vec{E}$ showing the contribution to the electric field of this representative piece. Drawing this vector helps you figure out the direction of the net field at the location of interest.

Step 2: Write an expression for the electric field due to one piece

- Pick an origin for your coordinate system, and show it on your diagram.

- Draw the vector $\vec{r}$ from the source piece to the observation loctaion. Write algebraic expressions for $\vec{r}$ and $\hat{r}$ (a unit vector in the direction of $\vec{r}$).

- Write an algebraic expression for the magnitude $|\Delta \vec{E}|$ contributed by the representative piece. Multiply by $\hat{r}$ to get a vector $\Delta \vec{E}$, from which you can read the components ΔE_x, ΔE_y, and ΔE_z. Your expressions should contain one or more "integration variables" related to the coordinates of the piece.

- Write the amount of charge on the piece, Δq, in terms of your variables.

- If a representative piece is small in size, your algebraic expressions should include small increments of the integration variable. For example, if your integration variable is y, your expressions must be proportional to Δy.

Step 3: Add up the contributions of all the pieces

- The net field is the sum of the contributions of all the pieces. To write the sum as a definite integral, you must include limits given by the range of the integration variable. If the integral can be done symbolically, do it. If not, choose a finite number of pieces and do the sum with a calculator or a computer.

Step 4: Check the result

- Check that the direction of the net field is qualitatively correct.

- Check the units of your result, which should be newtons per coulomb.

- Look at special cases. For example, if the net charge is nonzero, your result should reduce to the field of a point charge when you are very far away. For a numerical integration on a computer, check that the computation gives the correct numerical result for special cases that can be calculated by hand.

15.4 A uniformly charged thin ring

Next we'll calculate the electric field of a uniformly charged thin ring. We'll encounter rings of charge later when we study electric circuits. We'll also use the results for a ring to find the electric field of a disk later in this chapter.

15.4.1 Field of a ring, at a point on the axis

We'll calculate the electric field due to a uniformly charged ring of radius R and total positive charge q. We'll do only the easiest case—the field at a location along the axis of the ring, which is a line going through the center and perpendicular to the ring. Finding the electric field at other locations is harder.

Step 1: Cut up the charge distribution into pieces and draw $\Delta\vec{E}$

See Figure 15.11.

Step 2: Write an expression for the electric field due to one piece

Origin: center of the ring. Axes shown in Figure 15.11.

Location of piece: described by θ, where $\theta = 0$ is along the x axis.

Vector from source to observation location: (*Figure 15.11*)

$$\vec{r} = \langle\text{obs. loc.}\rangle - \langle\text{source}\rangle = \langle 0, 0, z\rangle - \langle R\cos\theta, R\sin\theta, 0\rangle$$

$$= \langle -R\cos\theta, -R\sin\theta, z\rangle$$

Magnitude of $\vec{r}$:

$$|\vec{r}| = \sqrt{(-R\cos\theta)^2 + (-R\sin\theta)^2 + z^2} = (R^2 + z^2)^{1/2}$$

Unit vector $\hat{r}$:

$$\hat{r} = \frac{\vec{r}}{|\vec{r}|} = \frac{\langle -R\cos\theta, -R\sin\theta, z\rangle}{(R^2 + z^2)^{1/2}}$$

Magnitude of $\Delta\vec{E}$:

$$\Delta E = \frac{1}{4\pi\varepsilon_0}\frac{\Delta q}{r^2} = \frac{1}{4\pi\varepsilon_0}\frac{\Delta q}{(R^2 + z^2)}$$

Δq in terms of variables:

$$\Delta q = q\left(\frac{\Delta\theta}{2\pi}\right) \text{ (There are } 2\pi \text{ radians in the complete ring.)}$$

Integration variable: θ

Expression for $\Delta\vec{E}$:

$$\Delta\vec{E} = (\Delta E)\hat{r} = \frac{1}{4\pi\varepsilon_0}\frac{q\left(\dfrac{\Delta\theta}{2\pi}\right)}{(R^2 + z^2)}\frac{\langle -R\cos\theta, -R\sin\theta, z\rangle}{(R^2 + z^2)^{1/2}}$$

$$\Delta\vec{E} = \frac{1}{4\pi\varepsilon_0}\frac{q}{2\pi}\frac{\Delta\theta}{(R^2 + z^2)^{3/2}}\langle -R\cos\theta, -R\sin\theta, z\rangle$$

Components: From the symmetry of the situation, we see that the x components will sum to zero, as will the y components. The z components will not cancel, so we need ΔE_z:

$$\Delta E_z = \frac{1}{4\pi\varepsilon_0}\frac{q}{2\pi}\frac{z}{(R^2 + z^2)^{3/2}}\Delta\theta$$

$$\text{or } dE_z = \frac{1}{4\pi\varepsilon_0}\frac{q}{2\pi}\frac{z}{(R^2 + z^2)^{3/2}}d\theta$$

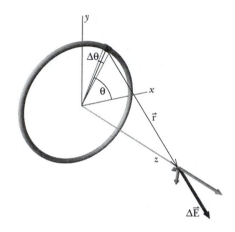

Figure 15.11 The contribution $\Delta\vec{E}$ to the total electric field of a charged ring, made by a segment of the ring of angular size $\Delta\theta$. We have chosen the center of the ring for the origin of our coordinate system. The x, y, and z components of $\Delta\vec{E}$ are shown, but not labeled.

Figure 15.12 Electric field of a uniformly charged ring.

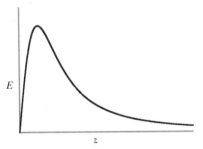

Figure 15.13 The electric field of a charged ring, along the z axis (perpendicular to the ring).

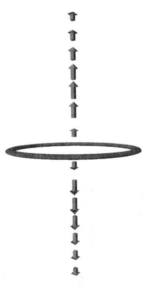

Figure 15.14 Electric field of a ring along the z axis, varying with distance. As the graph in Figure 15.13 indicates, as z increases the magnitude of the field first increases, then decreases.

Step 3: Add up the contributions of all the pieces

In this case we can evaluate the integral (sum) analytically:

$$E_z = \int_0^{2\pi} dE_z$$

$$= \int_0^{2\pi} \frac{1}{4\pi\varepsilon_0} \frac{q}{2\pi} \frac{z}{(R^2 + z^2)^{3/2}} d\theta$$

$$= \frac{1}{4\pi\varepsilon_0} \frac{q}{2\pi} \frac{z}{(R^2 + z^2)^{3/2}} \int_0^{2\pi} d\theta$$

ELECTRIC FIELD OF A UNIFORMLY CHARGED THIN RING

$$E = \frac{1}{4\pi\varepsilon_0} \frac{qz}{(R^2 + z^2)^{3/2}}$$

along the axis, for a ring of radius R and charge q (Figure 15.12).

Step 4: Check the result

Direction: Correct, by symmetry.

Units: $\left(\dfrac{\text{N} \cdot \text{m}^2}{\text{C}^2}\right)\left(\dfrac{\text{C} \cdot \text{m}}{(\text{m}^2)^{3/2}}\right) = \dfrac{\text{N}}{\text{C}}$ correct

Special cases

Exact center of the ring:

$$z = 0 \Rightarrow E = 0$$

This is correct, since all contributions to E will cancel at this location.

$z \gg R$:

$$(R^2 + z^2)^{3/2} \approx (z^2)^{3/2} = z^3, \text{ so } E \propto z/z^3 = 1/z^2$$

This is correct; at locations far from the ring, the ring should look like a point charge.

Distance dependence of the electric field of a ring

Since the field is zero at the center of the ring, but falls off like $1/z^2$ far from the ring, the field must first increase, then decrease, with distance from the ring. With a graphing calculator you can easily show that a plot of E vs. z looks like Figure 15.13.

Summary: Electric field of a uniformly charged ring

The electric field of a thin uniformly charged ring, a distance z from the midpoint of the ring along a line perpendicular to the plane of the ring is:

$$E = \frac{1}{4\pi\varepsilon_0} \frac{qz}{(R^2 + z^2)^{3/2}}$$

Note that we have calculated the field of a ring only along the axis perpendicular to the ring (Figure 15.14). As shown in Figure 15.15, the pattern of field at other locations is much more complex.

Figure 15.15 is the result of a numerical integration, in which the ring was cut up into short segments and the summation was made at many observation locations. We calculated the field analytically along the z axis not just because this is one of the rare cases where an analytical solution is possible, but also because this result will be useful to us later, including in the next section.

Ex. 15.3 Two rings of radius 5 cm are 20 cm apart and concentric with a common horizontal axis. The ring on the left carries a uniformly distributed charge of +35 nC, and the ring on the right carries a uniformly distributed charge of –35 nC. What is the magnitude and direction of the electric field on the axis, halfway between the two rings?

Ex. 15.4 In the previous exercise, if a charge of –5 nC were placed midway between the rings, what would be the magnitude and direction of the force exerted on this charge by the rings?

Ex. 15.5 What is the magnitude and direction of the electric field midway between the rings if both rings carry a charge of +35 nC?

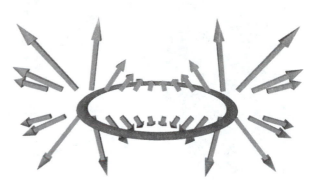

Figure 15.15 The electric field of a charged ring at locations across the ring, slightly out of the plane of the ring.

15.5 A uniformly charged disk

One reason that a uniformly charged disk of charge is important is that two oppositely charged metal disks can form a "capacitor," a device that is important in electric circuits. Before discussing metal disks, we will consider a glass disk that has been rubbed with silk in such a way as to deposit a uniform density of positive charge all over the surface.

15.5.1 Field along the axis of a uniformly charged disk

The electric field of a uniformly charged disk of course varies in both magnitude and direction at observation locations near the disk, as illustrated in Figure 15.16, which shows the computed pattern of electric field at many locations near a uniformly charged disk (done by numerical integration, with the surface of the disk divided into small squares). Note, however, that near the center of the disk the field is quite uniform, both in direction and magnitude. Even near the edge of the disk the magnitude of the electric field isn't very different, though the direction is no longer nearly perpendicular to the disk.

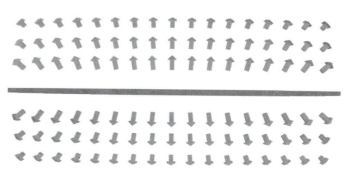

Figure 15.16 A uniformly charged disk viewed edge on. The electric field is shown at locations across the diameter of the disk.

Again, we'll pick an easy location for an analytical solution—the field at any location along the axis of the disk, which is a line going through the center and perpendicular to the disk, as shown in Figure 15.17. This is a more useful choice than one might expect, since the field turns out to nearly uniform in regions far from the edge of the disk; our result will be applicable to a variety of situations.

We consider a disk of radius R, with a total charge Q uniformly distributed over the front surface of the disk.

Step 1: Cut up the charge distribution into pieces; draw $\Delta\vec{E}$

Use thin concentric rings as pieces, as shown in Figure 15.17, since we already know the electric field of a uniform ring. Approximate each ring as having some average radius r.

Step 2: Write an expression for the $\Delta\vec{E}$ due to one piece

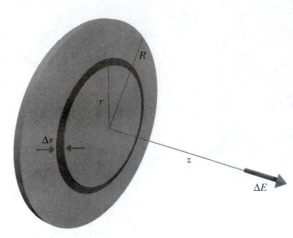

Figure 15.17 A ring of width Δr makes a contribution ΔE to the total electric field.

Origin: center of ring.

Location of piece: given by radius r of ring.

Vector from center of ring to observation location:

$$\vec{r} = \langle 0, 0, z \rangle$$

Magnitude of $\vec{r}$: $|\vec{r}| = z$

Unit vector $\hat{r}$:

$$\hat{r} = \frac{\vec{r}}{|\vec{r}|} = \langle 0, 0, 1 \rangle$$

Magnitude of $\Delta\vec{E}$ *contributed by this ring-shaped piece:*

$$\Delta E = \frac{1}{4\pi\varepsilon_0} \frac{(\Delta q)z}{(r^2 + z^2)^{3/2}} \text{ (from Section 15.4.1)}$$

Note that both Δq and r will be different for each piece.

Δq in terms of variables:

$$\Delta q = Q \frac{(\text{area of ring})}{(\text{area of disk})} = Q \frac{2\pi r \Delta r}{\pi R^2} \text{ (see Figure 15.18)}$$

Vector $\Delta\vec{E}$:

Figure 15.18 A ring of radius r and thickness Δr cut and rolled out straight.

$$\Delta\vec{E} = \Delta E \hat{r} = \frac{1}{4\pi\varepsilon_0} \frac{\left(Q\dfrac{2\pi r \Delta r}{\pi R^2}\right)z}{(r^2 + z^2)^{3/2}} \langle 0, 0, 1 \rangle$$

Integration variable: r

Components to calculate: only ΔE_z is nonzero.

$$\Delta E_z = \frac{1}{2\varepsilon_0}\left(\frac{Q}{\pi R^2}\right)\frac{z}{(r^2 + z^2)^{3/2}} \; r\Delta r$$

or, for infinitesimally thin rings:

$$dE_z = \frac{1}{2\varepsilon_0}\left(\frac{Q}{\pi R^2}\right)\frac{z}{(r^2 + z^2)^{3/2}} \; rdr$$

Step 3: Sum all contributions

$$E_z = \int_0^R \frac{1}{2\varepsilon_0}\left(\frac{Q}{\pi R^2}\right)\frac{z}{(r^2 + z^2)^{3/2}} \; rdr$$

Many of these quantities have the same values for different values of r, and these can be taken out of the integral as common factors:

$$E_z = \frac{1}{2\varepsilon_0}\left(\frac{Q}{\pi R^2}\right)z\int_0^R \frac{r}{(r^2 + z^2)^{3/2}} dr$$

This particular integral can be done by a change of variables, letting $u = (r^2 + z^2)$. You can work it out yourself, or look up the result in a table of integrals, or use a symbolic math package to evaluate it. The result is:

$$E_z = \frac{1}{2\varepsilon_0}\left(\frac{Q}{\pi R^2}\right)\left[1 - \frac{z}{(R^2 + z^2)^{1/2}}\right]$$

for a uniformly charged disk of charge Q and radius R, at locations along the axis of the disk. This is often written in terms of the area A of the disk ($A = \pi R^2$):

$$E_z = \frac{(Q/A)}{2\varepsilon_0}\left[1 - \frac{z}{(R^2 + z^2)^{1/2}}\right]$$

Step 4: Check

Direction:

Away from the disk, as expected.

Special location:

$0 < z \ll R$ (very close to the disk, but not touching it. See Figure 15.20.)

$$E \approx \frac{Q/A}{2\varepsilon_0}\left[1 - \frac{z}{R}\right]$$

If z/R is extremely small, $[1 - z/R]$ reduces to 1, so we have approximately

$$E \approx \frac{Q/A}{2\varepsilon_0}$$

Interestingly, this field is independent of distance! This result is approximately true near any large uniformly charged plate, not just a circular one.

ELECTRIC FIELD OF A UNIFORMLY CHARGED DISK

$$E = \frac{(Q/A)}{2\varepsilon_0}\left[1 - \frac{z}{(R^2 + z^2)^{1/2}}\right] \quad \text{(along axis; Figure 15.19)}$$

Approximation: close to the disk ($z \ll R$, Figure 15.20)

$$E \approx \frac{Q/A}{2\varepsilon_0}\left[1 - \frac{z}{R}\right]$$

Approximation: extremely close to the disk ($z \ll R$),

$$E \approx \frac{Q/A}{2\varepsilon_0}$$

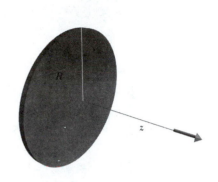

Figure 15.19 Electric field of a uniformly charged disk, at a location on the axis.

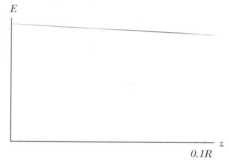

Figure 15.20 Electric field along the axis of a disk, for $z < 0.1R$.

Ex. 15.6 Explain qualitatively how it is possible for the electric field at locations near the center of the disk not to vary with distance away from the disk.

Ex. 15.7 Suppose that the radius of the disk $R = 20$ cm, and the total charge distributed uniformly all over the disk is $Q = 6\times10^{-6}$ C. Use the exact result to calculate the electric field 1 mm from the center of the disk, and also 3 mm from the center of the disk. Does the field decrease significantly?

Ex. 15.8 For a disk of radius $R = 20$ cm and $Q = 6\times10^{-6}$ C, calculate the electric field 2 mm from the center of the disk using all three formulas:

$$E = \frac{(Q/A)}{2\varepsilon_0}\left[1 - \frac{z}{(R^2 + z^2)^{1/2}}\right], \; E \approx \frac{Q/A}{2\varepsilon_0}\left[1 - \frac{z}{R}\right], \text{ and } E \approx \frac{Q/A}{2\varepsilon_0}$$

How good are the approximate formulas at this distance?

Ex. 15.9 For the same disk, calculate E at a distance of 5 cm (50 mm) using all three formulas. How good are the approximate formulas at this distance?

Ex. 15.10 Coulomb's law says that electric field falls off like $1/z^2$. How can E depend on $[1 - z/R]$, or be independent of distance?

Figure 15.21 The two charged metal plates of a capacitor are separated by a very small gap s (exaggerated in this figure).

15.6 Two uniformly charged disks: A capacitor

Consider two uniformly charged metal disks placed very near each other (separation or gap distance s), carrying charges of $-Q$ and $+Q$ (Figure 15.21). This arrangement is called a "capacitor," and we will work with such devices later in electric circuits.

A single charged metal disk cannot have a truly uniform charge density, because the mobile charges tend to push each other to the edge of the disk. Nevertheless, our results for disks made of insulating material are approximately correct for metal disks, especially in this two-disk configuration if the disks are very close together ($s << R$). Due to the attraction by the neighboring disk, almost all of the charge is nearly uniformly distributed on the inner surfaces of the disks, with very little charge on the outer surfaces of the disks (Figure 15.22).

We'll calculate the strength of the electric field at locations near the center of the disks, both inside and outside the capacitor.

Step 1: Cut up the charge distribution into pieces and draw $\Delta\vec{E}$

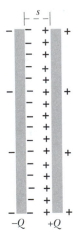

We know the electric field made by a single uniformly charged disk, so we can use this as a "piece." Consider the locations labeled 1, 2, and 3 in Figure 15.23, which shows a side view of the region near the center of the disks, blown up so there is room to draw. The disks are very close together ($s << R$), and extend up and down beyond the boundaries of the drawing.

? Before reading farther, predict the direction and relative magnitude (that is, whether the net field will be large or small) of $\vec{E}_{net}$ at locations 1, 2, and 3 in Figure 15.23.

At locations 1 and 3, the contributions of the two disks are in opposite directions, and nearly equal, so the net field is very small. However, the distances to the two disks are slightly different, so $E_+ \neq E_-$, and the net field is to the right at locations 1 and 3 (Figure 15.24 on facing page). At location 2, both disks contribute $\Delta\vec{E}$ in the same direction, so $\vec{E}_{net}$ is large, and to the left.

Figure 15.22 Schematic side view of a charged capacitor, with the gap distance s exaggerated for clarity.

Step 2: Write an expression for the electric field due to one piece

Origin: at surface of left plate; z axis runs to right (Figure 15.24)

Location of a piece given by: z (pieces are at $z = 0$ and $z = s$)

Distance from piece to observation location: z

Δq in terms of variables: Q (all the charge on one disk)

Assumptions: uniform charge density on inner surface of each disk (ignore the small charge on the outer surfaces of the disks); ($s << R$); locations 1 and 3 are very near the disks ($z << R$).

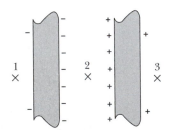

Figure 15.23 Expanded view of a region near the center of a capacitor. Only a small section of the plates is shown; the radius R of a plate is actually much larger than the distance s between plates.

E_- *(contribution of left piece):*

$$E_- \approx \frac{Q/A}{2\varepsilon_0}\left[1 - \frac{z}{R}\right] \text{ where } A = \pi R^2 \text{ is the area of one disk.}$$

Step 3: Add up the contributions of all the pieces

Location 2: Consider a point a distance z from the negative plate (and a distance $s - z$ from the positive plate):

$$E_2 \approx \frac{Q/A}{2\varepsilon_0}\left[1 - \frac{z}{R}\right] + \frac{Q/A}{2\varepsilon_0}\left[1 - \frac{s-z}{R}\right]$$

$$= \frac{Q/A}{2\varepsilon_0}\left[1 - \frac{z}{R} + 1 - \frac{s-z}{R}\right]$$

$$= \frac{Q/A}{\varepsilon_0}\left[1 - \frac{s/2}{R}\right]$$

$$\approx \frac{Q/A}{\varepsilon_0}$$

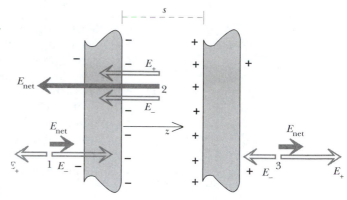

Figure 15.24 Electric field at three locations near the center of a capacitor.

The electric field between the plates is essentially twice the field due to one plate. The magnitude hardly depends on z, so the field is remarkably uniform. As you move away from the negative plate, the very slightly smaller contribution of the negative plate is nearly compensated by the very slightly larger contribution of the positive plate, as shown in Figure 15.25.

Fringe field

The electric field at locations outside the capacitor is called the "fringe field."

Location 3: $\vec{E}_3$ points to the right (see Figure 15.24) and its magnitude is the difference of the magnitudes of the fields of the two disks:

$$E_3 \approx \frac{Q/A}{2\varepsilon_0}\left[1 - \frac{z-s}{R}\right] - \frac{Q/A}{2\varepsilon_0}\left[1 - \frac{z}{R}\right] = \frac{Q/A}{2\varepsilon_0}\left(\frac{s}{R}\right)$$

Location 1: $\vec{E}_1$ also points to the right, and

$$E_1 \approx \frac{Q/A}{2\varepsilon_0}\left[1 - \frac{z}{R}\right] - \frac{Q/A}{2\varepsilon_0}\left[1 - \frac{z+s}{R}\right] = \frac{Q/A}{2\varepsilon_0}\left(\frac{s}{R}\right)$$

The fringe field is very small compared to the field inside the capacitor. Inside the gap, the fields of the two disks are in the same direction, but outside the gap the fields of the two disks are in opposite directions.

? Calculate the ratio of the outside field to the inside field, and show that this ratio is very small if $s \ll R$.

We have calculated the fringe field at a location outside the capacitor but very close to it, where the fringe field hardly changes with distance z. (If however you are very far away, so $z \gg R$, the capacitor looks like an electric dipole, and the fringe field falls off like $1/z^3$.)

Step 4: Check the result

Units: inside the capacitor

$$\frac{C/m^2}{N \cdot m^2/C^2} = \frac{N}{C}$$

Units: fringe field

$$\left(\frac{C/m^2}{N \cdot m^2/C^2}\right)\left(\frac{m}{m}\right) = \frac{N}{C}$$

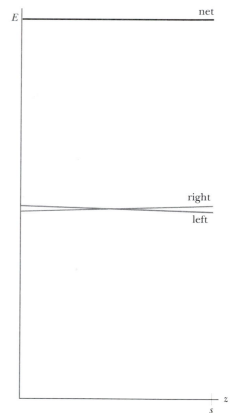

Figure 15.25 E inside a capacitor. The gray lines represent contributions from the left and right plates; the black line is the net electric field.

Applicability

Because we have considered locations far from the edges of the plates, our results apply not only to circular capacitors but also to capacitors with rectangular or other shapes of plates, as long as the same conditions are satisfied—the plate separation s must be much smaller than the width or height of a plate.

Although we cannot prove it at this time (we will need Gauss's Law—see Chapter 21), it is also true that the magnitude and direction of the electric field are practically the same everywhere in the gap, not just near the center of the plates. Only if you get near the outer edges of the plates does the electric field vary much from its center value.

ELECTRIC FIELD OF A CAPACITOR

Near center of a two-plate capacitor (plate has charge $+Q$ or $-Q$ and area A):

$$E \approx \frac{Q/A}{\varepsilon_0}$$

Fringe field (just outside the plates):

$$E_{\text{fringe}} \approx \frac{Q/A}{2\varepsilon_0}\left(\frac{s}{R}\right)$$

Ex. 15.11 If the magnitude of the electric field in air exceeds roughly 3×10^6 N/C, the air breaks down and a spark forms. For a two-disk capacitor of radius 50 cm with a gap of 1 mm, what is the maximum charge (plus and minus) that can be placed on the disks without a spark forming (which would permit charge to flow from one disk to the other)?

Ex. 15.12 Under these conditions, what is the strength of the fringe field just outside the center of the capacitor?

Ex. 15.13 Consider a capacitor made of two rectangular metal plates of length L and width W, with a very small gap s between the plates. There is a charge $+Q$ on one plate and a charge $-Q$ on the other. Assume that the electric field is nearly uniform throughout the gap region and negligibly small outside. Calculate the attractive force that one plate exerts on the other.

15.7 The electric field of a spherical shell of charge

As mentioned in Chapter 13, a sphere with charge spread uniformly over its surface produces a surprisingly simple pattern of electric field. For brevity, we will use the phrase "a uniform spherical shell" to refer to "a sphere with charge spread uniformly over its surface," since the charge itself forms a thin, shell-like spherical layer. As shown in Figure 15.26, a spherical object of radius R with charge Q uniformly spread out over its surface produces an electric field with the following distance dependence, if $\hat{\mathbf{r}}$ is a vector from the center of the sphere to the observation location:

ELECTRIC FIELD OF A UNIFORMLY CHARGED SPHERICAL SHELL

$$\vec{E}_{\text{sphere}} = \frac{1}{4\pi\varepsilon_0}\frac{Q}{r^2}\hat{\mathbf{r}} \quad \text{for } r > R \text{ (outside the sphere)}$$

$$\vec{E}_{\text{sphere}} = 0 \quad \text{for } r < R \text{ (inside the sphere)}$$

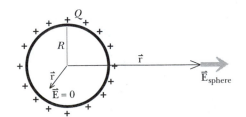

Figure 15.26 The electric field of a uniformly charged sphere (shown in cross-section), at locations outside and inside the sphere.

This can be shown by setting up and evaluating an integral, as we did in the case of the rod, the ring, and the disk, following the procedure that we used in those cases. The geometry involved is somewhat complex, and the details of the process are given in an appendix at the end of this chapter. In later chapters we will see alternative ways to prove this, one involving Gauss's law and one involving electric potential.

In this section we will try to develop a qualitative understanding of how this surprising result can be true.

15.7.1 Outside a uniform spherical shell of charge

Let's divide the sphere into six regions, as shown in Figure 15.27. $\vec{E}_2$ means "the electric field due to the charges in region 2 of the spherical shell." As long as we are rather far from a region of distributed charge we can approximate the electric field of that region as being due to a point particle with the total charge of the region.

Direction

Consider the field at location A outside the shell, a distance r from the center of the sphere. By symmetry, the net electric field is horizontal and to the right, because the vertical component of the net field cancels for pairs of fields such as $\vec{E}_2$ and $\vec{E}_6$, or $\vec{E}_3$ and $\vec{E}_5$. It is clear that the net field (the superposition of the contributions of all the source charges) is radially outward from the center of the sphere.

Magnitude

It is somewhat surprising that the magnitude of the net field turns out to be

$$E = \frac{1}{4\pi\varepsilon_0} \frac{Q}{r^2}$$

where Q is the total charge on the sphere and r is the distance to location A from the center of the sphere (not from the surface of the sphere!).

In Figure 15.28, notice that the distance from location A to some regions is less than r, while the distance from location A to other regions is greater than r. Because of the $1/r^2$ distance dependence for point charges, a smaller amount of charge closer than r makes as big a contribution to the net field as the larger amount of charge that is farther away than r.

It is an extraordinary aspect of the $1/r^2$ behavior of Coulomb's law that the net effect of all these charges comes out to be simply equivalent to placing one point charge Q at the center of the sphere. This is not an obvious result!

Force on a uniformly charged sphere

A uniformly charged spherical shell not only makes a field outside that looks as though it were made by a point charge, but the shell also reacts to outside charges as though it were a point charge. Consider the interaction between a uniform spherical shell and an outside point charge. The shell exerts a force on the point charge as though both were point charges. By the reciprocity principle for electric forces (Newton's third law), the real point charge must exert an equal and opposite force on the shell, and so it exerts a force on the shell as though the shell were a point charge. Hence from the outside a uniform spherical shell "looks" just like a point charge, both as a source of electric field and when it reacts to external fields.

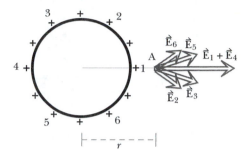

Figure 15.27 Contributions to the net electric field outside a charged sphere.

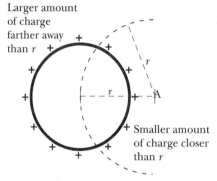

Figure 15.28 Some charge is closer than r to point A, but more of the charge is farther away.

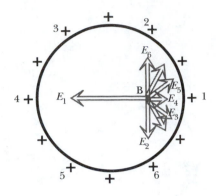

Figure 15.29 Contributions to the net electric field at a location inside a uniformly charged sphere add up to zero.

15.7.2 Inside a uniform spherical shell of charge

Now consider a location B inside the shell, as indicated in Figure 15.29. A small number of charges near location B (region 1) make a large contribution to the left, while a large number of charges far away from location B (regions 3, 4, and 5) each make small contributions to the right. Because the surface area of a portion of the sphere is proportional to r^2 while the electric field contributed by this region is proportional to $1/r^2$ these contributions exactly cancel each other. The electric field at location B due to charges on the surface of the sphere turns out to be exactly zero. Although intuitively plausible, this is not an obvious result.

Warning: *E* is not always zero inside a charged sphere

Do not over generalize this result. Other charges in the Universe may make a nonzero electric field inside the shell. It is only the electric field due to charges uniformly distributed on the surface of the sphere that is zero inside the sphere.

Implications

The fact that the electric field due to uniform charges on the shell is zero at any location inside the shell has interesting implications. For example, if the uniformly charged shell is filled with plastic, the charges on the surface do not polarize the molecules in the plastic, because the surface charges contribute zero field. Note that even if the location of a molecule is very close to the charged surface of the sphere, *E* inside the sphere is still zero at that location. (Charges external to the sphere may contribute to a nonzero electric field inside the sphere, however.)

Since a solid metal sphere is a conductor, any excess charge on the sphere arranges itself uniformly on the outer surface. Therefore, inside the metal there is no excess charge and no field. This is a special case of a more general behavior of metals, that in static equilibrium the electric field is zero inside a metal.

What is the electric field right at the surface?

The electric field is zero just inside the surface of a uniformly charged spherical shell, but the field is nonzero just outside the surface. You might wonder what the field is right at the surface. The physical reality is that the electric field is highly variable in direction and magnitude right at the surface, because the individual excess point charges (electrons and ions) on the surface are many atomic diameters away from each other, as you found for the charges on a tape (Problem 14.1). It is only when you get some thousands of atomic diameters away from the surface (about 10^{-7} m) that it is appropriate to consider the surface charge as approximately a uniform, continuous sheet of charge. As long as you are this far away from the surface, the field outside the surface looks like the field of a point charge at the center of the sphere, and the field inside the surface is zero.

Ex. 15.14 What is wrong with Figure 15.30 and this associated incorrect student explanation? "The electric field at location D inside the uniformly charged sphere points in the direction shown, because the charges closest to this location have the largest effect." (Spheres provide the most common exception to the normally useful rule that the nearest charges usually make the largest contribution to the electric field.)

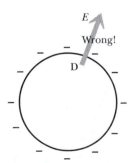

Figure 15.30 A student's incorrect prediction of the electric field inside a uniformly charged sphere (Ex 15.14).

Ex. 15.15 A thin plastic spherical shell of radius 5 cm has a uniformly distributed charge of –25 nC on its outer surface. A concentric thin plastic spherical shell of radius 8 cm has a uniformly distributed charge of +64 nC on its outer surface. Find the magnitude and direction of the electric field at distances of 3 cm, 7 cm, and 10 cm from the center. See Figure 15.31.

Ex. 15.16 A solid metal ball of radius 1.5 cm bearing a charge of –17 nC is located near a solid plastic ball of radius 2 cm bearing a uniformly distributed charge of +7 nC (Figure 15.32) on its outer surface. The distance between the centers of the balls is 9 cm. Show the approximate charge distribution in and on each ball. What is the electric field at the center of the metal ball?

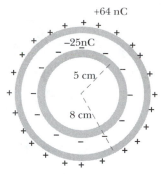

Figure 15.31 Two concentric charged plastic shells (Ex 15.15).

15.8 A solid sphere of charge

Until this point, all of the charge distributions we have considered involved charge distributed over the surface of an object. There are some cases, however, in which charge can be distributed throughout an object. One such case is the nucleus of an atom, which is composed of protons and neutrons packed into a sphere of radius on the order of $\sim 10^{-15}$ m. We can treat the charge density inside the nucleus as approximately uniform. Another such case is the electron cloud in an atom, in which the negative charge of the electrons appears to be distributed throughout a spherical region of radius on the order of $\sim 10^{-10}$ m (although not necessarily uniformly so).

We will consider a solid sphere with radius R, having charge Q uniformly distributed throughout its volume.

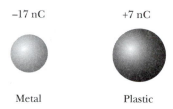

Figure 15.32 Charged metal and plastic balls (Ex 15.16).

Step 1: Cut up the charge distribution into pieces and draw $\Delta \vec{E}$

We can model the sphere as a series of concentric spherical shells.

Outside a solid sphere of charge

At a location outside the sphere, and hence outside all the shells, (Figure 15.33) each shell "looks" like a point charge at the center of the sphere. Hence, outside the sphere:

$$E = \frac{1}{4\pi\varepsilon_0}\frac{Q}{r^2} \text{ for } r > R$$

Inside a solid sphere of charge

At a location inside the sphere, we are inside some of the spherical shells (gray shells in Figure 15.34); the contribution of these shells to $\vec{E}_{net}$ is therefore zero.

We are outside the rest of the shells (white shells as in Figure 15.34); each of these shells "looks" like a point charge at the center of the sphere. To get $\vec{E}_{net}$ at this location we must add the contributions of all the inner shells. Since all the inner shells together will contribute

$$E = \frac{1}{4\pi\varepsilon_0}\frac{\Delta Q}{r^2}$$

our only remaining task is to find ΔQ.

$$\Delta Q = Q\frac{(\text{volume of inner shells})}{(\text{volume of sphere})} = Q\frac{((4/3)\pi r^3)}{((4/3)\pi R^3)}$$

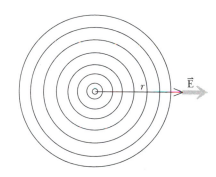

Figure 15.33 A charged solid sphere modeled as a series of concentric spherical shells, all charged.

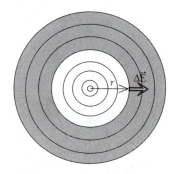

Figure 15.34 A location inside the gray shells, but outside the white shells. At this location only the inner shells make a nonzero contribution to the electric field.

Assembling the result:

$$E = \frac{1}{4\pi\varepsilon_0}\frac{Q}{r^2}\frac{\left(\frac{4}{3}\pi r^3\right)}{\left(\frac{4}{3}\pi R^3\right)} = \frac{1}{4\pi\varepsilon_0}\frac{Q}{R^3}r \text{ for } r < R$$

So the electric field inside the sphere is directly proportional to r.

? Does this make sense?

Yes. While inside the sphere, as r increases, the amount of charge "inside" the observation location increases. The larger volume ($V \propto r^3$) more than compensates for the $1/r^2$ field dependence. By symmetry, $E = 0$ at $r = 0$.

Note that in the special case $r = R$, $E = \frac{1}{4\pi\varepsilon_0}\frac{Q}{R^2}$, which is correct.

15.9 Infinitesimals and integrals in science

In pure mathematics, an infinitesimal quantity is conceived to be arbitrarily close to zero. In science, however, this may not make sense. For example, in the case of the uniformly charged thin rod it makes no physical sense to choose dz to be smaller than an atomic diameter. It would enormously complicate the summation if we had to take into account the detailed distribution of charged particles inside each atom. All that is required is that dz be small compared to the significant dimensions of the situation (r and L). We want each slice dz to be an adequate approximation of a point charge, and $dz = 0.02$ m may be good enough! We can consider the mathematical infinitesimal dz to be an idealization of what a scientist means by an infinitesimal quantity, which is a quantity that is "small enough" within the desired and possible precision of measurements and analysis.

Oddly, the result of doing an integral is approximate, not exact, because at the microscopic level the actual charge distribution is not in fact uniformly distributed. You found that on a charged piece of tape the individual electrons or ions were quite far from each other on an atomic scale. However, if you are many atomic diameters away from a "uniformly" charged rod, the mathematical integral gives an excellent approximation to the actual electric field.

There is another important sense in which our analysis is approximate. If a rod is an insulator, and has been charged by rubbing it with a cloth, the charge distribution will be only approximately uniform, even on a macroscopic scale, since charge transfer cannot be accurately controlled. If the rod is a metal, charge will tend to pile up at the ends rather than being uniformly distributed. Nevertheless, modeling a charged rod as a uniformly charged rod in practice can be a useful model. We just need to remember that our analytical solution is approximate, not exact.

15.10 *Integrating the spherical shell

In this section, we'll sketch the proof that a uniformly charged spherical shell looks from the outside like a point charge but on the inside has a zero electric field. Two quite different proofs are given in Chapters 16 and 21.

Step 1: Divide the sphere into pieces

Divide the spherical shell into rings of charge, each delimited by the angle θ and the angle $\theta+\Delta\theta$, and carrying an amount of charge ΔQ (Figure 15.35).

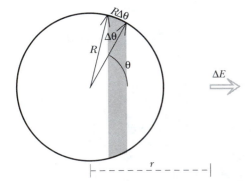

Figure 15.35 Divide the spherical shell (shown here in cross-section) into rings, each carrying a (variable) amount of charge ΔQ.

Each ring contributes ΔE at an observation point a distance r from the center of the spherical shell.

Step 2: Write an expression for ΔE

Origin: At the center of the sphere.

Use polar coordinates (r, θ, ϕ) As shown in Figure 15.36, θ is an angle measured from the r (horizontal) axis; ϕ refers to rotation about the r (horizontal) axis.

Location of one piece (ring): Given by angle θ.

Components: Only horizontal component.

Distance from center of one ring to observation location:

$$d = (r - R\cos\theta)$$

Amount of charge on each ring:

$$\Delta Q = Q\frac{(\text{surface area of ring})}{(\text{surface area of sphere})} = Q\left(\frac{2\pi(R\sin\theta)(R\Delta\theta)}{4\pi R^2}\right)$$

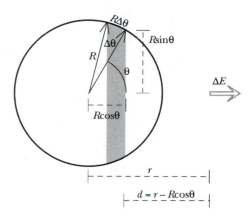

Figure 15.36 The sphere may be divided into ring-shaped segments.

As before, we calculate the surface area ΔA of the ring by laying the ring out flat (Figure 15.37), noting that the radius of the ring is $R\sin\theta$ and its width is $R\Delta\theta$ (since arc length is radius times angle, with angle measured in radians).

Integration variable: θ

Magnitude of ΔE

$$\Delta E = \frac{1}{4\pi\varepsilon_0}\frac{(\Delta Q)d}{(d^2 + (R\sin\theta)^2)^{3/2}}$$

$$= \frac{1}{4\pi\varepsilon_0}\frac{(r - R\cos\theta)}{[(r - R\cos\theta)^2 + (R\sin\theta)^2]^{3/2}}Q\left(\frac{2\pi(R\sin\theta)}{4\pi R^2}\right)(R\Delta\theta)$$

Figure 15.37 We can calculate the area of the ring of charge by "unrolling" it.

Step 3: Sum

$$E = \frac{1}{4\pi\varepsilon_0}\frac{Q}{2}\int_0^\pi \frac{(r - R\cos\theta)}{[(r - R\cos\theta)^2 + (R\sin\theta)^2]^{3/2}}\sin\theta\, d\theta$$

The limits on the integral are determined by the fact that if we let θ range between 0 and π radians (180 degrees), we add up rings that account for the entire surface of the spherical shell.

Again, evaluating the integral is just math (in this case, it is rather difficult math). The results are:

Outside the shell ($r > R$):

$$E = \frac{1}{4\pi\varepsilon_0}\frac{Q}{r^2} \text{ for } r > R$$

as though all the charge were concentrated into a point at the center of the spherical shell.

Inside the shell ($r < R$):

$$E = 0 \text{ for } r < R$$

15.11 Summary

Fundamental Principles

No new fundamental principles were introduced in this chapter.

New Techniques

A procedure for calculating the electric field due to a distribution of electric charges.

Step 1: Divide the charge distribution into pieces and draw $\Delta \vec{E}$.

Choose pieces whose electric field you know.

Step 2: Write an expression for the electric field due to one piece.

This involves selecting an origin, finding $\vec{r}$ and $\hat{r}$, describing the location of one piece in terms of an integration variable, expressing Δq in terms of your variables, and writing expressions for ΔE_x, ΔE_y, and ΔE_z.

Step 3: Add up the contributions of all the pieces.

Sometimes the sum can be written as a definite integral which can be symbolically evaluated; otherwise you must calculate a finite sum numerically.

Step 4: Check the result.

Checks should include a qualitative check of field direction, a check of the units of the answer, and checks of special cases.

Results

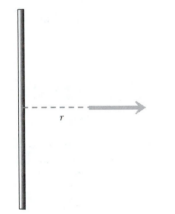

Rod

Electric field of a uniformly charged thin rod of length L, a distance r from the midpoint along a line perpendicular to the rod (Figure 15.38):

$$E = \frac{1}{4\pi\varepsilon_0}\left[\frac{Q}{r\sqrt{r^2 + (L/2)^2}}\right]$$

Approximation: for a very long rod ($r \ll L$), $E \approx \dfrac{1}{4\pi\varepsilon_0}\dfrac{2(Q/L)}{r}$

Figure 15.38 Electric field of a uniformly charged thin rod.

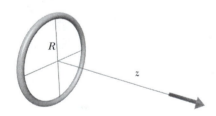

Ring

Electric field of a thin uniformly charged ring of radius R, a distance z from the midpoint of the ring along a line perpendicular to the plane of the ring (Figure 15.39):

$$E = \frac{1}{4\pi\varepsilon_0}\frac{qz}{(R^2 + z^2)^{3/2}}$$

Figure 15.39 Electric field of a uniformly charged ring.

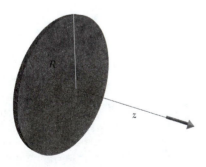

Disk

Electric field of a uniformly charged disk of radius R, a distance z from the center of the disk along a line perpendicular to the disk (Figure 15.40):

$$E = \frac{(Q/A)}{2\varepsilon_0}\left[1 - \frac{z}{(R^2 + z^2)^{1/2}}\right], \text{ where } A = \pi R^2$$

Approximation: close to the disk ($z \ll R$), $E \approx \dfrac{Q/A}{2\varepsilon_0}\left[1 - \dfrac{z}{R}\right]$

Approximation: extremely close to the disk ($z \ll R$), $E \approx \dfrac{Q/A}{2\varepsilon_0}$

Figure 15.40 Electric field of a uniformly charged disk.

Capacitor

Inside a capacitor (Figure 15.41), near its center: $E \approx \dfrac{Q/A}{\varepsilon_0}$

Fringe field just outside the plates $E_{\text{fringe}} \approx \dfrac{Q/A}{2\varepsilon_0}\left(\dfrac{s}{R}\right)$

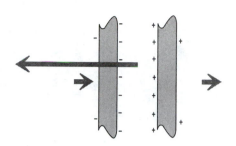

Figure 15.41 Electric field inside and outside a capacitor.

Spherical shell

Inside a uniformly charged thin spherical shell:

$$E = 0 \text{ due to charges on the shell}$$

Outside a uniformly charged thin spherical shell (Figure 15.42):

$$E = \frac{1}{4\pi\varepsilon_0}\frac{Q}{r^2} \text{ due to charges on the shell}$$

where r is the distance from the center of the sphere to the observation location.

Figure 15.42 Electric field at a location outside a uniformly charged spherical shell.

Sphere charged throughout its volume:

Inside a solid sphere of radius R, charged throughout its volume, at a location a distance r from the center of the sphere:

$$E = \frac{1}{4\pi\varepsilon_0}\frac{Q}{R^3}r$$

Warning: do not confuse this result with the previous one. The words "solid" and "shell" refer to the location of the charge (on the surface or throughout the object). If a plastic ball is not hollow, but has charge uniformly distributed over its surface (but not inside), the electric field inside the plastic, due to the charge on the surface, is zero—this is an example of a spherical shell of charge.

Figure 15.43 A charged hollow 3/4 cylinder. The observation location is at the center of the cylinder.

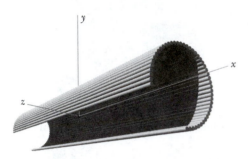

Figure 15.44 We can consider the cylinder to be composed of a large number of very thin charged rods.

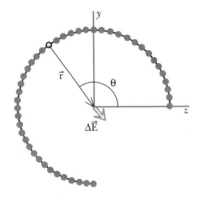

Figure 15.45 An end view of the cylinder, looking along the *x* axis in the +*x* direction.

15.12 Example problem: A hollow cylinder

A hollow 3/4 cylinder (Figure 15.43) carries a negative charge of Q spread uniformly over it surface. The radius of the cylinder is R, and its length is L, where $L \gg R$. Calculate the electric field at the center of the cylinder, whose orientation is shown in Figure 15.43.

Solution

1) Since we know the electric field near the center of a long rod, we could consider the cylinder to be made up of many thin charged rods, as shown in Figure 15.44. We will place the origin at the observation location, as shown. Each rod contributes $\Delta\vec{E}$ as shown in Figure 15.45.

2) The location of the center of a particular rod can be described in terms of the angle θ, defined as shown in Figure 15.45.

$$\vec{r} = \langle\text{obs. location}\rangle - \langle\text{source}\rangle = \langle 0, 0, 0\rangle - \langle 0, R\sin\theta, R\cos\theta\rangle$$
$$= \langle 0, -R\sin\theta, -R\cos\theta\rangle$$

$$r = \sqrt{(-R\cos\theta)^2 + (-R\sin\theta)^2} = R$$

$$\hat{r} = \frac{\vec{r}}{r} = \langle 0, -\sin\theta, -\cos\theta\rangle$$

The amount of charge on one rod is given by:

$$dQ = Q\left(\frac{d\theta}{\theta_{\text{total}}}\right) = \frac{Q\, d\theta}{(3/2)\pi}$$

where $d\theta$ is the angular width of one rod, and θ_{total} is the angular extent of the cylinder.

The contribution of this rod to the net electric field is:

$$\Delta\vec{E} = |\Delta\vec{E}|\hat{r} \approx \frac{1}{4\pi\varepsilon_0}\frac{2\,dQ}{L}\frac{1}{r}\langle 0, -\sin\theta, -\cos\theta\rangle$$

$$= \frac{1}{4\pi\varepsilon_0}\frac{2}{LR}\frac{Q\, d\theta}{(3/2)\pi}\langle 0, -\sin\theta, -\cos\theta\rangle$$

because in this case $r = R$; the center of every rod is the same distance R from the observation location.

3) We need to consider the x, y, and z components of electric field separately. To add up the contributions of all of the rods, we can integrate:

$$E_x = \int_0^{\frac{3}{2}\pi}\frac{1}{4\pi\varepsilon_0}\frac{2Q}{LR(3/2)\pi}\cdot 0\, d\theta = 0$$

$$E_y = \int_0^{\frac{3}{2}\pi}\frac{1}{4\pi\varepsilon_0}\frac{2Q}{LR(3/2)\pi}(-\sin\theta)\, d\theta = \frac{1}{4\pi\varepsilon_0}\frac{2Q}{LR(3/2)\pi}\cos\theta\Big|_0^{\frac{3}{2}\pi}$$

$$= -\frac{1}{4\pi\varepsilon_0}\frac{4Q}{3\pi LR}$$

$$E_z = \int_0^{\frac{3}{2}\pi} \frac{1}{4\pi\varepsilon_0} \frac{2Q}{LR(3/2)\pi}(-\cos\theta)\,d\theta = \frac{-1}{4\pi\varepsilon_0} \frac{2Q}{LR(3/2)\pi}\sin\theta\Big|_0^{\frac{3}{2}\pi}$$

$$= \frac{1}{4\pi\varepsilon_0} \frac{4Q}{3\pi LR}$$

So the net electric field is:

$$\vec{E} = \langle 0, -\frac{1}{4\pi\varepsilon_0}\frac{4Q}{3\pi LR}, \frac{1}{4\pi\varepsilon_0}\frac{4Q}{3\pi LR}\rangle$$

4) The direction is reasonable, since the y component of $\vec{E}$ should be negative, and the z component should be positive.

Units are $(Nm^2/C^2)(C/(m\cdot m)) = N/C$, which is correct.

15.13 Review questions

Rings, disks, capacitors, and spherical shells

RQ 15.1 Explain briefly how knowing the electric field of a ring helps in calculating the field of a disk.

RQ 15.2 Two very thin circular plastic sheets are close to each other and carry equal and opposite uniform charge distributions. Explain briefly why the field between the sheets is much larger than the field outside. Illustrate your argument on a diagram.

RQ 15.3 Define "fringe field."

RQ 15.4 A solid spherical plastic ball was rubbed with wool in such a way that it acquired a uniform negative charge all over the surface. Make a sketch showing the polarization of molecules inside the plastic ball, and explain briefly.

RQ 15.5 A student said "the electric field inside a uniformly charged sphere is always zero." Criticize this statement.

RQ 15.6 A capacitor consists of two large metal disks of radius 2.5 meters placed parallel to each other, a distance of 1.2 millimeters apart. The capacitor is charged up to have an increasing amount of charge $+Q$ on one disk and $-Q$ on the other. At about what value of Q does a spark appear between the disks?

Checking results

RQ 15.7 A student claimed that the formula for the electric field outside a cube of edge length L, carrying a uniformly distributed charge Q, at a distance x from the center of the cube, was

$$E = \frac{Q}{\varepsilon_0} \frac{L}{x^{1/2}}$$

Explain how you know that this cannot be the right formula.

RQ 15.8 A student claimed that the formula for the electric field outside a cube of edge length L, carrying a uniformly distributed charge Q, at a distance x from the center of the cube, was

$$\frac{1}{4\pi\varepsilon_0} \frac{50\,QL}{x^3}$$

Explain how you know that this cannot be the right formula.

Dependence on distance

RQ 15.9 Give an example of a configuration of charges that yields an electric field whose magnitude varies approximately with distance as specified:

- independent of distance
- $1/r$
- $1/r^2$
- $1/r^3$
- force (not field) that goes like $1/r^5$

15.14 Homework problems

Problem 15.1 Charged tape
A strip of invisible tape 12 cm by 2 cm is charged uniformly with a total net charge of 4×10^{-8} coulomb and hangs vertically.

(a) Divide the strip into three sections each 4 cm high by 2 cm wide, and use numerical summation to calculate the magnitude of the electric field at a perpendicular distance of 3 cm from the center of the tape. Show your calculations in a legible and understandable format.

(Hint: Recall that the Pythagorean theorem in three dimensions has the form $d^2 = x^2 + y^2 + z^2$.)

(b) Divide the strip into eight sections each 3 cm high by 1 cm wide, and use numerical summation to calculate the magnitude of the electric field at a perpendicular distance of 3 cm from the center of the tape. Show your calculations in a legible and understandable format.

(c) Which of your two answers do you think is closer to the true value of the electric field? Why?

(d) Write a computer program (or use a spreadsheet or math package) to calculate the field, dividing the tape into a large number of small sections, and compare with your other results. Calculate the electric field at several locations, including the one given above, and if possible display your results graphically.

Problem 15.2 Glass rod
Consider a thin glass rod of length L lying along the x axis with one end at the origin (Figure 15.46). The rod carries a uniformly distributed positive charge Q.

At a location $d > L$, on the x axis to the right of the rod, what is the electric field due to the rod? Follow the standard four steps.

(a) Use a diagram to explain how you will cut up the charged rod, and draw the $\Delta \vec{E}$ contributed by a representative piece.

(b) Express algebraically the contribution each piece makes to the electric field. Be sure to show your integration variable and its origin on your drawing.

(c) Write the summation as an integral, and simplify the integral as much as possible. State explicitly the range of your integration variable. Evaluate the integral.

(d) Show that your result is reasonable. Apply as many tests as you can think of.

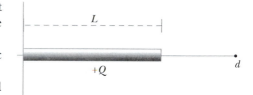

Figure 15.46 Glass rod (Problem 15.2).

Problem 15.3 Bent plastic rod
Consider a thin plastic rod bent into an arc of radius R and angle α (Figure 15.47). The rod carries a uniformly distributed negative charge $-Q$.

What are the components E_x and E_y of the electric field at the origin? Follow the standard four steps.

(a) Use a diagram to explain how you will cut up the charged rod, and draw the $\Delta \vec{E}$ contributed by a representative piece.

(b) Express algebraically the contribution each piece makes to the x and y components of the electric field. Be sure to show your integration variable and its origin on your drawing. (Hint: an arc of radius R and angle $\Delta\theta$ measured in radians has a length $R\Delta\theta$.)

(c) Write the summation as an integral, and simplify the integral as much as possible. State explicitly the range of your integration variable. Evaluate the integral.

(d) Show that your result is reasonable. Apply as many tests as you can think of.

Figure 15.47 Bent plastic rod (Problem 15.3).

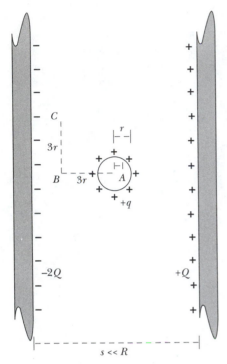

Figure 15.48 Plates and sphere (Problem 15.4).

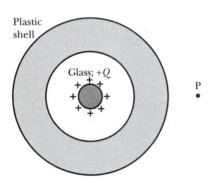

Figure 15.49 Qualitative fields with spheres (Problem 15.5).

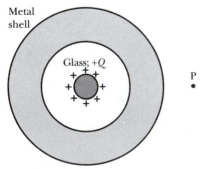

Figure 15.50 Qualitative fields with spheres (Problem 15.5).

Problem 15.4 Plates and sphere

Two large, thin charged plastic circular plates each of radius R are placed a short distance s apart; s is much smaller than the dimensions of a plate (Figure 15.48). The right-hand plate has a positive charge of $+Q$ evenly distributed over its inner surface (Q is a positive number). The left-hand plate has a negative charge of $-2Q$ evenly distributed over its inner surface. A very thin plastic spherical shell of radius r is placed midway between the plates (and shown in cross section). It has a uniformly distributed positive charge of $+q$. You can ignore the contributions to the electric field due to the polarization of the thin plastic shell and the thin plastic plates.

(a) Calculate the x and y components of the electric field at location A, a horizontal distance $r/2$ to the right of the center of the sphere.

(b) Calculate the x and y components of the electric field at location B, a horizontal distance $3r$ to the left of the center of the sphere.

(c) Calculate the x and y components of the electric field at location C, a horizontal distance $3r$ to the left and a vertical distance $3r$ above the center of the sphere.

Problem 15.5 Qualitative fields with spheres

(a) A glass sphere carrying a uniformly-distributed charge of $+Q$ is surrounded by an initially neutral spherical plastic shell (Figure 15.49). Qualitatively, indicate the polarization of the plastic.

(b) Qualitatively, indicate the polarization of the inner glass sphere. Explain briefly.

(c) Is the electric field at location P outside the plastic shell larger, smaller, or the same as it would be if the plastic weren't there? Explain briefly.

(d) Now suppose that the glass sphere carrying a uniform charge of $+Q$ is surrounded by an initially neutral metal shell (Figure 15.50). Qualitatively, indicate the polarization of the metal.

(e) Now be *quantitative* about the polarization of the metal sphere and prove your assertions.

(f) Is the electric field at location P outside the metal shell larger, smaller, or the same as it would be if the metal shell weren't there? Explain briefly.

Problem 15.6 Metal sphere in air

Breakdown field strength for air is roughly 3×10^6 N/C. If the electric field is greater than this value, stray free electrons (due to ionization of air molecules by cosmic rays) will be accelerated to high enough energies to trigger a chain reaction.

(a) There is a limit to the amount of charge that you can put on a metal sphere in air. If you slightly exceed this limit, why would breakdown occur, and why would the breakdown occur very near the surface of the sphere, rather than somewhere else?

(b) How much excess charge can you put on a metal sphere of 10 cm radius without causing breakdown in the neighboring air, which would discharge the sphere?

(c) How much excess charge can you put on a metal sphere of only 1 mm radius? These results hint at the reason why a highly charged piece of metal tends to spark at places where the radius of curvature is small, or at places where there are sharp points.

(d) Suppose that a positively charged sphere hangs from a thread in the center of a small, closed room. If breakdown occurs, describe qualitatively the final charge distribution, and discuss the chemical mechanisms involved in the redistribution of the charge.

Problem 15.7 Cathode ray tube

In a cathode-ray tube, an electron travels in a vacuum and enters a region between two deflection plates where there is an upward electric field of magnitude 10^5 N/C (Figure 15.51).

a) Sketch the trajectory of the electron, continuing on well past the deflection plates (the electron is going fast enough that it does not strike the plates).

(b) Calculate the acceleration of the electron while it is between the deflection plates.

(c) The deflection plates measure 12 cm by 3 cm, and the gap between them is 2.5 mm. The plates are charged equally and oppositely. What are the magnitude and sign of the charge on the upper plate?

(d) What is the direction and approximate magnitude of the electric field just above the center of the upper plate?

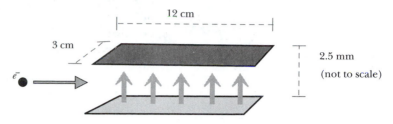

Figure 15.51 Cathode ray tube (Problem 15.7).

Problem 15.8 Two plastic spheres

Figure 15.52 shows two thin plastic spherical shells (shown in cross section) that are uniformly charged. The center of the larger sphere is at (0,0); it has a radius of 12 cm and a uniform positive charge of $+4\times10^{-9}$ C. The center of the smaller sphere is at (25 cm, 0); it has a radius of 3 cm and a uniform negative charge of -10^{-9} C.

(a) What are the components $E_{A,x}$ and $E_{A,y}$ of the electric field $\Delta\vec{E}$ at location A (6 cm to the right of the center of the large sphere)? Neglect the small contribution of the polarized molecules in the plastic, because the shells are very thin and don't contain much matter.

(b) What are the components $E_{B,x}$ and $E_{B,y}$ of the electric field at location B (15 cm above the center of the small sphere)? Neglect the small contribution of the polarized molecules in the plastic, because the shells are very thin and don't contain much matter.

(c) What are the components F_x and F_y of the force on an electron placed at location B?

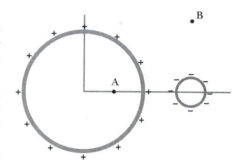

Figure 15.52 Metal sphere in air (Problem 15.6).

Problem 15.9 Concentric spheres

A solid plastic sphere of radius R_1 has a charge $-Q_1$ on its surface. A concentric spherical metal shell of inner radius R_2 and outer radius R_3 carries a charge Q_2 on the inner surface and a charge Q_3 on the outer surface. Q_1, Q_2, and Q_3 are positive numbers, and the total charge $Q_2 + Q_3$ on the metal shell is greater than Q_1.

At an observation location a distance r from the center, determine the magnitude and direction of the electric field in the following regions, and explain briefly in each case:

(a) $r < R_1$ (inside the plastic sphere)
(b) $R_1 < r < R_2$ (in the air gap)
(c) $R_2 < r < R_3$ (in the metal)
(d) $r > R_3$ (outside the metal)

For parts (a)–(d), be sure to give both the direction and the magnitude of the electric field, and explain briefly.

(e) Suppose $-Q_1 = -5$ nC. What is Q_2? Explain fully on the basis of fundamental principles.

(f) What can you say about the molecular polarization in the plastic? Explain briefly. Include a drawing if appropriate.

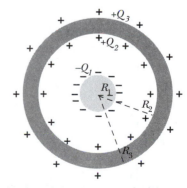

Figure 15.53 Concentric spheres (Problem 15.9).

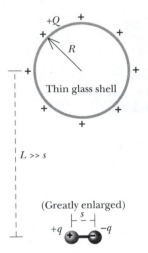

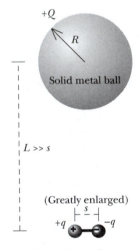

Figure 15.54 Dipole with glass sphere (Problem 15.10).

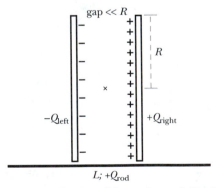

Figure 15.55 Dipole with metal sphere (Problem 15.10).

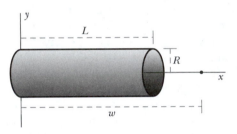

Figure 15.57 Two disks and a rod (Problem 15.12).

Problem 15.10 Dipole with glass or metal sphere
A small, thin, hollow spherical glass shell of radius R carries a uniformly distributed positive charge $+Q$ (Figure 15.54). Below it is a horizontal permanent dipole with charges $+q$ and $-q$ separated by a distance s (s is shown greatly enlarged for clarity). The dipole is fixed in position and is not free to rotate. The distance from the center of the glass shell to the center of the dipole is L.

(a) Calculate the magnitude and direction of the electric field at the center of the glass shell, and explain briefly, including showing the electric field on the diagram. Your results must not contain any symbols other than the given quantities R, Q, q, s, and L (and fundamental constants), unless you define intermediate results in terms of the given quantities. What simplifying assumption do you have to make?

(b) If the upper sphere were a solid metal ball with a charge $+Q$ (Figure 15.55), what would be the magnitude and direction of the electric field at its center? Explain briefly. Show the distribution of charges everywhere, and at the center of the ball accurately draw and label the electric field due to the ball charges $\vec{E}_{ball}$ and the electric field $\vec{E}_{dipole}$ of the dipole.

Problem 15.11 A hollow glass tube
A thin-walled hollow circular glass tube, open at both ends, has a radius R and length L (Figure 15.56). The axis of the tube lies along the x axis, with the left end at the origin. The outer sides are rubbed with silk and acquire a net positive charge Q distributed uniformly. Determine the electric field at a location on the x axis, a distance w from the origin. Carry out all steps including checking your result. Explain each step. (You may have to refer to a table of integrals.)

Figure 15.56 A hollow glass tube (Problem 15.11).

Problem 15.12 Two disks and a rod
In Figure 15.57 are two uniformly charged disks of radius R which are very close to each other (gap $\ll R$). The disk on the left has a charge of $-Q_{left}$ and the disk on the right has a charge of $+Q_{right}$ (Q_{right} is greater than Q_{left}). A uniformly charged thin rod of length L lies at the edge of the disks, parallel to the axis of the disks and centered on the gap. The rod has a charge of $+Q_{rod}$.

(a) Calculate the magnitude and direction of the electric field at the point marked $\times$ at the center of the gap region, and explain briefly, including showing the electric field on a diagram. Your results must not contain any symbols other than the given quantities R, Q_{left}, Q_{right}, L, and Q_{rod} (and fundamental constants), unless you define intermediate results in terms of the given quantities.

(b) If an electron is placed at the center of the gap region, what is the magnitude and direction of the electric force that acts on the electron?

Problem 15.13 A rod and two balloons

A very thin glass rod 4 meters long is rubbed all over with a silk cloth. It gains a uniformly distributed charge $+1.3\times10^{-6}$ C. Two small spherical rubber balloons of radius 1.2 cm are rubbed all over with wool. They each gain a uniformly distributed charge of -2×10^{-8} C. The balloons are near the midpoint of the glass rod, with their centers 3 cm from the rod. The balloons are 2 cm apart (4.4 cm between centers).

Length 4 m (drawing not to scale)

2 cm

3 cm

×

Radius 1.2 cm

0.6 cm

(a) Find the magnitude of the electric field at the location marked by the ×, 0.6 cm to the right of the center of the left balloon. Also calculate the angle the electric field makes with the horizontal. Show all your work, including showing vectors. State any approximations you are making.

(b) Next a neutral hydrogen atom is placed at that same location (marked by the ×). Draw a diagram showing the effect on the hydrogen atom while it is at that location. The polarizability of atomic hydrogen has been measured to be $\alpha = 7.4\times10^{-41}$ m^2/(N·m^2/C^2). What is the distance between the center of the proton and the center of the electron cloud in the hydrogen atom?

Problem 15.14 A plastic disk and a piece of aluminum foil

A large, thin plastic disk with radius $R = 1.5$ meter carries a uniformly distributed charge of $-Q = -3\times10^{-5}$ C (Figure 15.58). A circular piece of aluminum foil is placed $d = 3$ mm from the disk, parallel to the disk. The foil has a radius of $r = 2$ cm and a thickness $t = 1$ millimeter.

(a) Show the charge distribution on the close-up of the foil.

(b) Calculate the magnitude and direction of the electric field at location × at the center of the foil, inside the foil.

(c) Calculate the magnitude q of the charge on the left circular face of the foil.

Problem 15.15 Charge on the outer surfaces of a capacitor

A capacitor made of two parallel uniformly charged circular metal disks carries a charge of $+Q$ and $-Q$ on the inner surfaces of the plates, and very small amounts of charge $+q$ and $-q$ on the outer surfaces of the plates. Each plate has a radius R and thickness t, and the gap distance between the plates is s. How much charge q is on the outside surface of the positive disk, in terms of Q?

Problem 15.16 Force on a neutral metal disk

A small plastic ball carrying a negative charge $-Q$ hangs from two threads (Figure 15.59). A small uncharged thin metal disk of radius R and thickness t hangs from two threads at a distance L from the plastic ball, facing the ball so that the center of the ball is located on the axis of the disk. The double threads keep the objects from twisting. The distance L is much greater than R ($L \gg R$), and the thickness t is much less than R ($t \ll R$). Calculate the force that the charged ball exerts on the neutral metal disk, in terms of Q, R, L, and t.

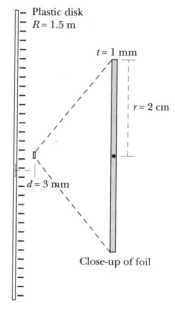

Plastic disk
$R = 1.5$ m

$t = 1$ mm

$r = 2$ cm

$d = 3$ mm

Close-up of foil

Figure 15.58 A plastic disk and a piece of aluminum foil (Problem 15.14).

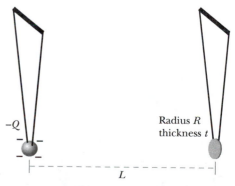

$-Q$

Radius R
thickness t

L

Figure 15.59 Force on a neutral metal disk (Problem 15.16).

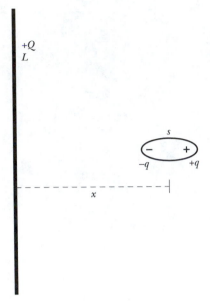

Figure 15.60 Force on a water molecule (Problem 15.17).

Problem 15.17 Force on a water molecule

A water molecule is a permanent dipole with a known dipole moment p (= qs). There is a water molecule in the air a very short distance x from the midpoint of a long glass rod of length L carrying a uniformly distributed positive charge $+Q$ (Figure 15.60). The axis of the dipole is perpendicular to the rod. Note that $s \ll x \ll L$. (The charged rod induces an increase in the dipole moment, but the induced portion of the dipole moment is completely negligible compared to p.)

(a) Find the magnitude and direction of the electric force acting on the water molecule. Your final result for the magnitude of the force must be expressed in terms of Q, L, x, and p. You can use q and s in your calculations, but your final result must not include q or s, since it is only their product $p = qs$ that is measurable for a water molecule. Explain your work carefully, including appropriate diagrams.

(b) If the electric field of the rod has the magnitude 1×10^6 N/C at the location of the water molecule, 1 cm from the rod, what is the magnitude of the horizontal component of the acceleration of the water molecule? The measured dipole moment for H_2O is about 6×10^{-30} C·m, and the mass of one mole is 18 grams (1+1+16). Be sure to check units in your calculation!

Problem 15.18 Charged pen and aluminum foil

This is a challenging problem which requires you to construct a model of a real physical situation, to make idealizations, approximations, and assumptions, and to work through a detailed analysis based on physical principles. Make sure to allow yourself enough time to think it through.

You may need to measure, estimate, or look up various quantities.

A clear plastic ball point pen is rubbed thoroughly with wool. The charged plastic pen is held above a small, uncharged disk-shaped piece of aluminum foil, smaller than a hole in a sheet of three-ring binder paper.

(a) Make a clear physics diagram of the situation, showing charges, fields, forces, and distances. Refer to this physics diagram in your analysis.

(b) Starting from fundamental physical principles, predict quantitatively how close you must move the pen to the foil in order to pick up the foil (i.e. predict an actual numerical distance). State explicitly all approximations and assumptions that you make.

(c) Try the experiment and compare your observation to your prediction.

Problem 15.19 An electrostatic dust precipitator

An electrostatic dust precipitator installed in a factory smokestack includes a straight metal wire of length $L = 0.8$ m that is charged approximately uniformly with a total charge $Q = 0.4 \times 10^{-7}$ C. A speck of coal dust (which is mostly carbon) is near the wire, far from both ends of the wire; the distance from the wire to the speck is $d = 1.5$ cm. Carbon has an atomic mass of 12 (6 protons and 6 neutrons in the nucleus). A careful measurement of the polarizability of a carbon atom gives the value

$$\alpha = 1.96 \times 10^{-40} \text{ m}^2 \frac{\text{C} \cdot \text{m}}{\text{N/C}}$$

(a) Calculate the initial acceleration of the speck of coal dust, neglecting gravity. Explain your steps clearly. Your final must be expressed in terms of Q, L, d, and α. You can use other quantities in your calculations, but your final result must not include them. Don't put numbers into your calculation until the very end, but then show the numerical calculation that you carry out on your calculator.

(b) If the speck of coal dust were initially twice as far from the charged wire, how much smaller would the initial acceleration of the speck be?

Problem 15.20 A semicircular arc

Consider a thin plastic rod bent into a semicircular arc of radius R with center at the origin (Figure 15.61). The rod carries a uniformly distributed negative charge $-Q$.

(a) Determine the electric field $\vec{E}$ at the origin contributed by the rod. Include carefully labeled diagrams, and be sure to check your result.

(b) An ion with charge $-2e$ and mass M is placed at rest at the origin. After a *very* short time Δt the ion has moved only a very short distance but has acquired some momentum $\vec{p}$. Calculate $\vec{p}$.

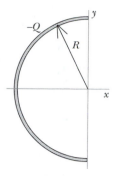

Figure 15.61 A semicircular arc (Problem 15.20).

Problem 15.21 Rod and neutral ball

A very long thin wire of length L carries a total charge $+Q$, nearly uniformly distributed along the wire. The center of a small neutral metal ball of radius r is located a distance d from the wire, where $d \ll L$. Determine the approximate magnitude of the force that the rod exerts on the ball. Show labeled physics information on the diagram. Explain how you model the situation.

$+Q, L$ (not to scale)

$d \ll L$

Radius r

Problem 15.22 Hydrogen atom

A simplified model of a hydrogen atom is that the electron cloud is a sphere of radius R with uniform charge density and total charge $-e$. (The actual charge density in the ground state is nonuniform.)

(a) For the uniform-density model, calculate the polarizability α of atomic hydrogen in terms of R. Consider the case where the magnitude E of the applied electric field is much smaller than the electric field required to ionize the atom. Suggestions for your analysis:

- Imagine that the hydrogen atom is inside a capacitor whose uniform field polarizes but does not accelerate the atom.
- Consider forces on the proton in the equilibrium situation, where the proton is displaced a distance s from the center of the electron cloud ($s \ll R$, Figure 15.62).

(b) For a hydrogen atom, R can be taken as roughly 10^{-10} m (the Bohr model of the H atom gives $R = 0.5 \times 10^{-10}$ m). Calculate a numerical value for the polarizability α of atomic hydrogen. For comparison, the measured polarizability of a hydrogen atom is $\alpha = 7.4 \times 10^{-41}$ C·m/(N/C); see note below.

(c) If the magnitude E of the applied electric field is 10^6 N/C, use the measured value of α to calculate the shift s.

(d) For some purposes it is useful to model an atom as though the nucleus and electron cloud were connected by a spring. Use the measured value of α to calculate the effective spring stiffness k_s for atomic hydrogen. For comparison, remember that last semester we found that the effective spring stiffness of the *interatomic* force in solid aluminum is about 16 N/m.

(e) If α were twice as large, what would k_s be?

Figure 15.62 In the polarized atom the center of the spherical electron cloud shifts a distance s from the center of the nucleus (Problem 15.22).

Note: Quantum-mechanical calculations agree with the experimental measurement of α reported in T. M. Miller and B. Bederson, "Atomic and molecular polarizabilities: a review of recent advances," *Advances in Atomic and Molecular Physics* **13**, 1-55, 1977. They use "cgs" units, so their value is $1/(4\pi\varepsilon_0)$ greater than the value given here.

Figure 15.63 Locations for calculating electric field; display E in circular patterns around the rod, in 3D (Problem 15.23).

Figure 15.64 End view of rod showing circular pattern of observation locations.

Problem 15.23 Electric field of a uniformly charged rod

Write a computer program to compute and display the electric field of a thin, uniformly charged rod at locations around the rod. Consider a rod of length $L = 1.0$ m, with a charge of $Q = 1 \times 10^{-8}$ C.

• Model the rod as a collection of N point charges, each with charge Q/N coulombs. First try very small values of N, then increase N until you are satisfied with the accuracy of your results. Explain what criterion you used to make this decision.

• Display arrows representing electric field vectors around the rod at locations 5 cm from the rod, near the ends of the rod and near the middle of the rod, as shown in Figure 15.63. At each height, display 6 vectors in a circular pattern around the rod, as shown in Figure 15.64. (You will of course have to scale the electric field vectors to fit in the display.)

• What is the magnitude E of the electric field at location A shown in Figure 15.63, for your most accurate calculation?

15.15 Answers to exercises

15.1 (page 517) 1.12×10^4 N/C

15.2 (page 517)

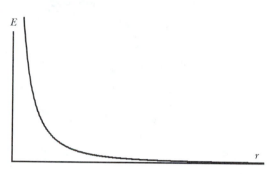

Yes, E decreases monotonically with r.

15.3 (page 521) 4.5×10^4 N/C, toward negative ring

15.4 (page 521) 2.25×10^{-4} N, toward positive ring

15.5 (page 521) 0

15.6 (page 523) As you move away from the center, the distant regions of the disk can contribute more to the net electric field, because the component perpendicular to the disk is larger. Apparently this nearly compensates for the larger distance from the nearby regions.

15.7 (page 523) 2.69×10^6 N/C; 2.66×10^6 N/C

15.8 (page 523) 2.67×10^6 N/C; 2.67×10^6 N/C; 2.70×10^6 N/C

15.9 (page 524) 2.05×10^6 N/C; 2.02×10^6 N/C; 2.70×10^6 N/C

15.10 (page 524) The superposition of the electric fields of many point charges can have a different distance dependence than the field of a single point charge.

15.11 (page 526) About 2×10^{-5} C

15.12 (page 526) About 3000 N/C

15.13 (page 526) $Q \left[\dfrac{Q/(LW)}{2\varepsilon_0} \right]$. If you calculated twice this amount, think again, carefully. One plate is affected by the electric field made by the other plate only.

15.14 (page 528) While the other charges are much farther away, there are a lot more of them, and in a sphere the two effects exactly cancel each other. The electric field at location D is zero.

15.15 (page 529)

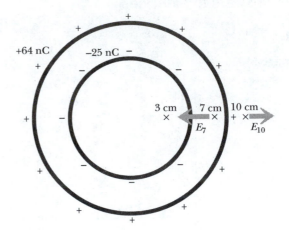

At 3 cm, $E_3 = 0$.

At 7 cm, $E_7 = 4.6 \times 10^4$ N/C, inward.

At 10 cm, $E_{10} = 3.5 \times 10^4$ N/C, outward.

15.16 (page 529)

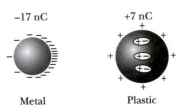

The electric field at the center of the metal ball is zero.

Chapter 16

Electric Potential

Chapter 16

Electric Potential

In analyzing the dynamics of moving objects, both at the macroscopic and the microscopic level, we found that it was frequently important to consider not just forces and momenta, but also work and energy, in trying to model the behavior of a physical system. Similarly, to complement our use of the concept of electric field, we need the concept of electric potential. Electric potential is defined as electric potential energy per unit charge.

The concept of electric potential is useful for some of the same reasons that the concept of electric field is useful. It allows us to reason about energy in a range of situations without having to worry about the details of some particular distribution of point charges. Electric potential has practical importance, in part because batteries and electric generators maintain a potential difference across themselves that is nearly independent of what is connected to them. The concept also provides significant theoretical power, enabling us to draw conclusions about a surprising range of issues, including, for example, what patterns of electric field in space are possible, and the magnitude and direction of the average electric field due to the polarized molecules in an insulator.

16.1 Review of electric potential energy

Stated briefly, electric potential is electric potential energy per unit charge, just as electric field is electric force per unit charge. We first review electric potential energy, which involves interactions of pairs of charged particles.

Energy of a single particle

The energy of a single particle with charge q_1 (Figure 16.1) consists solely of its particle energy

q_1

Figure 16.1 No electric potential energy is associated with a single charged particle.

$$E = \frac{mc^2}{\sqrt{1 - v^2/c^2}} = mc^2 + K$$

where the kinetic energy K is approximately $K \approx \frac{1}{2}mv^2$ for $v \ll c$.

There is no potential energy, because this is not a multiparticle system, and there can be no interactions inside this single-particle system.

Energy of two charged particles

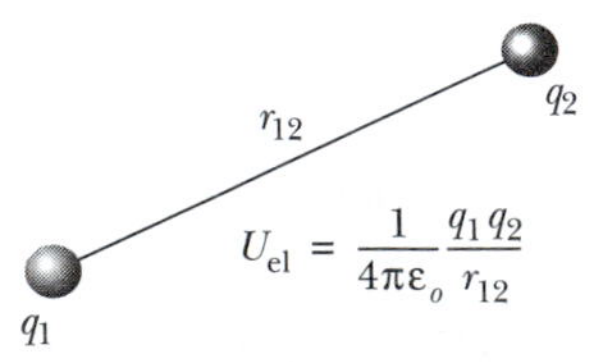

r_{12} q_2

$$U_{el} = \frac{1}{4\pi\varepsilon_o}\frac{q_1 q_2}{r_{12}}$$

q_1

Figure 16.2 Electric potential energy of two charges.

Next, we add to our system a second particle with charge q_2 at a distance r_{12} from q_1 (Figure 16.2). The total energy of this two-particle system includes not only the individual particle energies E_1 and E_2 but also their interaction energy U_{el}, so that the total energy of the system is $(E_1 + E_2 + U_{el})$, where the particle energies E_1 and E_2 are given by the formula shown above.

The potential energy U_{el} for the two-particle system is this:

ELECTRIC POTENTIAL ENERGY OF TWO PARTICLES

$$U_{el} = \frac{1}{4\pi\varepsilon_o}\frac{q_1 q_2}{r_{12}} \text{ (joules)}$$

We'll quickly review the main points in the derivation in Chapter 4 that led to this important result. Consider the system composed of the two particles and analyze a process in which work is done to change the particle energies.

We separate this work into work done by external forces (forces exerted by objects not part of the two-particle system), and work done by the forces that the two objects exert on each other:

$$\Delta(E_1 + E_2) = W_{\text{ext. forces}} + W_{\text{int. forces}}$$

Move $W_{\text{int. forces}}$ to the other side of the equation:

$$\Delta(E_1 + E_2 + (-W_{\text{int. forces}})) = W_{\text{ext. forces}}\text{ext}$$

We define the change in electric potential energy as $\Delta U_{\text{el}} = -W_{\text{int. forces}}$:

$$\Delta(E_1 + E_2 + U_{\text{el}}) = W_{\text{ext. forces}}$$

We define the total energy of the two-particle system to be the sum of the particle energies and the electric potential energy, so now the energy equation takes the simple form $\Delta E_{\text{sys}} = W$, where W is the work done by the surroundings.

Work done by internal forces in one short displacement $\Delta\vec{r}$ is $\vec{F}_{\text{int}} \bullet \Delta\vec{r}$, and to sum up lots of small contributions we integrate:

$$\Delta U = -W_{\text{int. forces}} = -\int_i^f \vec{F}_{\text{int}} \bullet d\vec{r}$$

An integral is required, because the electric force $\vec{F}_{\text{int}}$ varies in magnitude and direction during the process. We can carry out this integral for the electric force (in Chapter 4 we looked for the antiderivative rather than calculating the integral, but the methods are equivalent):

$$\Delta U_{\text{el}} = -\int_i^f \frac{1}{4\pi\varepsilon_o}\frac{q_1 q_2}{r_{12}^2}\hat{r}_{12} \bullet d\vec{r}_{12}$$

$$\Delta U_{\text{el}} = -\int_i^f \frac{1}{4\pi\varepsilon_o}\frac{q_1 q_2}{r_{12}^2}dr_{12} \quad (\text{since } \hat{r}_{12} \text{ is parallel to } \vec{r}_{12})$$

$$\Delta U_{\text{el}} = -\frac{q_1 q_2}{4\pi\varepsilon_o}\left[-\frac{1}{r_{12}}\right]_i^f$$

$$\Delta U_{\text{el}} = \Delta\left(\frac{1}{4\pi\varepsilon_o}\frac{q_1 q_2}{r_{12}}\right)$$

Therefore we define the electric potential energy of a pair of particles as

$$U_{\text{el}} = \frac{1}{4\pi\varepsilon_o}\frac{q_1 q_2}{r_{12}}$$

If the two charges have the same sign, the potential energy is positive, corresponding to repulsion (Figure 16.3). If the two charges have opposite signs, the potential energy is negative, corresponding to attraction (Figure 16.4), just as the attractive gravitational force leads to negative gravitational potential energy $U_{\text{grav}} = -Gm_1 m_2 / r_{12}$.

More than two charges

If there are more than two charges, there is an interaction energy term U_{ij} for each pair of charges q_i and q_j. For example, consider a system of three charged particles (Figure 16.5). The electric potential energy of this three-charge system involves three interaction pairs:

$$U = U_{12} + U_{13} + U_{23}$$

$$U = \frac{1}{4\pi\varepsilon_o}\frac{q_1 q_2}{r_{12}} + \frac{1}{4\pi\varepsilon_o}\frac{q_1 q_3}{r_{13}} + \frac{1}{4\pi\varepsilon_o}\frac{q_2 q_3}{r_{23}}$$

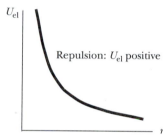

Figure 16.3 Electric potential energy of two like-sign point charges is positive for repulsion.

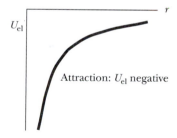

Figure 16.4 Electric potential energy of two unlike-sign point charges is negative for attraction.

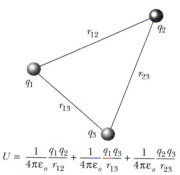

$$U = \frac{1}{4\pi\varepsilon_o}\frac{q_1 q_2}{r_{12}} + \frac{1}{4\pi\varepsilon_o}\frac{q_1 q_3}{r_{13}} + \frac{1}{4\pi\varepsilon_o}\frac{q_2 q_3}{r_{23}}$$

Figure 16.5 Electric potential energy of three point charges.

16.2 Defining electric potential

Electric potential is defined as electric potential energy per unit charge:

DEFINITION OF ELECTRIC POTENTIAL *V*

$$V = \frac{U_{el}}{q} \quad \text{(joules per coulomb; also called volts)}$$

And as a consequence,

$$U_{el} = qV$$

It is unfortunate that "electric potential" (or just "potential") and "electric potential energy" have such very similar names. One way to keep them straight is through their units: potential energy, like any other kind of energy, has units of joules, while potential has units of joules/coulomb, or volts. Often a difference in potential is called a "voltage," for obvious reasons.

V due to one particle

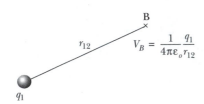

Figure 16.6 Electric potential near a single point charge.

A particularly important special case is the potential due to a single charged point particle of charge q_1. At location B, a distance r_{12} from charge q_1 (Figure 16.6), the electric potential is:

POTENTIAL AT LOCATION B DUE TO A SINGLE POINT CHARGE

$$V_B = \frac{1}{4\pi\varepsilon_o}\frac{q_1}{r_{12}} \quad \text{(J/C, or volts)}$$

A single charged particle doesn't have electric potential energy, but it does create electric potential all around itself. If we place a second charge q_2 at location B (Figure 16.7), the two-particle system would have electric potential energy U_{12} given by

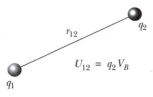

Figure 16.7 Place charge q_2 at location B, and we have $U_{12} = q_2 V_B$.

$$U_{12} = q_2 V_B = q_2\left(\frac{1}{4\pi\varepsilon_o}\frac{q_1}{r_{12}}\right) = \frac{1}{4\pi\varepsilon_o}\frac{q_1 q_2}{r_{12}}$$

which is just what we expect for this two-particle system.

V due to two particles

Next consider the electric potential produced by these two charges at location C (Figure 16.8). Since potential is a scalar, we simply add the potentials due to each charge:

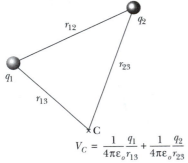

$$V_C = \frac{1}{4\pi\varepsilon_o}\frac{q_1}{r_{13}} + \frac{1}{4\pi\varepsilon_o}\frac{q_2}{r_{23}}$$

Figure 16.8 Electric potential near two point charges.

$$V_C = \frac{1}{4\pi\varepsilon_o}\frac{q_1}{r_{13}} + \frac{1}{4\pi\varepsilon_o}\frac{q_2}{r_{23}}$$

The electric potential energy for this two-particle system is just U_{12}. But if we place a third charge q_3 at location C (Figure 16.9), this introduces additional potential energy $q_3 V_C$:

$$U = \frac{1}{4\pi\varepsilon_o}\frac{q_1 q_2}{r_{12}} + q_3 V_C$$

$$U = \frac{1}{4\pi\varepsilon_o}\frac{q_1 q_2}{r_{12}} + q_3\left(\frac{1}{4\pi\varepsilon_o}\frac{q_1}{r_{13}} + \frac{1}{4\pi\varepsilon_o}\frac{q_2}{r_{23}}\right)$$

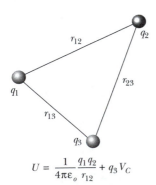

$$U = \frac{1}{4\pi\varepsilon_o}\frac{q_1 q_2}{r_{12}} + q_3 V_C$$

Figure 16.9 Place charge q_3 at location C, and we have added electric potential energy $q_3 V_C$.

This is indeed the electric potential energy of the three-particle system, involving three interaction pairs: $U = U_{12} + U_{13} + U_{23}$.

As is the case with electric field, we often split the Universe into two parts: a set of charges which contribute a potential V everywhere in space, and a different charge q that is affected by this potential.

V at infinity

A special case to consider is the potential at a location very far away from a single point charge q. We find that infinitely far from a point charge:

$$V_\infty = \frac{1}{4\pi\varepsilon_o}\frac{q}{\infty} = 0 \quad (\infty \text{ is shorthand for "extremely large"})$$

? Does this make sense?

The electric potential energy U_{el} for two interacting particles must be zero at infinitely large separation (because, as discussed in Chapter 4, the total energy of the system must then be the sum of the rest energies and kinetic energies of the particles). Therefore the electric potential V should also be zero at a location infinitely distant from a single point charge, as shown in Figure 16.10 and Figure 16.11.

Figure 16.10 Potential due to a positive point charge.

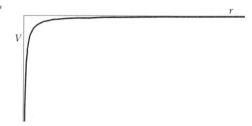

Figure 16.11 Potential due to a negative point charge.

Example: What is the potential at a location 10^{-10} m from a proton, including sign? If you were to place an electron at that location, what would the electric potential energy be, including sign?

Solution: The proton charge is $+e$:

$$V = \frac{1}{4\pi\varepsilon_o}\frac{e}{r}$$

$$V = \left(9\times10^9\frac{\text{N·m}^2}{\text{C}^2}\right)\frac{(1.6\times10^{-19}\text{ C})}{(10^{-10}\text{ m})} = 14.4\text{ J/C} \quad (\text{or } 14.4 \text{ volts})$$

This shows that typical atomic or molecular potentials are in the range of a few volts. If we place an electron at that location, the electric potential energy will be negative:

$$U_{el} = qV = (-1.6\times10^{-19}\text{ C})(14.4\text{ J/C}) = -2.3\times10^{-18}\text{ J}$$

Ex. 16.1 What is the change in potential ΔV in going from a location 10^{-10} m from a proton to a location 2×10^{-10} m from the proton? The result should be in joules/coulomb, or volts, with an appropriate sign.

Ex. 16.2 Use the result of the previous exercise to calculate the change in electric potential energy ΔU associated with moving an electron from a location 10^{-10} m from a proton to a location 2×10^{-10} m from the proton. The result should be in joules, with an appropriate sign. Does the sign make physical sense?

Ex. 16.3 Calculate the potential at location A in Figure 16.12. (1 nC is 10^{-9} C.)

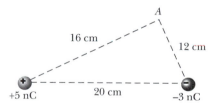

Figure 16.12 Calculate the potential at location A (Exercise 16.3).

16.3 Electric field is negative gradient of potential

The definitions of electric field and potential are similar. Electric field is force per unit charge, and electric potential is potential energy per unit charge:

$$\text{Per unit charge: } \vec{E} = \frac{\vec{F}}{q} \text{ and } V = \frac{U_{el}}{q}$$

Recall that force can be expressed as the negative gradient of potential energy. The x component of force is the negative derivative of U with respect to x:

$$F_x = -\frac{dU}{dx}$$

Write the derivative on a per-charge basis:

$$F_x/q = -\frac{d(U/q)}{dx}$$

Since F_x/q is E_x, and U/q is V, we find the following important relation between electric field and potential (Figure 16.13):

ELECTRIC FIELD IS NEGATIVE GRADIENT OF POTENTIAL

$$E_x = -\frac{dV}{dx}, \; E_y = -\frac{dV}{dy}, \; E_z = -\frac{dV}{dz}$$

Note that the units work out correctly: joules/coulomb divided by meters is N/C, since a joule is a N · m . Also, we see that the units of electric field can be given as volts per meter, written V/m.

For example, if the potential rises from 90 volts to 100 volts in a distance of 1 mm, the x component of the electric field is

$$E_x = -\frac{(100-90) \text{ volts}}{0.001 \text{ m}} = -10^4 \text{ V/m}$$

More technically, we'll mention that the gradient should be written using "partial derivatives," like this:

$$E_x = -\frac{\partial V}{\partial x}, \; E_y = -\frac{\partial V}{\partial y}, \; E_z = -\frac{\partial V}{\partial z}$$

The meaning of $\partial V/\partial x$ is simply "take the derivative of V with respect to x while holding y and z constant." For example, if $V = 3xy + 2z^2$, then we have for the partial derivatives $\partial V/\partial x = 3y$, $\partial V/\partial y = 3x$, and $\partial V/\partial z = 4z$.

We'll also mention that the gradient is often written in the form $\vec{E} = -\vec{\nabla} V$, where the inverted delta ("nabla") represents taking the x, y, and z partial derivatives of the potential to find the x, y, and z components of the electric field.

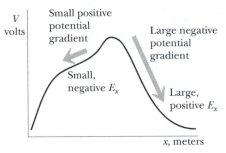

Figure 16.13 The negative slope of potential *vs.* x is the x component of the electric field.

Ex. 16.4 Suppose the potential at $z = 3.50$ m is 40 volts, and the potential at $z = 3.52$ m is 44 volts. What is the approximate value of E_z in this region? Include the appropriate sign.

Ex. 16.5 If throughout a particular region of space the potential can be expressed as $V = 3xz + 5y - 8z$, what are the vector components of the electric field at location $<x,y,z>$?

16.4 Potential difference

We will often be interested in the difference in potential between two locations in space. For example, if we know the potential at location A and the potential at location B, we can calculate how much work we would have to do in order to move a charged particle from an initial location to a final location (without changing its kinetic energy):

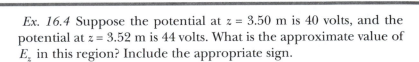

$$W = \Delta E_{\text{sys}} = \Delta U_{\text{el}} = qV_f - qV_i = q\,\Delta V$$

We can calculate ΔV, the potential difference between two locations, in two ways. If we know the potential at each location, we can simply take the difference. If we don't know the potential at each location, but we do know the electric field in the intervening region, we can still find ΔV, as we will show.

16.4.1 Potential difference and electric field

In the preceding section, we found that electric field is the negative spatial derivative of the potential (the negative gradient). Conversely, we can show that the difference in potential in moving from one location to another is the negative integral of the electric field along a path between the locations.

The change in potential energy is equal to the negative of the work done by internal forces:

$$\Delta U = -W_{\text{int. forces}} = -\int_i^f \vec{F}_{\text{int}} \bullet d\vec{l}$$

where $d\vec{l}$ represents a small step along the chosen path.

For electric forces, write this on a per-unit-charge basis:

$$\Delta\left(\frac{U_{\text{el}}}{q}\right) = -\int_i^f \left(\frac{\vec{F}_{\text{int}}}{q}\right) \bullet d\vec{l}$$

But U_{el}/q is V, and $\vec{F}/q$ is $\vec{E}$, so we have the following important relation between ΔV and electric field:

POTENTIAL DIFFERENCE AND ELECTRIC FIELD

$$V_f - V_i = \Delta V = -\int_i^f \vec{E} \bullet d\vec{l}$$

If $\vec{E}$ does not change very much over a short total path $\Delta\vec{l}$, we can write

$$\Delta V = -\int_i^f \vec{E} \bullet d\vec{l} \approx -\vec{E} \bullet \Delta\vec{l} \quad \text{(very short path)}$$

For example, if E_x at $x = 5.00$ m is 1000 N/C and we move from $x = 5.00$ m to $x = 5.01$ m, the potential difference is

$$\Delta V \approx -\vec{E} \bullet \Delta\vec{l} = -(1000 \text{ V/m})(0.01 \text{ m}) = -10 \text{ volts}$$

If E_x is constant in this region, this result is exact. If E_x is varying, our result is approximate.

Potential near a point charge

Let's apply the integral formulation to the familiar case of a point charge q:

$$V_f - V_i = -\int_i^f \vec{E} \bullet d\vec{l} = -\int_i^f \frac{1}{4\pi\varepsilon_0}\frac{q}{r^2}\,dr = \frac{1}{4\pi\varepsilon_0}\left[\frac{q}{r}\right]_i^f = \frac{1}{4\pi\varepsilon_0}\left(\frac{q}{r_f} - \frac{q}{r_i}\right)$$

This is consistent with the fact that $V = \dfrac{1}{4\pi\varepsilon_0}\dfrac{q}{r}$ due to a point charge.

Potential inside a uniformly charged hollow sphere

Since the electric potential is zero at a location infinitely far from a charge, we can write V at a location A as:

$$V_A = V_A - V_\infty = -\int_\infty^A \vec{E} \bullet d\vec{l}$$

Using this idea, we can find the potential at a location B inside a uniformly charged hollow sphere with charge Q and radius R by considering a path

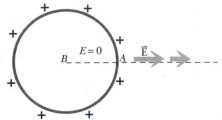

Figure 16.14 A path from infinity to location B inside a uniformly charged hollow sphere, going through location A.

from infinity to location *B*, as shown in Figure 16.14. We can consider the path to consist of two segments: a path from infinity to location *A*, just outside the sphere, and from *A* to location *B* inside the sphere. Along the first segment of the path the electric field is the same as that of a point charge located at the center of the sphere, so

$$V_A = V_A - V_\infty = \frac{1}{4\pi\varepsilon_0}\frac{Q}{R}$$

Inside the sphere, the electric field due to the charges on the surface of the sphere is zero, so

$$V_B - V_A = -\int_A^B \vec{E} \cdot d\vec{l} = 0 \text{, and}$$

$$V_B = (V_B - V_A) + (V_A - V_\infty) = \frac{1}{4\pi\varepsilon_0}\frac{Q}{R}$$

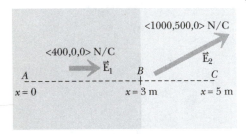

Figure 16.15 The electric field differs in the two regions.

Example: Suppose that from $x = 0$ to $x = 3$ the electric field is given by $\vec{E} = <400, 0, 0>$ N/C, and that from $x = 3$ to $x = 5$ the electric field is given by $\vec{E} = <1000, 500, 0>$ N/C. See Figure 16.15. What is the potential difference $\Delta V = V_C - V_A$?

Solution: Split the path into two parts, *A* to *B* and *B* to *C*. In each part the electric field is constant in magnitude and direction. We have the following:

$$\Delta V = -\sum \vec{E} \cdot \Delta \vec{l}$$

$$\Delta V = -<400, 0, 0>V/m \bullet <3, 0, 0>m - <1000, 500, 0>V/m \bullet <2, 0, 0>m$$

$$\Delta V = (-1200 - 2000) \text{ volts} = -3200 \text{ volts}$$

In evaluating ΔV we used the definition of the dot product:

$$\vec{E} \bullet \Delta \vec{l} = E_x\Delta x + E_y\Delta y + E_z\Delta z$$

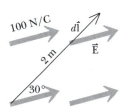

Figure 16.16 Moving at a 30° angle to the electric field.

Example: Suppose that in a certain region of space (Figure 16.16) there is a nearly uniform electric field of magnitude 100 N/C (that is, the electric field is the same in magnitude and direction throughout this region). If you move 2 meters at an angle of 30° to this electric field, what is the change in potential?

Solution: $\Delta V \approx -\vec{E} \bullet \Delta \vec{l} = E\Delta l \cos\theta$, using another way to evaluate the dot product. So we have

$$\Delta V = -(100 \text{ N/C})(2 \text{ m})\cos(30°) = -173 \text{ volts}$$

Ex. 16.6 You move from location *i* at <2,5,4> m to location *f* at <3,5,9> m. All along this path there is a nearly uniform electric field $\vec{E} = <1000, 200, -500>$ N/C. Calculate $\Delta V = V_f - V_i$, including sign and units.

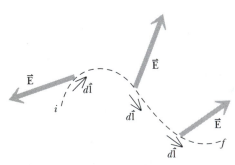

Figure 16.17 The potential difference for each step along this path is $-\vec{E} \bullet \Delta \vec{l}$. The potential difference $\Delta V = V_f - V_i$ is the sum of all these contributions.

16.4.2 Limit on mathematical complexity in this introductory course

Using $V_f - V_i = \Delta V = -\int_i^f \vec{E} \bullet d\vec{l}$ to calculate the potential difference along a path that is not straight, such as that shown in Figure 16.17, along which the electric field is varying in magnitude and direction, can be mathematically challenging. In many cases the best way to do this is to do it numerically, using a computer. In this course we will not ask you to do arbitrarily complex integrations of this kind analytically.

In this introductory course we will typically consider relatively simple situations. Often the electric field is uniform (same magnitude and direction) all along a straight path, which makes the calculation very simple. Another relatively simple situation is one where the magnitude varies but not the direction along a straight path. A third simple situation that we will encounter is one where we move along a curved path, but the electric field is always in the same direction as this curved path, and often constant in magnitude. In each of these simple cases it can be quite easy to evaluate the potential difference, as was seen in the preceding exercises.

Ex. 16.7 If the electric field exceeds about 3×10^6 N/C in air a spark occurs. Approximately, what is the maximum possible potential difference between the plates of a capacitor whose gap is 3 mm, without causing a spark in the air between them?

Ex. 16.8 A capacitor with a gap of 2 mm has a potential difference from one plate to the other of 30 volts. What is the magnitude of the electric field between the plates?

16.4.3 Electron volt—a unit of energy

From an earlier exercise you saw that potential differences near atoms are of the order of a few volts. If an electron moves through a potential difference of one volt there is a change in the electric potential energy whose magnitude is

$$|\Delta U| = (e)(1 \text{ volt}) = (1.6 \times 10^{-19} \text{ C})(1 \text{ J/C}) = 1.6 \times 10^{-19} \text{ joule}$$

This amount of energy is called one "electron volt" and abbreviated 1 eV. It is a convenient unit for measuring energies of atomic processes in physics and chemistry. Note that an electron volt is a unit of energy, not potential (potential is measured in volts, not electron volts).

Ex. 16.9 An electron starts from rest in a vacuum, in a region of strong electric field. The electron moves through a potential difference of 35 volts. What is the kinetic energy of the electron in electron volts (eV)? What if the particle were a proton?

16.4.4 Utility of electric potential

The utility of the concept of potential, energy per unit charge, is similar to the utility of the concept of electric field, force per unit charge. These concepts help us to analyze and discuss phenomena that would be more difficult to analyze directly in terms of energy and force. In particular, we can split a problem into two parts. First, we determine the electric potential and/or electric field produced by a set of charges. Second, we determine how some other charge is affected by this electric potential and/or electric field.

Just as it is sometimes more useful to analyze a system in terms of energy rather than force, it is sometimes more useful to analyze a system in terms of potential rather than field. It is often the case that an analysis in terms of energy (or electric potential, energy per unit charge) is simpler or clearer than an analysis in terms of force (or electric field, force per unit charge).

Because potential is a scalar rather than a vector, it is often simpler to calculate than electric field. On occasion it may be easier to find a formula for potential and then take its gradient to find electric field, than to calculate the electric field directly.

16.4.5 The sign of the potential difference

The potential difference ΔV can be positive or negative, and the sign is extremely important. For example, an increase in potential energy $q \, \Delta V$ is associated with a decrease in kinetic energy, while a decrease in potential energy $q \, \Delta V$ is associated with an increase in kinetic energy.

To show an easy way to get the right sign, we'll consider the situation in a capacitor with two large plates, close together, carrying charges $+Q$ and $-Q$ uniformly distributed over the area of the plates (Figure 16.18). As we saw in the previous chapter, the electric field is nearly the same in magnitude and direction everywhere between the plates, as long as you're not too close to the edge of the plates. In Figure 16.18 the electric field points to the left, away from the positive plate and toward the negative plate.

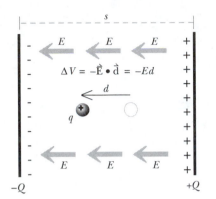

Figure 16.18 $\Delta V < 0$ in this case, moving in the direction of $\vec{E}$ so that $\cos(0°) = +1$.

If you release a positive charge from rest it will move toward a position with a lower electric potential (negative ΔV), and the decrease in electric potential energy, $\Delta U_{el} = q \, \Delta V$, goes into increasing the kinetic energy (Figure 16.18). Note that the positive charge is heading in the direction of the electric field. Alternatively, we can slowly ease the positive charge down to the lower potential and extract useful work from it, letting it push on us as it goes.

If you push a positive charge to a higher electric potential (positive ΔV) there is an increase in the electric potential energy, $\Delta U_{el} = q \, \Delta V$. This increase is associated with moving in the direction opposite to the electric field (Figure 16.19), due to the minus sign in $\Delta V = -\int \vec{E} \bullet d\vec{l}$.

It is important to be able to identify whether a potential difference is positive or negative, because the sign governs whether a charged particle will gain or lose energy in moving from one place to another.

It is easy to make sign errors in formal calculations using potential, and we need an independent check on the sign. The following rule is extremely useful:

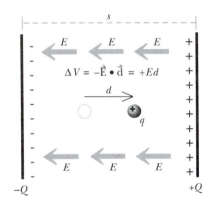

Figure 16.19 $\Delta V > 0$ in this case, moving opposite the direction of $\vec{E}$ so that $\cos(180°) = -1$.

SIGN OF ΔV AND DIRECTION OF $\vec{E}$

Moving in the direction of $\vec{E}$ means that the potential is decreasing.

Moving opposite to $\vec{E}$ means that the potential is increasing.

It may help to think of holding a positive charge in your hand. If you simply release the positive charge, it starts to move toward lower potential, in the direction of the electric field. But if you push it against the electric field, you are moving to higher potential.

? Suppose you let go of a negative charge. Will it head toward a region of lower potential or toward a region of higher potential? Why?

A negative charge $-q$ that is free to move will accelerate in the direction opposite to the direction of the electric field, since $\vec{F} = (-q)\vec{E}$. Therefore a negative charge heads toward a region of higher potential (moving against $\vec{E}$ means moving to higher potential).

Because you can determine the sign of a potential difference on purely physical grounds, you should never get the wrong sign! In the following exercises you are asked to give the correct sign for the potential difference between two locations. Use the rule given above.

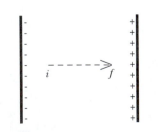

Figure 16.20 Exercise 16.10.

Ex. 16.10 In Figure 16.20, what is the direction of the electric field? Is $\Delta V = V_f - V_i$ positive or negative?

Ex. 16.11 In Figure 16.21, what is the direction of the electric field? Is $\Delta V = V_f - V_i$ positive or negative?

Ex. 16.12 In Figure 16.22, what is the direction of the electric field? Is $\Delta V = V_f - V_i$ positive or negative?

Figure 16.21 Exercise 16.11.

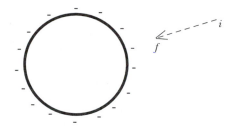

Figure 16.22 Exercise 16.12.

16.4.6 Application: Potential difference in a metal in static equilibrium

In Section 14.5.5 we proved that inside a conductor in static equilibrium, the electric field is zero (because otherwise mobile charges would shift until they contribute a field large enough to cancel the applied field). Therefore in static equilibrium the electric field is zero at all locations along any path through a metal. This means that the potential difference

$$\Delta V = V_f - V_i = -\int_i^f \vec{\mathbf{E}} \bullet d\vec{\mathbf{l}}$$

is zero between any two locations inside the metal, and the potential at any location must be the same as the potential at any other location (Figure 16.23).

FOR A METAL IN STATIC EQUILIBRIUM

$$\Delta V = V_f - V_i = -\int_i^f \vec{\mathbf{E}} \bullet d\vec{\mathbf{l}} = 0$$

In a metal in static equilibrium the potential is exactly the same everywhere inside that piece of metal, because $E = 0$ inside the metal.

? Does this mean that the potential is zero at every location inside a metal in static equilibrium?

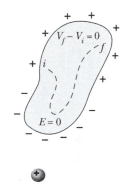

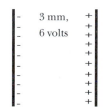

Figure 16.23 Potential difference is zero inside a metal in static equilibrium.

No. The potential inside a metal object in static equilibrium is constant (the same at every location), but it need not be zero. Charges on the surface of the metal object or on other objects may contribute to a nonzero but uniform potential inside the metal.

Let's see a concrete example of the potential inside a metal. Suppose a capacitor with large plates and a small gap of 3 mm has a potential difference of 6 volts from one plate to the other (Figure 16.24).

? What is the direction and magnitude of the electric field in the gap?

The direction of the electric field in the gap is toward the left, away from the positive plate and toward the negative plate. The magnitude can be found from noting that the potential difference $\Delta V = Es = 6$ volts, so that $E = (6 \text{ volts})/(0.003 \text{ m}) = 2000 \text{ volts/meter}$.

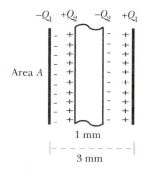

Figure 16.24 A capacitor with large plates and small gap.

A more complex situation

Next we insert into the center of the gap a 1-mm-thick metal slab with the same area as the capacitor plates, as shown in Figure 16.25. We are careful not to touch the charged capacitor plates as we insert the metal slab. We want to figure out what is the new potential difference between the two outer plates

First we need to understand the new pattern of electric field that comes about when the metal slab is inserted. The charges on the outer plates are $-Q_1$ and $+Q_1$, both distributed approximately uniformly over a large plate area A. The metal slab of course polarizes and has charges of $+Q_2$ and $-Q_2$ on its surfaces, as indicated in Figure 16.25.

The electric field inside a capacitor is approximately $E = (Q/A)/\varepsilon_0$, where Q is the charge on one plate and A is the area of the plate, if the plate separation s is small compared to the size of the plates.

Figure 16.25 Insert a metal slab in the middle of the capacitor gap.

? Since the electric field inside the metal slab must be zero, we can conclude that Q_2 is equal to Q_1. Why?

The plates have the same area A. The outer charges $-Q_1$ and $+Q_1$ produce an electric field in the metal slab $E_1 = (Q_1/A)/\varepsilon_0$ to the left. The inner charges $+Q_2$ and $-Q_2$ are arranged like the charges on a capacitor, so they produce an electric field in the metal slab $E_2 = (Q_2/A)/\varepsilon_0$ to the right. The sum of these two contributions must be zero, because the electric field inside a metal in static equilibrium must be zero. Hence Q_2 is equal to Q_1.

We can consider that the left pair of charges produces a field like that of a capacitor, and the right pair of charges also produces a field like that of a capacitor. The effect is that after inserting the metal slab, the electric field remains 2000 V/m in the air gaps but is now zero inside the metal slab. (The fringe fields are small if the gap is small.)

? Now that you know the electric field everywhere, what are the potential differences across each of the three regions between the plates (air gap, metal slab, air gap).

The electric field in the air gap is essentially unchanged, so

$$\Delta V_{\text{left}} = \Delta V_{\text{right}} = (2000\text{V/m})(0.001\text{m}) = 2\text{ V}$$

ΔV inside the metal slab must be zero, because E = 0 inside the slab. The potential difference between the plates of the capacitor is now 4 V, not 6 V as it was originally. (Note that there is no such thing as "conservation of potential"; the potential difference changed when we inserted the slab.)

A powerful reasoning tool

When you know that the electric field inside a metal must be zero, you know that the sum of all the contributions to that electric field must be zero. This is a powerful tool for reasoning about fields and charges in and on metals in static equilibrium.

Ex. 16.13 Explain qualitatively why the new potential difference is less than the original 6 volts.

Ex. 16.14 If the potential of the negative plate were 0 and we measured potential relative to that location, what would be the potential at the left edge of the metal slab? In the middle of the metal slab? At the right edge of the metal slab?

Ex. 16.15 A metal sphere of radius R in static equilibrium has a uniformly distributed charge Q on its surface. There are no other charged objects around. What is the potential at the center of the sphere? What is the potential at a radius $0.85R$ from the center of the sphere?

16.4.7 Application: A metal not in static equilibrium

During the (extremely) brief time when a metal is polarizing due to an external charge, there is a nonzero electric field inside the metal. Before static equilibrium is reached, the electric field in the metal is not zero.

As we will study in detail in a later chapter, in an electric circuit the battery maintains a nonzero electric field inside the wires, as a kind of continuing polarization process that doesn't lead to static equilibrium. The electric field continually pushes the sea of mobile electrons through the wires. The electric field is not zero in this situation, which is not static equilibrium.

? In Figure 16.26 there is a nonzero electric field of uniform magnitude E throughout the interior of a wire of length L, and the direction of the electric field follows the direction of the wire. What is the potential difference $V_B - V_A$?

In this case the potential difference is simply $-EL$, the sum of all the $-\vec{E} \bullet \Delta\vec{l} = -E\Delta l$ terms along the wire. So if a metal is not in static equilibrium, the potential isn't constant in the metal. However, in a circuit a thick copper wire may have such a small electric field in it that there is very little potential difference from one end to the other, in which case the potential is almost (but not quite) constant in the wire.

16.5 Path independence

In Chapter 5 (page 169) we showed that potential energy differences depend only on the initial and final states and are independent of path: different path, different process, but same change of state. Similarly, the potential difference $\Delta V = V_B - V_A$ between two locations A and B depends solely on the positions of all the charges that contribute to V_A and V_B. The potential at A is the sum of all of the $(1/(4\pi\varepsilon_0))(q/r)$ contributions by all of the charges; similarly for the potential at B.

However, we can also calculate the potential difference between A and B as the integral of $-\vec{E} \bullet d\vec{l}$, and to do this we must first choose a path to follow, along which we add up all the $-\vec{E} \bullet d\vec{l}$ contributions. What if we take some other path between the two locations? We better get the same result no matter what path we take. This property of potential, "path independence," allows us to reason about a variety of different situations.

A simple example of two different paths

To see how the integral for potential difference between two locations can be the same along different paths, we'll consider two different paths through a capacitor. Suppose you move from the positive plate of a capacitor to the negative plate, moving at an angle θ to the electric field E (Figure 16.27).

? If the gap distance is s, calculate the potential difference in going from A to C: $\Delta V = V_f - V_i = V_C - V_A$.

Since the electric field is the same in magnitude and direction all along the path, we can write

$$\Delta V = -(E_x \Delta x + E_y \Delta y + E_z \Delta z) = -Es$$

Alternatively, since $\Delta l = s/\cos\theta$, we can calculate

$$\Delta V = -\vec{E} \bullet d\vec{l} = -E[s/(\cos\theta)]\cos\theta = -Es$$

Suppose instead we move along the two-part path shown in Figure 16.28.

? $\Delta V = V_f - V_i = V_C - V_A = ?$

Along the path from B to C, $-\vec{E} \bullet \Delta\vec{l} = -\langle E, 0, 0 \rangle \bullet \langle 0, 0, \Delta l \rangle = 0$ (alternatively, since the cosine of 90° is zero, $V_C - V_B = 0$).

Therefore we again find that $V_C - V_A = V_B - V_A = -Es$, which is what we found for going directly from A to C.

? Does it make sense that $V_C - V_B = 0$?

Yes. Just as no work is required to move at right angles to a force, it takes no effort to move at right angles to the electric field, so the electric potential energy and the electric potential don't change. The potential difference

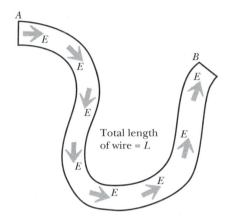

Figure 16.26 Inside a wire in a circuit, there can be an electric field with uniform magnitude E but which follows the direction of the wire. The wire's length is L.

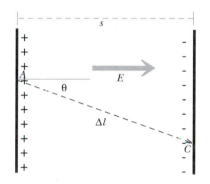

Figure 16.27 Moving at an angle to the electric field.

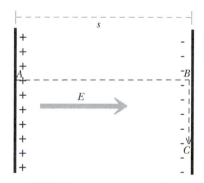

Figure 16.28 Moving at right angles to the electric field.

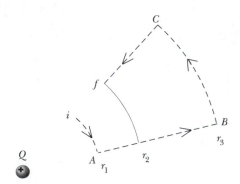

Figure 16.29 What is the potential difference along this complicated path?

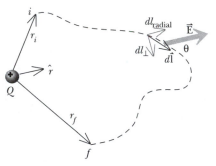

Figure 16.30 Calculating the potential difference along an arbitrary path.

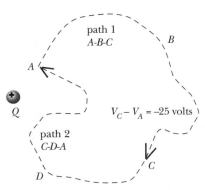

Figure 16.31 Round-trip potential difference on a path from *A* back to *A* again is zero.

doesn't depend on the angle of our path in Figure 16.27 because the sideways components of our steps don't affect ΔV.

Two different paths near a point charge

Along a path straight away from a stationary charge *Q* we know this:

$$\Delta V = \frac{1}{4\pi\varepsilon_0} Q\left[\frac{1}{r_f} - \frac{1}{r_i}\right] \text{ near a positive or negative point charge } Q$$

It is instructive to work out the potential difference along a different path.

? In the situation shown in Figure 16.29, calculate the potential difference for each branch of the path from location *i* to *A* to *B* to *C* to location *f*, in terms of *Q* and the radii r_1, r_2, and r_3, then add these up to get the potential difference along this path. As a first step, draw the electric field at various locations along the path.

You should find that you get exactly the same result that you would get by going directly in a straight line from *i* to *f*. We see here that it is only when $d\vec{l}$ has a radial component that there is a nonzero contribution to ΔV. Along the curved arcs the electric field is perpendicular to the path, so the arcs contribute zero to the potential difference. Going along the straight lines is equivalent to just going from r_1 to r_2. Evidently it doesn't matter what path we take. We say that potential difference is "path independent."

An arbitrary path near a point charge

We can see more generally how this works out for *any* path near a stationary point charge. Follow an arbitrary path from initial location *i* to final location *f* (Figure 16.30). At each location along the path draw the component dl_{radial} of $d\vec{l}$ in the direction of $\vec{E}$, pointing away from the charge *Q*.

At every location along the path we have

$$-\vec{E} \bullet d\vec{l} = -\left|\vec{E}\right|\hat{r} \bullet d\vec{l} = -E dl\cos\theta$$

But $dl\cos\theta$ is the radial component of $d\vec{l}$ and is equal to dl_{radial}, the change in distance from the charge. Therefore along this arbitrary path we are simply integrating $-E dl_{radial}$, so

$$\Delta V = \frac{1}{4\pi\varepsilon_0} Q\left[\frac{1}{r_f} - \frac{1}{r_i}\right]$$

All that matters are the initial and final distances from the charge, not the path we happen to follow. In fact, we already knew that the potential at any distance from the charge is $(1/(4\pi\varepsilon_0))(Q/r)$, so the potential difference between location *i* and location *f* should indeed be the expression given above. What is interesting is seeing how it worked out that the integral of $-\vec{E} \bullet d\vec{l}$ didn't depend on what path we took.

16.5.1 Round trip potential difference is zero

A consequence of path independence is that the integral of $-\vec{E} \bullet d\vec{l}$ along a round-trip path has a potential difference of zero, as we can now show (Figure 16.31). Move along path 1 (*A-B-C*), and then back to location *A* along path 2 (*C-D-A*).

? Suppose the potential difference along path 1 is −25 volts. What would be the potential difference in going backward along path 2 (*A-D-C*)? Forward along path 2 (*C-D-A*)?

Going backward along path 2 is the same as going forward along path 1, so the potential difference along the path (*A-D-C*) must be −25 volts. Going in the opposite direction, along path 2 (*C-D-A*), reverses the sign of the parallel component of the electric field, so the potential difference along path 2 must be +25 volts.

? So if we walk all the way around the closed path, *A-B-C-D-A*, what is the potential difference?

The round-trip potential difference is zero (−25 volts plus +25 volts). Note that this result is the direct consequence of the path independence of electric potential. It is also not surprising, since we do expect $V_A - V_A = 0$!

> Potential difference due to a stationary point charge
> is independent of the path.
> Potential difference along a round trip is zero.

Since potential difference can be calculated as an integral involving the electric field, and the electric field is the sum of all the contributions of many stationary point charges, potential difference is the sum of the contributions of all the point charges. The potential difference integral is independent of the path for one point charge, and since all charge distributions are made up of atomic point charges, the round trip potential difference must be zero for any configuration of stationary charges whatsoever.

PATH INDEPENDENCE

ΔV is always independent of the path taken between two locations.
Along a round-trip path, ΔV is *zero*.

These are extremely important general results, and in the future we will often have occasion to refer to these properties of electric potential due to charges. Path independence is important because when we use potential we are reasoning about the relationship between initial and final states, independent of the intervening process. We are free to analyze the most convenient process that takes the system from the initial state to the final state.

A puzzle

In a previous section (Section 16.4.7) we noted that in a metal object that is not in static equilibrium, such as a current-carrying wire in a circuit, there can be a nonzero electric field, so there is a difference in potential between two locations in the wire. The path independence of ΔV, however, implies that we should get the same potential difference if we take a path through the air, rather than a path through the wire (Figure 16.32). We are forced to conclude that near an electric circuit there must be an electric field in the air! We will see how this comes about in Chapter 18.

Ex. 16.16 Calculate the potential difference along the closed path consisting of two radial segments and two circular segments centered on the charge q (Figure 16.33). Show that the four ΔV's add up to zero. It is helpful to draw electric field vectors at several locations on each path segment to help keep track of signs.

16.5.2 Reasoning about patterns of electric field

We can use our results from the previous section to reason about particular patterns of electric field. For example, consider the pattern of electric field shown in Figure 16.34, where the electric field is tangent to the circle at every point on the circle.

? Try to think of an arrangement of stationary charges that could produce the pattern of electric field shown in Figure 16.34.

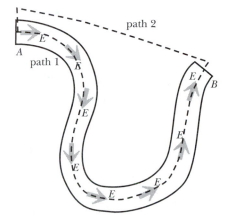

Figure 16.32 The potential difference between two points in a wire in a circuit must be the same for a path through the air and for a path through the wire.

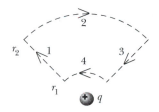

Figure 16.33 Exercise 16.16.

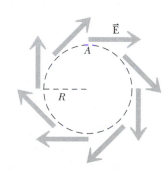

Figure 16.34 Is it possible to produce this pattern of electric field ($\vec{E}$ tangent to a circle at every point on the circle) with some arrangement of point charges?

We can demonstrate very simply that it is impossible to produce this pattern of electric field by any arrangement of stationary point charges! Consider a path that starts at location A and goes clockwise around the circle, ending back at location A. Since at every location along the path $\vec{E}$ is parallel to $d\vec{l}$, the round trip potential difference along this path is:

$$\Delta V = -\int \vec{E} \bullet d\vec{l} = -(2\pi R)E$$

But we showed previously that for a single point charge, and hence for any assemblage of point charges, $\Delta V = 0$ for a round trip. So this "curly" pattern of electric field must be impossible to produce by arranging any number of stationary point charges. We would not have gained this insight without introducing the concept of potential.

Although we have shown that we cannot produce this pattern of electric field with any arrangement of stationary point charges, it might be possible if there were some other way to produce electric fields. We'll return to this issue later in the course.

16.6 Shifting the zero of potential

As we saw in Chapter 4, electric potential energy (and therefore potential) must go to zero at large distances, to satisfy the requirements of relativity. Nevertheless, we are often interested only in differences of potential or potential energy, in which case we can choose a different zero location as a matter of convenience. This is exactly analogous to approximating change in gravitational potential energy as $\Delta(-GMm/r) \approx \Delta(mgy)$ near the surface of the Earth, and then choosing the zero of potential energy to be at or near the surface, rather than at infinity.

The electric field $\vec{E}$ is given by the gradient of the potential, not by the value of the potential. This means that only potential difference ΔV matters in determining the electric field, not the value V of the potential.

? Find the magnitude of the electric field in the 0.2-mm gap of a narrow-gap capacitor whose negative plate is at a potential of 35 volts and whose positive plate is at a potential of 135 volts (Figure 16.35).

The magnitude of the electric field in this case is $(100 \text{ volts})/(0.2\times10^{-3} \text{ m}) = 5\times10^{5} \text{ volts/m}$. The direction is from the positive plate toward the negative plate (direction of decreasing potential).

? Suppose instead that the potential of the negative plate is 75000 volts and the potential of the positive plate is 75100 volts (Figure 16.36). Now what is the electric field in the 0.2-mm gap?

This makes no difference in the calculation, because it is only the 100-volt difference in the potential that counts, not the particular values of the potential.

? Now suppose that the potential of the negative plate is –500 volts and the potential of the positive plate is –400 volts (Figure 16.37). What is the electric field in the 0.2-mm gap?

The magnitude of the electric field is again 5×10^{5} volts/m. These calculations show that what counts in calculating the electric field is the gradient of potential, or potential *difference* per unit distance, not the value of the potential. This is exactly analogous to the fact that it is no harder to walk up from the 75th floor to the 76th floor of a building than it is to walk up from the 1st floor to the 2nd floor. To put it another way, adding a constant V_0 to all the potential values changes nothing as far as potential difference and electric field are concerned:

$$(V_f + V_0) - (V_i + V_0) = V_f - V_i$$

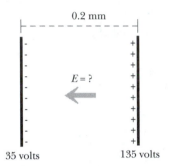

Figure 16.35 What is the magnitude of the electric field?

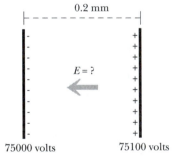

Figure 16.36 What is the magnitude of the electric field?

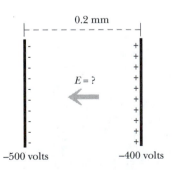

Figure 16.37 What is the magnitude of the electric field?

16.7 Potential of distributed charges

We can calculate the potential due to a charge distribution in two ways: either by dividing the distribution into point-like pieces and adding the potential due to each piece, or by calculating $-\int \vec{E} \cdot \vec{dl}$ along a path from infinity to the location of interest.

16.7.1 Potential along the axis of a ring

Adding *V* of point charges

We can use superposition to calculate the potential at a location z from the center of a thin ring, along the axis. The ring carries a total charge Q. The total charge Q is made up of point charges, each of charge q, and each of these point charges is a distance $\sqrt{z^2 + R^2}$ from the location of interest (Figure 16.38).

The potential contributed by any one of these point charges q is

$$V = \frac{1}{4\pi\varepsilon_0}\frac{q}{\sqrt{z^2 + R^2}} \quad \text{for one point charge } q$$

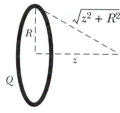

Figure 16.38 The potential due to a uniformly charged ring.

Adding up the potential contributed by all of the point charges, we have the following result:

$$V_{\text{ring}} = \sum \frac{1}{4\pi\varepsilon_0}\frac{q}{\sqrt{z^2 + R^2}} = \frac{1}{4\pi\varepsilon_0}\frac{1}{\sqrt{z^2 + R^2}}\sum q = \frac{1}{4\pi\varepsilon_0}\frac{Q}{\sqrt{z^2 + R^2}}$$

Integrating

This same result can be obtained by integrating the electric field of the ring along a path, using the expression for this field which we found in the previous chapter.

$$V(z) = -\int_{\infty}^{z} \frac{1}{4\pi\varepsilon_0}\frac{qw}{(R^2 + w^2)^{3/2}}\hat{w} \cdot \vec{dl}$$

Finding field from potential

Conversely, from the result for the potential we could determine the electric field by differentiation, since the electric field is the negative gradient of the potential (see Problem 16.9).

Ex. 16.17 Show that if you are very far from the ring ($z \gg R$), the potential is approximately equal to that of a point charge. (This is to be expected, because if you are very far away, the ring appears to be nearly a point.)

16.7.2 Potential along the axis of a uniformly charged disk

Consider a disk of radius R (area $A = \pi R^2$) with charge Q uniformly distributed over its surface. To calculate the potential due to this disk we follow a procedure similar to the one we used in finding the electric field of a charge distribution. We divide the disk into rings as we did in finding the electric field of a disk in the previous chapter (Figure 16.39). At location z along the axis of a disk, the potential contributed by one ring is given by the result found above:

$$V_{\text{ring}} = \frac{1}{4\pi\varepsilon_0}\frac{\Delta q}{(z^2 + r^2)^{1/2}} = \frac{1}{4\pi\varepsilon_0}\frac{Q}{A}\frac{2\pi r \Delta r}{(z^2 + r^2)^{1/2}}$$

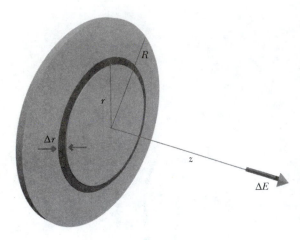

Figure 16.39 A ring of radius r and width Δr makes a contribution V_{ring} to the potential.

We have used the result from the previous chapter for the charge on one of the rings, $\Delta q = (Q/A)2\pi r\Delta r$. Add up the contributions of all the rings:

$$V = \frac{1}{2\varepsilon_0} \frac{Q}{A}\int_0^R \frac{r\,dr}{(z^2 + r^2)^{1/2}}$$

$$V = \frac{1}{2\varepsilon_0} \frac{Q}{A}[(z^2 + r^2)^{1/2}]_0^R$$

$$V = \frac{1}{2\varepsilon_0} \frac{Q}{A}[(z^2 + R^2)^{1/2} - z]$$

The units for this result check: meters in the bracket, and m^2 in the A term in the denominator, so the units are those of $Q/(\varepsilon_0 R)$, which are indeed units of potential.

We can find the electric field from the negative gradient of the potential. We want the z component of electric field, and taking the derivative of the potential we find the following:

$$E = -\frac{\partial V}{\partial z} = \frac{(Q/A)}{2\varepsilon_0}\left[1 - \frac{z}{(z^2 + R^2)^{1/2}}\right]$$

This is the same result obtained in the previous chapter by summing the contributions to the electric field.

Ex. 16.18 What is the potential at the center of a spherical shell of radius R which has a charge Q uniformly distributed over its surface? Don't do an integral of the electric field; just add up the contributions to the potential by the charges. (Afterwards, compare with the alternative analysis starting on page 553.)

16.8 Reflection: Potential and potential difference

We have seen that there are two different ways to find the potential at a particular location:

• Add up the contributions of all point charges at all other locations:

$$V_A = \sum \frac{1}{4\pi\varepsilon_0}\frac{q_i}{r_i}$$

• Travel along a path from a point very far away to the location of interest, adding up $-\vec{E} \bullet \vec{dl}$ at each step:

$$V_A = -\int_\infty^A \vec{E} \bullet \vec{dl}$$

One way to think of the meaning of the potential at location A is to consider how much work per unit charge you would have to do to move a charged particle from a location very far away to location A. This work would of course depend on the electric field in the region through which you moved the charge.

We have also seen that there are two ways to find the potential difference between two locations A and B:

• Subtract the potential at the initial location A from the potential at the final location B:

$$\Delta V = V_B - V_A$$

• Travel along a path from A to B, adding up $-\vec{E} \bullet d\vec{l}$ at each step:

$$\Delta V = V_B - V_A = -\int_A^B \vec{E} \bullet d\vec{l}$$

Again, note that the potential difference between two locations depends on the electric field in the region between the two locations. If you were to move a charge from location A to location B, the electric field in the intervening region would determine how much work per unit charge you would have to do.

Avoiding a common pitfall

A common error is to assume that the electric field at a location determines the potential at a location. In fact, the electric field at location A has very little to do with the potential at location A! Similarly, the electric field at A and the electric field at B have very little to do with the potential difference between locations A and B; it is the electric field in the intervening region that determines ΔV.

For example, consider Ex. 16.18 on page 564. Although the electric field due to a charged spherical shell is zero at the center of the shell, the potential at that location is not zero!

16.9 Energy density associated with electric field

Until now we have thought of energy (that is, electric potential energy) as associated with interacting charged particles. There is an alternative view, however, which considers energy to be stored in electric fields themselves. To see how this view works out quantitatively, we'll consider moving one plate of a capacitor.

The force that one capacitor plate exerts on the other (Figure 16.40) is equal to the charge Q on one plate times the field made by the other plate, which is half the total field in the gap:

$$E_{\text{one plate}} = \frac{(Q/A)}{2\varepsilon_0} \quad \text{(for very small gap)}$$

So the force on one plate is $F = Q\dfrac{(Q/A)}{2\varepsilon_0}$

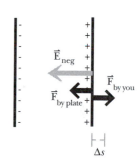

Figure 16.40 To move a capacitor plate you must exert a force equal to the force on the plate by the other plate.

Suppose you pull the positive plate of the capacitor slowly away from the negative plate, exerting a force only infinitesimally larger than the force exerted by the other plate. The work that you do in moving the plate a distance Δs goes into increasing the electric potential energy U:

$$\Delta U = W = Q\left(\frac{Q/A}{2\varepsilon_0}\right)\Delta s$$

We can rearrange this expression in the following way:

$$\Delta U = \tfrac{1}{2}\varepsilon_0\left(\frac{Q/A}{\varepsilon_0}\right)^2 A\,\Delta s$$

The expression in parentheses is the electric field E inside the capacitor. The quantity $A\,\Delta s$ is the change in the volume occupied by electric field inside the capacitor (length times width times increase in the plate separation). Therefore we can write

$$\frac{\Delta U}{\Delta(\text{volume})} = \tfrac{1}{2}\varepsilon_0 E^2$$

We ascribe an *energy density* (joules per m^3) to the electric field. By pulling the capacitor plates apart we increased the volume of space in which there is a sizable electric field. We say that the energy expended by us was converted into energy stored *in the electric field.*

This view is numerically equivalent to our earlier view of electric potential energy, but it turns out to be a more fundamental view. For example, the energy carried by electromagnetic radiation propagating through space far from any charges can best be expressed in terms of the energy density associated with the field.

FIELD ENERGY DENSITY

$$\tfrac{1}{2}\varepsilon_0 E^2 \quad (\text{J}/\text{m}^3)$$

Although we derived the field energy density for the particular case of a capacitor, the result is general. Anywhere there is electric field, there is energy density given by this formula.

16.9.1 An electron and a positron

In Chapter 4 we stated the principle of conservation of energy in this way: The change in energy of a system plus the change in energy of its surroundings must be zero. In the following example we will see that in order to understand energy conservation in quite a simple situation it is necessary to invoke the idea that energy is stored in fields.

Consider an electron and a positron which are released from rest some distance from each other. We will take the electron to be the system under consideration, so therefore the positron and everything else in the Universe are the "surroundings."

Because of the attractive electric force between the particles, the electron accelerates toward the positron, gaining kinetic energy. By the principle of conservation of energy, the energy of the surroundings must therefore decrease. (Remember that the system can have no potential energy, since it consists of a single particle, and potential energy is a property of pairs of particles.)

? Does the energy of the positron decrease?

No, the energy of the positron also increases, since it accelerates toward the electron, gaining kinetic energy.

? Where is there a decrease of energy in the surroundings?

Evidently the energy stored in the fields surrounding the two particles must decrease. Clearly, the electric field at any location in space does change as the positions of the particles change. The electric field in the region between the particles gets larger, but the electric field everywhere else in space decreases (since E_{dipole} is proportional to s, the distance between the particles). It would be a somewhat daunting task to integrate E^2 over the volume of the Universe, with the additional complication that close to a charged particle E approaches infinity. However, we do not actually need to do this integral to figure out the change in energy of the electric field throughout space. Since

$$\Delta(\text{Field energy}) + \Delta K_{\text{positron}} + \Delta K_{\text{electron}} = 0$$

then $\Delta(\text{Field energy}) = -2(\Delta K_{\text{electron}})$

In this example, the principle of conservation of energy leads us directly to the idea that energy must be stored in electric fields, since there is no other way to account for the decrease of energy in the surroundings.

Figure 16.41 An electron and a positron are released from rest, some distance apart. We choose the electron as the system, and everything else (including the positron) as the surroundings.

If we had chosen the electron plus the positron as our system, we would have found that ΔU_{el} is equal to $-2(\Delta K_{\mathrm{electron}})$. The change in potential energy for the two-particle system is the same as the change in the field energy. Evidently in a multiparticle system we can either consider a change in potential energy or a change in field energy (but not both); the quantities are equal.

The idea of energy stored in fields is a general one. It is not only electric fields that carry energy, but magnetic fields and gravitational fields as well.

Ex. 16.19 The energy density inside a certain capacitor is $10\,\mathrm{J/m^3}$. What is the magnitude of the electric field inside the capacitor?

Ex. 16.20 What is the energy density associated with an electric field of 3×10^6 V/m (large enough to initiate a spark)?

16.10 Potential difference in an insulator

Reasoning from the definition of static equilibrium and the existence of a sea of mobile electrons in a metal, we were able to conclude that in static equilibrium the net electric field everywhere inside a metal is zero. Given this, we were able to conclude that the potential difference between any two locations inside a metal object in static equilibrium must be zero.

The situation inside a polarized insulator is more complex. An applied field, such as the uniform field inside a a capacitor (Figure 16.42), polarizes the molecules in the insulator. These polarized molecules themselves contribute to the net field inside the material. Presumably, however, the electric field inside the plastic due to the polarized molecules varies depending on the observation location.

For example, consider locations A and B inside the polarized plastic shown in Figure 16.43, where we show electric field contributed solely by the induced dipoles in the polarized plastic. If we consider columns of polarized molecules to be similar to capacitors (consisting of two vertical sheets of charge), then location A is inside "capacitor" 1, while location B is between "capacitors" 2 and 3. At location A the dominant contribution by the dipoles will be from the molecules in "capacitor" 1, and the electric field due to the dipoles will be large and to the left.

At location B, the fringe fields of the nearest two "capacitors" represent the dominant contributions by the dipoles, and the field due to the dipoles will point to the right, and be smaller than the field due to the dipoles at A. If we follow a path from the left side of the plastic to the right side, $\vec{E}$ due to the dipoles will sometimes point in the direction of $d\vec{l}$ and sometimes opposite to the direction of $d\vec{l}$.

It would be useful to consider an average electric field due to the polarized molecules inside the plastic, but it isn't clear how we would calculate such a field. It is not even clear from simply inspecting the situation what the direction of this average field will be. We can construct a surprisingly simple argument based on potential to help us answer this question.

Round-trip potential difference

Although the pattern of E_{dipoles} is complex inside the material, it is much less complex at locations outside the plastic. Intuitively, you may be able to see that at locations outside the plastic the electric field due to the polarized plastic will have a similar pattern to the electric field near a single dipole.

In Figure 16.44 the electric field due just to the polarized plastic is shown at locations along a path outside the plastic. If we travel clockwise around

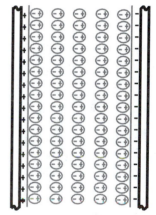

Figure 16.42 Polarization of molecules in a piece of plastic inside a capacitor.

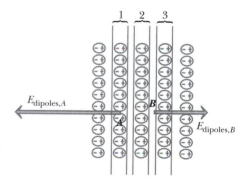

Figure 16.43 Electric field contributed solely by the induced dipoles in the plastic.

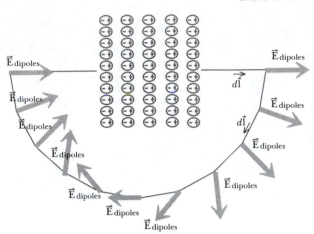

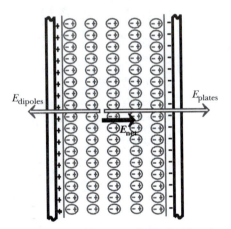

Figure 16.44 Electric field due solely to the induced dipoles in the plastic, along a path outside the plastic.

Figure 16.45 The net field inside the polarized insulator is smaller in magnitude than the applied field.

the path, it is clear from the diagram that $\vec{E} \bullet d\vec{l}$ is always a positive quantity: there is never a component of $\vec{E}$ opposite to $d\vec{l}$.

Since the molecular dipoles consist of stationary point charges, the round-trip path integral of the electric field (the round-trip potential difference) due to the molecular dipoles must sum to zero.

? Explain why this implies that the average field inside the plastic must point to the left and cannot point to the right.

If the average field inside the plastic pointed to the right, the round-trip path integral of electric field in Figure 16.44 would be nonzero, which is impossible. So a very general argument concerning the nature of potential lets us deduce the direction of the average field inside the plastic.

16.10.1 Dielectric constant and net field

The net electric field inside the plastic is the sum of the field due to the capacitor plates and the field due to the induced dipoles in the plastic, which, as we have just demonstrated, points in a direction opposite to the capacitor's field (Figure 16.45). The net field inside the plastic is therefore smaller than the field due to the capacitor alone. We can say informally that the electric field is "weakened" inside an insulator. Note carefully that the *net* field is in the same direction as the field made by the plates, but smaller in magnitude.

We define a constant K, called the "dielectric constant" (dielectric is another word for insulator), as the factor by which the net electric field is weakened inside an insulator:

$$\text{Inside an insulator } \vec{E}_{\text{net}} = \frac{\vec{E}_{\text{applied}}}{K}$$

For a capacitor, the applied field is $(Q/A)/\varepsilon_0$ and the net field inside a capacitor which is filled with an insulator is $(Q/A)/\varepsilon_0/K$. The dielectric constant is related to the atomic polarizability, but the details are beyond the scope of this introductory course. (The difficulty in making the connection is in accounting for the electric fields produced by the other polarized molecules.)

Note that the dielectric constant K is always bigger than 1, since polarization always weakens the net electric field inside the insulator. The easier it is to polarize a molecule, the bigger is the field-weakening effect and the bigger the value of K.

Here are representative values of the dielectric constant K for various insulators:

vacuum	1 (by definition)
air	1.0006 (approximately like vacuum)
carbon tetrachloride	2.2
typical plastic	5
sodium chloride	6.1
water	80 at 20° C
strontium titanate	310 (huge!)

If there is no insulator in the gap, the potential difference across the capacitor is $\Delta V = Es$ (since $\vec{E}$ is uniform, and parallel to $d\vec{r}$). If the insulator fills the gap and we maintain the same charges $+Q$ and $-Q$ on the metal plates, both the electric field E and the potential difference ΔV are reduced by the same factor K:

$$\Delta V_{\text{insulator}} = \frac{\Delta V_{\text{vacuum}}}{K}$$

Since we can't actually get inside the insulator and measure the field there, we determine K for a particular insulator by measuring the effect on the potential difference ΔV between the plates for a fixed capacitor charge Q.

In summary, placing an insulator between the plates of a capacitor

- decreases the electric field inside the insulator
- decreases the potential difference across the insulator

If the insulator doesn't fill the gap, the electric field inside the insulator is reduced by the factor K, but the electric field at other places in the gap is hardly affected, because the electric field of the insulator is small outside the insulator. Outside the gap, however, the fringe field of the plates is small, and the small field of the plastic further reduces the fringe field.

In Section 16.4.6 starting on page 557 we examined the electric field and potential difference between the two metal plates of a capacitor, then inserted a metal slab. In the following example we will start with the same device, but insert a glass slab.

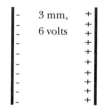

Ex. 16.21 A capacitor with a 3 mm gap has a potential difference of 6 volts (Figure 16.46). A disk of glass 1 mm thick, with area the same as the area of the metal plates, has a dielectric constant of 2.5. It is inserted in the middle of the gap between the metal plates. Now what is the potential difference of the two metal disks? (It helps to make a diagram showing the electric field along a path.)

Figure 16.46 Capacitor before inserting a glass disk 1 mm in thickness between the plates.

16.11 *Integrating the spherical shell

In this advanced section, we'll use potential to prove that a uniformly charged spherical shell looks from the outside like a point charge but on the inside has a zero electric field. We did this by integrating the electric field directly in the previous chapter, and a very different kind of proof using Gauss's law is given in a later chapter.

Divide the spherical shell into rings of charge, each delimited by the angle θ and the angle $\theta+\Delta\theta$, and carrying an amount of charge ΔQ (Figure 16.47). The angle θ will be the integration variable used in summing up the contributions of the various rings.

Each ring contributes V_{ring} at an observation point a distance r from the center of the spherical shell. The center of each ring is a distance $(r - R\cos\theta)$ from the observation location (we will eventually choose $\Delta\theta$ so small that it makes no difference whether we measure to the edge of the ring or to the center of the ring). Therefore the distance d from the observation location to each charge on the ring is $\sqrt{[(r - R\cos\theta)^2 + (R\sin\theta)^2]}$:

$$V_{\text{ring}} = \frac{1}{4\pi\varepsilon_0} \frac{\Delta Q}{[(r - R\cos\theta)^2 + (R\sin\theta)^2]^{1/2}}$$

The total amount of charge on each ring is

$$\Delta Q = \left(\frac{Q}{4\pi R^2}\right)\Delta A$$

where ΔA is the surface area of the ring, since the total charge Q is spread uniformly all over the spherical surface, whose total area is $4\pi R^2$. To calculate the surface area ΔA of the ring, as before we lay the ring out flat (Figure 16.48), noting that the radius of the ring is $R\sin\theta$ and its width is $R\Delta\theta$ (since arc length is radius times angle, with angle measured in radians).

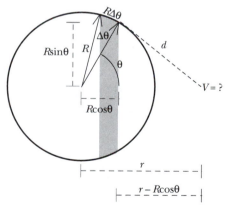

Figure 16.47 The sphere may be divided into ring-shaped segments.

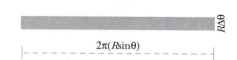

Figure 16.48 We can calculate the area of the ring of charge by "unrolling" it.

Putting these elements together, we have for the potential due to the ring

$$V_{ring} = \frac{1}{4\pi\varepsilon_0}\frac{1}{[(r-R\cos\theta)^2 + (R\sin\theta)^2]^{1/2}}\left(\frac{Q}{4\pi R^2}\right)2\pi(R\sin\theta)R\Delta\theta$$

We can add up all these contributions in the form of a definite integral:

$$V = \frac{1}{4\pi\varepsilon_0}\frac{Q}{2}\int_0^\pi \frac{\sin\theta\, d\theta}{[(r-R\cos\theta)^2 + (R\sin\theta)^2]^{1/2}}$$

The limits on the integral are determined by the fact that if we let θ range between 0 and π radians (180 degrees), we add up rings that account for the entire surface of the spherical shell.

A change of variables lets us evaluate this integral. Let

$$u = (r-R\cos\theta)^2 + (R\sin\theta)^2 = r^2 - 2rR\cos\theta + (R\cos\theta)^2 + (R\sin\theta)^2$$

But $(R\cos\theta)^2 + (R\sin\theta)^2 = R^2$, from the Pythagorean theorem. Therefore

$$u = r^2 - 2rR\cos\theta + R^2$$

$$du = 2rR\sin\theta\, d\theta \quad \text{and} \quad \sin\theta\, d\theta = \frac{1}{2rR}du$$

$$V = \frac{1}{4\pi\varepsilon_0}\frac{Q}{2}\frac{1}{2rR}\int_{\theta=0}^{\theta=\pi}\frac{du}{u^{1/2}}$$

$$V = \frac{1}{4\pi\varepsilon_0}\frac{Q}{2}\frac{1}{2rR}[2\sqrt{(r-R\cos\theta)^2 + (R\sin\theta)^2}]_0^\pi$$

$$V = \frac{1}{4\pi\varepsilon_0}\frac{Q}{2rR}[\sqrt{(r+R)^2} - \sqrt{(r-R)^2}]$$

$$V = \frac{1}{4\pi\varepsilon_0}\frac{Q}{2rR}[(r+R)\mp(r-R)]$$

This gives two results, depending on which sign is taken for the square root. Take the "−" sign, which turns out to correspond to being outside the shell:

$$V = \frac{1}{4\pi\varepsilon_0}\frac{Q}{r} \quad \text{(outside the shell; } r > R)$$

This says that the potential outside a uniformly charged spherical shell is exactly the same as the potential due to a point charge located at the center of the shell, as though the shell were collapsed to a point. Since the electric field is the negative gradient of the potential, the electric field outside the shell is the same as if the shell were collapsed to its center.

If we take the "+" sign, we have

$$V = \frac{1}{4\pi\varepsilon_0}\frac{Q}{R} \quad \text{(inside the shell; } r < R)$$

This says that the potential inside the shell is constant, and equal to the potential just outside the shell. A constant potential means that the electric field inside the shell is zero. Note that for locations inside the spherical shell, some of the rings give a field to the left, and some give a field to the right. It turns out that these contributions exactly cancel each other.

It is the $1/r^2$ behavior of the electric field of point charges that leads to these unusual results. If the electric field of a point charge were not exactly proportional to $1/r^2$, the electric field outside a uniform spherical shell would not look like the field of a point charge at the center, and the electric field would not be zero inside the shell.

16.12 Summary

Fundamental concepts

ELECTRIC POTENTIAL

Electric potential $V = \dfrac{U_{el}}{q}$, units are J/C or volts

POTENTIAL OF A POINT CHARGE

$$V = \frac{1}{4\pi\varepsilon_0}\frac{q}{r}$$

FIELD ENERGY DENSITY

$$\tfrac{1}{2}\varepsilon_0 E^2 \quad (\text{J}/\text{m}^3)$$

Results

$$\Delta V = V_f - V_i = -\int_i^f \vec{\mathbf{E}} \bullet d\vec{\mathbf{l}}$$

$\vec{\mathbf{E}}$ is gradient of V: $E_x = -\dfrac{\partial V}{\partial x}$, $E_y = -\dfrac{\partial V}{\partial y}$, $E_z = -\dfrac{\partial V}{\partial z}$

$\Delta U_e = q\Delta V$

Electric potential properties:

V is the superposition of V's due to individual charges
Independent of the path taken from i to f
$\Delta V = 0$ around a closed path ($a \rightarrow b \rightarrow c \rightarrow d \rightarrow ...... \rightarrow a$)
Electric field unchanged if we shift all the V's:
$$(V_f + V_0) - (V_i + V_0) = V_f - V_i$$
$\vec{\mathbf{E}}$ points toward *lower* potential

Metal in static equilibrium: $\Delta V = -\int_i^f \vec{\mathbf{E}} \bullet d\vec{\mathbf{l}} = 0$ along *any* path,

so potential is *uniform* (constant everywhere)
throughout a metal in static equilibrium

Stationary point charges cannot produce a "curly" field (Figure 16.49).

Dielectric constant

Inside an insulator $\vec{\mathbf{E}}_{net} = \dfrac{\vec{\mathbf{E}}_{applied}}{K}$; K is "dielectric constant"; $K > 1$

so $\Delta V_{insulator} = \dfrac{\Delta V_{vacuum}}{K}$

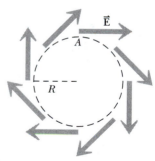

Figure 16.49 An impossible pattern of electric field made by stationary point charges.

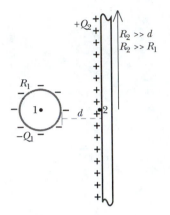

Figure 16.50 A uniformly charged plastic shell and a uniformly charged glass disk.

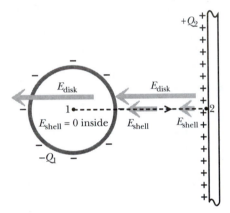

Figure 16.51 Electric field contributions of the shell and the disk.

16.13 Example problem: A disk and a spherical shell

A thin spherical shell made of plastic carries a uniformly distributed negative charge $-Q_1$. A thin circular disk made of glass carries a uniformly distributed positive charge $+Q_2$. The radius R_1 of the plastic spherical shell is very small compared to the large radius R_2 of the glass disk. The distance from the surface of the spherical shell to the plastic sheet is d, and d is much smaller than R_2. Find the potential difference $V_2 - V_1$. Location 1 is at the center of the plastic sphere, and location 2 is just outside the circular glass sheet. State what approximations or simplifying assumptions you make.

Solution

Fundamental principle: Superposition. We can find the potential difference due to the spherical shell and the potential difference due to the disk, and add these two contributions (or we could add the two electric field contributions, then integrate $\vec{E}_{net} \cdot d\vec{l}$ along a path going from 1 to 2).

Path: A straight line path from 1 to 2 is simplest. The path and the electric field contributions at various locations along the path are shown in Figure 16.51.

Simplifying assumption: We can neglect the polarization of the plastic and glass, because there is little matter in the thin shell and thin disk, so the field of the polarized molecules is negligible compared to the contributions of $-Q_1$ and $+Q_2$.

ΔV due to shell:

$$V_{\text{surface of shell}} - V_1 = 0 \text{ because } E_{\text{shell}} = 0 \text{ inside shell.}$$

$$\text{Outside the shell } \vec{E}_{\text{shell}} = \frac{1}{4\pi\varepsilon_0}\frac{-Q_1}{r^2}\hat{r}, \text{ so}$$

$$V_2 - V_{\text{surface of shell}} = \Delta\!\left(\frac{1}{4\pi\varepsilon_0}\frac{-Q_1}{r}\right) = \frac{1}{4\pi\varepsilon_0}\left(\frac{-Q_1}{R_1 + d} - \frac{-Q_1}{R_1}\right)$$

Check sign: Moving opposite to the field, so potential should increase. Result agrees, since $+Q_1/R_1$ is the larger term.

ΔV due to disk:

$$\text{Approximation: since } d \ll R_2 \text{ and } R_1 \ll R_2, \ E_{\text{disk}} \approx \frac{Q_2/(\pi R_2^2)}{2\varepsilon_0} \text{ (to the left)}$$

$$\text{So } V_2 - V_1 = -\int_i^f \vec{E}\cdot d\vec{l} \approx +\frac{Q_2/(\pi R_2^2)}{2\varepsilon_0}(R_1 + d)$$

Check sign: Moving opposite to the field, so potential should increase. Result agrees.

ΔV due to both shell and disk:

$$V_2 - V_1 = \frac{1}{4\pi\varepsilon_0}\left(\frac{-Q_1}{R_1 + d} - \frac{-Q_1}{R_1}\right) + \frac{Q_2/(\pi R_2^2)}{2\varepsilon_0}(R_1 + d)$$

16.14 Review questions

Terminology and units

RQ 16.1 What is the difference between electric potential energy and electric potential?

RQ 16.2 What are the units of electric potential energy, of electric potential, and of electric field?

Potential gradient

RQ 16.3 Two capacitors have the same size of plates and the same distance (2 mm) between plates. The potentials of the two plates in capacitor 1 are −10 volts and +10 volts. The potentials of the two plates in capacitor 2 are 350 volts and 370 volts. What is the electric field inside capacitor 1? Inside capacitor 2?

Potential at a location

RQ 16.4 A particle with charge $+q_1$ and a particle with charge $-q_2$ are located as shown in Figure 16.52. What is the potential at location A?

RQ 16.5 Four point charges, q_1, q_2, q_3, and q_4, are located at the corners of a square with side d, as shown in Figure 16.53. What is the potential at location A, at the center of the square? What is the potential at location B, in the middle of one side,?

RQ 16.6 What is the potential at location B, a distance h from a ring of radius a with charge $-Q$, as shown in Figure 16.54?

RQ 16.7 A student said "The electric field at the center of a charged spherical shell is zero, so the potential at that location is also zero." What is wrong with this statement?

Potential vs. potential difference

RQ 16.8 For each of the following statements, say whether it is true or false and explain why it is true or false. Be complete in your explanation, but be brief. **Pay particular attention to the distinction between potential V and potential difference ΔV.**
   (a) "The electric potential inside a metal in static equilibrium is always zero."
   (b) "If there is a constant large positive potential throughout a region, the electric field in that region is large."
   (c) "If you get close enough to a negative point charge, the potential is negative, no matter what other charges are around."
   (d) "Near a point charge, the potential difference between two points a distance L apart is $-EL$."
   (e) "In a region where the electric field is varying, the potential difference between two points a distance L apart is $-(E_f - E_i)L$."

Dielectric constant

RQ 16.9 We discussed a method for measuring the dielectric constant by placing a slab of the material between the plates of a capacitor. Using this method, what would we get for the dielectric constant if we inserted a slab of metal (not quite touching the plates, of course)?

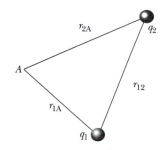

Figure 16.52 What is the potential at location A? (RQ 16.4)

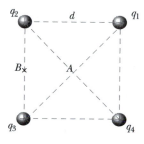

Figure 16.53 What is the potential at locations A and B? (RQ 16.5)

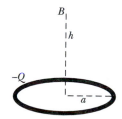

Figure 16.54 What is the potential at location B, a distance R from a negatively charged ring? (RQ 16.6)

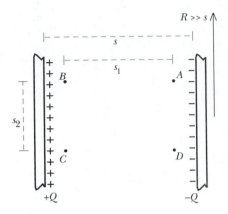

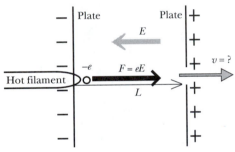

Figure 16.55 Different paths in a capacitor
(Problem 16.1).

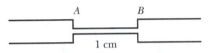

Figure 16.56 $V_B - V_A = 1.5$ volts (Problem
16.2).

Figure 16.57 A television tube (Problem
16.3).

16.15 Homework problems

Problem 16.1 Potential along different paths in a capacitor

A capacitor consists of two charged disks of radius R separated by a distance
s, where $R \gg s$ (Figure 16.55). The magnitude of the charge on each disk is
Q.

Consider points A, B, C, and D inside the capacitor, as shown in the diagram.

(a) Show that $\Delta V = V_C - V_A$ is the same for these paths by evaluating ΔV
along each path:

        Path 1: A - B - C

        Path 2: A - C

        Path 3: A - D - B - C

(b) If $Q = 43$ μC, $R = 4.0$ m, $s_1 = 1.5$ mm, and $s_2 = 0.7$ mm, what is the value
of $\Delta V = V_C - V_A$?

(c) Choose two different paths from point A back to point A again, and
show that $\Delta V = 0$ for a round trip along both of these paths.

Problem 16.2 Potential difference along a wire

The potential difference from one end of a 1-cm-long wire to the other in a
circuit is $\Delta V = V_B - V_A = 1.5$ volts, as shown in Figure 16.56.

Which end of the wire is at the higher potential? What are the magnitude
and direction of the electric field E inside the wire?

Problem 16.3 Electrons in a television picture tube

In a television picture tube electrons are boiled out of a very hot metal filament placed near a negative metal plate (Figure 16.57). These electrons
start out nearly at rest and are accelerated toward a positive metal plate.
They pass through a hole in the positive plate on their way toward the picture screen.

If the high-voltage supply in the television set maintains a potential difference of 15000 volts between the two plates, what speed do the electrons
reach?

Problem 16.4 Potential difference near a dipole

We know the magnitude of the electric field at a location on the x axis and
at a location on the y axis, if we are far from the dipole.

(a) Find $\Delta V = V_B - V_A$ along a line perpendicular to the axis of a dipole.
Do it two ways: from the superposition of V due to the two charges, and from
the integral of the electric field.

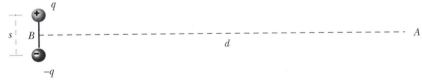

(b) Find $\Delta V = V_C - V_D$ along the axis of the dipole. Include the correct
sign. Do it two ways: from the superposition of V due to the two charges, and
from the integral of the electric field.

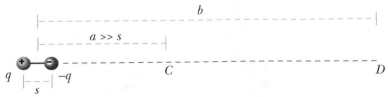

(c) What is the change in potential energy ΔU in moving an electron from
D to C?

Problem 16.5 Field and potential in a capacitor

Two very large disks of radius R are carrying uniformly distributed charges Q_A and Q_B. The plates are parallel and 0.1 millimeters apart (Figure 16.58). The potential difference between the plates is $V_B - V_A = -10$ volts.

(a) What is the direction of the electric field between the disks?

(b) Invent values of Q_A, Q_B, and R that would make $V_B - V_A = -10$ volts.

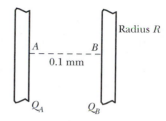

Figure 16.58 Field and potential in a capacitor (Problem 16.5).

Problem 16.6 Potential near a charged spherical shell

A uniform spherical shell of charge $+Q$ is centered at point B (Figure 16.59). Show that $\Delta V = V_C - V_A$ is independent of path by calculating ΔV for each of these two paths (actually do the integrals):

Path 1: A - B - C (along a straight line through the shell).

Path 2: A - D - C (along a circular arc around point B).

Problem 16.7 Field and potential of a dipole

A dipole is oriented along the x axis. The dipole moment is $p\ (= qs)$.

(a) Calculate exactly the potential V at a location $<x,0,0>$ on the x axis and at a location $<0,y,0>$ on the y axis, by superposition of the individual $1/r$ contributions to the potential.

(b) What are the approximate values of V at the locations in part (a) if these locations are far from the dipole?

(b) Using the approximate results of part (b), calculate the gradient of the potential along the x axis, and show that the negative gradient is equal to the magnitude of the electric field E_x.

(c) Along the y axis, $dV/dy = 0$. Why isn't this equal to the magnitude of the electric field E along the y axis?

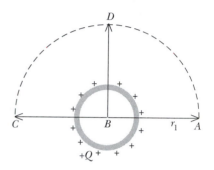

Figure 16.59 A charged spherical shell (Problem 16.6).

Problem 16.8 Potential difference in a capacitor

A capacitor consists of two large metal disks placed a distance s apart (Figure 16.60). The radius of each disk is R ($R \gg s$), and the thickness of each disk is t. The disk on the left has a net charge of $+Q$, and the disk on the right has a net charge of $-Q$. Calculate the potential difference $V_2 - V_1$, where location 1 is inside the left disk at its center, and location 2 is in the center of the air gap between the disks. Explain briefly.

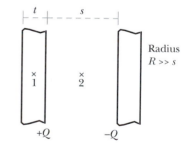

Figure 16.60 A capacitor (Problem 16.8).

Problem 16.9 Electric field due to a charged ring

Refer to Section 16.7.1 on page 563 for the potential along the axis of a ring of radius R carrying a charge Q. Let the axis of the ring be the z axis of the coordinate system, and determine E_z at any location z along the axis. (If the charge is nearly uniformly distributed around the ring, at these locations there is no E_x or E_y, due to the symmetry of the situation.) Compare with the result obtained in the previous chapter by integration.

Problem 16.10 Maximum possible potential in air

What is the maximum possible potential of a metal sphere of 10 cm radius in air? What is the maximum possible potential of a metal sphere of only 1 mm radius? These results hint at the reason why a highly charged piece of metal (with uniform potential throughout) tends to spark at places where the radius of curvature is small, or at places where there are sharp points. Remember that the breakdown electric field strength for air is roughly 3×10^6 V/m.

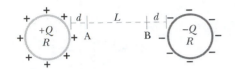

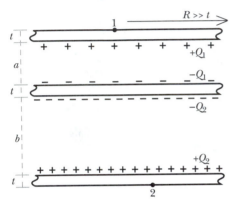

Figure 16.61 Two spherical shells (Problem 16.11).

Figure 16.62 Three charged disks (Problem 16.13).

Problem 16.11 Two spherical shells

A thin spherical glass shell of radius R carries a uniformly distributed charge $+Q$, and a thin spherical plastic shell of radius R carries a uniformly distributed charge $-Q$ (Figure 16.61). The surfaces of the spheres are a distance $L+2d$ from each other, and locations A and B are a distance d from the surfaces of the spheres. Calculate the potential difference $V_B - V_A$.

Problem 16.12 Solid glass ball

A small solid glass ball of radius r is irradiated by a beam of positive ions, and gains a charge $+q$ distributed uniformly throughout its volume. What is the potential at location A, inside the ball, a distance $r/2$ from the center of the ball?

Problem 16.13 Three disks

Figure 16.62 shows three very large metal disks (seen edgewise), carrying charges as indicated. On each surface the charges are distributed approximately uniformly. Each disk has a very large radius R and a small thickness t. The distances between the disks are a and b, as shown; they also are small compared to R. Calculate $V_2 - V_1$, and explain your calculation briefly.

Problem 16.14 Electron deflection in an oscilloscope

This problem deals with several aspects of an oscilloscope. You have an 18000-volt supply for accelerating electrons to a speed adequate to make the front phosphor-coated screen glow when the electrons hit it. Once the electron has emerged from the accelerating region, it coasts through a vacuum at nearly constant speed.

You can apply a potential difference of plus or minus 40 volts across the deflection plates to steer the electron beam up or down on the screen to paint a display (other deflection plates not shown in the diagram are used to steer the beam horizontally).

Each of the two deflection plates is a thin metal plate of length $L = 8$ cm and width (into the diagram) 4 cm. The distance between the deflection plates is s = 3 mm. The distance from the deflection plates to the screen is $d = 30$ cm.

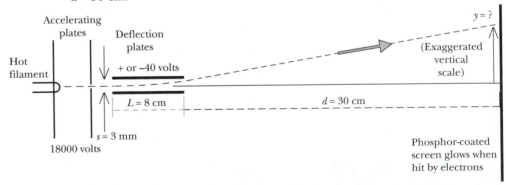

When there is a 40 volt potential difference between the deflection plates, what is the deflection y of the electron beam where it hits the screen? An approximate treatment is fine, but state your assumptions. As is usually the case, it pays to carry out all of your calculations algebraically and only evaluate the final algebraic result numerically. Note the exaggerated vertical scale: the deflection is actually small compared to the distance to the screen.

Problem 16.15 Charge transfer from one sphere to another

A small metal sphere of radius r initially has a charge q_0. Then a long copper wire is connected from this small sphere to a distant large uncharged metal sphere of radius R. Calculate the final charge q on the small sphere and the final charge Q on the large sphere. You may neglect the small amount of charge on the wire. What other approximations did you make? (Hint for this problem: think about potential...)

Problem 16.16 Bent charged rod
A rod uniformly charged with charge $+Q$ is bent into a semicircular arc of radius b, as shown in Figure 16.63. What is the potential at location A, at the center of the arc?

Problem 16.17 Potential difference near a rod
A long rod of length L carries a uniform charge $-Q$ (Figure 16.64). Calculate the potential difference $V_A - V_C$. All of the distances are small compared to L. Explain your work carefully.

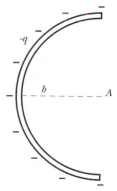

Figure 16.63 A charged rod bent into a semicircle (Problem 16.16).

Problem 16.18 Potential difference for nested metal spheres
A metal sphere of radius r_1 carries a positive charge of amount Q. A concentric spherical metal shell with inner radius r_2 and outer radius r_3 surrounds the inner sphere and carries a total positive charge of amount $4Q$, with some of this charge on the outer surface (at r_3) and some on the inner surface (at r_2).

(a) How is the charge of $4Q$ distributed on the two surfaces of the outer shell? Prove this!

(b) What is the potential just outside r_3, half-way between r_2 and r_3, just inside r_2, just outside r_1, and at the center?

Problem 16.19 Van de Graaff generator
In a Van de Graaff generator, a rubber belt carries electrons up through a small hole in a large hollow spherical metal shell (Figure 16.65). The electrons come off the upper part of the belt and drift through a wire to the outer surface of the metal shell, so that the metal shell acquires a sizable negative charge, approximately uniformly distributed over the sphere. At a time when the sphere has acquired a sizable charge $-Q$, approximately how much work must be done by the motor to move one more electron from the base (a distance h below the sphere) to the upper pulley (located a distance $R/2$ from the center of the hollow sphere)? Explain your work, and state explicitly what approximations you had to make.

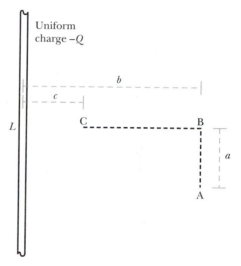

Figure 16.64 Potential difference near a charged rod (Problem 16.17).

Problem 16.20 Nuclear potential
A nucleus contains Z protons which on average are uniformly distributed throughout a tiny sphere of radius R.

(a) Calculate the potential at the center of the nucleus. Assume that there are no electrons or other charged particles in the vicinity of this bare nucleus.

(b) Suppose that in an accelerator experiment a positive pion is produced at rest at the center of a nucleus containing Z protons. The pion decays into a positive muon (essentially a heavy positron) and a neutrino. The muon has initial kinetic energy K_i. How much kinetic energy does the muon have by the time it has been repelled very far away from the nucleus? (The muon interacts with the nucleus only through Coulomb's law and is unaffected by nuclear forces. The massive nucleus hardly moves and gets negligible kinetic energy.)

(c) If the nucleus is gold, with 79 protons, what is the numerical value of $K_f - K_i$ in electron volts?

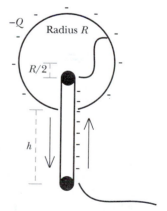

Figure 16.65 Van de Graaff generator (Problem 16.19).

Problem 16.21 Electric potential near an HCl molecule

An HCl molecule in the gas phase has an internuclear separation s (Figure 16.66). For the purposes of this problem, assume:

- The molecule can be considered as two point charges, H^+ and Cl^-, located at the centers of the nuclei (i.e., ignore polarization of one ion by the other, sharing of electrons, etc.);
- The molecule remains fixed in position;
- The distances a and b shown in the diagram are much greater than ($a > b >> s$).

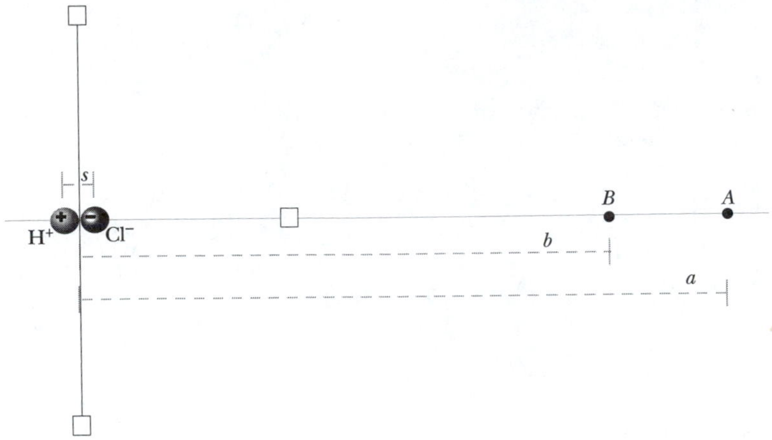

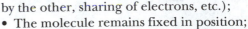

(a) Draw vectors showing the relative magnitude and direction of the electric field at each of the three points indicated by a square box. Your drawing should be accurate enough to show clearly whether one vector is equal in magnitude, greater in magnitude, or less in magnitude than another vector, but need not be more accurate than that.

Figure 16.66 An HCl molecule (Problem 16.21).

(b) What is the potential difference $\Delta V = V_B - V_A$, in terms of the given quantities?

(c) An electron passes point A traveling along the x axis in the negative x direction, toward the Cl^-. Its kinetic energy is K_A. What is its kinetic energy at point B?

Problem 16.22 A disk and a rod

A very long, thin glass rod of length $2R$ carries a uniformly distributed charge $+q$, as shown in Figure 16.67. A very large plastic disk of radius R, carrying a uniformly distributed charge $-Q$, is located a distance d from the rod, where $d << R$. Calculate the potential difference $V_B - V_A$ from the center of the surface of the disk (location A) to a location a distance h from the center of the disk (location B). If you have to make any approximations, state what they are.

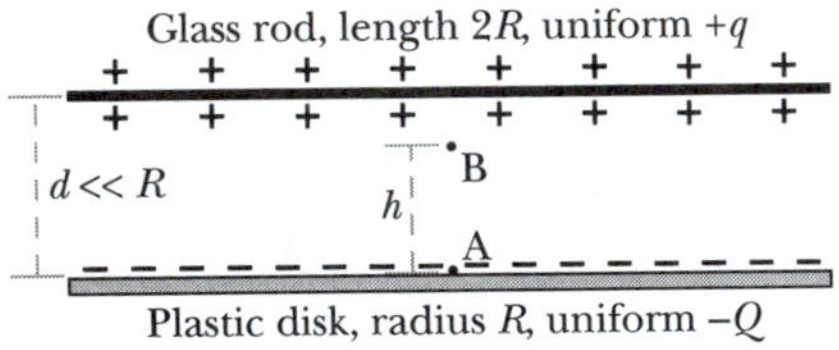

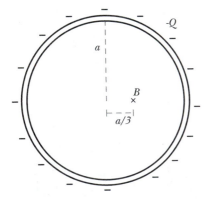

Figure 16.67 A disk and a rod (Problem 16.22).

Problem 16.23 Connect two capacitors together

An isolated parallel-plate capacitor of area A_1 with an air gap of length s_1 is charged up to a potential difference ΔV_1. A second parallel-plate capacitor, initially uncharged, has an area A_2 and a gap of length s_2 filled with plastic whose dielectric constant is K. You connect a wire from the positive plate of the first capacitor to one of the plates of the second capacitor, and you connect another wire from the negative plate of the first capacitor to the other plate of the second capacitor. What is the final potential difference across the first capacitor?

Problem 16.24 Inside a charged spherical shell

A thin plastic spherical shell of radius a is rubbed all over with wool, and gains a charge of $-Q$. What is the potential at location B, a distance $a/3$ from the center of the sphere, as shown in Figure 16.68?

Problem 16.25 A solid plastic sphere

A solid plastic sphere of radius R has charge $-Q$ distributed uniformly over its surface. It is far from all other objects.

(a) Sketch the molecular polarization in the interior of the sphere and explain briefly.

(b) Calculate the potential at the center of the sphere.

Figure 16.68 A spherical shell with charge $-Q$ (Problem 16.24).

Problem 16.26 Disks and a sphere

A thin spherical shell made of plastic carries a uniformly distributed negative charge $-Q_1$. Two large thin disks made of glass carry uniformly distributed positive and negative charges $+Q_2$ and $-Q_2$ (Figure 16.69). The radius R_1 of the plastic spherical shell is very small compared to the radius R_2 of the glass disks. The distance from the center of the spherical shell to the positive disk is d, and d is much smaller than R_2.

(a) Find the potential difference $V_2 - V_1$ in terms of the given quantities (Q_1, Q_2, R_1, R_2, and d). Point 1 is at the center of the plastic sphere, and point 2 is just outside the sphere.

(b) Find the potential difference $V_3 - V_2$. Point 2 is just below the sphere, and point 3 is right beside the positive glass disk.

(c) Suppose the plastic shell is replaced by a solid metal sphere with radius R_1 carrying charge $-Q_1$. State whether the absolute magnitudes of the potential differences would be greater than, less than, or the same as they were with the plastic shell in place. Explain briefly, including an appropriate diagram.

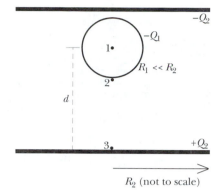

Figure 16.69 Disks and a sphere (Problem 16.26).

Problem 16.27 Insert a plastic slab into a capacitor

An isolated large-plate capacitor (not connected to anything) originally has a potential difference of 1000 volts with an air gap of 2 mm. Then a plastic slab 1 mm thick, with dielectric constant 5, is inserted into the middle of the air gap as shown in Figure 16.70. Calculate the following potential differences, and explain your work.

$$V_1 - V_2 = ? \qquad V_2 - V_3 = ?$$

$$V_3 - V_4 = ? \qquad V_1 - V_4 = ?$$

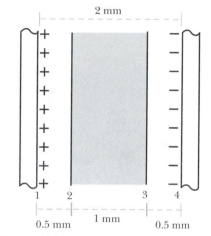

Figure 16.70 Plastic slab in a capacitor (Problem 16.27).

Problem 16.28 Pattern of electric field

Figure 16.71 shows a pattern of electric field in which the electric field is horizontal throughout this region but is larger toward the top of the region and smaller toward the bottom. If this pattern of electric field can be produced by some arrangement of stationary charges, sketch such a distribution of charges. If this pattern of electric field cannot be produced by any arrangement of stationary charges, prove that it is impossible.

Figure 16.71 Pattern of electric field for Problem 16.28.

Problem 16.29 Two spherical shells

A thin spherical shell of radius R_1 made of plastic carries a uniformly distributed negative charge $-Q_1$ (Figure 16.72). A thin spherical shell of radius R_2 made of glass carries a uniformly distributed positive charge $+Q_2$. The distance between centers is L.

(a) Find the potential difference $V_B - V_A$. Location A is at the center of the glass sphere, and location B is just outside the glass sphere.

(b) Find the potential difference $V_C - V_B$. Location B is just outside the glass sphere, and location C is a distance d to the right of B.

(c) Suppose the glass shell is replaced by a solid metal sphere with radius R_2 carrying charge $+Q_2$. Would the magnitude of the potential difference $V_B - V_A$ be greater than, less than, or the same as it was with the glass shell in place? Explain briefly, including an appropriate physics diagram.

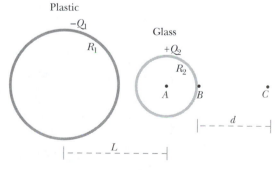

Figure 16.72 A plastic shell and a glass shell (Problem 16.29).

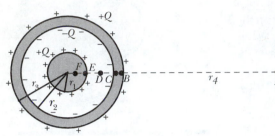

Figure 16.73 Nested metal spheres (Problem 16.30).

Problem 16.30 Nested metal spheres

A solid metal sphere of radius r_1 has a charge $+Q$ (Figure 16.73). It is surrounded by a concentric spherical metal shell with inner radius r_2 and outer radius r_3, that has a charge $-Q$ on its inner surface and $+Q$ on its outer surface. In the diagram, point A is located at a distance r_4 from the center of the spheres. Points B and C are inside the metal shell, very near the outer and inner surfaces respectively. Point E is just inside the surface of the solid sphere. Point D is halfway between C and E. Point F is a distance $r_1/2$ from the center.

(a) Is each of the following potential differences > 0, $= 0$, or < 0? Briefly explain why in terms of electric field.

$V_B - V_A$; $V_C - V_B$; $V_D - V_C$; $V_F - V_E$

(b) Calculate the potential at location F, V_F. Explain your work.

Problem 16.31 A Geiger tube

A long thin metal wire with radius r and length L is surrounded by a concentric long narrow metal tube of radius R, where $R \ll L$ (Figure 16.74). Insulating spokes hold the wire in the center of the tube and prevent electrical contact between the wire and the tube. A variable power supply is connected to the device as shown. There is a charge $+Q$ on the inner wire and a charge $-Q$ on the outer tube. As we will see when we study Gauss's law in a later chapter, the electric field inside the tube is contributed solely by the wire; the outer tube does not contribute as long as we are not near the ends of the tube.

(a) In terms of the charge Q, length L, inner radius r, and outer radius R, what is the potential difference $V_{tube} - V_{wire}$ between the inner wire and the outer tube? Explain, and include checks on your answer.

(b) The power-supply voltage is slowly increased until you see a glow in the air very near the inner wire. Calculate this power-supply voltage (give a numerical value), and explain your calculation. The length $L = 80$ cm, the inner radius $r = 0.7$ mm, and the outer radius $R = 3$ cm.

This device is called a "Geiger tube" and was one of the first electronic particle detectors. The voltage is set just below the threshold for making the air glow near the wire (part b). A charged particle that passes near the center wire can trigger breakdown in the air, leading to a large current that can be easily measured.

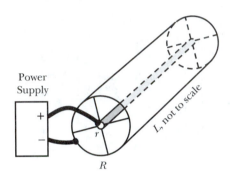

Figure 16.74 A Geiger tube (Problem 16.31).

16.16 Answers to exercises

16.1 (page 551):　　-7.2 volts

16.2 (page 551):　　$+1.15 \times 10^{-18}$ J ; yes, it takes energy to move the electron farther away from the proton

16.3 (page 551):　　56 volts

16.4 (page 552):　　$E_z = -200$ V/m (or -200 N/C)

16.5 (page 552):　　$(-3z, -5, 8 - 3x)$

16.6 (page 554):　　-1500 volts

16.7 (page 555):　　9000 volts

16.8 (page 555):　　15000 volts/m

16.9 (page 555):　　35 eV ; also 35 eV

16.10 (page 556):　　To the left; positive

16.11 (page 557):　　To the right; negative

16.12 (page 557):　　Toward the sphere; negative

16.13 (page 558):　　Zero field in metal slab, so there is no longer a contribution to the potential difference along that part of the path.

16.14 (page 558):　　Left edge of slab: 2 volts
Middle of slab: 2 volts
Right edge of slab: 2 volts

16.15 (page 558):　　$\dfrac{1}{4\pi\varepsilon_0}\dfrac{Q}{R}$; $\dfrac{1}{4\pi\varepsilon_0}\dfrac{Q}{R}$; the same throughout the metal, and equal to the potential just outside the surface

16.16 (page 561):　　$\Delta V_1 = \dfrac{1}{4\pi\varepsilon_0}\left(\dfrac{q}{r_2} - \dfrac{q}{r_1}\right),\ \Delta V_2 = 0$

$\Delta V_3 = \dfrac{1}{4\pi\varepsilon_0}\left(\dfrac{q}{r_1} - \dfrac{q}{r_2}\right),\ \Delta V_4 = 0$

$\Delta V_{\text{round trip}} = 0$

16.17 (page 563):　　If $z \gg R$, $\sqrt{z^2 + R^2} \approx z$, and $V \approx \dfrac{1}{4\pi\varepsilon_0}\dfrac{Q}{z}$

16.18 (page 564):　　$\dfrac{1}{4\pi\varepsilon_0}\dfrac{Q}{R}$

16.19 (page 567)　　1.5×10^6 V/m

16.20 (page 567)　　40 J/m^3

16.21 (page 569):　　4.8 volts

Chapter 17

Magnetic Field

Chapter 17

Magnetic Field

Electric fields are not the only kind of field associated with charged particles. When a compass needle turns and points in a particular direction, we say that there is a "magnetic field" pointing in that direction, which forces the needle to line up with it (Figure 17.1). Initially we'll simply define magnetic field as "whatever it is that is detected by a compass," and in accordance with convention, we will use the symbol *"B"* to refer to magnetic field. The twist of a compass needle is an indicator of magnetic fields, just as the twist of a suspended electric dipole is an indicator of electric fields (Section 14.2.3). Magnetic fields are made by moving charges, so we will need to have some moving charges at hand. We will assemble some simple electric circuits in which currents can run steadily, providing a convenient source of moving charges. In this chapter we will study magnetic field, and we will also develop an atomic-level description of magnets.

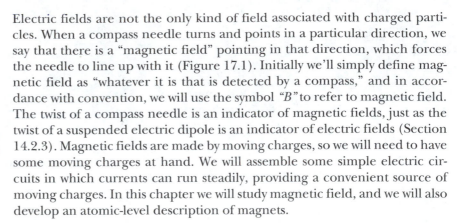

Figure 17.1 A compass needle points in the direction of the net magnetic field at its location.

Equipment

In the study of magnetic field you will need the following equipment:
- two flashlight batteries in a battery holder
- light bulbs of two kinds
- screw-in bulb sockets
- several short copper wires with clips on the ends ("connecting wires")
- a long wire (about 2 m in length)
- a liquid-filled magnetic compass
- unmagnetized nails for an experiment on page 610

17.1 Electron current

Magnetic fields are made by moving charges. A current in a wire provides a convenient source of moving charges, and allows us to experiment with producing and detecting magnetic fields.

In static equilibrium, there is no net motion of the sea of mobile electrons inside a metal. In the electric circuits we will construct in this chapter, we can arrange things so the electron sea does keep moving continuously. This continuous flow of electrons is called an "electric current," and is an indication that the system is not in static equilibrium. In order to be able to talk about what things affect electric current, we need a precise definition:

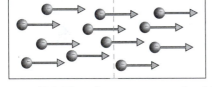

Figure 17.2 The electron current in this conductor is the number of electrons per second passing the dashed line.

DEFINITION OF ELECTRON CURRENT

The electron current is the number of electrons per
second that enter a section of a conductor.

We will use the symbol *i* for electron current.

If we could count the number of electrons per second passing a particular point in a circuit (Figure 17.2), we could measure electron current directly. This is difficult to do, so we use indirect measurements to determine the magnitude of electron current in a wire. One such indirect measurement involves measuring the magnetic field created by the moving electrons.

As electrons drift through a wire, they collide with the atomic cores, and this "friction" heats the wire (and prevents the electrons from going faster and faster). Both the heating and the magnetic effects are proportional to the "electron current"—the number of electrons that enter the wire every second.

Ex. 17.1 If 1.8×10^{16} electrons enter a light bulb in 3 milliseconds, what is the magnitude of the electron current at that point in the circuit?

Ex. 17.2 If the electron current at a particular location in a circuit is 9×10^{18} electrons/s, how many electrons pass that point in 10 minutes?

17.1.1 Simple circuits

To observe the magnetic effects of electric currents, we will construct simple circuits containing wires, light bulbs, and batteries. These are the simplest examples of systems in which we can observe the fundamental electric and magnetic properties of continuous electric currents. In our first activities we will examine our equipment systematically, and observe how it behaves under different conditions. As you do these experiments keep in mind the definition of electron current.

Light bulbs and sockets

The filament of a light bulb (the very thin metal wire that glows) is made of tungsten, a metal that does not melt until reaching a very high temperature. A glowing tungsten wire would rapidly oxidize and burn up in air, so there is a vacuum or an inert gas such as argon inside the bulb.

The thin tungsten filament in the bulb strongly resists the passage of electrons. When the electron sea is forced to move through the tungsten, the mobile electrons collide with the positive cores (nuclei plus inner electrons), and this "friction" makes the metal get hot and glow.

Figure 17.3 A battery and a round bulb.

Figure 17.4 Cutaway sketch of a light bulb.

Experiment 17.1 Experiments with simple circuits

(a) Using one battery, some connecting wires (the insulated wires with clips on the ends, not the bare Nichrome wire), and a round bulb (#14) *but no socket*, make the bulb light up (Figure 17.3). If the bulb glows with a steady light, this is a "steady state." Make a diagram showing how you connected the circuit.

(b) Examine a light bulb carefully, and imagine slicing the bulb in half lengthwise. A cutaway sketch is shown in Figure 17.4. On a copy of the sketch, label the important parts and connections, indicating which parts of the bulb you think are metal conductors, and which parts are insulators. Using a different color if available, show the conducting path that electric current follows through the bulb. What happens if you switch the connections to the battery? Does the light bulb still light?

(c) Closely examine the long #48 bulb and the round #14 bulb (Figure 17.5). The tungsten filaments in both bulbs are about the same length, but perhaps you can see even with the naked eye that the filament in the long bulb is extremely thin—thinner than the filament in the round bulb. Through which bulb would you guess it would be easier to push electric current, through a thin filament (as in the long bulb) or through a thick filament (as in the round bulb)? Why? (At this point this is mostly just a guess, but we'll study this in detail later.)

(d) Examine a bulb *socket* (the receptacle into which a bulb is screwed), and imagine slicing the socket in half lengthwise. Make a cutaway sketch,

Figure 17.5 A long #48 bulb (left) and a round #14 bulb (right).

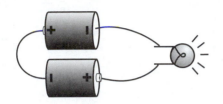

Figure 17.6 Two batteries and a round bulb in series. The socket is not shown.

and label the important parts and connections. Indicate which parts are metal conductors and which parts are insulators. On your sketch, trace the path (in a different color if available) along which electrons will move through the socket when there is a bulb in it. Screw a bulb into the socket and connect the socket to the battery with two connecting wires. Make sure the bulb lights.

(e) Connect a round bulb in a socket to two batteries "in series" (that is, one after the other) using connecting wires as shown in Figure 17.6. To connect two batteries in series, put them in the battery holder in opposite directions, and connect them as shown (note that "+" is connected to "–"). Compare the brightness of the bulb with one battery and with two batteries in series. Because of this difference we'll usually use two batteries in series in our experiments.

17.2 Detecting magnetic fields

We will use a magnetic compass as a detector of magnetic fields. Just as the deflection of a charged piece of invisible tape or the twisting of a permanent electric dipole indicates the presence of an electric field, the twisting of a compass needle indicates the presence of a magnetic field.

? Observe the interactions of your compass with nearby objects. How can you be sure that a compass needle is not simply responding to electric fields?

Among the reasons you may have listed are the following:
- The compass needle is affected by the proximity of objects made of iron or steel, even if these objects are electrically neutral (and therefore attract both positively and negatively charged tapes). It is also affected by the presence of nickel or cobalt, though these are less readily available.
- The compass needle is unaffected by objects made of most other elements, including aluminum, copper, zinc, and carbon, whereas charged tapes interact with these objects.
- If it is not near objects made of iron, the compass needle points toward the Earth's magnetic North pole, while neither electrically charged objects nor electric dipoles do this.

Figure 17.7 Wire aligned with the compass needle.

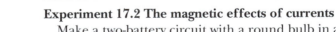

Experiment 17.2 The magnetic effects of currents

Make a two-battery circuit with a round bulb in a socket, as shown in Figure 17.6. Place your magnetic compass on a flat surface under one of the wires (Figure 17.7). Keep the compass away from steel objects, such as the steel-jacketed batteries, and the alligator clips on the ends of your wires. If you are working on a steel table, you may need to put the compass on a thick book. (For flexibility in placement, you may find it useful to make a long wire by connecting two of your wires together.)

Connect the circuit so that the bulb glows, and do the following:
- Lift the wire up above the compass.
- Orient the wire to be horizontal and lined up with the compass needle. (Using a long wire may make it easier to do this.)
- Bring the aligned wire down onto the compass (Figure 17.7).

(a) What is the effect on the compass needle as you bring the wire down on top of the compass?

(b) What happens when the wire is initially aligned perpendicular instead of parallel to the needle (Figure 17.8)?

Figure 17.8 Wire perpendicular to the compass needle.

(c) Reverse the connections to the batteries, or reverse the direction of the wire over the compass, in order to force electrons through the circuit in the opposite direction. Again make the compass needle deflect. How is the deflection of the compass needle affected by changing the direction of the current?

(d) Run the wire under the compass instead of over the compass (Figure 17.9). What changes?

(e) To make sure that it is the current in the wire, and not the metal wire itself, that affects the compass, disconnect the batteries. When you bring the wire down on the compass, is there a deflection?

(f) Record the observed compass deflection in the following cases:

    1: Two batteries and the (bright) round bulb.

    2: Two batteries and the (dim) long bulb.

    3: Two batteries, and just a long wire (no bulb). This is called a "short circuit", and it puts a large drain on the batteries, so you should not leave this connected for many minutes.

Figure 17.9 Wire under compass.

The effect you have just observed was discovered by Oersted in 1820, and is sometimes called "the Oersted effect." From your experiments, you should have drawn the following conclusions:

- the magnitude of the magnetic field produced by a current of moving electrons depends on the amount of current
- a wire with no current running in it produces no magnetic field
- the magnetic field due to the current appears to be perpendicular to the direction of the current
- the direction of the magnetic field due to the current under the wire is opposite to the direction of the magnetic field due to the current above the wire

A model for our observations

The moving electrons in the wire create a magnetic field at various locations in space, including at the location of the compass. The vector sum of the Earth's magnetic field $\vec{B}_{Earth}$ plus the magnetic field $\vec{B}_{wire}$ of a current-carrying wire makes a net magnetic field $\vec{B}_{net}$ (Figure 17.10). Since a compass needle points in the direction of the net magnetic field at its location, the needle turns away from its original north-south direction to align with the net field. Because the magnetic field made by the wire is perpendicular to the wire and is in opposite directions above and below the wire, the pattern of magnetic field made by the wire must look like Figure 17.11.

It happens (conveniently) to be the case that the magnitudes of $\vec{B}_{Earth}$ and $\vec{B}_{wire}$ are similar. If B_{Earth} were much larger than B_{wire}, then we would see almost no response from the compass.

? What would we observe if B_{wire} were much larger than B_{Earth}?

If B_{wire} were much larger than B_{Earth}, then the compass deflection would be nearly $90°$, since the net magnetic field would be primarily due to the magnetic field made by the current in the wire.

There is hardly any observable electric interaction of the current-carrying wire with nearby materials, because the wires have no net charge (actually, we'll see later that there are tiny amounts of excess charge on the surface, but too little to observe easily).

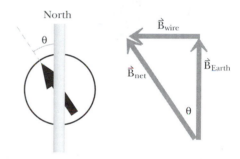

Figure 17.10 A compass needle points in the direction of the net magnetic field, which is the superposition of the magnetic field of the Earth and the magnetic field of the current-carrying wire.

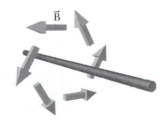

Figure 17.11 The magnetic field made by a current-carrying wire.

Example: A current-carrying wire is oriented north-south and laid on top of a compass. At the location of the compass needle the magnetic field due to the wire points west and has a magnitude of 3×10^{-6} T. The horizontal component of Earth's magnetic field has

Figure 17.12 The compass points in the direction of the net magnetic field.

a magnitude of about 2×10^{-5} tesla. What compass deflection will you observe?

Solution: The compass needle will point in the direction of the net magnetic field, so from the diagram in Figure 17.12:

$$\theta = \tan^{-1}\left(\frac{B_{\text{wire}}}{B_{\text{Earth}}}\right) = \tan^{-1}\left(\frac{3 \times 10^{-6}\text{T}}{2 \times 10^{-5}\text{T}}\right) = 8.5° \text{ west}$$

Ex. 17.3 A current-carrying wire oriented north-south and laid over a compass deflects the compass 15° east. What are the magnitude and direction of the magnetic field made by the current? The horizontal component of Earth's magnetic field is about 2×10^{-5} tesla.

17.3 The Biot-Savart law for a single moving charge

Careful experimentation has shown that a point charge makes an electric field given by

$$\vec{E} = \frac{1}{4\pi\varepsilon_0}\frac{q}{r^2}\hat{r}$$

and this relation between charge and electric field is called Coulomb's law.

Similarly, careful experimentation shows that a moving point charge not only makes an electric field but also makes a magnetic field (Figure 17.13), which curls around the moving charge. This curly pattern is characteristic of magnetic fields. (In contrast, we saw in Chapter 16 that it is impossible to produce a curly electric field by arranging stationary point charges.) Magnetic field is measured in "tesla" and its magnitude and direction are given by the "Biot-Savart law" (pronounced bee-oh sah-VAR):

THE BIOT-SAVART LAW FOR A SINGLE CHARGE

$$\vec{B} = \frac{\mu_0}{4\pi}\frac{q\vec{v} \times \hat{r}}{r^2}, \text{ where}$$

$$\frac{\mu_0}{4\pi} = 10^{-7}\frac{\text{tesla}\cdot\text{m}^2}{\text{coulomb}\cdot\text{m/s}} \text{ } exactly$$

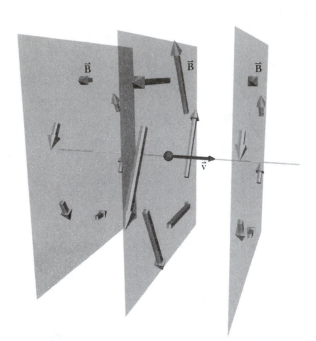

Figure 17.13 The magnetic field made by a moving positive charge, shown in three planes normal to $\hat{v}$.

$\vec{v}$ is the velocity of the point charge q, and $\hat{r}$ is a unit vector that points from the source charge toward the observation location.

17.3.1 The cross product

The magnetic field involves a "cross product" $\vec{v} \times \hat{r}$ which you encountered in the study of angular momentum. There are two ways of calculating the cross product of two vectors $\vec{A}$ and $\vec{B}$. The first is the complete vector calculation; the second involves finding the magnitude and direction separately. Both are useful.

CROSS PRODUCT $\vec{A} \times \vec{B}$

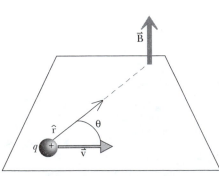

Figure 17.13b The magnetic field made by a moving charge is perpendicular to the plane defined by $\hat{v}$ and $\hat{r}$.

Vector: $\vec{A} \times \vec{B} = \langle (A_yB_z - A_zB_y), (A_zB_x - A_xB_z), (A_xB_y - A_yB_x)\rangle$

Magnitude of vector: $|\vec{A} \times \vec{B}| = AB\sin\theta$

Direction of vector: given by "right-hand rule" (see below)

The cross product $\vec{A} \times \vec{B}$ is a vector in a direction perpendicular to the plane defined by $\vec{A}$ and $\vec{B}$. In Figure 17.13b (on the previous page), both $\vec{v}$ and $\hat{r}$ lie in the plane, and the vector $\vec{v} \times \hat{r}$ is perpendicular to the plane. In the use of the cross product in the Biot-Savart law, $\vec{v} \times \hat{r}$ is a vector in the direction of the magnetic field produced by the moving charge.

Right-hand rule for vector cross product

- point fingers of right hand in direction of first vector $\vec{v}$ (Figure 17.14)
- rotate wrist, if necessary, to make it possible to
- curl fingers of right hand through angle θ toward second vector $\hat{r}$ (Figure 17.15)
- stick out thumb, which points in direction of cross product $\vec{v} \times \hat{r}$ (Figure 17.15).

Wrist rotation

The rotation of the wrist is an important part of the right-hand rule. Consider the situation in Figure 17.16. With your right hand in this position, you can't bend your fingers backwards from the first vector $\vec{v}$ toward the second vector $\hat{r}$—it is a physical impossibility. You need to rotate your wrist into a position from which it is possible to bend the fingers, as shown in Figure 17.17. Pay attention to the size of the angle through which you bend your fingers. This is the angle whose sine is part of the definition of the magnitude of the cross product. This angle should never be more than 180°. If it is, you have made a mistake in orienting your hand; probably you need to rotate your wrist.

Two-dimensional projections

Because it is more difficult to sketch a situation in three dimensions, whenever possible we will work with two-dimensional projections onto the *x-y* plane. If $\vec{v}$ and $\hat{r}$ lie in the *x-y* plane, the cross product vector $\vec{v} \times \hat{r}$ points in the +*z* direction (out of the page) or in the −*z* direction (into the page). A special notation is used to indicate "out of the page" ($\odot$) and "into the page" ($\otimes$). The symbol $\odot$ can be thought of as showing the tip of an arrow (a vector) pointing out at you, while the symbol $\otimes$ represents the feathers of an arrow (a vector) pointing away from you.

Ex. 17.4 Given that $\vec{v} = \langle v_x, v_y, v_z \rangle$ and $\hat{r} = \dfrac{\langle x, y, z \rangle}{\sqrt{x^2 + y^2 + z^2}}$, write out $\vec{v} \times \hat{r}$ as a vector.

Ex. 17.5 This is an important exercise. At the locations marked with × in the accompanying figure, determine the direction of the

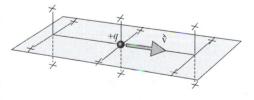

magnetic-field vectors due to a positive charge +*q* moving with a velocity $\vec{v}$. For each observation location, draw the unit vector $\hat{r}$ *from* the charge *to* that location, then consider the cross product $\vec{v} \times \hat{r}$. Pay attention not only to the directions of the magnetic field but also to the relative magnitudes of the vectors.

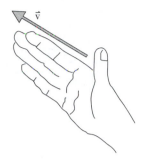

Figure 17.14 Open right hand, fingers point in direction of $\vec{v}$.

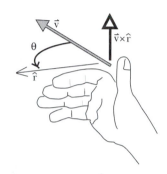

Figure 17.15 Bend fingers toward lining up with $\hat{r}$. Thumb points in direction of $\vec{v} \times \hat{r}$.

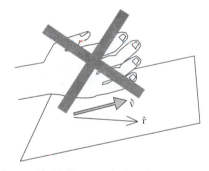

Figure 17.16 You can't bend your fingers backward. You must rotate the wrist into a position that lets you bend the fingers.

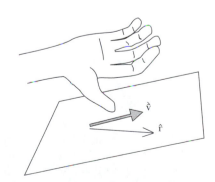

Figure 17.17 After rotating the wrist, it is possible to bend the fingers.

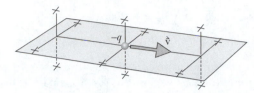

Ex. 17.6 If the charge is negative $(-q)$ as in the adjacent figure, how does this change the pattern of magnetic field?

Ex. 17.7 Briefly explain why the magnetic field is zero straight ahead of and straight behind a moving charge.

Ex. 17.8 How does the magnetic field of a moving point charge fall off with distance at a given angle: like $1/r$, $1/r^2$, or $1/r^3$?

Ex. 17.9 Describe the magnetic field made by a charge that is not moving.

Ex. 17.10 To get an idea of the size of magnetic fields at the atomic level, consider the magnitude of the magnetic field due to the electron in the simple Bohr model of the hydrogen atom. In the ground state the Bohr model predicts that the electron speed would be 2.2×10^6 m/s, and the distance from the proton would be 0.5×10^{-10} m. What is B at the location of the proton?

Figure 17.18 The magnetic field due to moving charges in the wire curls around the wire. Here the moving charges are assumed to be negative.

17.3.2 Sign of the moving charge

Assume that the moving charged particles in a current-carrying wire are negatively charged, and move in the direction shown in Figure 17.18.

? Use the Biot-Savart law and the right hand rule to predict the direction of the magnetic field at several locations around the wire (Figure 17.18).

Convince yourself that the cross product $\hat{v} \times \hat{r}$ does correctly give the direction and pattern of magnetic field near a wire, as observed in your experiment where you saw a change of direction for the wire above and below the compass.

? Try the analysis again, but assume that the moving charges are positively charged particles moving in a direction opposite to that shown. Do you still predict the curly magnetic field as shown in Figure 17.18?

You should have concluded that for the purpose of predicting the direction of magnetic field, it does not matter whether we assume that the moving charges are negative and moving in a given direction, or positive and moving in the opposite direction. It is common to describe current flow in terms of "conventional current," which means assuming that the moving charges are positive, even if this is known not to be the case in a particular situation.

Figure 17.19 If the moving particles in the wire are electrons, in what direction are they moving? If we assume that the particles are positively charged, in which direction would they be moving?

Ex. 17.11 In Figure 17.19 a current-carrying wire lies on top of a compass. Judging from the deflection of the compass away from North, what is the direction of the electron current in the wire? If the current were due to the motion of positive charges, which way would they be moving?

Ex. 17.12 Observe the compass deflection in a circuit. From which end of the battery, + or −, does electron current emerge? From which end does conventional current emerge?

17.4 Relativistic effects

17.4.1 Magnetic field depends on your frame of reference

Electric fields are made by charges, whether at rest or moving. Magnetic fields seem to be made solely by moving charges. But consider the following odd "thought experiment."

Suppose Jack sits in the classroom with a charged piece of invisible tape stuck to the edge of his desk. He can of course observe an electric field due to the charged tape, but he doesn't observe any magnetic field. His compass is unaffected by the charged tape, since those charges aren't moving:

$$\vec{B} = \frac{\mu_0}{4\pi} \frac{q\vec{v} \times \hat{r}}{r^2} = 0 \text{, since } v = 0$$

Jill runs at high speed through the classroom past the charged tape, carrying her own very sensitive compass. She observes an electric field due to the charged tape. But in addition, in her frame of reference the charged tape is moving, so she observes a small magnetic field due to the moving charges, which affects her compass! As far as Jack is concerned, the charged tape just makes an electric field. But apparently Jill sees a mixture of electric and magnetic fields made by what is for her a moving charged tape.

Up until now we have implied that electric fields and magnetic fields are fundamentally different, but this "thought experiment" shows that they are in fact closely related. Moreover, this connection raises questions about the Biot-Savart law:

$$\vec{B} = \frac{\mu_0}{4\pi} \frac{q\vec{v} \times \hat{r}}{r^2}$$

Just what velocity are we supposed to use in this formula? The velocity of the tape relative to Jack (which is zero) or the velocity of the tape relative to Jill (opposite her own velocity)?

The correct answer is that you use the velocities of the charges as you observe them in your frame of reference. Using these velocities in the Biot-Savart law you calculate the magnetic field; in your frame of reference you observe a magnetic field that agrees with your prediction. Observers in a different reference frame use the velocities observed in their frame to calculate the magnetic field using the Biot-Savart law, and in their frame of reference they observe a magnetic field that agrees with their prediction. You and they make different predictions, and observe different magnetic fields, but both you and they find agreement between theory and experiment.

Evidently there is a deep connection between electric fields and magnetic fields, and this connection is made explicit in Einstein's special theory of relativity. Later in the course we will see further aspects of this connection.

17.4.2 Retardation

Remember that when you move a charge, the electric field of that charge at a distance from the charge doesn't change instantaneously (Section 13.6.1). The electric field doesn't change until a time sufficient for light to reach the observation location, and you measure a change in the electric field at the same instant that you see the charge move.

The same retardation effect is observed with magnetic fields. If you suddenly change the current in a wire, the magnetic field at some distance from the source of the magnetic field stays the same until enough time has elapsed for light to reach the observation location. So magnetic field has some reality in its own right, independent of the moving charges that originally produced it.

The Biot-Savart law does not contain any reference to time, so it cannot be relativistically correct. Like Coulomb's law, the Biot-Savart law is only approximately correct and will give accurate results only if the speeds of the moving source charges are small compared to the speed of light.

17.5 Electron current and conventional current

An easy way to observe the magnetic fields made by moving charged particles is to initiate and sustain a current–a continual flow of charged particles in one direction. In order to do this, we need to find a way to produce and sustain an electric field inside a wire, since we know that if the electric field inside a metal becomes zero, the metal object will be in static equilibrium, and no current will flow. How exactly it is possible to arrange charges in order to create such an electric field everywhere in a wire is the subject of Chapter 18. For the moment, we'll assume that we have somehow accomplished this by assembling a circuit from batteries, wires, and perhaps light bulbs, since evidently currents do flow in such circuits.

In order to apply the Biot-Savart law to predict the magnitude and direction of the magnetic field associated with this current, we need to know the number of moving charges making the field, and how fast they are moving.

17.5.1 A formula for electron current

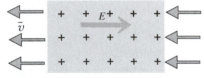

Figure 17.20 The electron sea drifts to the left with a speed $\bar{v}$.

Consider a section of a metal wire through which the mobile-electron sea is continuously shifting, under the influence of a a nonzero electric field, as indicated in Figure 17.20. There can be no excess charge anywhere inside the wire. To every mobile electron there corresponds a singly charged positive atomic core— an atom minus one electron which has been released to roam freely in the mobile-electron sea. Averaged over a few atomic diameters, the interior of the metal is neutral, and the repulsion between the mobile electrons is balanced out on the average by the attractions of the positive cores.

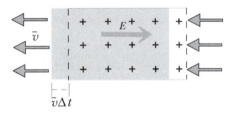

Figure 17.21 During a time Δt the electron sea shifts a distance $\bar{v}\Delta t$.

Suppose that in a metal wire the mobile-electron sea as a whole has an average "drift speed" $\bar{v}$. Assume that the current is evenly distributed across the entire cross section of the wire (in a later chapter we will be able to show that this is true.) After a short time Δt, this section of the electron sea will drift to the left a distance $\bar{v}\Delta t$, as indicated in Figure 17.21.

As shown in Figure 17.22, if the cross-sectional area of this wire is A, the volume of the disk-shaped portion of the sea that has flowed out of the left end of this section of the wire (and into the adjoining section of wire) in the time Δt is $A\bar{v}\Delta t$.

Figure 17.22 The volume of the electron sea that has flowed past a point in the wire during a time Δt is $A\bar{v}\Delta t$.

? If the density of mobile electrons (that is, the number of mobile electrons per unit volume) everywhere in the metal is n, how many electrons are in this disk?

$$\left(n\,\frac{\text{electrons}}{\text{m}^3}\right)(A\text{ m}^2)(\bar{v}\,\frac{\text{m}}{\text{s}})(\Delta t\text{ s}) \;=\; nA\bar{v}\Delta t \text{ electrons}$$

The number of electrons in this disk is the number of electrons that have flowed out of this section of the wire in the short time Δt.

? What is the number of mobile electrons per second that flow past the left end of this piece of the wire?

We divide by the time Δt to calculate the number of mobile electrons per second that flow past the left end of this piece of the wire:

$$\frac{\#}{s} = \frac{nA\bar{v}\Delta t}{\Delta t} = nA\bar{v}$$

This result, that the number of electrons passing some section of the wire per second is $nA\bar{v}$, is sufficiently important to be worth remembering (though it is even better to remember the simple derivation that gives this result).

ELECTRON CURRENT

The rate i at which electrons pass a section of a wire:

$$i = nA\bar{v} \quad (\# \text{ of electrons per second})$$

n is the mobile electron density (the number of mobile electrons per unit volume), A is the cross-sectional area of wire, and $\bar{v}$ is the average drift speed of electrons.

Ex. 17.13 You have used copper wires in your circuits. Let's calculate the mobile electron density n for copper. A mole of copper has a mass of 64 g (0.064 kg), and one mobile electron is released by each atom in metallic copper. The density of copper is about 9 grams per cubic centimeter, or $9\times10^3 \text{kg/m}^3$. Show that the number of mobile electrons per cubic meter in copper is $8.4\times10^{28} \text{m}^{-3}$.

Ex. 17.14 Suppose that $i = 3.4\times10^{18}$ electrons/s are drifting through a copper wire. (This is a typical value for a simple circuit.) The cross-sectional area of the wire is $8\times10^{-7} \text{m}^2$ (corresponding to a diameter of 1 mm), and the density of mobile electrons in copper is $8.4\times10^{28} \text{m}^{-3}$. What is the drift speed of the electrons?

Ex. 17.15 At the drift speed found in the previous exercise, about how many minutes would it take for a single electron in the electron sea to drift from one end to the other end of a wire 30 cm long, about one foot? (A puzzle: if the drift speed is so slow, how can a lamp light up as soon as you turn it on? We'll come back to this in the next chapter.)

17.5.2 Conventional current

In most metals, current consists of drifting electrons. However there are a few materials in which the moving charges are positive "holes" in the sea of mobile electrons. The positive holes act in every way like real positive particles. For example, holes drift in the direction of the electric field, and they have a charge of $+e$.

As you could see in Exercise 17.12 (page 590), electron current moves from the negative end of a battery through a circuit to the positive end of the battery; hole current would go in the opposite direction. Given this information, consider the following question:

? From your own observations of the direction of magnetic field around a copper wire, can you tell whether the current in copper consists of electrons, or holes?

The observations you have made are not sufficient to tell the difference, because when you change the sign of the moving charge, you also change the direction of the drift velocity: $(+e)(+\bar{v}) \rightarrow (-e)(-\bar{v})$ (see Section 17.3.2). Therefore the prediction of the Biot-Savart law is exactly the same in either case.

In copper, as in most metals, the current is due to moving electrons. The few metals that have hole current include aluminum and zinc. In the doped semiconductors important in electronics, "p-type" material (positive type) involves hole current, and "n-type" (negative type) involves electron current. Later in this course we will study the Hall effect, which can be used to determine whether the current in a wire is due to electrons or to holes.

Because most effects (other than the Hall effect) are the same for electron current and hole current, it is traditional to define "conventional current" to run in the direction of hole current, even if the actual current consists of moving electrons, in which case the "conventional current" runs in the opposite direction to the electrons. The advantage to this approach is that it eliminates the minus sign associated with electrons. (This sign ultimately goes back to a choice made by Benjamin Franklin when he arbitrarily assigned "positive" charge to be the charge we now know to be carried by protons.)

In addition, conventional current I is defined not as the number of holes passing some point per second but rather as the amount of charge (in coulombs) passing that point per second. This is the number of holes per second multiplied by the (positive) charge $|q|$ associated with one hole:

CONVENTIONAL CURRENT

$$I = |q|\,nA\bar{v} \text{ (coulombs per second, or amperes)}$$

In a metal, $|q| = e$, but in an ionic solution the moving ions might have charges that are a multiple of e.

Ex. 17.16 In Exercise 17.13 you calculated an electron current of $i = 3.4 \times 10^{18}$ electrons/s. What was the conventional current, including units?

17.6 The Biot-Savart law for currents

We don't often observe the magnetic field of a single moving charge. Usually we are interested in the magnetic field produced by a large number of charges moving through a wire in a circuit. The superposition principle is valid for magnetic fields, so we need to add up the magnetic-field contributions of the individual charges.

Let's calculate the magnetic field due to a bunch of moving positive charges contained in a small thin wire of length Δl and cross-sectional area A (Figure 17.23). If there are n moving charges per unit volume, there are $nA\Delta l$ moving charges in this small volume. We will measure the magnetic field at a location far enough away from this small volume that each moving charge produces approximately the same magnetic field at that location.

? Show that the net magnetic field of all the moving charges in this small volume, far from the volume, is

$$\Delta\vec{B} = \frac{\mu_0}{4\pi}\frac{I\Delta\vec{l} \times \hat{r}}{r^2},$$

where $\Delta\vec{l}$ is a vector with magnitude Δl, pointing in the direction of the conventional current I.

This results from the fact that the volume $A\Delta l$ contains $n(A\Delta l)$ moving charges, so the magnetic field is $n(A\Delta l)$ times as large as the magnetic field of one of the charges. Each moving charge has a charge q, and the conven-

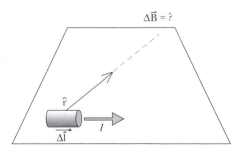

Figure 17.23 Magnetic field contributed by a short thin length of current-carrying wire.

tional current is $I = |q|(nA\bar{v})$. Collecting terms, we find the result given above.

So we have an alternative form of the Biot-Savart law for the magnetic field contributed by a short thin length of current-carrying wire:

THE BIOT-SAVART LAW FOR A SHORT THIN LENGTH OF WIRE

$$\Delta\vec{B} = \frac{\mu_0}{4\pi} \frac{I\Delta\vec{l} \times \hat{r}}{r^2}$$

where $\dfrac{\mu_0}{4\pi} = 10^{-7} \dfrac{\text{tesla·m}^2}{\text{coulomb·m/s}}$ *exactly*,

$\Delta\vec{l}$ is a vector in the direction of the conventional current I, and

$\hat{r}$ is a unit vector that points *from* the source charge *toward* the observation location.

? Explain why this formula also gives the right results if the moving charges are (negative) electrons, as long as we interpret $\Delta\vec{l}$ as a vector in the direction of the conventional current.

The law works for moving electrons because while they have the opposite charge, they also move in the direction opposite to the direction of the conventional current. These two changes in sign cancel each other.

It is important to keep in mind that there is really only one Biot-Savart law, not two. Always remember that

$$\Delta\vec{B} = \frac{\mu_0}{4\pi} \frac{I\Delta\vec{l} \times \hat{r}}{r^2}$$

is simply the result of adding up the effects of many moving charges in a short thin length of wire, each of which contributes a magnetic field

$$\vec{B} = \frac{\mu_0}{4\pi} \frac{q\vec{v} \times \hat{r}}{r^2}$$

The key point is that there are $nA\Delta l$ electrons in a short length of wire, each moving with average speed $\bar{v}$, so that the sum of all the $q\vec{v}$ contributions is $(nA\Delta l)|q|\bar{v} = (|q|nA\bar{v})\Delta l = I\Delta l$.

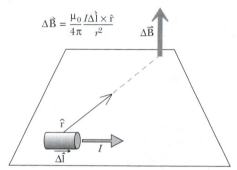

Figure 17.24 The Biot-Savart law for the magnetic field contributed by a short thin length of wire carrying conventional current I.

17.7 The magnetic field of current distributions

In the next sections we will apply the Biot-Savart law to find the magnetic field of various distributions of currents in wires. We will use the same four-step approach we used to find the electric field of distributed charges in Chapter 15:

1) Cut up the current distribution into pieces; draw $\Delta\vec{B}$ for a representative piece.
2) Write an expression for the magnetic field due to one piece.
3) Add up the contributions of all the pieces.
4) Check the result.

17.7.1 Applying the Biot-Savart law: A long straight wire

Using the procedure outlined above, we can calculate the magnetic field near a long straight wire. This will make it possible to predict the compass needle deflection that you observe when you bring a wire near your compass, and you can compare your prediction with your experimental observation.

The long straight wire is one of the few cases that we can calculate completely by hand. Except for a few special cases, calculating the magnetic field due to distributed currents is often best done by computer. The basic concepts you would use in a computer program are the same as we will use here, but the computer would carry out the tedious arithmetic involved in the summation step of the procedure.

Step 1: Cut up the distribution into pieces and draw $\Delta \vec{B}$

We will consider a wire of length L (Figure 17.25). We cut up the wire into very short sections each of length Δy, where Δy is a small portion of the total length. We will calculate the magnetic field a perpendicular distance x from the center of the wire. Using the right hand rule, we can draw the direction of $\Delta \vec{B}$ for this "piece" of current.

Step 2: Write an expression for the magnetic field due to one piece

origin: Center of wire

vector $\vec{r}$: $\vec{r} = \langle x, 0, 0 \rangle - \langle 0, y, 0 \rangle = \langle x, -y, 0 \rangle$ from source to observation location

magnitude of $\vec{r}$: $r = [x^2 + (-y)^2]^{1/2}$

unit vector $\hat{r} = \dfrac{\vec{r}}{r} = \dfrac{\langle x, -y, 0 \rangle}{[x^2 + (-y)^2]^{1/2}}$

location of piece: depends on y (so y will be the integration variable)

$\Delta \vec{l}$ *in terms of* y: $\Delta \vec{l} = \Delta y \langle 0, 1, 0 \rangle = \Delta y \hat{j}$

magnetic field due to one piece:

$$\Delta \vec{B} = \frac{\mu_0}{4\pi} \frac{I \Delta y \hat{j} \times \hat{r}}{(x^2 + y^2)} = \frac{\mu_0}{4\pi} \frac{I \Delta y}{(x^2 + y^2)} \langle 0, 1, 0 \rangle \times \frac{\langle x, -y, 0 \rangle}{[x^2 + (-y)^2]^{1/2}}$$

evaluate cross product:

$$\langle 0, 1, 0 \rangle \times \langle x, -y, 0 \rangle = \langle 0, 0, -x \rangle$$

expression for $\Delta \vec{B}$:

$$\Delta \vec{B} = -\frac{\mu_0}{4\pi} \frac{I x \Delta y}{(x^2 + y^2)^{3/2}} \langle 0, 0, 1 \rangle = -\frac{\mu_0}{4\pi} \frac{I x \Delta y}{(x^2 + y^2)^{3/2}} \hat{k}$$

components: Every piece of the straight wire contributes some magnetic field in the $-z$ direction, so we need to calculate only the z component.

Step 3: Add up the contributions of all the pieces

By letting $\Delta y \to 0$, we can write the sum as an integral:

$$B = \frac{\mu_0}{4\pi} I x \int_{y = -L/2}^{y = +L/2} \frac{dy}{(x^2 + y^2)^{3/2}}$$

Fortunately, this integral can be found in standard tables of integrals:

$$B = \frac{\mu_0}{4\pi} I x \left[\frac{y}{x^2 \sqrt{x^2 + y^2}} \right]_{y = -L/2}^{y = +L/2}$$

$$B = \frac{\mu_0}{4\pi} I x \left[\frac{L/2}{x^2 \sqrt{x^2 + (L/2)^2}} - \frac{-L/2}{x^2 \sqrt{x^2 + (L/2)^2}} \right]$$

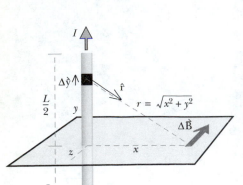

Figure 17.25 Cut the wire into short pieces and draw the magnetic field contributed by one of the pieces.

$$B = \frac{\mu_0}{4\pi} \frac{LI}{x\sqrt{x^2 + (L/2)^2}}$$

An extremely important case is that of a very long wire ($L \gg x$) or, equivalently, the magnetic field very near a short wire ($x \ll L$).

? Show that you get

$$B \approx \frac{\mu_0}{4\pi} \frac{2I}{x} \text{ if } L \gg x$$

If $L \gg x$, $\sqrt{x^2 + (L/2)^2} \approx L/2$, which leads to the stated result.

In summary, Figure 17.26 shows the pattern of magnetic field a distance x from the center of a straight wire of length L. Because of the symmetry of this field, it does not matter where we draw our x axis—the field will be the same all around the rod. To indicate this, we replace x in our formula with r, the radial distance from the rod.

Although the magnitude of the field is constant at constant r, the direction of the field is different at each angle, since the field curls around the wire. To express this as a vector equation we would have to include a cross product in the expression; it's simpler to get the direction using the right hand rule.

MAGNETIC FIELD OF A STRAIGHT WIRE

$$B_{\text{wire}} = \frac{\mu_0}{4\pi} \frac{LI}{r\sqrt{r^2 + (L/2)^2}}$$

(length L, conventional current I, a distance r from center of wire)

$$B_{\text{wire}} \approx \frac{\mu_0}{4\pi} \frac{2I}{r} \text{ if } L \gg r$$

Historically, the result for a very long straight wire was first obtained by the French physicists Biot and Savart. Their names have come to be associated with the more fundamental principle (the "Biot-Savart law") that leads to this result.

Step 4: Check the result

direction: In Figure 17.26 the right-hand rule is consistent with the diagram.
units:

$$\left(\frac{\text{T} \cdot \text{m}}{\text{A}}\right)\frac{(\text{m} \cdot \text{A})}{(\text{m} \cdot \text{m})} = \text{T}$$

far away ($r \gg L$):

$$B_{\text{wire}} \approx \frac{\mu_0}{4\pi} \frac{LI}{r^2} = \frac{\mu_0}{4\pi} \frac{I\Delta l}{r^2}$$

which is indeed the magnetic field contributed by a short wire of length Δl.

Another right-hand rule

There is another right-hand rule which is often convenient to use with current-carrying wires. As you can see in Figure 17.27, if you grasp the wire in your right hand with your thumb pointing in the direction of conventional current, your fingers curl around in the direction of the magnetic field. Clearly this right-hand rule is consistent with the one we used for the Biot-Savart law.

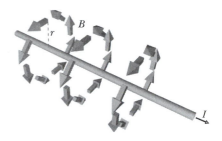

Figure 17.26 Magnetic field of a straight current-carrying wire.

Figure 17.27 The thumb of the right hand points in the direction of conventional current flow I, and the fingers curl around in the direction of the magnetic field.

17.7.2 Experimental observations of a long straight wire

Experiment 17.3 The magnetic field of a long straight wire

Use clip leads to connect a wire about 2 meters long to a battery, but don't make the final connection yet. Lay the wire down on the floor or a table, making a straight length as long as possible, heading north and south, as far from the steel battery as possible.

- Hold the compass above the wire at a height where the compass deflection is 20° when you turn the current on and off. Mark on a vertically held piece of paper this height $r_{20°}$ of the compass needle above the wire, and record this height.

- Mark the paper at twice this height and use the second mark to place the compass at a height $2(r_{20°})$ above the paper, and record the compass deflection.

(a) Are your data consistent with the theoretical prediction that the magnitude of the magnetic field near a long straight wire is proportional to $1/r$, where r is the perpendicular distance to the wire?

(b) Using your value of r_{20}, determine the amount of current I in the wire.

17.8 Applying the Biot-Savart law: A circular loop of wire

Next we'll calculate the magnetic field of a circular loop of wire that carries a conventional current I. We'll do only the easiest case—the magnetic field at any location along the axis of the loop, which is a line going through the center and perpendicular to the loop.

This calculation is important for two reasons. First, many scientific and technological applications of magnetism involve circular loops of current-carrying wire. Second, the calculation will also lead into an analysis of atomic current loops in your bar magnet. After calculating the magnetic field we will compare our predictions with experiments.

17.8.1 Magnetic field of a circular loop of wire

Step 1: Cut up the distribution into pieces and draw $\Delta\vec{B}$

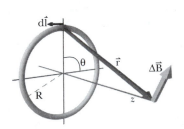

Figure 17.28 Cut the loop into short pieces and draw the magnetic field contributed by one of the pieces. Note that $d\vec{l} \perp \hat{r}$ for every piece of the loop.

See Figure 17.28. We cut up the loop into very short sections each of length Δl (a small portion of the total circumference $2\pi R$) (Figure 17.28; side view in Figure 17.29). Determine direction of $\Delta\vec{B}$ with right-hand rule. Note that the angle between $\Delta\vec{l}$ and $\hat{r}$ is 90° for every location on the ring.

Step 2: Expression for $\Delta\vec{B}$ due to one piece

To work out $\Delta\vec{B}$ for an arbitrary $\Delta\vec{l}$ will be algebraically messy, since

$$\Delta\vec{l} = \langle R\cos(\theta + d\theta), R\sin(\theta + d\theta), 0 \rangle - \langle R\cos\theta, R\sin\theta, 0 \rangle, \text{ and}$$

$$\hat{r} = \langle 0, 0, z \rangle - \langle R\cos\theta, R\sin\theta, 0 \rangle$$

However, we can simplify the problem considerably by noticing the symmetry of the situation. Because of the circular symmetry of the ring, the ΔB_x and ΔB_y contributed by one piece will be canceled by the contributions of a piece located on the other side of the loop.

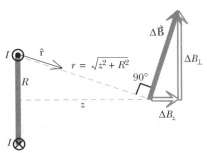

Figure 17.29 Side view of the current-carrying loop. Again note that $d\vec{l} \perp \hat{r}$ for every piece of the loop.

Furthermore, the ΔB_z contributed by each piece of the ring will be exactly the same. This allows us to select one piece for which ΔB_z is easy to calculate, and use this value for every piece in the sum. We will select the piece shown in the diagram, located on the y axis.

origin: Center of loop

vector $\vec{r}$: $\vec{r} = \langle$ obs. loc.$\rangle - \langle$ source $\rangle = \langle 0, 0, z \rangle - \langle 0, R, 0 \rangle = \langle 0, -R, z \rangle$

magnitude of $\vec{r}$: $r = [R^2 + z^2]^{1/2}$

unit vector $\hat{r} = \dfrac{\vec{r}}{r} = \dfrac{\langle 0, -R, z \rangle}{[R^2 + z^2]^{1/2}}$

location of piece: depends on θ (so θ will be the integration variable)

$\Delta\vec{l}$: $|\Delta\vec{l}| = \langle -R\Delta\theta, 0, 0 \rangle$ the magnitude of $\Delta\vec{l}$ is $R\Delta\theta$.

magnetic field due to one piece:

$$\Delta\vec{B} = \frac{\mu_0}{4\pi}\frac{I\,d\vec{l} \times \hat{r}}{r^2} = \frac{\mu_0}{4\pi}I\frac{\langle -R\Delta\theta, 0, 0 \rangle \times \langle 0, -R, z \rangle}{[R^2 + z^2]^{3/2}}$$

evaluate cross product:

$$\langle -R\Delta\theta, 0, 0 \rangle \times \langle 0, -R, z \rangle = \langle 0, -zR\Delta\theta, R^2\Delta\theta \rangle$$

We need only the z component, since the others will add up to zero: *expression for* ΔB_z:

$$\Delta B_z = \frac{\mu_0}{4\pi}\frac{IR^2\Delta\theta}{[R^2 + z^2]^{3/2}}$$

In Figure 17.29 we show the component ΔB_z along the axis and the component $\Delta B_\perp$ perpendicular to the axis.

Step 3: Sum the contributions of all the pieces

We can express the sum as an integral, where θ, which specifies the location of a piece of the ring, runs all the way around the ring. Remember that ΔB_z contributed by each piece is the same.

$$B_z = \int_0^{2\pi} \frac{\mu_0}{4\pi}\frac{IR^2\,d\theta}{(z^2 + R^2)^{3/2}}$$

The integral $\int_0^{2\pi} d\theta$ is just 2π. Here is the result (Figure 17.30):

MAGNETIC FIELD OF A LOOP

$$B_{\text{loop}} = \frac{\mu_0}{4\pi}\frac{2\pi R^2 I}{(z^2 + R^2)^{3/2}}$$

for a circular loop of radius R and conventional current I
at a distance z from the center, along the axis

Step 4: Check the result

units:

$$\left(\frac{\text{T} \cdot \text{m}}{\text{A}}\right)\frac{(\text{m}^2 \cdot \text{A})}{(\text{m}^2)^{3/2}} = \text{T}$$

direction: Check several pieces with the right hand rule.

special case: Center of the loop (Figure 17.31), where the magnetic field is especially easy to calculate from scratch. By the right-hand rule, $\Delta\vec{l} \times \hat{r}$ points out of the page for every piece of the loop, and its magnitude is simply Δl, since $\Delta\vec{l} \perp \hat{r}$—the magnitude of $\hat{r}$ is 1, and $\sin(90°)=1$.

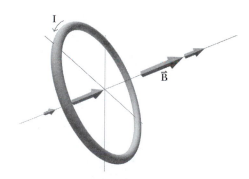

Figure 17.30 Magnetic field of a circular loop of current-carrying wire.

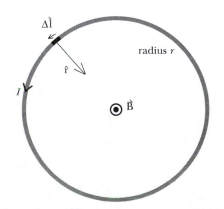

Figure 17.31 Magnetic field at the center of a loop of current-carrying wire.

So we have this:

$$B = \sum \frac{\mu_0}{4\pi} \frac{I\Delta l}{R^2} = \frac{\mu_0}{4\pi} \frac{I(2\pi R)}{R^2} = \frac{\mu_0}{4\pi} \frac{2\pi I}{R}$$

because the sum of all the Δl's is just $2\pi R$, the radius of the ring.

? Show that the general formula for the magnetic field of a loop reduces to this result if you let $z = 0$, which is another kind of check.

17.8.2 Qualitative features of the magnetic field of a loop of wire

It is interesting to see what the magnetic field is *far* from the loop, along the axis of the loop.

? Show that if z is very much larger than the radius R, the magnetic field is approximately equal to

$$\frac{\mu_0}{4\pi} \frac{2\pi R^2 I}{z^3}$$

The key to this result is that if $z \gg R$, $(z^2 + R^2)^{3/2} \approx (z^2)^{3/2} = z^3$. We see that the magnetic field of a circular loop falls off like $1/z^3$.

Ex. 17.17 Figure 17.32 shows a coil of wire containing many loops. Sketch the predicted magnetic field direction and relative magnitude at the locations indicated in Figure 17.32. (Hint: for the locations to the side of the coil, apply the Biot-Savart law qualitatively to the near and far halves of the circular loop.) Compare this pattern of magnetic field with the pattern of *electric* field around an electric dipole.

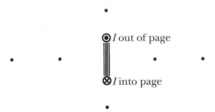

Figure 17.32 Sketch the predicted magnetic field of the coil at the indicated locations.

17.8.3 Experimental observations of a coil of wire

Next we will make measurements of a coil of wire, consisting of multiple loops to make a larger magnetic field than a single loop would provide.

Experiment 17.4 The magnetic field of a coil of wire

Take an insulated wire about 2 meters in length and wrap a coil of about $N = 20$ turns loosely around two fingers. Twist the ends together to help hold the coil together, and remove the coil from your fingers. Use clip leads to connect the coil to a battery.

By appropriate placement and orientation of your compass relative to your N-turn current loop, measure the experimental magnetic field direction and relative magnitude at the locations indicated in Figure 17.33. Remember that you must always position the compass in such a way that the coil's magnetic field is perpendicular to the Earth's magnetic field.

On a diagram like Figure 17.33, draw the magnetic field vectors at all marked locations. Record the number of turns N, the approximate radius of the coil R, the distances to the measurement locations, and the compass deflections at those locations.

(a) Do your measurements agree with the predicted pattern of magnetic field around a coil of wire?

(b) Determine the amount of current I in the wire.

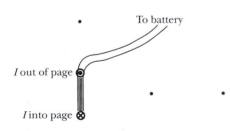

Figure 17.33 Measure the magnetic field at the indicated locations.

17.8.4 A right-hand rule for current loops

There is another "right-hand rule" that is often used to get the direction of the magnetic field along the axis of a loop. Let the fingers of your right hand curl around in the direction of the conventional current, and your thumb will point in the direction of the magnetic field.

? Try using this right-hand rule to determine the direction of the magnetic field at the indicated observation location in Figure 17.34.

You should find that the magnetic field points down. This right-hand rule should of course give the same result as applying the more general right-hand rule to the cross product $\Delta \vec{l} \times \hat{r}$ and adding up the contributions of the various parts of the loop, as called for by the Biot-Savart law.

? On the diagram, consider $\Delta \vec{l} \times \hat{r}$ for two short pieces of the loop, on opposite sides of the loop. Show that the two pieces together contribute a magnetic field in the downward direction at the indicated location.

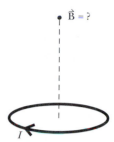

Figure 17.34 Curl the fingers of your right hand in the direction of the conventional current, and your thumb will point in the direction of the magnetic field.

17.8.5 Magnetic field at other locations outside the loop

The magnetic field at locations outside the loop is more difficult to calculate analytically, but as your experimental measurements suggest, the magnetic field has a characteristic dipole pattern (Figure 17.35).

17.9 Magnetic dipole moment

Recall the formula for the electric field along the axis of an electric dipole, at a distance r far from the dipole:

$$E_{\text{axis}} \approx \frac{1}{4\pi\varepsilon_0} \frac{2p}{r^3}, \text{ where the "electric dipole moment" } p = qs$$

Similarly, in the formula for the magnetic field along the axis of a current-carrying coil, at a distance r far from the coil, we can write this:

$$B_{\text{axis}} \approx \frac{\mu_0}{4\pi} \frac{2\mu}{r^3}, \text{ where the "magnetic dipole moment" } \mu = IA$$

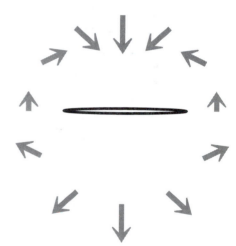

Figure 17.35 The magnetic field of a current loop (viewed edge-on), at locations outside the loop, in a plane containing the axis of the loop.

Here A is the area of the loop (πR^2 for circular loops). This formula for magnetic field is approximately valid even if the loop is not circular. The magnetic dipole moment $\vec{\mu}$ is considered to be a vector pointing in the direction of the magnetic field along the axis (Figure 17.36). This means that the direction of the magnetic dipole moment can be obtained by curling the fingers of your right hand in the direction of the conventional current, and your thumb points in the direction of the magnetic dipole moment.

We will see later that the concept of magnetic dipole moment also applies to magnets and provides a way of characterizing the strength of a magnet.

Ex. 17.18 What is the magnetic dipole moment of a 3000-turn rectangular coil that measures 3 cm by 5 cm and carries a current of 2 amperes?

Figure 17.36 The magnetic dipole moment $\vec{\mu}$ is considered to be a vector pointing in the direction of the magnetic field along the axis.

17.9.1 Twisting of a magnetic dipole

For reasons that we will discuss in a later chapter, the magnetic dipole moment vector $\vec{\mu}$ acts just like a compass needle. In an applied magnetic field, a current-carrying loop rotates so as to align the magnetic dipole moment $\vec{\mu}$ with the field.

It is hard to observe this twisting with your own hanging coil, because the rather stiff wires prevent the coil from freely rotating, unless you suspend the entire apparatus from a thread, batteries and all. Or do this:

Experiment 17.5 A coil is a compass

If you have access to kitchen equipment, you may be able to float your batteries and coil in a bowl or large glass or aluminum pan, or on a block of wood. (Avoid steel containers!) Then you may be able to observe the axis of the coil line up with the Earth's magnetic field, just as though the magnetic dipole moment of the coil were a compass needle.

17.10 The magnetic field of a bar magnet

From its name, you might guess that a magnet also makes a magnetic field. We say that a magnetic field is present if we see a compass needle twist, so let's see whether your bar magnet does make your compass needle deflect.

Experiment 17.6 Directions of the magnetic field of a bar magnet

Place your compass to the right of your bar magnet, with the magnet oriented in such a way as to make the compass deflect to the east as shown in Figure 17.37. Make a similar diagram in your notebook.

We define the direction of a magnetic field $\vec{B}$ at a particular location as the direction that the "north" end of a compass points to when placed at that location.

(a) Assuming that the bar magnet makes a magnetic field along the axis of the magnet, draw a vector in your notebook to show the direction of the compass needle, and another arrow to show the direction of the magnetic field $\vec{B}_m$ of your bar magnet at the present location of the compass. For future reference, write "N" for "north" on the end of the magnet that the magnetic field points out of (attach tape to write on if necessary).

(b) Move the compass to the left of your magnet, above your magnet, and below your magnet, as shown in Figure 17.37. For each location, record the direction of the compass needle, and also record on your diagram the direction of the magnetic field due to the bar magnet at that location.

(c) Does this pattern of field look familiar? Where have we seen a similar pattern of field directions in space before?

(d) Suspend your magnet from a thread or a hair, using a piece of tape, or float the magnet on a dish or plate in water. Note which end points toward the north. Evidently a compass consists simply of a magnet in the shape of a needle, mounted on a pivot.

Experiment 17.7 Magnets and matter

Obtain another magnet by working with a partner or finding a kitchen magnet. Make sure that you have labeled the north and south ends of both magnets.

(a) Describe briefly the interactions that the two magnets have with each other.

(b) You probably know that magnets don't interact strongly with anything but iron or steel objects (steel is mostly iron). Magnets also interact with nickel or cobalt objects, but these aren't so readily available. Check to see for yourself that your magnet doesn't interact with aluminum (aluminum foil) or copper (a penny, or an electrical wire), or any nonmetal. Also notice that your magnet is strongly attracted to steel parts of your electricity kit.

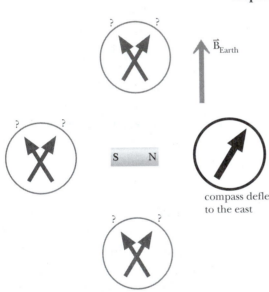

compass defle
to the east

Figure 17.37 Place your compass at each of the locations shown above, relative to your bar magnet, and note the deflection of the compass needle. What is the direction of the magnet's magnetic field at these locations?

(c) Check what happens with charged invisible tape, and see that while there is the usual attraction of a charged tape for any uncharged object, there doesn't seem to be any magnetic interaction, because either end of the magnet acts the same.

(d) Can you give any evidence that magnetic interactions pass through matter, as electric interactions do?

(e) Place two magnets in various positions near your compass to demonstrate to your satisfaction that the superposition principle holds for the magnetic fields of magnets; that is, the net field of the two magnets is the vector sum of the two fields. Give one example of your observation of the superposition principle for magnetic fields.

You have seen that magnetic interactions have some similarity to electric interactions: there is both attraction and repulsion, the superposition principle is valid, and the interactions pass through matter. However, there are also some major differences. Your bar magnet interacts only with iron or steel objects, while a charged invisible tape interacts with *all* objects. Your magnet is permanently magnetic, while a charged invisible tape or plastic pen loses its charge within a relatively short time. Also, the magnetic field around a current-carrying wire has a "curly" pattern that we don't see with electric fields.

17.10.1 The magnetic field of the Earth

The magnetic field of the Earth has a pattern that looks like that of a bar magnet (Figure 17.38). In the Northern Hemisphere, the Earth's magnetic field dips downward toward the Earth. A horizontally held compass is sensitive just to the horizontal component of the magnetic field, and this horizontal component gets smaller as you move closer to the magnetic poles.

For example, the horizontal component is smaller in Canada than it is in Mexico, despite the fact that the magnitude of the magnetic field is larger in Canada. The horizontal component is zero right at the magnetic poles, where the magnitude of the magnetic field is largest.

The Earth is a big magnet, but its type "S" pole is in the Arctic, and the "N" end of a compass needle points toward this type "S" pole.

The Earth's magnetic poles are not located exactly at the geographic poles. The "S" pole is located in northern Canada, about 1300 km (800 miles) from the geographic north pole, and the "N" pole is on the Antarctic continent, about 1300 km from the geographic south pole.

17.10.2 Dependence on distance

The horizontal component of the Earth's magnetic field, which is the component that affects a horizontally held compass, is different at different latitudes, depending on the distance from the "magnetic poles" of the Earth. Here are measurements of the magnitude of the horizontal component of the Earth's magnetic field at a few selected locations in the United States:

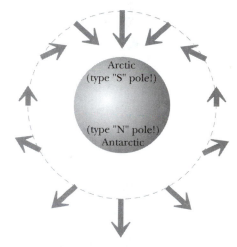

Figure 17.38 The Earth acts like a magnet. Its type "N" pole is in the Antarctic, and its type "S" pole is in the Arctic.

Location	Horizontal component of Earth's magnetic field
Maine	about 1.5×10^{-5} tesla
Much of the United States	about 2×10^{-5} tesla
Florida, Hawaii	about 3×10^{-5} tesla

Knowing the horizontal component of the Earth's magnetic field, we can use the deflection of the compass needle to measure the magnitude of the

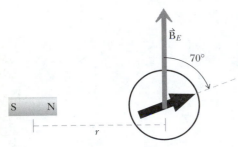

Figure 17.39 Place the magnet at a center-to-center distance r such that the compass needle deflection is 70°.

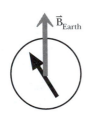

Figure 17.40 Place the compass north or south of the bar magnet and note the deflection of the compass needle.

magnetic field of the magnet. We can then study how the magnetic field of the magnet varies with distance from the magnet.

Experiment 17.8 Dependence on distance for a bar magnet

Again place the compass to the right of your bar magnet, oriented in such a way as to make the compass deflect 70° to the east (Figure 17.39). *Do not work on a table that contains steel components*, which can interfere with the measurements. The best place to make the measurements is outdoors, away from iron and steel objects.

(a) At the location where you get a 70° deflection, measure the distance r from the center of the magnet to the center of the compass. A centimeter ruler is provided at the back of the book.

(b) Use vector analysis and the magnitude of the Earth's field to calculate the magnitude B_r of the magnetic field of the magnet at this distance r from the center of the magnet.

(c) Now move the magnet farther away by a factor of two; that is, place the magnet so that the distance from the center of the magnet to the center of the compass is $2r$. Record the new distance and the new compass deflection angle. Calculate the magnitude B_{2r} of the magnetic field of the magnet at this new distance from the magnet.

(d) By what ratio did the magnetic field of the magnet decrease when the distance from the magnet was doubled?

(e) The magnetic field of the magnet gets smaller with distance, and it is plausible to guess that the magnetic field of the magnet might vary as $1/r^n$, where n is initially unknown. According to your measurements, what is n? (You may wish to make measurements at some other distances to give further support to your analysis.)

(f) Extrapolating to a location near one end of your bar magnet, approximately how strong is the magnetic field in tesla near one end of your magnet? (Remember, measure r to the center of the magnet.)

(g) Place the compass north or south of the magnet, at the original distance r from the center of the magnet, and determine the magnitude of the magnetic field at this location (Figure 17.40). How does this magnitude compare to your result in (b)?

17.10.3 A bar magnet is a magnetic dipole

The pattern of directions of magnetic field around a bar magnet is very similar to the pattern of directions of the magnetic field around a current loop, (Section 17.8.2 on page 600) and to the pattern of the electric field around a permanent electric dipole seen in Chapter 13. Also, along the axis of magnet or a current loop or a permanent electric dipole, the magnetic or electric field varies like $1/r^3$. Moreover, the magnitude of the field off to the side of an electric dipole or a magnet is half as large as it is at the same distance along the axis. Because of these strong similarities, a magnet is often called a "magnetic dipole."

Problem 17.1 Magnetic dipole moment of your bar magnet

The magnetic field along the axis of a bar magnet can be written like this:

$$B_{\text{axis}} \approx \frac{\mu_0}{4\pi} \frac{2\mu}{r^3}$$

Use your own data from Experiment 17.8 on page 604 to determine the magnetic moment μ of your bar magnet. If you have mislaid those data, quickly repeat the measurements now. You will use the value of your magnetic moment in later work. The magnetic moment describes how strong a magnet is: the bigger the magnetic moment, the bigger the magnetic field at some distance r.

Experiment 17.9 Coils attract and repel like magnets
There is yet another way in which the behavior of a magnet and of a current-carrying coil of wire is very similar. Hang your coil over the side of the table and bring your bar magnet near the hanging coil, along the axis of the coil. Reverse the magnet. Repeat from the other side of the coil. Is this behavior similar to the interaction you have observed between two bar magnets?

17.10.4 Magnetic monopoles?

There is one dramatic difference between electric and magnetic dipoles. The individual positive and negative electric charges ("monopoles") making up an electric dipole can be separated from each other, and these point charges make outward-going or inward-going electric fields (which vary like $1/r^2$). One might expect a magnetic dipole to be made of positive and negative magnetic monopoles, but no one has ever found an individual magnetic monopole. Such a magnetic monopole would presumably make an outward-going or inward-going magnetic field (which would vary like $1/r^2$), but such a pattern of magnetic field has never been observed.

Figure 17.41 is an example of the sort of magnetic field pattern that many scientists have diligently looked for but not found. If you cut a magnet in two, you don't get two magnetic monopoles—you just get two magnets!

Figure 17.41 This pattern of magnetic field has never been observed.

Experiment 17.10 "Cutting a magnet in two"
Without actually cutting a magnet into two pieces, you can simulate such an operation by putting two bar magnets together north end to south end, then taking them apart (Figure 17.42).

Do you in fact find that the put-together longer magnet acts just like a single magnet, and that after pulling them apart the two pieces still act like magnets?

If you have the equipment available, you can magnetize a soft iron nail by stroking it with a magnet; then use a hacksaw to cut it in half and investigate the properties of each half.

Figure 17.42 Place two magnets together end-to-end: do they behave like a single magnet?

17.11 The atomic structure of magnets

We have seen that the magnetic field of a circular current loop looks suspiciously similar to the magnetic field of a magnet. The pattern of directions of the magnetic field looks the same, and it is even the case that for both magnets and current loops the magnitude of the field at large distances along the axis falls off like $1/r^3$. Both magnets and current loops can be described in terms of magnetic dipole moment $\vec{\mu}$.

A possible explanation for this curious similarity is that in each atom of a magnet there might be electrons moving in circular orbits, and each of these orbiting electrons forms a current loop (Figure 17.43). Each atomic current loop would contribute an amount of magnetic field

$$\frac{\mu_0}{4\pi}\frac{2\mu}{r^3} = \frac{\mu_0}{4\pi}\frac{2\pi R^2 I}{r^3}$$

Figure 17.43 A simple model for a magnet might involve electrons in circular atomic orbits.

at a distance r, with $R =$ the radius of the orbit. (Note that $r \gg R$, since R is the radius of an atom, so this approximation for the field is a very good approximation.) The net magnetic field would be the sum of these individual atomic contributions.

> **?** In Figure 17.43, what is the direction of the resulting magnetic field to the left and right of the atoms, and of the magnetic dipole moment of one atom? (Remember that electrons are negative.)

In Figure 17.43, you should be able to use the loop form of the right-hand rule to show that the magnetic field points to the left (if the orbiting charges were positive, the magnetic field would point to the right). This means that the magnetic dipole moment of each atom points to the left. Note too that the "N" pole of this multiatom magnet is to the left, and the "S" pole to the right.

> **?** If an electron with charge $-e$ travels in a circular orbit of radius R at speed v, what is the magnitude of the average current I in one electron current loop? (Hint: The time T it takes to go around once can be determined from the fact that $v = 2\pi R / T$.)

The current I is a measure of how much charge passes one location per second, and the time to go around once is $T = 2\pi R / v$, so

$$I = \frac{e}{\left(\dfrac{2\pi R}{v}\right)} = \frac{ev}{2\pi R}$$

> **?** What is the magnetic dipole moment $\mu = IA$ for one electron current loop?

The area A of the current loop is πR^2, so

$$\mu = \left(\frac{ev}{2\pi R}\right)(\pi R^2) = \tfrac{1}{2}eRv \quad \text{(magnetic dipole moment of one atom)}$$

The next two short sections offer two different ways to estimate the magnitude of this atomic magnetic dipole moment.

Estimating the magnetic dipole moment: Quantized angular momentum

If you have not studied angular momentum, and the quantization of angular momentum, you should skip to the next section to estimate a value for the atomic magnetic dipole moment.

We can rewrite the magnetic dipole moment in terms of the orbital angular momentum $L = Rmv$:

$$\mu = \tfrac{1}{2}eRv = \tfrac{1}{2}\frac{e}{m}(Rmv) = \tfrac{1}{2}\frac{e}{m}L$$

This is an important result in its own right: the magnetic dipole moment is proportional to the angular momentum. It turns out that this is true not only for translational angular momentum but also for rotational angular momentum, although the proportionality factor in the latter case is 1 instead of $1/2$.

In quantum-mechanical systems, orbital angular momentum is quantized in multiples of $\hbar = 1.05 \times 10^{-34}$ J·s . Assuming one quantum of angular momentum, we have the following:

$$\mu \approx \tfrac{1}{2}\frac{e}{m}\hbar$$

Evaluating this expression, we find that:

$$\mu \approx (\tfrac{1}{2})\frac{(1.6\times10^{-19}\ \text{C})}{(9\times10^{-31}\ \text{kg})}(1.05\times10^{-34}\ \text{J·s})$$

$$\mu \approx 0.9\times10^{-23}\ \text{ampere·m}^2\ \text{per atom}$$

Estimating the magnetic dipole moment: A simple model of the atom

If you worked through the previous section using quantized momentum, you can skip this section.

Electrons don't really move in well-defined orbits in an atom, but let's nevertheless pursue this crude model a bit further to see whether the model gives a magnetic field with roughly the right magnitude. In this model (Figure 17.44), an outer electron is held in a circular orbit of radius R by the electric attraction to the rest of the atom (which has a net charge of $+e$).

The momentum principle is $d\vec{p}/dt = \vec{F}_{net}$, where $\vec{p}$ is the momentum of the electron and the net force is the electric force that the rest of the atom exerts on the electron. In circular motion the magnitude of the momentum doesn't change, but the direction does, and the magnitude of the rate of change of the rotating momentum vector can be calculated from the angular speed $\omega = v/R$:

$$\left|\frac{d\vec{p}}{dt}\right| = \omega p = \left(\frac{v}{R}\right)(mv) = F_{net}$$

Alternatively, we can write the momentum principle in the form mass times acceleration equals force. For circular motion with radius R, the magnitude of the acceleration is v^2/R, so we have

$$\left|m\frac{d\vec{v}}{dt}\right| = m\left(\frac{v^2}{R}\right) = F_{net}$$

Either way, we obtain

$$\frac{mv^2}{R} = \frac{1}{4\pi\varepsilon_0}\frac{e^2}{R^2}$$

$$v = \sqrt{\frac{1}{4\pi\varepsilon_0}\frac{e^2}{Rm}}$$

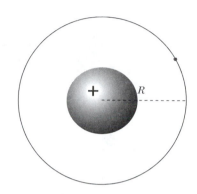

Figure 17.44 Simple model of an atom in which an outer electron orbits a positive inner core.

Plugging in the electron charge and mass, and assuming an approximate atomic radius of $R \approx 10^{-10}$ m, we get $v \approx 1.6\times10^6$ m/s (you can check this calculation if you like). Inserting this value of the electron orbital speed v into our earlier formulas, we find that the magnetic dipole moment of one orbiting electron should be roughly

$$\mu = \tfrac{1}{2}eRv \approx \tfrac{1}{2}(1.6\times10^{-19}\ \text{C})(10^{-10}\ \text{m})(1.6\times10^6\ \text{m/s})$$

$$\mu \approx 1.3\times10^{-23}\ \text{ampere·m}^2\ \text{per atom}$$

This differs somewhat from the estimate in the preceding section. The former estimate is more faithful to the full quantum-mechanical analysis that is necessary to obtain accurate predictions. The key point is that both estimates are approximately 10^{-23} ampere·m^2.

Comparing with experiment

Assume that the magnetic dipole moment of your bar magnet is due to orbiting electrons, one per atom, all aligned in the same direction, and see what value of magnetic dipole moment μ you would predict for your magnet. You need to work out how many atoms are in your magnet.

Although your magnet is probably made of some alloy such as Alnico V (51% iron, 8% aluminum, 14% nickel, 24% cobalt, and 3% copper), for simplicity assume it is made just of iron, which has a density of 8 grams/cm^3 and an atomic mass of 56 (that is, 6×10^{23} atoms weigh a total of 56 grams).

Problem 17.2 Predicting the magnetic dipole moment of your magnet
Determine the mass of your magnet and thereby determine the number N of atoms in your magnet, assuming it is made of iron. Assuming that each of the atoms has a magnetic moment with a value estimated above, see how well the atomic model for a magnet fits the measured value of the magnetic moment of your bar magnet (Problem 17.1 (page 604) Magnetic dipole moment of your bar magnet).

You presumably found that the predicted and measured values of the magnetic dipole moment of your bar magnet differ by a factor of only two or so. Given our crude approximations, this is pretty good (instead of being off by a factor of, say, a thousand!). It is at least possible that we have partially explained magnets in terms of orbiting electrons. All we put into the theoretical prediction was the assumption that each of the N atoms has one orbiting electron, the assumption that the magnetic dipole moments in all the atoms are aligned in the same direction, and either (a) angular momentum quantization or (b) an estimate of 10^{-10} m for the radius of an atom.

While explaining a magnet in terms of atomic current loops captures much of the flavor of what is believed to be the true explanation, a great deal of experimental and theoretical work has shown that this explanation is not the whole story. We will summarize the presently accepted analysis of the atomic structure of magnets.

17.11.1 The modern theory of magnets

Orbital motion

One oversimplification in our simple calculation is our assumption that electrons in an atom orbit the nucleus in well-defined circular paths (the Bohr model of an atom). In more sophisticated atomic models, there is only a probability for finding the electron at a particular place. Moreover, many electrons in atoms have spherically symmetric probability distributions ("s orbitals"), and such a symmetric distribution has a zero average magnetic dipole moment (and zero angular momentum). However, electrons in non spherically symmetric probability distributions ("p, d, or f orbitals") can contribute a non-zero magnetic dipole moment. Also, we assumed that in every atom there was just one electron that contributed a magnetic field, but in some materials two or more atomic electrons may contribute.

Spin

The orbital motion of atomic electrons is not the only contribution to the magnetic field. Each electron acts as though it were a little spinning ball of charge, rotating at very high speed, and therefore producing magnetic field due to the bits of charge that are going around in circles about the center of the ball (Figure 17.45). The amount of magnetic field produced by a single spinning electron is numerically about the same as the magnetic field we calculated for a single orbiting outer electron. So our calculation is numerically roughly right, but the magnetic field can be due to spin motion (about the center of the electron) or orbital motion (about the nucleus), or both.

Protons and neutrons in nuclei also have spin, but their magnetic dipole moments are much smaller than that of the electron. The quantized rota-

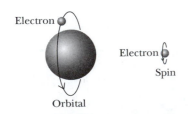

Figure 17.45 Electrons have spin as well as orbital motion that can contribute to the magnetic dipole moment.

tional angular momentum of the proton or neutron is the same as that of the electron, but since the magnetic dipole moment is $\mu \approx \frac{1}{2}(e/m)\hbar$, the large mass makes the magnetic dipole moment be very small. We can ignore nuclear contributions to the magnetic field of your bar magnet, though nuclear magnetic dipole moments play an important role in the phenomenon of nuclear magnetic resonance (NMR), and in the technology of magnetic resonance imaging (MRI) used in medicine, which is based on NMR.

The "spinning ball" model of the electron isn't really adequate, because the most sensitive experiments are consistent with the notion that an electron is a true point charge and has zero radius! It is nevertheless an experimental fact that an electron does make a magnetic field as though it were a spinning ball of charge, and it does little harm to think of it literally as a spinning ball.

Alignment of the atomic magnetic dipole moments

In many materials, an atom has no net orbital or spin magnetism. In most materials whose individual atoms do make a magnetic field, the orbital and spin motions in different atoms don't line up with each other, so the net field of the many atoms in a piece of the material averages out to zero. But in a few materials, notably iron, nickel, cobalt, and some alloys of these elements, the orbital and spin motions in neighboring atoms line up with each other and therefore produce a sizable magnetic field. These unusual materials are called "ferromagnetic." The reason for the alignment can be adequately discussed only within the framework of quantum mechanics, but it is worth mentioning that the alignment is due to electric interactions between the atoms, not to the much weaker magnetic interactions.

Magnetic domains

In an ordinary piece of iron that isn't a magnet, the iron is a patchwork of small regions, called "magnetic domains," within which the alignment of all the atomic magnetic dipole moments is nearly perfect. But normally these domains are oriented in random directions, so the net effect is that this piece of iron doesn't produce a significant net magnetic field. Figure 17.46 shows a picture of the situation, in which the arrows indicate the magnetic dipole moments of the atoms.

If you use a coil to apply a large magnetic field to the iron, domains that happen to be nearly aligned with the applied field tend to grow at the expense of domains that have a different orientation (Figure 17.47). Also, with a sufficiently large applied magnetic field, the magnetic dipole moments of the atoms in a domain tend to rotate toward aligning with the applied field, just as a compass needle turns to align with an applied field. The result of both of these effects is to partially align the magnetic dipole moments of most of the atoms, which produces a magnetic field $\vec{B}_{iron}$ which may be much larger than the applied magnetic field. There is a kind of multiplier effect.

In the case of very pure iron, when you turn off the current in the coil, the piece of iron goes back nearly to its original disordered patchwork of domains and doesn't act like a magnet any more. But in some other ferromagnetic materials, including some alloys of iron, when you remove the applied field the domains remain nearly aligned. In that case you end up with a permanent magnet like the one you have, which is probably made of the alloy Alnico. Hitting a magnet a hard blow may disorder the domains and make the metal no longer act like a magnet. Heating above a critical temperature also destroys the alignments.

Figure 17.46 Disordered magnetic domains—the net magnetic field produced by the iron is nearly zero.

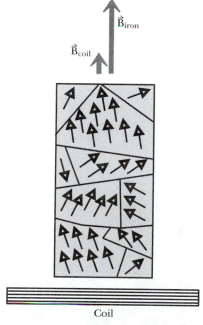

Figure 17.47 The magnetic field of a coil causes a partial ordering of the magnetic domains, which produces a significant nonzero magnetic field contributed by the iron.

17.11.2 Iron inside a coil

You can observe this multiplier effect. A small magnetic field created by current-carrying loops of wire wrapped around a piece of iron can lead to a large observed magnetic field being contributed by the iron, due to the alignment of the magnetic dipole moments in the iron.

Experiment 17.11 The magnetic multiplier effect

Note: You need unmagnetized nails for this experiment. If the nails have ever been near a magnet, they may have become strongly magnetized. Check this by bringing a nail near the compass, first one end, then the other. If the nail is unmagnetized, both ends of the nail will affect the compass equally. If however the nail is strongly magnetized, the two ends of the nail will affect the compass quite differently.

If a nail is strongly magnetized, bring your bar magnet near it (but not touching the nail) to magnetize the nail in the opposite direction. Check again with the compass and repeat as necessary, trying to reduce the magnetization to zero. Do not proceed until you have at least three unmagnetized nails.

Now you can study the magnetic multiplier effect. Take an insulated wire about 2 meters in length and wrap a coil of about 20 turns loosely around two fingers. Twist the ends together to help hold the coil together and remove it from your fingers. Connect the coil to a battery and position your compass to the east or west of the coil so that the compass needle deflects away from north by about 5 degrees. Turn off the current, so the compass deflection goes to zero.

(a) Insert an unmagnetized iron nail into the coil and record the compass deflection (if any). Now start the current again and observe what happens to the compass deflection. Add additional nails and observe and record what happens to the compass deflection.

(b) Explain briefly why each nail produces a sizable increase in the compass deflection when you turn on the current.

(c) To check whether the nails are permanently magnetized after being inserted into the coil, turn off the current. If the compass still deflects significantly without any current in the coil, the nails have become strongly magnetized. Are they?

(d) Bring one of these nails near the compass, first one end and then the other. Is the nail slightly magnetized?

Experiment 17.12 An electromagnet

Next, construct an "electromagnet." Take an insulated wire about 2 meters in length and wrap it tightly around an iron nail, all along its length. You can make multiple layers of windings, back and forth along the nail, to increase the effects, but leave the ends of the nail sticking out so that you can touch the nail to other objects. Twist the ends of the wire together to help hold the coil together. Connect the coil to a battery and bring it near another nail.

Can you pick up the other nail? (If the battery is not fresh, you may not get enough current to lift the other nail, but you should be able to lift a paper clip!) Why does the other nail fall when you cut the current? Is there any evidence that the electromagnet nail retains some magnetization?

17.11.3 Why are there multiple domains?

Within one domain of a ferromagnetic material such as iron, strong electric forces between neighboring atoms make the atomic magnetic dipole moments line up with each other. Why don't all the atoms in a piece of iron spontaneously align with each other? Why does the piece divide into small magnetic domains with varying magnetic orientations in the absence of an applied magnetic field?

There is a simple way to get some insight into why this happens. Consider two possible patterns of magnetic domains in a bar of iron in the absence of an applied magnetic field, a single domain or two opposed domains (Figure 17.48).

You can simulate this situation by taking two permanently magnetized bar magnets and holding them in one of these two positions, as shown in Figure 17.49.

? Which of these two magnet configurations do you find to be more stable, the parallel or the antiparallel alignment? By analogy, which is more likely for a bar of iron, that it have just one domain or that it split into two domains in the absence of an applied magnetic field?

You will find that if you start with the parallel alignment the magnets tend to flip into the antiparallel alignment. The domain structure of iron can be a compromise between a tendency for neighboring atoms to have their magnetic dipole moments line up with each other (a strong but short-range electric interaction), and a competing tendency for magnetic dipole moments to flip each other (a weaker but longer-range magnetic interaction).

However, the net effect depends critically on the details of the geometric arrangement. Although two long bar magnets that are side by side tend to line up with their magnetic dipole moments in opposite directions, two thin disk-shaped magnets tend to line up with their magnetic dipole moments in the same direction. What happens inside a magnetic material depends on the details of the geometrical arrangement of the atoms and on the details of the electric interactions between neighboring atoms.

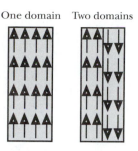

Figure 17.48 Possible arrangements—one magnetic domain or two.

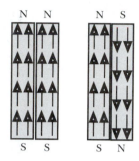

Figure 17.49 Arrange two bar magnets side by side, aligned parallel or antiparallel.

17.12 *Applying the Biot-Savart law to a solenoid

The analysis in this section offers one more example of how to apply the Biot-Savart law. It shows that the magnetic field is nearly uniform along the axis of a solenoid, far from the ends.

Step 1: Cut up the distribution into pieces and draw $\Delta\vec{B}$

We consider a solenoid of length L that is made up of N circular loops wound tightly right next to each other, each of radius R (Figure 17.50). We consider each loop as one piece. The conventional current in the loops is I.

? Given what you know about individual current loops, what will be the direction of the magnetic field $\Delta\vec{B}$ contributed by each of the loops at *any* location along the axis of the solenoid in Figure 17.50?

All of the contributions inside the solenoid will point to the right in Figure 17.50.

Step 2: Write an expression for the magnetic field due to one piece

origin: Center of solenoid
location of one piece: Given by z, so integration variable is z
distance from loop to observation location: $d - z$ (Figure 17.51).

? How many closely packed loops are contained in a short length Δz of the solenoid?

There are N/L loops per meter, so the number of loops in a length Δz is $(N/L)\Delta z$. We know the magnetic field made by each loop along its axis.

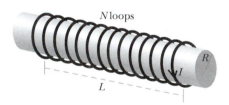

Figure 17.50 A solenoid of length L with N loops of radius R carrying current I.

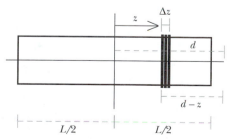

Figure 17.51 Defining an integration variable z for summing the contributions of the many loops.

? Show that the loops contained in the section Δz of the solenoid contribute a magnetic field at the observation location in the z direction of this amount:

$$\Delta B_z = \frac{\mu_0}{4\pi}\frac{2\pi R^2 I}{[(d-z)^2 + R^2]^{3/2}}\frac{N}{L}\Delta z$$

Step 3: Add up the contributions of all the pieces

The net magnetic field lies along the axis and is the summation of all the ΔB_z contributed by all the loops:

$$B_z = \sum \Delta B_z = \sum \frac{\mu_0}{4\pi}\frac{2\pi R^2 I N}{L}\frac{\Delta z}{[(d-z)^2 + R^2]^{3/2}}$$

Many of these quantities are the same for every piece and can be taken outside the summation as common multiplicative factors:

$$B_z = \frac{\mu_0}{4\pi}\frac{2\pi R^2 I N}{L}\sum \frac{\Delta z}{[(d-z)^2 + R^2]^{3/2}}$$

This can be turned into an integral, with z ranging from $-L/2$ to $+L/2$:

$$B_z = \frac{\mu_0}{4\pi}\frac{2\pi R^2 I N}{L}\int_{z=-L/2}^{z=+L/2} \frac{dz}{[(d-z)^2 + R^2]^{3/2}}$$

Let $u = d - z$, in which case $dz = -du$, and therefore the limits on the integration run from $u = d - (-L/2)$ to $u = d - (+L/2)$:

$$B_z = \frac{\mu_0}{4\pi}\frac{2\pi R^2 I N}{L}\int_{u=d+L/2}^{u=d-L/2} \frac{-du}{[u^2 + R^2]^{3/2}} = \frac{\mu_0}{4\pi}\frac{2\pi R^2 I N}{L}\int_{u=d-L/2}^{u=d+L/2} \frac{du}{[u^2 + R^2]^{3/2}}$$

Fortunately, this integral can be found in standard tables of integrals:

$$B_z = \frac{\mu_0}{4\pi}\frac{2\pi R^2 I N}{L}\left[\frac{u}{R^2\sqrt{u^2 + R^2}}\right]_{u=d-L/2}^{u=d+L/2}$$

$$B_z = \frac{\mu_0}{4\pi}\frac{2\pi I N}{L}\left[\frac{d+L/2}{\sqrt{(d+L/2)^2 + R^2}} - \frac{d-L/2}{\sqrt{(d-L/2)^2 + R^2}}\right]$$

In this expression, d is the distance from the center of the solenoid.

Here is a plot of this expression for B_z for a particular solenoid, as a function of the distance from the center of the solenoid. Notice that the magnetic field is nearly uniform inside the solenoid, as long as you are not too near the ends. The field is even more uniform for solenoids that are longer and/or thinner than this one.

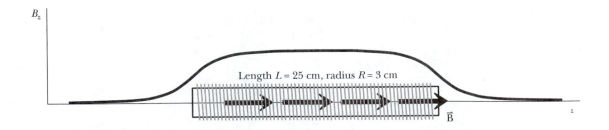

? If the radius R is small compared with the length L ($R \ll L$), show that at the center of the solenoid the magnetic field has a remarkably simple value:

$$B_z \approx \frac{\mu_0 NI}{L}$$

With $d = 0$ and $R \ll L$, the quantity in square brackets above reduces to

$$\frac{L/2}{\sqrt{(L/2)^2}} - \frac{-L/2}{\sqrt{(-L/2)^2}}$$

which is equal to 2, from which follows the simple form $B_z \approx \mu_0 NI/L$.

The function plotted above shows that this is also the approximate magnetic field inside the solenoid along much of the axis, not too close to the ends. Outside the solenoid, the magnetic field falls off rapidly. At a large distance the field falls off like $1/r^3$, since the solenoid then looks like a simple collection of current loops.

MAGNETIC FIELD INSIDE A LONG SOLENOID

$$B_z \approx \frac{\mu_0 NI}{L} \quad \text{inside a long solenoid (radius of solenoid} \ll L)$$

The same result can be obtained with much less effort by using Ampere's law, which we will study in a later chapter.

Step 4: Check the result

Does our final result make sense? In particular, do we have the right units? Comparing with the magnetic field for a single current element,

$$\frac{\mu_0}{4\pi} \frac{I\Delta \vec{\mathbf{l}} \times \hat{\mathbf{r}}}{r^2}$$

we easily verify that our answer does have the right units, since

$$\frac{NI}{L} \quad \text{has the same units as} \quad \frac{I\Delta l}{r^2}$$

(Remember that $\hat{\mathbf{r}}$ is dimensionless.)

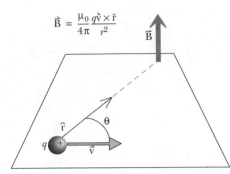

Figure 17.52 The Biot-Savart law.

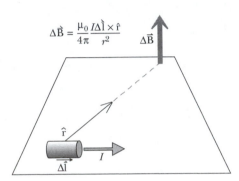

Figure 17.53 The Biot-Savart law for currents.

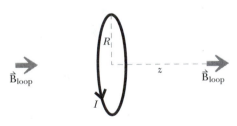

Figure 17.54 Magnetic field of a wire.

Figure 17.55 Magnetic field of a loop.

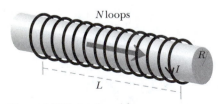

Figure 17.56 A solenoid.

17.13 Summary

Fundamental principles

The Biot-Savart law

$$\vec{B} = \frac{\mu_0}{4\pi} \frac{q\vec{v} \times \hat{r}}{r^2} \text{ (single moving particle, Figure 17.52)}$$

$$\Delta\vec{B} = \frac{\mu_0}{4\pi} \frac{I\Delta\vec{l} \times \hat{r}}{r^2} \text{ (current distribution, Figure 17.53)}$$

$$\frac{\mu_0}{4\pi} = 10^{-7} \frac{\text{tesla·m}^2}{\text{coulomb·m/s}} \text{ } exactly,$$

$\hat{r}$ is a unit vector that points *from* the source charge
toward the observation location.

Results

Electron current: $i = nA\bar{v}$

Conventional current: $I = |q|nA\bar{v}$

A magnetic field can be detected by a compass; horizontal component of $B_{\text{Earth}} \approx 2\times10^{-5}$ tesla in much of the continental United States.

Magnetic field of a wire (Figure 17.54)

$$B_{\text{wire}} = \frac{\mu_0}{4\pi} \frac{LI}{r\sqrt{r^2 + (L/2)^2}} \approx \frac{\mu_0}{4\pi}\frac{2I}{r} \text{ for } r << L$$

for a straight wire of length L and conventional current I at a perpendicular distance r from the center of the wire

Magnetic field of a loop (Figure 17.55)

$$B_{\text{loop}} = \frac{\mu_0}{4\pi}\frac{2\pi R^2 I}{(z^2 + R^2)^{3/2}} \approx \frac{\mu_0}{4\pi}\frac{2\mu}{z^3} \text{ for } z >> R$$

for a circular loop of radius R and conventional current I at a distance z from the center, along the axis

Magnetic dipole moment of a loop: $\mu = (\pi R^2)I$

Magnetic field inside a long solenoid (Figure 17.56)

$$B_z \approx \frac{\mu_0 NI}{L} \text{ inside a long solenoid (radius of solenoid} << L)$$

Note that a solenoid is very long compared to its radius—it is not the same thing as a coil composed of a few loops close together.

Ferromagnetic material

Organized into domains of aligned atomic magnetic dipole moments, and an applied field can orient these domains. Removal of the applied field may leave the material partially aligned, forming a permanent magnet.

17.14 Example problem: A circuit in the Antarctic

In a research station in the Antarctic, a circuit containing a partial loop of wire, of radius 5 cm, lies on a table. A top view of the circuit (looking down on the table) is shown in Figure 17.57. A current of 5 amperes runs in the circuit in the direction shown. You have a bar magnet with magnetic moment 1.2 A·m². How far above location A, at the center of the loop, would you have to hold the bar magnet, and in what orientation, so that the net magnetic field at location A would be zero? The magnitude of the Earth's magnetic field in the Antarctic is about 6×10^{-5} T.

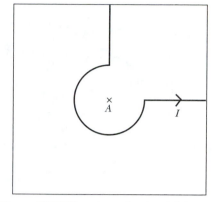

Figure 17.57 Top view of a circuit lying on a table (looking down at the table).

Solution

Direction: The right hand rule for current loops indicates that the circuit shown makes a magnetic field pointing up (out of the table). In the Antarctic the Earth's magnetic field also points approximately up (out of the ground). Therefore the bar magnet must be held with its north pole downwards, to make a magnetic field down (into the ground).

Magnitude of B due to current:

The straight wires contribute nothing to the magnetic field at A, because for each wire $d\vec{l} \times \hat{r} = 0$.

To find B at the center of the 3/4 loop we can integrate the Biot-Savart law:

$$\vec{B} = \int d\vec{B} = \int \frac{\mu_0}{4\pi} \frac{I d\vec{l} \times \hat{r}}{r^2}$$

We choose an origin at the center of the loop. The magnitude of $\hat{r}$ is constant, and equal to R, the radius of the loop. At all locations $d\vec{l} \perp \hat{r}$, so $\left| d\vec{l} \times \hat{r} \right| = dl \sin 90° = dl = R d\theta$, as shown in Figure 17.58. So

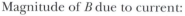

$$\left| \vec{B}_{loop} \right| = \int_{\pi/2}^{2\pi} \frac{\mu_0}{4\pi} \frac{I(R d\theta)}{R^2} = \frac{\mu_0}{4\pi} \frac{IR}{R^2} \int_{\pi/2}^{2\pi} d\theta = \frac{\mu_0}{4\pi} \frac{I}{R} \theta \Big|_{\pi/2}^{2\pi} = \frac{\mu_0}{4\pi} \frac{3\pi I}{2R}$$

$$\left| \vec{B}_{loop} \right| = \left(1 \times 10^{-7} \frac{\text{T} \cdot \text{m}}{\text{A}} \right) \frac{(3\pi)(5 \text{ A})}{2(0.05 \text{ m})} = 4.7 \times 10^{-5} \text{ T}$$

Magnitude of B due to bar magnet must therefore be:

$$B_{mag} = B_{loop} + B_{Earth} = 4.7 \times 10^{-5} \text{ T} + 6 \times 10^{-5} \text{ T} = 1.1 \times 10^{-4} \text{ T}$$

Since $B_{mag} \approx \frac{\mu_0}{4\pi} \frac{2\mu}{z^3}$ we have

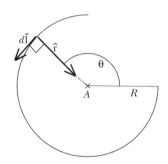

Figure 17.58 $d\vec{l} \perp \hat{r}$ for every step around the loop. The magnitude of r is constant and equal to the radius of the loop.

$$z = \left[\frac{\frac{\mu_0}{4\pi}(2\mu)}{B_{mag}} \right]^{1/3} = \left[\frac{\left(1 \times 10^{-7} \frac{\text{T} \cdot \text{m}}{\text{A}} \right)(2)(1.2 \text{ A} \cdot \text{m}^2)}{(1.1 \times 10^{-4} \text{ T})} \right]^{1/3} = 0.13 \text{ m}$$

So we hold the magnet with its center about 13 cm above the circuit, with its north pole toward the ground.

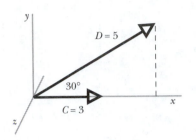

Figure 17.59 The angle between two vectors is 30 degrees (RQ 17.1).

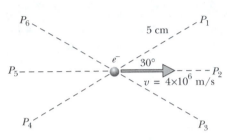

Figure 17.60 An electron moves to the right with speed 4×10^6 m/s (RQ 17.2).

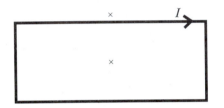

Figure 17.61 A rectangular loop of wire carrying current I (RQ 17.3).

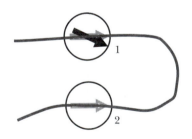

Figure 17.62 The wire rests on top of two compasses (RQ 17.5).

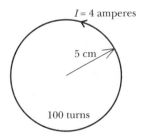

Figure 17.63 A current of 4 A runs through a thin circular coil of wire with 100 turns (RQ 17.8).

17.15 Review questions

Cross product

RQ 17.1 A vector $\vec{C}$ of magnitude 3 lies along the x axis, and a vector $\vec{D}$ of magnitude 5 lies in the xy plane, 30 degrees from the x axis (Figure 17.59). What is the magnitude and direction of the cross product $\vec{C} \times \vec{D}$? What is the magnitude and direction of the cross product $\vec{D} \times \vec{C}$? Draw both vectors on a diagram.

The Biot-Savart law

RQ 17.2 An electron is moving horizontally to the right with speed 4×10^6 m/s. What is the magnetic field due to this moving electron at the indicated locations in Figure 17.60? (Each location is 5 cm from the electron.) Give both magnitude and direction of the magnetic field at each location.

RQ 17.3 What is the direction of the magnetic field at the indicated locations inside and outside this current-carrying rectangular coil of wire shown in Figure 17.61? Explain briefly. (Direction of conventional current is shown.)

Measuring magnetic field with a compass

RQ 17.4 A current-carrying wire oriented north-south is laid on top of a compass. If the compass deflection is $17°$, what is the magnitude of the magnetic field due to the current?

RQ 17.5 Consider the portion of a circuit shown in Figure 17.62. When no current is running, both compasses point north (direction shown by the gray arrows). When current runs in the circuit, the needle of compass 1 deflects as shown. What direction will the needle of compass 2 point? Draw a sketch indicating its deflection.

Field of a straight wire

RQ 17.6 In a circuit consisting of a long bulb and two flashlight batteries in series the conventional current is about 0.1 ampere. What is the magnetic field 5 mm from the wire? (This is about how far away the compass needle is when you place the wire on top of the compass.) Is this a big or a small field?

Two straight wires

RQ 17.7 A battery is connected to a Nichrome wire and a conventional current of 0.3 ampere runs through the wire. The wire is laid out in the form of a rectangle 50 cm by 3 centimeters. What is the magnetic field at the center of the rectangle? Give the direction as well as the magnitude.

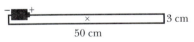

Two straight wires (RQ 17.7).

Coil of wire

RQ 17.8 A thin circular coil of wire of radius 5 cm consists of 100 turns of wire, as shown in Figure 17.63. If the conventional current in the wire is 4 amperes, what are the magnitude and direction of the magnetic field at the center of the coil? (Direction of conventional current is shown.)

Dependence on distance

RQ 17.9 (a) How can you produce a magnetic field that is nearly uniform in a region?

(b) How can you produce a magnetic field that falls off like $1/r$?

(c) How can you produce a magnetic field that falls off like $1/r^2$? (Note that you *cannot* use just a short piece of current-carrying wire, because the other parts of the wire also contribute.)

(d) How can you produce a magnetic field that falls off like $1/r^3$?

Magnetic materials

RQ 17.10 An iron bar magnet makes a pattern of magnetic field that looks just like the pattern of magnetic field outside a long current-carrying coil of wire. Are there currents in the iron? Explain briefly.

RQ 17.11 Suppose you have two Alnico bar magnets, one with a mass of 100 grams and one with a mass of a kilogram. At a distance of a meter from the center of either one, how would the magnetic field differ? Why?

17.16 Homework problems

17.16.1 In-line problems from this chapter

Problem 17.1 (page 604) Magnetic dipole moment of your bar magnet
The magnetic field along the axis of a bar magnet can be written like this:

$$B_{\text{axis}} \approx \frac{\mu_0}{4\pi} \frac{2\mu}{r^3}$$

Use your own data from Experiment 17.8 on page 604 to determine the magnetic moment μ of your bar magnet. If you have mislaid those data, quickly repeat the measurements now. You will use the value of your magnetic moment in later work. The magnetic moment describes how strong a magnet is: the bigger the magnetic moment, the bigger the magnetic field at some distance r.

Problem 17.2 (page 608) Predicting the magnetic dipole moment of your magnet
Determine the mass of your magnet and thereby determine the number N of atoms in your magnet, assuming it is made of iron. Assuming that each of the atoms has a magnetic moment with a value estimated above, see how well the atomic model for a magnet fits the measured value of the magnetic moment of your bar magnet (Problem 17.1 (page 604) Magnetic dipole moment of your bar magnet).

17.16.2 Additional problems

Problem 17.3 Fields of an electron
The electron in Figure 17.64 is traveling with a speed of 3×10^6 m/s.

(a) At location A, what are the directions of the electric and magnetic fields contributed by the electron?

(b) Calculate the magnitudes of the electric and magnetic fields at location A.

Problem 17.4 Wire on compass
You place a long straight wire on top of your compass, and the wire is a height of 5 millimeters above the compass needle. If the conventional current in the wire is $I = 0.2$ ampere and runs left to right as shown, calculate

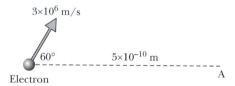

Figure 17.64 An electron makes electric and magnetic fields (Problem 17.3).

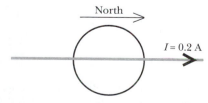

Figure 17.65 Wire on top of a compass (Problem 17.5).

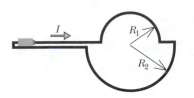

Figure 17.66 Two arc-shaped wires (Problem 17.5).

the approximate angle the needle deflects away from North and draw the position of the compass needle.

Problem 17.5 Magnetic field of arcs

Two long wires lie very close together and carry a conventional current I as shown in Figure 17.66, and each wire has a semi-circular kink, one of radius R_1 and the other of radius R_2. Calculate the magnitude and direction of the magnetic field at the common center of the two semi-circular arcs

Problem 17.6 Deflecting a compass needle

When you bring a current-carrying wire down onto the top of a compass, aligned with the original direction of the needle and 5 mm above the needle, the needle deflects by 10 degrees (Figure 17.67).

(a) Show on a diagram the direction of conventional current in the wire, and the direction of the additional magnetic field made by the wire underneath the wire, where the compass needle is located. Explain briefly.

(b) Calculate the amount of current flowing in the wire. The measurement was made at a location where the horizontal component of the Earth's magnetic field is $B_{Earth} \approx 2 \times 10^{-5}$ tesla.

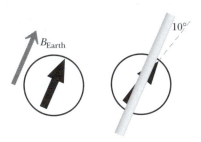

Figure 17.67 Deflecting a compass needle (Problem 17.6).

Problem 17.7 Using magnetic field to measure current

You can use measurements of the magnetic field of a coil to determine how much current your battery is supplying to the coil. Using your value of B (page 600), determine the conventional current I through your coil. If this current is less than 3 ampere, you should replace the battery.

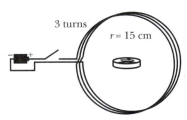

Figure 17.68 Wire with a loop (Problem 17.8).

Problem 17.8 Wire with a loop in it

A very long wire carrying a conventional current I is straight except for a circular loop of radius R (Figure 17.68). Calculate the magnitude and direction of the magnetic field at the center of the loop.

Problem 17.9 Deflecting a compass needle with a coil

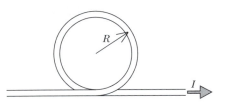

A thin circular coil of radius $r = 15$ cm contains $N = 3$ turns of Nichrome wire. A small compass is placed at the center of the coil, as shown in Figure 17.69. With the battery disconnected, the compass needle points to the right, in the plane of the coil. Assume that the horizontal component of the Earth's magnetic field is about $B_{Earth} \approx 2 \times 10^{-5}$ tesla.

When the battery is connected, a current of 0.25 ampere runs through the coil. Predict the deflection of the compass needle. If you have to make any approximations, state what they are. Is the deflection outward or inward as seen from above? What is the magnitude of the deflection?

Figure 17.69 Deflecting a compass needle with a coil (Problem 17.9).

Problem 17.10 Magnetic moment of a bar magnet

In Figure 17.70 a bar magnet is aligned east-west, with its center 16 cm from the center of a compass. The compass is observed to deflect 50° away from north as shown, and the horizontal component of the Earth's magnetic field is known to be 2×10^{-5} tesla .

(a) Label the N and S poles of the bar magnet and explain your choice.

(b) Determine the magnetic dipole moment of this bar magnet, including correct units.

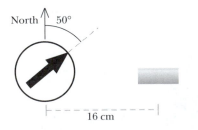

Figure 17.70 Determine the magnetic dipole moment of this bar magnet (Problem 17.10).

Problem 17.11 Magnetic field in a circuit

Figure 17.71 shows a circuit consisting of a battery and a Nichrome wire, through which runs a current I.

(a) At the location marked × (the center of the semicircle), what is the direction of the magnetic field?

(b) At the location marked × (the center of the semicircle), what is the magnitude of the magnetic field? If you have to make any approximations, state what they are.

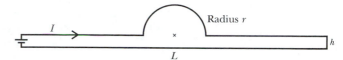

Figure 17.71 Magnetic field in a circuit (Problem 17.11).

Problem 17.12 Magnetic field of coils

Two thin coils of radius 3 cm are 20 cm apart and concentric with a common axis. Both coils contain 10 turns of wire with a conventional current of 2 amperes that runs counter-clockwise as viewed from the right side (Figure 17.72).

(a) What is the magnitude and direction of the magnetic field on the axis, halfway between the two loops, without making the approximation $z \gg r$? (For comparison, remember that the horizontal component of magnetic field in the United States is about 2×10^{-5} tesla).

(b) In this situation, the observation location is not very far from either coil. How bad is it to make the $1/z^3$ approximation? That is, what percentage error results if you calculate the magnetic field using the approximate formula for a current loop instead of the exact formula?

(c) What is the magnitude and direction of the magnetic field midway between the two coils if the current in the right loop is reversed to run clockwise?

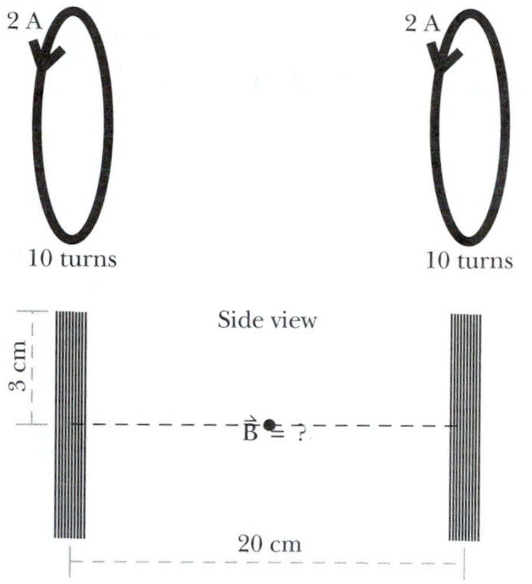

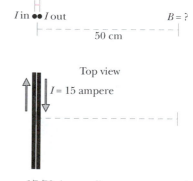

Figure 17.72 Magnetic field of coils (Problem 17.12).

Problem 17.13 Magnetic fields in the home

At one time, concern was raised about the possible health effects of the small alternating (60-hertz) magnetic fields created by electric currents, in houses and near power lines. In a house, most wires carry a maximum of 15 amperes (there are 15-ampere fuses that melt and break the circuit if this current is exceeded). The two wires in a home power cord are about 3 millimeters apart, as shown in Figure 17.73, and at any instant they carry currents in opposite directions (both of which change direction 60 times per second).

(a) Calculate the maximum magnitude of the alternating magnetic field, 50 cm away from the center of a long straight power cord that carries a current of 15 amperes. Both wires are at the same height as the observation location.

(b) Explain briefly why twisting the pair of wires into a braid as shown would minimize the magnetic field at the location discussed in (a).

Figure 17.73 An appliance power cord consists of two wires side by side (Problem 17.4).

(The magnitude of the field that you calculate is very small compared to the Earth's magnetic field, but there were questions as to whether a very small *alternating* magnetic field might have health effects. After many detailed studies, the consensus of most scientists now seems to be that these small alternating magnetic fields are not a hazard after all.)

Problem 17.14 A bent wire

A conventional current I runs in the direction shown in Figure 17.74. Determine the magnitude and direction of the magnetic field at point C, the center of the circular arcs.

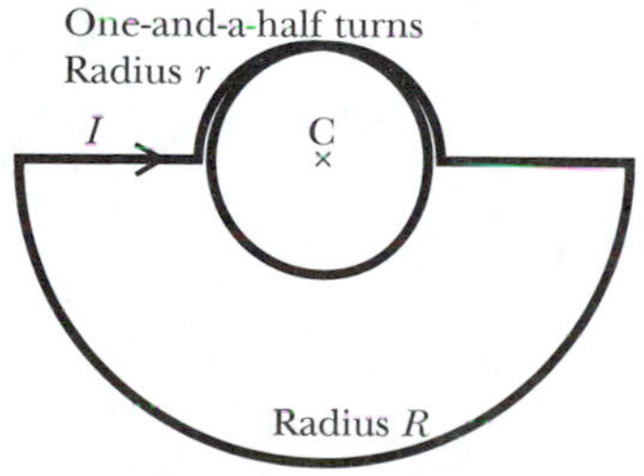

Figure 17.74 Conventional current I runs in this circuit (Problem 17.7).

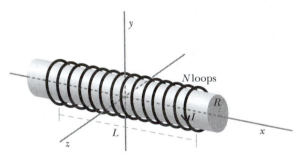

Figure 17.75 A thin wire is part of an electrical circuit (Problem 17.15).

Problem 17.15 Magnetic field of a current

A thin wire is part of a complete electrical circuit which carries a current I. For this problem consider only the piece of wire of length d as shown in the figure below. Answer the following questions based on this figure.

(a) What are the magnitude and direction of the magnetic field due to the wire at Q (location $<-w,0,0>$?

(b) Set up the integrals necessary to determine the x, y, and z components of the magnetic field at P (location $<-w,h,0>$). The integrals must be in a form which can be evaluated (no cross products in the integrand), but you do not need to evaluate them.

(c) What is the direction of the magnetic field at location P?

Problem 17.16: Calculating the magnetic field of a solenoid

This problem requires calculations in three dimensions and will familiarize you with the details of the magnetic field made by a solenoid (a long coil).

A solenoid of length $L = 0.5$ meter and radius $R = 3$ cm is wound with $N = 50$ turns of wire carrying a current $I = 1$ ampere. Its center line lies on the x axis, with the origin at the center of the solenoid (Figure 17.76).

(a) Calculate and display magnetic field vectors at the locations in the xy plane, inside and outside of the solenoid, shown in Figure 17.77. It is acceptable to approximate the helix as simple loops, if you find that is easier to do. (If you have a helix instead of loops, $\Delta \vec{l}$ also has a z component.) Do the pattern and direction of magnetic field make sense?

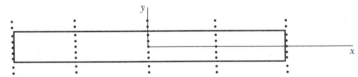

Figure 17.77 Locations at which to calculate the magnetic field of a solenoid.

(b) Display the numerical value of the magnitude of the magnetic field at one location, the center of the solenoid. Also display the theoretical numerical value of the magnetic field, in the approximation that this is a very long solenoid ($B \approx \mu_0 NI/L$). What is the minimum number of steps around one loop (or around one turn of the helix) that are necessary to obtain good agreement between the theoretical value and your numerical integration? What is your criterion for "good agreement"?

(c) Vary the number of loops. What is the minimum number of loops (or turns of the helix) that are necessary to get an approximately uniform field inside the solenoid?

(d) How do the magnitude and direction of the magnetic field outside the solenoid compare to the magnitude and direction of the magnetic field inside the solenoid?

Suggestions

Consider $\vec{r}$ to be a vector from the midpoint of $\Delta \vec{l}$ to the observation location.

One way to find $\Delta \vec{l}$ is to find two vectors $\vec{a}$ and $\vec{b}$ to the endpoints of $\Delta \vec{l}$, then subtract to get $\Delta \vec{l}$, as shown in Figure 17.78.

One way to get started is to calculate the magnetic field for a single loop located at the origin, then extend the program to include 50 loops.

In debugging your program, you may find it useful to display the $\Delta \vec{l}$ vector for each step.

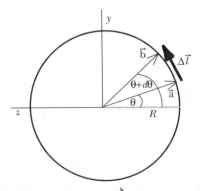

Figure 17.78 Finding $\Delta \vec{l}$.

17.17 Answers to exercises

17.1 (page 585) 6×10^{18} electrons/s

17.2 (page 585) 5.4×10^{21} electrons

17.3 (page 588) 5.4×10^{-6} T

17.4 (page 589) $\dfrac{\langle (v_y z - v_z y), (v_z x - v_x z), (v_x y - v_y x) \rangle}{\sqrt{x^2 + y^2 + z^2}}$

17.5 (page 589)

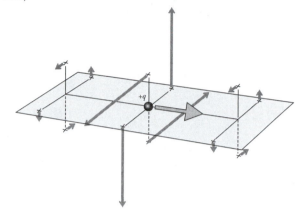

17.6 (page 590)

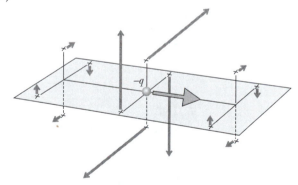

17.7 (page 590)   Cross product involves $\sin\theta$, which is zero.

17.8 (page 590) $1/r^2$

17.9 (page 590)   zero magnetic field

17.10 (page 590)   14 tesla, which is an extremely large magnetic field

17.11 (page 590)   Electron current flows to the left, conventional current to the right.

17.12 (page 590)   Electron current emerges from the − end of the battery; conventional current emerges from the + end.

17.14 (page 593) 5×10^{-5} m/s

17.15 (page 593)   100 minutes

17.16 (page 594)   0.54 ampere

17.17 (page 600)

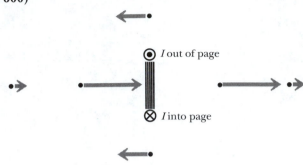

17.18 (page 601) $9 \, \text{A} \cdot \text{m}^2$

Chapter 18

A Microscopic View of Electric Circuits

Chapter 18

A Microscopic View of Electric Circuits

In Chapter 17 we studied the magnetic fields made by moving charges. However, we did not discuss the effect of those magnetic fields on other charges. It turns out that only moving charges are affected by magnetic fields, so one situation in which we can easily observe these effects is in electric circuits where currents are flowing. Before we can study magnetic forces in detail, we need to know more about the motion of charges in electric circuits.

This chapter presents a somewhat different view of electric circuits from the macroscopic view you may have seen in a previous physics course, because we are interested in constructing models of how things work at a microscopic level. Our goal in this chapter is to construct a microscopic model of what happens in an electric circuit, using familiar physics principles (the energy principle, the momentum principle), the concepts of electric field and potential, and the microscopic model of a metal that we developed in Chapter 14. We will be interested in questions such as:

Are charges used up in a circuit?

How is it possible to create and maintain a nonzero electric field inside a wire?

What is the role of the battery in a circuit?

Nonequilibrium systems

When an electric field is applied to a conductor, the mobile charges in the conductor experience forces, and begin to move in the direction of those forces. We have previously studied situations in which the flow of charges lasts for a very short time, until the buildup of charge at the surface of the conductor produces an electric field that is equal and opposite to the applied field, the net electric field inside the conductor becomes zero, and the system reaches static equilibrium.

In an electric circuit, the system does not reach equilibrium. Despite the motion of charges, the net field inside the conductor does not go to zero, and charge flow continues for a long period in this nonequilibrium system.

Conventional current and electron current

In Chapter 17 we talked about both electron current and conventional current. Remember that:

$$\text{electron current } i = \# \text{ electrons/s passing a point}$$

$$\text{electron current flows in a direction opposite to } \vec{E}$$

$$\text{conventional current I} = \# \text{ coulombs/s} = |q|\,i$$

$$\text{conventional current flows in the direction of } \vec{E}$$

18.1 Current in different parts of a circuit

As a first step in constructing a microscopic model to explain current flow in a circuit, we need to address a crucial question, both experimentally and theoretically: What happens to the charges that flow through the circuit? Is there a different amount of current in different parts of a series circuit?

18.1.1 The steady state

You have seen that once you have assembled a circuit, current keeps flowing at about the same rate for quite a long time. The brightness of a light bulb or the amount of compass deflection does not change noticeably over several minutes. In fact, since it takes quite a while for the batteries to run down under these circumstances, the current remains about the same for several hours. We say that these circuits are in a "steady state." The steady state is not the same as static equilibrium.

DEFINITIONS OF "STATIC EQUILIBRIUM" AND "STEADY STATE"

"Static equilibrium" means that no charges are flowing.

"Steady state" means that charges are moving, but their velocities at any location do not change with time, and there is no change in the deposits of excess charge anywhere.

18.1.2 Current in different parts of a simple circuit

We will measure the amount of current flowing in different parts of a circuit in the steady state. We will use compass deflections as an indicator of current, since the magnetic field made by a wire is proportional to the current in the wire.

Before making the measurement, it is instructive to think about what we expect to see. Consider a circuit consisting of two batteries and a round bulb, as shown in Figure 18.1. Electron current flows from the negative terminal of the battery toward the bulb.

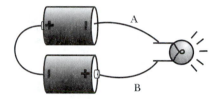

Figure 18.1 A circuit containing a round bulb and two batteries in series.

? How would you expect the amount of electron current at location A to compare to the electron current at location B? Make a clear prediction—write it down. (It is ok to guess.)

Most people choose one of the following possibilities:
(1) There should be no current at all at B, because all of the electron current coming from the negative end of the battery is used up in the bulb.
(2) The current at B should be less than the current at A, because some of the current is used up to make the bulb give off light and heat.
(3) The current should be the same at A and at B.

Before doing the experiment, let's consider these alternatives from a theoretical viewpoint, using what we know about electric interactions and the properties of metals.

Can current be used up in the bulb?

It is reasonable to expect that something will be used up in the bulb in order to produce heat and light. Let's consider whether electric current can be the thing that is used up.

? Current in a metal wire is simply the flow of electrons past a point. Can electrons be used up in the bulb?

Electrons alone cannot be destroyed, because this would violate the fundamental principle of conservation of charge. If negative particles were destroyed, the universe would become more and more positive!

We do know that electrons can react with positrons (positively charged anti-electrons), annihilating both particles and producing large amounts of energy. However, ordinary matter cannot coexist with antimatter for longer than a tiny fraction of a second, so there cannot be a supply of positrons in the circuit to annihilate electrons.

? Could electrons just accumulate in the bulb, so that the current at B would be less than at A?

If electrons accumulated in the bulb, the bulb would become negatively charged. The negative charge of the bulb would become large enough to repel incoming electrons and stop the current. Since in the steady state current keeps flowing, this can't be happening. (Moreover, an increasingly negatively charged bulb would strongly attract nearby objects, which you don't observe.)

We have used proof by contradiction to rule out the possibility that electrons are annihilated or get stuck in the bulb. So we must conclude that current cannot be used up in a light bulb. Since we have ruled out both possibilities (1) and (2) by this argument, we expect that alternative (3) is correct: the current should be the same at A and at B.

Measuring current all around a circuit

We have predicted on theoretical grounds that current cannot be used up in a circuit. Therefore, we expect to measure the same current everywhere in a series circuit (a circuit with no parallel branches). The following important experiment will check our theoretical conclusions. Be careful in your experimentation, and be sure to compare your results to those of several other groups of students, and check them with your instructor.

Experiment 18.1 Current in different parts of a circuit

Assemble a circuit containing a round bulb and two batteries in series, as shown in Figure 18.2. Use the following procedure to make careful measurements of the magnetic field produced by each wire. (Note that equal currents will produce equal magnetic fields, and hence equal compass deflections.)

- Disconnect one wire.
- Make sure the batteries and the steel alligator clips are not near the compass.
- Lay the wire in which you wish to measure current over the compass, carefully lining up the wire with the compass needle.
- Carefully reconnect the circuit without moving anything.
- Read the compass deflection to within 2 degrees or better.

(a) Is the circuit in a steady state? Leave the compass under the wire at A, and observe it for 15-20 seconds. Does the deflection remain constant?

(b) Using the compass, measure the current at A and B, and record the observed compass deflections.

(c) Use this sensitive "null" test to compare the current in the two wires: run wires A and B side-by-side in opposite directions over the compass, then connect the circuit. The magnetic fields produced by the two wires are in opposite directions. If the currents are equal and the wires are very close together, the magnetic fields in the two wires will cancel each other and give zero deflection.

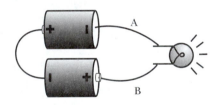

Figure 18.2 A circuit containing a round bulb and two batteries in series.

What is used up in the light bulb?

As you should have observed, the amount of current entering and leaving a lighted bulb in a steady state circuit is the same. You may still feel uneasy about all this. You might ask, "If electrons don't get used up in the bulb, what makes the light that we see? You can't get something for nothing!"

? Rub your hands together as hard as you can and as fast as you can for several seconds. What change do you observe in your hands?

Your hands get hot, but they are certainly not used up! However, the feeling that something must be used up to light the bulb is entirely reasonable.

? What does get used up to warm your hands, or to make a light bulb hot enough to emit visible light?

In both cases energy is being "used up," or rather transformed from one form of stored energy into thermal energy (and energy in the form of emitted light, in the case of the bulb). In order to move your hands you must use up some of the stored chemical energy in your body. Similarly, to force electrons through the filament, some of the chemical energy stored in the battery must be used up.

Forcing electrons through the filament heats the metal with a kind of friction, not unlike the friction that heats your hands when you rub them. In both cases chemical energy is converted into thermal energy, but without destroying the objects that rub against each other.

Since energy flows out of the system (the circuit) in the form of heat and light, we will need to take this into account when we study the energetics of circuits later in this chapter.

18.1.3 Current at a node

We can extend these considerations to more complicated arrangements of wires. In general we have the following important principle, based on charge conservation in the steady state (a node is a junction of two or more conductors in a circuit):

THE CURRENT NODE RULE

In the steady state, the electron current entering a node in a circuit is equal to the electron current leaving that node.

This rule is not really a fundamental principle, but rather a consequence of the fundamental principle of conservation of charge and the definition of steady state. It is also called "the Kirchhoff node rule." According to the current node rule, in the portion of a circuit shown in Figure 18.3, $i_1 = i_2$, and $i_2 = i_3 + i_4$.

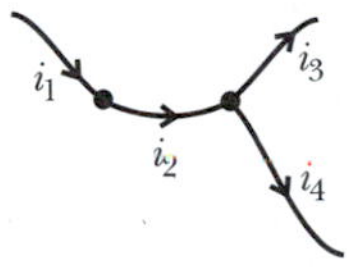

One can of course express this rule in terms of conventional current. Figure 18.4 shows a portion of a steady-state circuit, with the magnitudes of the conventional currents given in amperes (A), and the directions given by arrows. The following exercises apply to this situation:

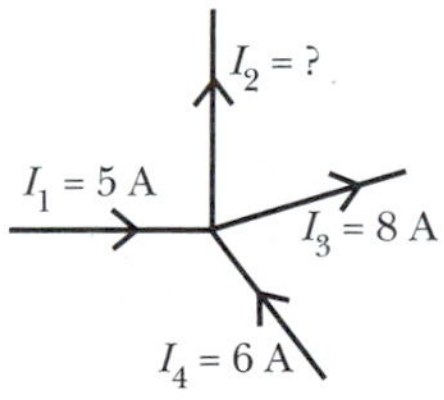

Ex. 18.1 Write the node equation for the circuit in Figure 18.4. What is the value of the outward-going current I_2?

Ex. 18.2 In Figure 18.4, if I_4 is 1 A (instead of 6 A), what is the value of the outward-going current I_2? What is the meaning of the minus sign?

Ex. 18.3 Looking at Figure 18.5, a student said, "Since the current everywhere in a circuit must be the same, $i_1 = i_2$ in this circuit." What is wrong with this statement?

18.2 Start-stop motion of electrons in a wire

We assert that in a current-carrying wire there must be an electric field to drive the sea of mobile electrons (or holes). In this section we will establish a relationship between the electric field and the current. For concreteness we will mainly discuss metals such as copper in which the current is electron

Figure 18.3 Electron current at four different locations in part of a circuit:
$$i_1 = i_2 \quad \text{and} \quad i_2 = i_3 + i_4$$

Figure 18.4 A simple example.

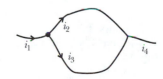

Figure 18.5 Circuit in Exercise 18.3.

current. First we'll examine the contention that the electric field must be nonzero inside a wire to produce a steady state current.

Why is a field necessary?

? According to the momentum principle, once an object is in motion, no force is required to keep it moving at constant speed. Given this, why is an electric field needed to keep a current flowing?

If the mobile electrons in a metal did not interact at all with the lattice of atomic cores, then once a current got started, it could flow forever. However, as we have observed, there is an interaction; the moving electrons do lose energy to the lattice, increasing the thermal motion of the atoms. We detect this by observing that the wire gets hot—in some cases hot enough to emit visible photons. So unless an electric field is present to increase the momentum of the mobile electrons, their energy will quickly be dissipated (as energy can be dissipated by friction), and the current flow will stop.

Electrons cannot push each other through the wire

? One might argue that in a circuit each mobile electron makes an electric field that affects the mobile electron to the left of it ("pushing" on its neighbor), thereby causing current to flow. Given our description of the mobile-electron sea in a metal, what is wrong with this analysis?

As we have stated previously, there can be no excess charge inside a conductor, so the number density of mobile electrons inside a metal wire must equal the number density of positive atomic cores—the inside of the wire is electrically neutral. In Figure 18.6 the repulsion of the mobile electron on the left by a mobile electron on the right is cancelled by the attraction of a neighboring positive core, so the total force on a mobile electron due to the other mobile electrons and the positive cores is zero. The result is that the mobile-electron sea behaves rather like an ideal gas, in which none of the particles interact with other particles. Electrons cannot continually push each other through the wire like peas pushed through a tube from one end of the tube. Rather, it must be other charges somewhere outside the wire that make an electric field throughout the wire that continually drives the electron current. At this point in our analysis it is not yet clear exactly where these charges are.

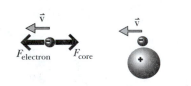

Figure 18.6 A mobile electron cannot push another mobile electron through a wire. The total force on a mobile electron due to the other mobile electrons and the positive cores is zero.

18.2.1 The Drude model

In a simple classical model of electron motion (called the "Drude model" after the physicist who first proposed it), a mobile electron in the metal, under the influence of the electric field inside the metal, does accelerate and gain energy, but then it loses that energy by colliding with the lattice of atomic cores, which is vibrating because of its own thermal energy. After a collision, an electron again gets accelerated, and again collides. This process is what makes the metal filament in a light bulb get hot. Figure 18.7 shows a graph of this start-stop motion for a single electron.

Drift speed and electron mobility

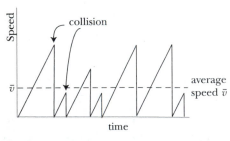

Figure 18.7 A mobile electron speeds up under the influence of the electric field inside the wire, then collides with an atomic core and loses energy.

The average speed of an electron in this start-stop motion is called the "drift" speed $\bar{v}$, and we say that the electron "drifts" through the metal. Actually, the slow drift motion is superimposed on high-speed motion of the electrons in all directions inside the metal, much as the wind is a slow drift motion superimposed on the high-speed motion of air molecules in all directions. A full treatment of electrons in a metal, including the reason for the high-speed motion in all directions, requires quantum mechanics, but the simple classical Drude model allows us to understand most of the important aspects of circuits on a microscopic level.

We express the momentum principle

$$\frac{\Delta \vec{p}}{\Delta t} = \vec{F}_{net}$$

in a form involving finite time steps,

$$\Delta p = F_{net} \Delta t = eE \Delta t$$

where E is the magnitude of the electric field inside the wire, and Δt is the time between collisions. If we make the simplifying assumption that the electron loses all its momentum during each collision, we have

$$\Delta p = p - 0 = eE \Delta t$$

The speed of the electron (of mass m_e) at the time of collision turns out to be small compared to the speed of light, so we have

$$v = \frac{p}{m_e} = \frac{eE \Delta t}{m_e}$$

However, the time between collisions is not the same for all electrons. Some experience longer times between collisions, some shorter times. To get an average, "drift" speed $\bar{v}$ for all electrons at a particular instant, we need the average time $\overline{\Delta t}$ between collisions:

$$\bar{v} = \frac{eE\overline{\Delta t}}{m_e}$$

$\overline{\Delta t}$, the average time between collisions of the electrons with the atomic cores, is determined by the high-speed motion of the electrons, and by the temperature of the metal (at a higher temperature the thermal motion of the atomic cores is greater, and the average time between collisions is reduced, leading to a smaller drift speed for the same field E).

? Is drift speed directly proportional to the magnitude of the electric field?

Assuming that increasing the electric field does not result in a significant change in temperature, then doubling the electric field E doubles the drift speed $\bar{v}$ attained in that time; hence the drift speed is directly proportional to the electric field. The proportionality factor is called the electron "mobility" and is denoted by u (or by μ in some books).

$$\bar{v} = uE$$

Evidently, $u = \dfrac{e}{m_e}\overline{\Delta t}$

Different metals have different electron mobilities. The higher the mobility the higher the drift speed for a given electric field:

MOBILITY *u*

$$\bar{v} = uE$$

Combining this with the expression for electron current $i = nA\bar{v}$,

we get $i = nAuE$

If, as in most metals, the mobile charges are negative, the direction of the drift velocity is opposite to the direction of the electric field. In a conductor whose mobile charges are positive (holes), the direction of the drift velocity will of course be the same as the direction of the electric field.

Magnitude of E in a circuit

Before continuing, it is useful to do the following calculation to become familiar with the approximate magnitude of the electric field in a circuit wire.

Ex. 18.4 In the previous chapter you calculated the drift speed in a copper wire to be 5×10^{-5} m/s for a typical electron current. Calculate the magnitude of the electric field E inside the copper wire. The mobility of mobile electrons in copper is

$$u = 4.5\times10^{-3}\frac{\text{m/s}}{\text{N/C}}$$

(Note that though the electric field in the wire is very small, it is adequate to push a sizeable electron current through the copper wire.)

18.2.2 Electric field and drift speed in different elements of a circuit

We confirmed by experiment that in the steady state, the current was the same everywhere in a series circuit (a circuit without parallel branches). However, we know that the electron current i depends on the cross sectional area A of a wire, since $i = nA\bar{v}$. The cross-sectional area of the tungsten filament of a light bulb is much less than the cross-sectional area of the copper connecting wires in the circuit (Figure 18.8).

In order to simplify thinking about this situation, let's first consider a circuit in which a wire leads into another, thinner wire of the same material.

? How must the drift speed of mobile electrons in the thin wire compare to the drift speed of mobile electrons in the thick wire?

Figure 18.8 The drift speed of electrons in the thin and thick wires.

Since the current must be the same in the thick and thin wires, the electron drift speed must be greater in the thin wire.

$$nA_{\text{thin}}\bar{v}_{\text{thin}} = nA_{\text{thick}}\bar{v}_{\text{thick}}$$

$$\text{and therefore } \bar{v}_{\text{thin}} = \frac{A_{\text{thick}}}{A_{\text{thin}}}\bar{v}_{\text{thick}}$$

Note that this is very different from "compressible" flows such as highway traffic. On a highway the density of cars (how close they are to each other) can vary. For example, when a highway narrows down from two lanes to one, cars usually go slower in the narrow region but are correspondingly closer together (higher density). Because the electron sea in a metal is nearly incompressible (its density cannot change significantly), in the steady state the electrons must go faster in the narrow region. The density n of mobile electrons in copper is the same in both wires.

? If the electron drift speed is greater in the thin wire, how must the electric field in the thin wire compare to the field in the thick wire?

Figure 18.9 Electric field in the thin and thick wires.

Since the drift speed must be greater in the thinner wire, the electric field in the thinner wire must also be larger, since $\bar{v} = uE$ (Figure 18.9).

Ex. 18.5 Suppose a wire leads into another, thinner wire of the same material which has only half the cross sectional area. In the "steady state," the number of electrons per second flowing through the thick wire must be equal to the number of electrons per second flowing through the thin wire. If the drift speed $\bar{v}_1$ in the thick wire is 4×10^{-5} m/s, what is the drift speed $\bar{v}_2$ in the thinner wire?

Ex. 18.6 If the electric field E_1 in the thick wire is 9×10^{-3} N/C, what is the electric field E_2 in the thinner wire?

Ex. 18.7 Suppose wire A and wire B are made of different metals, and are subjected to the same electric field in two different circuits. Wire B has four times the cross-sectional area, 1.5 times as many mobile electrons per cubic centimeter, and twice the mobility of wire A. In the steady state, 1×10^{18} electrons enter wire A every second. How many electrons enter wire B every second?

18.2.3 Direction of electric field in a wire

According to our measurements of the magnetic field made by the moving charges in our circuits, the current is the same in every part of a series circuit. Since the amount of current is proportional to E, and since conventional current flows in the direction of $\vec{E}$, this must mean that E is the same in every part of a wire in the circuit, and that $\vec{E}$ must be parallel to the wire at every location—even if the wire twists and turns, as indicated in Figure 18.10.

18.2.4 Proof that current fills the wire

Not only must the electric field follow the wire, but it must be uniform across a cross section of the wire. Suppose for a moment that this isn't true, that instead at different locations in the cross section of a straight wire the drift speed and therefore the electric field are different (Figure 18.11).

Consider the potential difference

$$\Delta V_{ABCDA} = -\int_A^A \vec{E} \cdot d\vec{l}$$

along the round-trip path *ABCDA* (from *A* to *B* to *C* to *D* and back to *A*).

? Calculate the contributions along each part of the path, ΔV_{AB}, ΔV_{BC}, ΔV_{CD}, and ΔV_{DA}, and the sum ΔV_{ABCDA}. What can you conclude about E_1 and E_2?

The requirement that the round-trip potential difference be zero means that E_1 and E_2 have to be equal. Therefore the electric field must be uniform both along the length of the wire and also across the cross-sectional area of the wire. Since the drift speed is proportional to E, we find that the current is indeed uniformly distributed across the cross section. This result is true only for uniform cross section and uniform material, in the steady state. The current is not uniformly distributed across the cross section in the case of high-frequency (non-steady-state) alternating currents, because time-varying currents can create non-Coulomb forces, as we will see in a later chapter on Faraday's law.

18.3 What charges make the electric field in the wires?

We have concluded that in a steady state circuit:
- There must be an electric field in the wires
- The magnitude of $\vec{E}$ must be the same throughout a wire of the same geometry and material

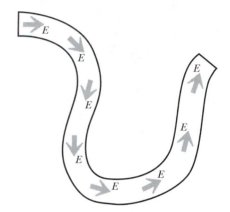

Figure 18.10 The electric field in a current-carrying wire must follow the wire.

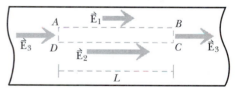

Figure 18.11 Could E vary at different locations in a uniform wire?

- The direction of the electric field at every location must be along the wire, since current flow follows the wire

Consider a very simple circuit consisting of a bulb connected by long wires to a battery (Figure 18.12). We have concluded that there must be an electric field inside the metal bulb filament, forcing electrons through the filament, heating it so much that it glows. Since electric fields are produced by charges, there must be excess charges somewhere to produce the electric field inside a wire in a circuit. Where might these charges be?

Are the excess charges inside the conductors?

The bulb filament and the wires are conductors, so there can be no excess charge in the interior of the bulb filament and wires. In these metal objects, there are equal numbers of mobile electrons and positive atomic cores; the excess charges must be somewhere else.

Are the excess charges on the battery?

Figure 18.12 There is an electric field in the bulb filament. Is it charges in and on the battery that produce this field?

One might reasonably assume that the charges that make the electric field are in and on the battery, since that's the active element in the circuit. Perhaps there are + charges on the + end of the battery, and − charges on the − end? In that case the field made by the battery will be something like the field of a dipole.

? Suppose the bulb is 10 cm from the battery, and shining brightly. Move the bulb to a distance of about 1 cm from the battery, allowing the wires to bend (Figure 18.13). Very roughly, about how much larger is the electric field made by the battery at this closer location?

Although we're too close to the battery for the electric field to be proportional to $1/r^3$, a rough estimate is that the electric field at 1 cm distance is about $10^3 = 1000$ times as large as the field at 10 cm.

Figure 18.13 Bend the wires so the bulb is much closer to the battery. Presumably the electric field in the bulb should get much larger?

? Now that the field made by the battery at the location of the bulb filament has increased by a factor of 1000 or so, what should happen to the electron current in the bulb, and the brightness of the bulb?

For an ordinary metal, the electron current is proportional to the electric field: $i = nA\bar{v} = nAuE$. Therefore the electron current should increase by about a factor of 1000. The bulb should be enormously brighter than before! Yet this doesn't happen. As you have seen in many experiments, moving the bulb has no effect at all on the bulb brightness, or on the amount of current indicated by your compass.

Another test of the "charge on the battery" hypothesis also fails. To drive current through the bulb filament, the electric field must have a component parallel to the filament. Yet if you rotate the bulb, you don't see any change in the bulb brightness (Figure 18.14). If the charges responsible for making the electric field were solely in and on the battery, rotating the bulb ought to make a big difference in the brightness—the bulb might even go out.

The fact that the bulb doesn't get brighter when moved toward the battery, and does not go out when rotated, proves that charges in and on the battery cannot be the only contributors to the electric field in the bulb filament. There must be other charges somewhere else (and these charges can't be inside the wires or bulb filament, either, where there is no net charge, averaged over a few atomic diameters). This is deeply puzzling.

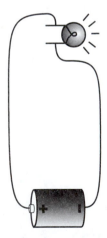

Figure 18.14 Rotate the bulb. Does the brightness change, due to a change in the parallel component of electric field?

18.3.1 A mechanical battery

To simplify our further analysis of circuits, including in particular our search for the location of the charges responsible for producing the electric field in a bulb filament, we introduce a "mechanical battery" that is much easier to understand in detail than a chemical battery yet behaves in a circuit in a way that is very similar to that of a chemical battery.

Our mechanical battery has a conveyor belt driven by a motor or a hand crank that pulls electrons out of a positive plate and pushes them onto a negative plate (Figure 18.15). This action replenishes electrons that leave the negative plate and move through the wire with drift speed $\bar{v}$, and it removes electrons that enter the positive plate after traveling through the wire. You may have seen a "Van de Graaff" generator which has a motor-driven conveyor belt like this that pumps lots of charge onto a metal sphere.

This combination of charged plates plus the motor-driven conveyor belt acts very much like a chemical battery. The main difference is that here we imagine moving the charges by mechanical means, and this mechanism is easier to understand than the mechanism for charge transfer in a chemical battery. As long as the motor is able to maintain the charge separation across the two plates, we can have a steady-state current running in the wire.

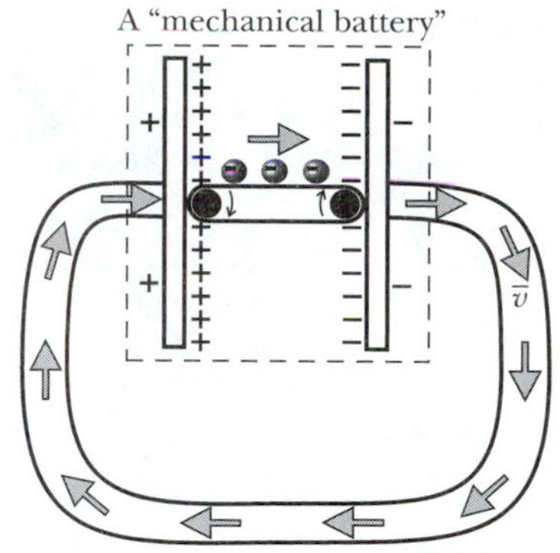

18.3.2 Field due to the battery

We will now consider a simple circuit consisting of a mechanical battery and a resistive Nichrome wire that has some twists and turns in it. (The bare wire in your experiment kit is Nichrome; the heating element in toasters are made of such wire.) You know from your own experiments that twists and turns do not seem to affect the amount of steady-state current, which seems odd. It will seem much odder in a moment!

? At the locations marked × on a copy of Figure 18.16 (locations 1 through 5), draw vectors representing the approximate electric field due solely to the charges on the metal plates of the mechanical battery. (The amount of charge on the belt is completely negligible compared to the charge on the plates.) In a different color or with a different-looking vector, indicate the drift velocity of the mobile-electron sea at those locations, assuming that the drift velocity is due solely to those electric fields.

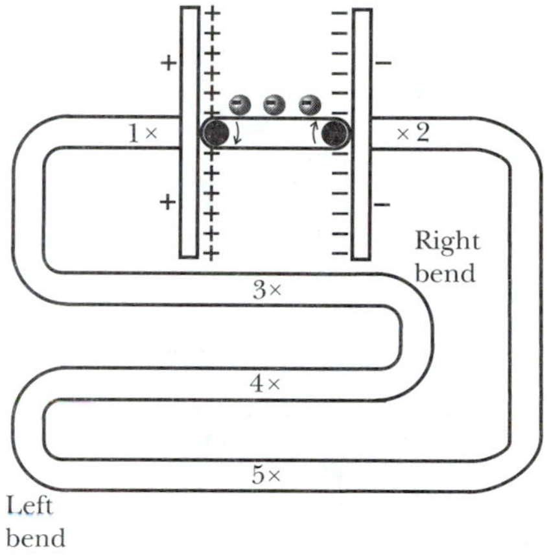

18.3.3 Charge buildup on the surface of the wire

Good grief! Figure 18.17 shows that we've got the electron current running *upstream* at location 4 in the wire! That can't be right in the steady state (the situation where charge distributions and currents are not changing).

> In the steady state there must be some other charges somewhere that contribute to the net electric field in such a way that the electric field points upstream everywhere (giving an electron drift velocity downstream everywhere).

Remember that the electrons in the mobile-electron sea inside the metal don't interact with each other, because their mutual repulsions are on the average canceled by attraction to the positive atomic cores, so they can't push each other through the wire. Some other charges must contribute to the electric field that is responsible for pushing the electrons through the wire.

Take a look at the left bend in the wire in Figure 18.17. If the only charges were on the plates, electron current would flow toward the left bend from both neighboring sections. Here we have a section of wire that has electron current flowing into it but not out.

? What effect will this have on this section of the wire? Be as specific as you can, bearing in mind that the wire is made of metal. Illustrate your thoughts on your copy of Figure 18.16.

Figure 18.15 A "mechanical battery": a conveyor belt maintains a charge separation that drives a steady current in the wire.

Figure 18.16 A circuit with a Nichrome wire and a mechanical battery.

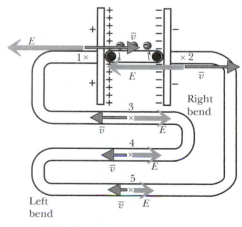

Figure 18.17 Electric field due solely to the mechanical battery, and electron drift velocities that would result from this pattern of electric field.

If electrons are flowing into a section of wire from both ends, that section will gain a net negative charge. Since the wire is made of metal, electrons are free to move, and all excess charge will move to the surface of the wire. In Chapter 14 we asserted that excess charge goes to the surface of a metal. This assertion is plausible, because excess charges in the interior of a metal repel each other toward the surface.

More precisely, excess negative charge expands the mobile-electron sea, so that some of the sea peeks out at the surface, giving a negative surface charge. Excess positive charge contracts the electron sea, allowing positive cores to peek out at the surface. A rigorous proof that all the excess charge goes to the surface of a metal requires Gauss's law, which we will study in a later chapter.

So negative charge accumulates on the surface of the left bend.

? What happens at the right bend in the wire? Illustrate your thoughts on a copy of Figure 18.16.

At the right bend electrons flow out of both ends, leaving a net positive charge, also on the surface of the wire.

? Focus your attention on location 4, a place where the electric field of the mechanical battery alone would drive current in the wrong direction. At location 4, what is the direction of the additional contribution to the electric field made by the excess + and − charges on the left and right bends?

The charges on the bends contribute an electric field to the left at location 4 that is opposite to the electric field of the mechanical battery. The "wrongly" directed net electric field is reduced in magnitude (Figure 18.18). But as long as the net field at location 4 still points to the right (the "wrong" direction), more and more excess charge will pile up on the bends, and the electric field contributed by these charges will grow.

? How far will this pileup process go? What will stop it?

The pileup will continue automatically until there is so much charge on the bends that the net field at 4 points to the left. Other charges must appear at various places on the surfaces of the wires in order that eventually the net field at locations 3, 4, and 5 all have the same magnitude. Only then will the electron current $i = nA\bar{v} = nAuE$ into a bend equal the electron current out of a bend, with no further change in the amount of charge on the surface of the bend.

We see examples of "feedback" in both the left bend and the right bend of the circuit. If the initial field is such that it drives current that is different in amount or direction from the steady-state current, surface charge automatically builds up in such a way as to alter the current to be more like the steady-state current. Moreover, once the steady state is established, there is "negative feedback": any deviation away from the steady state will produce a change in the surface charge that tends to restore the steady-state conditions.

We have deliberately emphasized the most dramatic aspect of this feedback, which is what happens at the bends in the wire. However, the resulting buildup of surface charge isn't limited to bends in the circuit. There must be positive and negative charge on the surface of the wires near the + and − plates of the mechanical battery. In the steady state the surface charge must vary smoothly from positive to negative around the circuit, possibly with a small amount of extra charge on the bends in the circuit. (Abrupt changes would imply very large electric fields.) The surface charge must arrange itself in such a way as to produce a pattern of electric field that follows the direction of the wire and has the same magnitude throughout the wire. Since

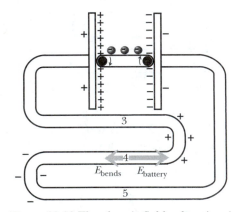

Figure 18.18 The electric field at location 4 is reduced by the contribution of the charges that have piled up on the bends.

the current is given by $i = nA\bar{v} = nAuE$, to have the same i everywhere in this series circuit we must have the same E.

In the steady state the distribution of surface charge, electric field, and electron current might look very roughly like Figure 18.19. At selected locations we indicate high density of surface charge (coulomb/m^2) with lots of +'s or −'s, and low density with few +'s or −'s. Regions between charge symbols have surface-charge density intermediate between the neighboring charge densities—the charge density is continuous.

18.3.4 The electric field of a simple surface-charge gradient

The distribution of excess surface charge in a circuit can be quite complicated, especially if there are bends and twists in the wires. It is useful to see in a simpler case how the surface charge contributes to the net electric field inside the wires.

Consider the simple circuit shown in Figure 18.20. Near the − end of the mechanical battery (location 1 in Figure 18.20) the surface charge on the wire must be negative. Near the + end of the mechanical battery (location 5) the surface charge on the wire must be positive. Somehow there has to be less and less negative surface charge the farther you go from the − end, symbolized on the diagram by fewer − signs (for example, at location 2). The symmetry of the situation means that the surface charge must be zero half-way around the circuit (at location 3). There must be more and more positive surface charge the closer you come to the + end at location 5 (symbolized by more + signs).

This charge distribution is only a very rough approximation to the actual distribution. For example, feedback in the circuit will tend to put some additional charge on the bends. The key point is that the surface charge at location 1 must be negative, and the surface charge at location 5 must be positive. When we connect a wire from one end of the battery to the other, there has to be some kind of smooth transition along the wire from − surface charge to + surface charge. There will even be some location where there is zero surface charge (location 3).

The surface charge has the shape of a hollow tube surrounding the metal wires, and we need to understand what kind of electric field this tube of charge creates inside the wires. Consider two very short slices of this tube of charge, two rings of charge, just to the right of location 3 in Figure 18.20. In Figure 18.21, the left-hand ring has less negative charge than the ring on the right (because of the variation of surface charge along the wire).

? At location ×, on the center line halfway between the two rings, what is the direction of the net electric field due to these two rings?

The net electric field points to the right, toward the ring with a larger amount of negative charge. Notice that if the rings had the same charge density, even if a large amount, they would produce zero field at location ×. It is the variation of surface charge density along the wire that produces electric field, not the amount of surface charge density. Evidently a charge distribution somewhat like that shown in Figure 18.20 will give a distribution of electric field in the wires that is appropriate for this series circuit.

Ex. 18.8 At locations 1, 2, 3, 4, 5 on a diagram like Figure 18.20, draw the net electric field. Also draw the resulting electron drift velocity. Pay close attention to direction and relative magnitudes.

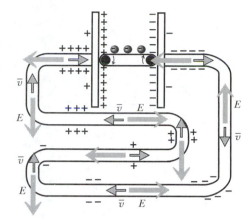

Figure 18.19 A possible approximate distribution of surface charge in the steady

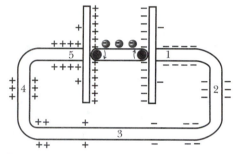

Figure 18.20 A geometrically simple circuit with approximate surface charge shown.

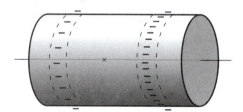

Figure 18.21 Two rings of surface charge on a wire, with different charge density. What is the field direction at location ×?

18.3.5 Amount of surface charge

Despite the large number of electrons moving around in a circuit, you will find that the wires in one of your circuits will not repel a hanging charged tape. Try it! (The tape is of course attracted to neutral matter, which masks the tiny repulsion due to surface charges.) The large number of electrons moving inside the wires are balanced by an equally large number of positively-charged atomic cores, so the net charge inside the wire is zero and cannot repel charges by electric interaction. (The current does of course produce magnetic effects.)

The surface charges and the electric field they make are essential to driving and guiding the current through the interiors of the wires, but the amount of charge present on the wires in a circuit powered by two 1.5 volt batteries is very small. Little surface charge is needed in such a circuit; it is very easy to move electrons through metal, and very small electric fields are sufficient to drive sizable currents. We will see that in a typical circuit the magnitude of the electric field inside a wire is small—around 5 V/m. It turns out that to create such a field in your two-battery circuits, the required amount of surface charge on a centimeter of wire near the negative end of the battery is very small—only about 10^6 electrons. Compare this with the very much larger number of electrons you found on a charged piece of invisible tape, which was about 10^{10} electrons.

It is possible to run a circuit with a 10,000-volt power supply and demonstrate how the surface charges can repel charged objects. The amount of surface charge is proportional to the voltage, and such circuits have much more surface charge than a 3-volt circuit. A description of such an experiment is given in Section 18.10 at the end of this chapter.

18.3.6 Summary: surface charge and electric field

To summarize, there is a variation of surface charge from negative to positive around a simple circuit that doesn't have big bends and twists, and the electric field inside the wire is to a significant extent due to variation in the charge density of nearby surface charge (Figure 18.22). The actual net field is due to all the charges, including faraway surface charges and charges in and on the battery.

An example of a circuit in which the surface charge distribution is definitely not simple is one in which a wire is wound into a tight coil. In that case the surface charge on one turn of the coil has a big effect on electrons flowing in other turns of the coil.

Generally speaking, the easy thing to draw accurately is the pattern of electric field and electron current. In the steady state, the charges arrange themselves on the surfaces of the wires in such a way as to produce the necessary electric field (and current). When you draw an approximate distribution of surface charge, you should make sure that it is at least roughly consistent with the pattern of electric field and current.

18.4 Connecting a circuit: The initial transient

When you complete a circuit by making the final connection, feedback forces a rapid rearrangement of surface charges leading to the steady state. This period of adjustment before establishing the steady state is called the initial transient. How does this work, and how long does the transient take? We'll look now in closer detail at how the steady state is established.

Consider the circuit in Figure 18.23, containing one chemical battery and two Nichrome wires with a gap between them. The system is in static equilibrium, so $E = 0$ everywhere inside the wires. An approximate surface-charge distribution is indicated, with one wire charged positive and the oth-

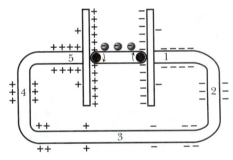

Figure 18.22 Simple circuit, with approximate surface charge distribution.

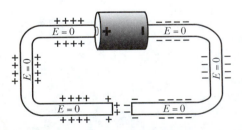

Figure 18.23 A circuit with a gap; approximate surface charge shown.

er charged negative, ignoring the exact details of how these charges are distributed. (In actual fact, there are larger concentrations of surface charge near the gap and near the ends of the battery, so this is a crude diagram.)

Let's look closely at the neighborhood of the gap. In Figure 18.24 we show the part of the electric field inside the wire that is contributed *solely* by the charges on the faces of the gap. You should convince yourself that these vectors are indeed drawn correctly.

? On a similar diagram, sketch the part of the electric field that is due to all the *other* charges (on the ends of the battery and along the wires). Why is the field due to all the other charges exactly opposite to the field due to the gap charges?

The net field inside the wire must be zero, because at this time the system is in static equilibrium. Therefore the field due to all other charges must be exactly equal in magnitude and opposite in direction to the field made by the charges on the faces of the gap (Figure 18.25).

Let's complete the circuit by connecting the Nichrome wires together. The charges on the facing ends of the wires neutralize each other, leaving a tube of surface charge on the outside of the wire, as shown in Figure 18.26 in a cross section of the region near the connection point. At this instant this surface charge distribution has a big, unstable "discontinuity" in it—closely neighboring areas on the outside of the wire have positive and negative charges.

? In a diagram similar to Figure 18.26, use vectors to indicate the electric field at the locations marked × inside the wire just after closing the gap. (Hint: The gap charges are now gone.)

Immediately after closing the gap we've lost the contribution to the net field formerly made by the charges on the faces of the gap. Therefore the net field will look like the field of the "other" charges that we drew before.

Electrons in the mobile-electron sea inside the metal will move due to these fields. Imagine that you could take a "snapshot" of the surface charge after the electrons inside the wire have moved just a little bit (that is, after a very short time). Electrons moving toward a positive region of the surface will make the surface charge in that region less positive, while electrons moving away from a negative region of the surface will make the surface charge in that region less negative.

? On your diagram corresponding to Figure 18.26, where there is a change in the surface charge, write what that change is (for example, increase of positive charge, decrease of negative charge, etc.). (Remember that a + sign on the surface does not represent some exotic positive charge. Rather, a + sign indicates a region where there is a deficiency of electrons, leaving partially exposed the metal's positive atomic cores.)

Figure 18.27 shows the new distribution of surface charge after a very short time (a fraction of a nanosecond!). The motion of electrons toward and away from the surface, driven by the new electric field, dilutes the surface charge near the location where the gap was, so that in the gap region the initial abrupt charge discontinuity turns into a more gradual change, indicated by placing fewer + or − symbols near the gap location. A nonzero electric field has come into being where the gap used to be, due to the + charges to the left and − charges to the right of that location.

In Figure 18.28 we see that the affected region, where there has been a rearrangement of the surface charge, grows rapidly outward from the gap location at approximately the speed of light, which is the speed with which the change in the distribution of charges is felt at places some distance away.

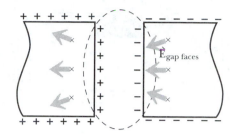

Figure 18.24 A close-up of the gap, showing the electric field contributed *solely* by the charges on the faces of the gap. At the locations marked ×, draw the electric field due to all *other* charges.

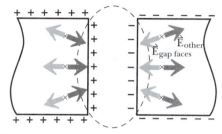

Figure 18.25 Field due to charges on the faces of the gap, and due to all others.

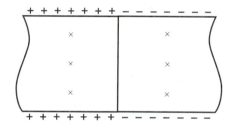

Figure 18.26 A cross-section near the former gap, just after closing the gap.

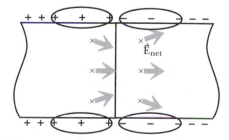

Figure 18.27 The dilution of the original surface charge. The circled regions at left become less positive, while the circled regions at right become less negative.

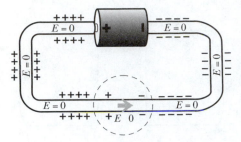

Figure 18.28 Parts of the circuit haven't yet been informed that the gap has been closed! The sphere enclosing the region of changed electric field expands at the speed of light.

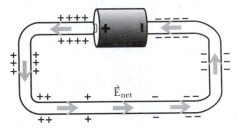

Figure 18.29 In the steady state the electric field has the same magnitude everywhere in the wire.

At the instant sketched in Figure 18.28, the surface charges and the electric field at distant locations haven't been affected yet, because those regions haven't yet received the information that the gap has been closed! The electric field is still zero inside the wires except near the location of the gap.

It is the speed of light that determines the minimum time required to establish the steady state. Light travels 30 centimeters (about one foot) in a nanosecond (10^{-9} s). The time required to move the electrons toward or away from the surface is completely negligible, because the electrons only have to move an infinitesimal distance in order to establish a significant amount of surface charge, through slight expansion or contraction of the mobile-electron sea (compare with Problem 14.17, where the molecular stretch associated with polarizing an insulator was shown to be less than the radius of a single proton!).

In just a few nanoseconds the rearrangement of the surface charges will extend all the way around the circuit. The initial transient leads to the establishment of a "steady state," in which the surface charge and current no longer vary with time. In the steady state there is a variation of surface charge around the circuit, and the electric field has uniform magnitude throughout the inside of the wire (Figure 18.29).

Important vocabulary:
- static equilibrium: nothing moving (no current)
- steady state: there is a constant (nonzero) current
- initial transient: short-time process leading to the steady state

18.4.1 Why a light comes on right away

As discussed earlier, the drift speed in one of your battery-and-bulb circuits is only about 5×10^{-5} m/s . However, a change in the circuit, such as making or breaking a connection, propagates its effects very rapidly through the circuit, at approximately the speed of light. It is not necessary for charges to move physically to a distant place to cause a change at that place: a rearrangement of charges due to tiny movements produces a change at the speed of light in the electric field at a distance. It is important to keep in mind that the effects do take a nonzero amount of time to propagate (nanoseconds). The propagation is not instantaneous, but it is very fast.

When you turn on the light switch in your room (making the final connection in a circuit), it could take hours on the average for electrons in the switch to drift to the overhead light. But the light comes on right away, because the rearrangement of surface charges in the circuit takes place at about the speed of light. The final steady state of the circuit is established in a few nanoseconds, after which the electron sea circulates slowly and majestically around and around, everywhere in the circuit, including through the switch and the wires and the light bulb. (Most lighting actually uses "alternating current," in which case the electron sea doesn't drift continuously but merely sloshes back and forth very short distances, everywhere in the circuit.)

The electrons don't have to move from the switch to the light bulb; there are already plenty of electrons in the bulb filament. All that is required is to establish the appropriate steady-state surface-charge distribution, very quickly, and then the electrons that are already in the filament start moving.

18.5 Feedback

Feedback during the initial transient produces surface charge of the right amount to create the appropriate steady-state electric field. It also maintains these steady-state conditions, as we will show next.

18.5.1 Feedback leads to current equalization

Consider a straight section of wire that at some moment has more incoming electron current than outgoing electron current, leading to a buildup of excess negative charge on the surface of this section of the wire (Figure 18.30).

This surface-charge buildup will tend to equalize the incoming and outgoing electron currents. The outer surface of this section of wire becomes more negative, and this negative surface charge tends to impede the incoming electrons and tends to speed up the outgoing electrons, thus making the incoming and outgoing electron currents $i = nA\bar{v}$ more nearly equal to each other. This feedback process will continue until the two currents are exactly equal to each other. At that point there will be no further change in the surface charge.

? If the incoming electron current is smaller than the outgoing current, what will happen on the outer surface of this section of the wire?

In this case (Figure 18.31) this section of wire becomes more positive, which speeds up the incoming electrons and slows down the outgoing electrons, so the incoming electron current increases and the outgoing electron current decreases. Again, the effect of this feedback mechanism is to equalize the current into and out of a section of wire.

So whichever way there is an imbalance between incoming and outgoing current, the two currents will quickly become equal to each other due to the automatic feedback mechanism of changes in the surface charge. Earlier you saw that in the steady state the current everywhere in a series circuit is the same. Now you can see how the feedback mechanism creates and maintains this steady state.

18.5.2 Feedback makes current follow the wire

It is amusing and instructive to see what happens if you bend a straight wire while the current is running (Figure 18.32). The oncoming electrons don't immediately know what has happened and continue in their original direction, but as they run into the bend, electrons pile up on the outside of the bend, which tends to turn the oncoming electrons.

? What determines how much charge piles up? (That is, what makes the pile-up process come to a stop?)

The electrons pile up on the bend until there are enough there to repel oncoming electrons just enough to make them turn the corner (follow the wire) without running into the side of the wire. A calculation by Rosser indicates that a single extra electron on the bend is sufficient to turn a sizable current! So not only does the feedback mechanism equalize incoming and outgoing current for any section of the wire, it also forces the electron current to follow the wire, no matter how the wire is bent or twisted.

18.5.3 Summary of feedback

We have learned that there must be some excess charge on the outside of the wire, all along the wire. Through the feedback mechanism, this surface charge will automatically arrange itself (or rearrange itself) in such a way as to ensure that the net electric field everywhere inside the wire points along the wire and has the appropriate magnitude to drive the appropriate amount of steady-state current. (The amount of this current is ultimately determined by the strength of the motor in our mechanical "battery," and by the electron mobility u of the material that the wire is made of.)

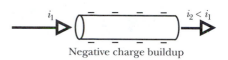

Figure 18.30 What if more electron current enters than leaves a section of wire?

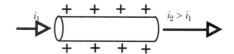

Figure 18.31 What if less electron current enters than leaves a section of wire?

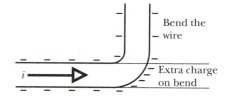

Figure 18.32 What happens if you bend a straight wire while current is running?

For details of the calculation about the amount of charge needed to turn the current, see W. G. V. Rosser, "Magnitudes of surface charge distributions associated with electric current flow," *American Journal of Physics* **38**, 265-266, (1970).

This is very similar to the feedback we have seen in static electricity. If an external field is applied to an isolated metal block, charges automatically arrange themselves on the surface of the metal block in such a way as to make the electric field inside the metal block have a particular form: the electric field must be zero throughout the inside of the block in static equilibrium.

To summarize:

Feedback in a circuit leads to surface charges and steady-state current: $\vec{E} \neq 0$ inside a metal.

Feedback in static electricity situations leads to static equilibrium: $\vec{E} = 0$ inside a metal.

18.6 Surface charge and resistors

We have used surface charges and electric fields to analyze a circuit made up just of a battery and a high-resistance Nichrome wire. In this section we will use the surface-charge model to analyze circuits involving "resistors"—sections of a circuit that resist the passage of electrons more than other parts of the circuit.

18.6.1 Narrow resistors and thick wires

We will analyze a circuit similar to a circuit you assembled, consisting of a battery, copper wires, and a bulb. The very thin tungsten filament of the light bulb is an example of a "resistor"—a component that has a much higher resistance to current flow than do the thick wires connecting the resistor to the battery.

For a simple example of a circuit containing a resistor, we'll again consider a battery and a high-resistance Nichrome wire, but we'll make part of the Nichrome wire very much thinner than the rest of the wire (Figure 18.33). We'll call the narrow section a "resistor," and there are two thick wires connecting the resistor to the battery.

Just after connecting the circuit but before the establishment of the steady-state surface charge, the electric field in the narrow resistor might temporarily be about the same as the electric field in the neighboring thick wires, leading to a comparable drift speed $\bar{v}$ in both sections. However, the number of electrons per second trying to enter the resistor is the large number $nA_{\text{thick}}\bar{v}$, whereas the number of electrons per second passing through the resistor is the small number $nA_{\text{thin}}\bar{v}$, so electrons pile up at location 3, the entrance to the resistor. (Remember that the mobile-electron density n is the same in the wires and the resistor, because they are both made of Nichrome in our example.)

Similarly, a deficiency of electrons, or + charge, accumulates at location 4, the exit from the resistor, because the number of electrons per second emerging from the resistor is small compared to the electron current in the thick wire that is removing electrons from the region.

? Later, in the steady state, how does the electron current in the thick wires compare with the electron current in the resistor? Why?

Once the steady state has been established, the currents must have become the same. Reason by contradiction: if the currents weren't the same in the steady state, surface charge would be building up on the resistor, which would mean we hadn't reached the steady state yet. Charge piles up until the resulting increased electric field in the resistor gives a high enough drift speed to make an electron current that equals the electron current in the thick wires. Once these two currents are the same, there will be no more pile-up.

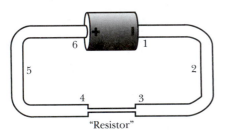

"Resistor"

Figure 18.33 A simple circuit with a "resistor"—a thin section of the Nichrome wire.

? In the steady state, how does the electric field in the resistor compare with the electric field in the thick wires? Why? On a diagram like Figure 18.33, draw the steady-state electric field in the resistor and in the neighboring thick wires, paying attention to relative magnitudes.

Since the steady-state current is the same in the thick and thin sections, we have the relation $nA_{thick}uE_{thick} = nA_{thin}uE_{thin}$. Therefore the electric field in the resistor must be quite a lot larger than the electric field in the thick sections. Quantitatively, $E_{thin} = (A_{thick}/A_{thin})E_{thick}$, since the mobile-electron density n and the mobility u are the same in the thick and thin wires (both made of Nichrome). The electric field has a uniform large magnitude throughout the resistor, and a uniform small magnitude throughout the thick wires.

? Now draw the approximate surface charge in the steady state. Be sure to make the distribution of surface charge be consistent with the distribution of electric field that you drew already.

In the steady state the surface charge varies only a little along the thick wires, producing a small electric field E_w and a small drift speed. Along the resistor there is a big variation from + at one end to − at the other, which makes a large field $E_{resistor}$ inside the thin resistor (Figure 18.34). This makes a large drift speed in the resistor (with the same electron current i as in the wires). We have an approximate charge distribution that is consistent with the distribution of electric field.

This resistor circuit offers another example of feedback in an electric circuit—an automatic adjustment leads to producing the electric fields that are needed in the steady state. Note that if the drift speed in the thick wires were momentarily to increase for any reason, extra charge would pile up at the resistor, which would tend to decrease the drift speed in the thick wires.

On the other hand, if the drift speed in the thick wires were to momentarily decrease, charge would drain through the resistor in such a way as to decrease the pile-up of charge at the ends of the resistor, which would tend to increase the drift speed in the thick wires.

Ex. 18.9 Sketch a graph of the magnitude of the electric field inside the wire, as a function of location around the circuit (Figure 18.35). In a different color, graph the drift speed of the electrons.

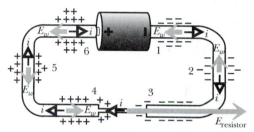

Figure 18.34 An approximate surface-charge distribution that is consistent with what we know about the electric field.

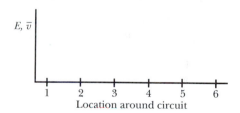

Figure 18.35 Make graphs of magnitude of electric field and of drift speed at locations around the circuit.

18.6.2 A good analysis strategy

The analysis in the preceding section is an example of a general strategy for understanding the relationship of charge and field in a circuit. Here is a summary of this strategy:

- Based on $i = nAuE$, draw the distribution of electric field in the wires.
- Next draw an approximate charge distribution that is consistent with the distribution of electric field.

The reason for carrying out the analysis in this order is that the distribution of electric field is fairly simple and can be analyzed nearly exactly, whereas the pattern of surface charge is complicated and can be determined only in a crude approximation without very difficult calculations.

What we know is that feedback will produce an arbitrarily complicated distribution of surface charge appropriate to the steady-state distribution of electric field. The pattern of electric field gives us quantitative information

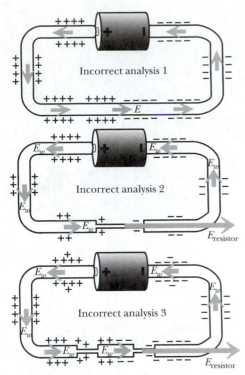

Figure 18.36 Circuits incorrectly analyzed: distribution of surface charge inconsistent with the distribution of electric field.

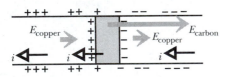

Figure 18.37 A portion of a circuit with a wide resistor.

about the electron current; the pattern of surface charge gives us a clear sense of the mechanism that produces the pattern of electric field.

Ex. 18.10 The circuits in Figure 18.36 have been analyzed incorrectly; the distribution of surface charge is inconsistent with the distribution of electric field. Briefly identify what is inconsistent in each case.

18.6.3 A wide resistor: Charges on the interface

Another kind of resistor is wide rather than narrow but has low mobility. Consider copper wires leading up to a short piece of carbon having the same, wide cross section (Figure 18.37; we don't show the entire circuit). Carbon has a much smaller electron mobility u than copper, so the steady-state electric field has to be much larger in the carbon than in the copper in order to have the same current in both materials.

In the transient leading up to the steady state, electrons pile up not only on the outer surfaces of the copper and carbon, but also on the interfaces between the copper and the carbon, and these interface charges make a major contribution to the large electric field inside the carbon resistor.

Our assertion that "excess charge is always on the surface of a metal" extends to this case, where the surface may be in contact with a different conducting material.

18.7 Energy in a circuit

Charge conservation plus the definition of a steady state led to the current node rule, which tells us that the current in a simple series battery and bulb circuit must be the same everywhere in the circuit. But how many amperes of current flow? What determines the magnitude of the current? The energy principle provides the answer, plus the properties of batteries.

Consider a circuit containing various circuit elements such as resistors and batteries. We could discuss energy conservation in the circuit by considering one electron as it makes a complete trip around the circuit. The energy gained by the electron as it is transported across the mechanical battery is dissipated in collisions with atomic cores as it travels through the wire. Alternatively, we can analyze the energy per unit charge gained and lost in a trip around the circuit. We know that over any path the round-trip potential difference must be zero. If we follow a round-trip path through a circuit, through circuit element 1, circuit element 2, etc., we find the following "loop rule":

ENERGY CONSERVATION (THE LOOP RULE)

$$\Delta V_1 + \Delta V_2 + ... = 0 \text{ along any closed path in a circuit}$$

This is essentially the energy principle, but on a per unit charge basis. Remember that electric potential is defined as potential energy per unit charge: $V = U/q$. The loop rule says that you can't gain energy in a round trip. In circuits this energy principle is often called the "Kirchhoff loop rule."

Along a wire or resistor of length L in which there is a uniform electric field of magnitude E, the magnitude of the potential difference is just EL. But what is the potential difference across a battery?

18.7.1 Potential difference across a battery

What determines how much potential difference the motor of a "mechanical battery" can maintain across the battery? Suppose you start with uncharged plates (Figure 18.38) and turn on the motor (with no wire connected to the plates), extracting electrons from the left-hand plate and transporting them to the right-hand plate. The conveyor belt, driven by the motor, exerts what we will call a "non-Coulomb" force $\vec{F}_{NC}$ on each electron. This being the only force, the belt accelerates electrons.

As the motor continues to transport electrons to the right, charge builds up on the plates. The charges on the plates exert a "Coulomb" force $F_C = eE_C$ on the electrons being transported, in the opposite direction to F_{NC}, the non-Coulomb force that the conveyor belt exerts (Figure 18.39). There is less acceleration of the electrons.

By "Coulomb force" and "Coulomb field" we mean the force and field due to point charges, as given by Coulomb's law. We use the term "non-Coulomb" to refer to other kinds of interactions. The Coulomb and non-Coulomb forces oppose each other inside the battery.

Eventually there is enough charge on the plates to make $F_C = eE_C = F_{NC}$. At that point the motor cannot pump any more charge, and the plates are charged up as much as they can be (Figure 18.40). We see that the function of a mechanical battery (or of a chemical battery) is to produce and maintain a charge separation (pull electrons out of the positive plate and push them onto the negative plate). The amount of charge separation is limited by and determined by the strength of the motor in a mechanical battery (or by the nature of the chemical reactions in a chemical battery).

If the distance between the (large) plates of a mechanical battery is s, and the electric field E_C of the charged plates is uniform between the plates, then the potential difference across the battery is

$$|\Delta V_{batt}| = E_C s = \frac{F_{NC}s}{e}$$

The quantity $F_{NC}s/e$, the energy input per unit charge, is a property of the battery and is called the "emf" (pronounced "e" "m" "f") of the battery. Historically, emf was an abbreviation for "electromotive force," which is a bad name, since emf is not a force at all but energy input per unit charge. We will avoid this terminology by just using the abbreviation emf. In a chemical battery, the emf is a measure of the chemical energy per unit charge expended by the battery in moving charge through the battery. The emf of a flashlight battery is 1.5 volts, and the associated charge separation is due to chemical reactions in which electrons are reactants or products.

The emf of a chemical battery turns out to be nearly constant, independent of how much current is running through the battery (as long as the current is not too large), and approximately independent of how long the battery has been used (though it does decline with time). It is the approximate constancy of the emf that makes this a useful concept, and which makes it useful to analyze circuits in terms of emf and the potential differences in the circuit elements attached to the battery.

It is important to keep in mind that although the units of emf are volts, the emf is not a potential difference. Potential difference is a path integral of the electric field made by real charges. The emf is the energy input per unit charge and might in principle be gravitational or nuclear in nature. For example, our mechanical battery could be run by a falling weight.

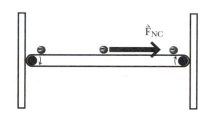

Figure 18.38 Start the motor and begin moving electrons from left to right.

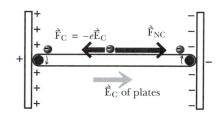

Figure 18.39 Charges build up on the plates and exert a "Coulomb" force on the electrons being transported.

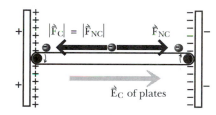

Figure 18.40 The Coulomb force grows to be as big as the non-Coulomb force.

ROLE OF A BATTERY

A battery maintains a potential difference across the battery.

This potential difference is numerically equal to the battery's emf.

Ex. 18.11 If the plates of a mechanical battery are very large compared to the distance s between the plates, and the area of a plate is A, show that the amount of charge Q on one of the plates of an isolated battery is determined by the battery's emf: $Q = A\varepsilon_0(\text{emf})/s$.

Ex. 18.12 For one of your batteries, emf = 1.5 volts. Using the result of the preceding exercise, make a rough estimate of the amount of charge in coulombs on the end plate of a battery like yours, whose length $s = 5$ cm and radius $r = 1$ cm. (This is a very rough estimate because 5 cm isn't small compared to 1 cm.) How does this compare to the charge on a U tape? Should you expect to be able to repel a U tape with one end of your battery?

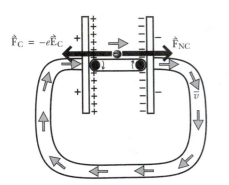

Figure 18.41 $F_C = F_{NC}$ if there is little internal resistance.

18.7.2 Internal resistance

If we now connect a Nichrome wire from one plate of the mechanical battery to the other (Figure 18.41), we can achieve a steady state if the motor transports as many electrons per second from the + plate to the − plate as are flowing through the wire. The larger the current, the more electrons per second the motor must transport to keep the plates resupplied. If there is no resistance to the flow of charge inside the battery, we will have $F_C = eE_C = F_{NC}$.

In any real battery there is some resistance to the flow of charge through the battery, and this resistance is called the "internal resistance" of the battery. If the mobility of an ion of charge e through the battery is u, the drift velocity through the battery will be proportional to the net force per unit charge acting on the ion: $\bar{v} = u[F_{NC}/e - E_C]$. Since the motor force F_{NC} is fixed, the maximum drift speed is obtained when $E_C = 0$, which means that there is no charge on the ends of the battery: the motor isn't keeping up, and $|\Delta V_{batt}| = 0$. Much of the time we will make the simplifying assumption that the mobility u of ions inside the battery is very high (low "internal resistance"), so that we get a sizable drift speed even if F_C (= eE_C) is nearly as large as F_{NC}. Correspondingly, we assume that $|\Delta V_{batt}| \approx \text{emf}$.

In Chapter 19 we will show how to model a battery when the internal resistance is not negligible. For now we will consider "ideal" batteries for which the internal resistance is negligible, and $|\Delta V_{batt}| = \text{emf}$.

18.7.3 Field and current in a simple circuit

In Figure 18.42, note that the electric field inside the mechanical battery points from the "+" end toward the "−" end, in the opposite direction to the electric field in the neighboring wires. The "+" end of a battery is at a higher potential than the "−" end.

If we start at the negative plate of the battery and go counter-clockwise in Figure 18.42, there is a potential increase of +emf across the battery and a potential drop of −EL along the wire of length L. The round-trip potential difference is zero:

Figure 18.42 The electric field inside the mechanical battery is opposite to the electric field in the neighboring wires.

$$\Delta V_{batt} + \Delta V_{wire} = 0$$

$$\text{emf} + (-EL) = 0$$

$$E = \frac{\text{emf}}{L}$$

We used the energy principle (round-trip potential difference is zero) to determine the magnitude of the electric field in the wire. If we know the properties of the wire (cross-sectional area A, electron density n, and electron mobility u), we can now determine the magnitude of the conventional current $I = enAuE$.

Two different paths

In Figure 18.43, start at the negative end of the battery and follow the dashed path clockwise through wires L_2 and L_3, and then through the battery to the starting location. There is a potential rise $+E_2L_2$ along the part of the path of length L_2, a potential rise $+E_3L_3$ along the rest of the path of length L_3, then a potential drop through the battery of amount $-$emf, yielding $E_2L_2 + E_3L_3 + (-\text{emf}) = 0$ for the round-trip potential difference. If we instead follow the path through L_1 and L_3, we find a round-trip potential difference $E_1L_1 + E_3L_3 + (-\text{emf}) = 0$.

These two equations imply (correctly) that $E_1L_1 = E_2L_2$, which makes sense, since with the same starting and ending points the two wires have the same potential differences. Also notice in Figure 18.43 that $i_3 = i_1 + i_2$, due to the current node rule (charge conservation in the steady state).

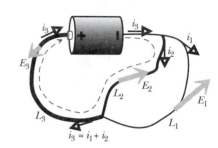

Figure 18.43 Follow the dashed path through wires L_2 and L_3.

18.7.4 Potential difference across connecting wires

In your own experiments you have seen that in a circuit containing a light bulb or a Nichrome wire, the number or length of the connecting wires used has a negligible effect on the amount of current in the circuit. In a series circuit with thick copper wires and a light bulb, the electric field is large in the bulb and quite small in the thick copper wires, because the steady-state electron current $nA\bar{v} = nAuE$ must be the same in both. The electron mobility in cool copper is much higher than the electron mobility in hot tungsten, and typically the copper wires are much thicker than the tungsten bulb filament, so that not only is it the case that $E_{\text{wires}} \ll E_{\text{bulb}}$, but even for rather long wires $E_{\text{wires}}L_{\text{wires}} \ll E_{\text{bulb}}L_{\text{bulb}}$. Therefore in the energy equation (emf $- E_{\text{bulb}}L_{\text{bulb}} - E_{\text{wires}}L_{\text{wires}} = 0$) the term $E_{\text{wires}}L_{\text{wires}}$ may be negligible; the work done by the battery goes mostly into energy dissipation in the bulb, with a small amount of energy dissipation in the thick copper wires.

Ex. 18.13 Explain briefly why the brightness of a bulb doesn't change noticeably when you use longer copper wires to connect it to the battery.

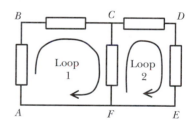

Figure 18.44 A circuit containing devices connected by thick copper wires.

18.7.5 General use of the loop rule

The loop rule can be applied along any path through an arbitrarily complicated circuit. Consider a circuit made up of five circuit elements connected together by thick copper wires that have high electron mobility (Figure 18.44). The circuit elements could be resistors, batteries, capacitors, or even more complex devices such as semiconductor diodes or even refrigerators!

In Figure 18.45 we show the left portion of the complete circuit (loop 1). Start at location A and walk along the path labeled $ABCFA$. From A to B the potential difference is $\Delta V_1 = V_B - V_A$; from B to C the potential difference is $\Delta V_2 = V_C - V_B$; and from C to F the potential difference is $\Delta V_3 = V_F - V_C$.

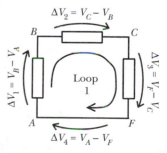

Figure 18.45 Walk around loop 1, following the path ABCFA.

From F to A there is negligible potential difference $\Delta V_4 = V_A - V_F \approx 0$, because the thick wire has high electron mobility and large cross-sectional area, and so requires a negligibly small electric field to drive the current.

Any round-trip potential difference must be zero, and the loop rule gives this:

$$\Delta V_1 + \Delta V_2 + \Delta V_3 + \Delta V_4 = 0$$
$$(V_B - V_A) + (V_C - V_B) + (V_F - V_C) + (V_A - V_F) = 0$$

We could write a similar equation corresponding to walking along loop 2 (path *FCDEF*). The idea is that you start anywhere you like on the circuit diagram and walk around a loop that brings you back to your starting location, adding up potential drops and rises as you go.

You can even walk along a round-trip path through the air, outside the circuit elements, but unless you know something about the electric field and potential differences outside the devices and connecting wires, this usually isn't very useful.

Ex. 18.14 In Figure 18.44, suppose $V_C - V_F = 5$ volts, and $V_D - V_E = 3.5$ volts. What is the potential difference $V_C - V_D$? If the element between C and D is a battery, is the + end of the battery at C or at D?

18.8 Application: Energy conservation in circuits

In the middle of Figure 18.46 is a graph of potential around the circuit shown at the top, along the path A-B-C-D-A. Below the potential graph is a graph of $E_\parallel$, the component of the electric field in the direction that we walk, counter-clockwise around the circuit. Inside the battery we are walking opposite to the direction of the electric field, so there is a potential rise equal in magnitude to the emf of the battery.

A big potential gradient (volts per meter) means a large field, so a steep slope on the potential graph corresponds to a large electric field (see bottom of Figure 18.46). Along the thick wires, the potential gradient is small and the field is small ($E = |\Delta V| / L$), yielding a small drift speed (wire AB, for instance). Along the thin wire (BC) there is a large potential gradient corresponding to a large field and a high drift speed.

For this circuit with its two thick wires and thin resistor connected to a battery, all the potential drops (negative potential differences) plus the potential rise across the battery (numerically equal to the emf) add up to 0:

$$\Delta V_1 \quad + \quad \Delta V_2 \quad + \quad \Delta V_3 \quad + \Delta V_{\text{battery}} = 0$$

$$(-E_1 L_1) + (-E_2 L_2) + (-E_3 L_3) + \text{ emf } = 0$$

Ex. 18.15 What would be the potential difference ΔV_2 across the thin resistor in Figure 18.46 (between locations B and C) if the battery emf is 1.5 volts? Do you have enough information to determine the current I in the circuit? Assume that the electric field in the thick wires is very small (so that the potential differences along the thick wires are negligible).

Ex. 18.16 A Nichrome wire 30 cm long and 0.25 mm in diameter is connected to a 1.5-volt flashlight battery. What is the electric field inside the wire? Why don't you have to know how the wire is bent? How would your answer change if the wire diameter were 0.35 mm? (Note that the electric field in the wire is quite small compared to the electric field near a charged tape.)

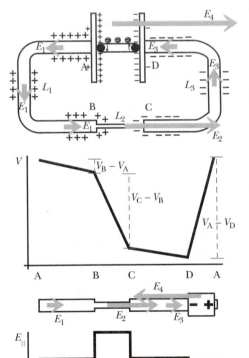

Figure 18.46 Potential rises and falls as you walk around a circuit. The electric field is the gradient of the potential.

18.9   Applications of the theory

We now have two principles on which to base the analysis of circuits:

THE CURRENT NODE RULE

In the steady state, for many electrons
flowing into and out of a node,

electron current: net i_{in} = net i_{out}, where $i = nAuE$

conventional current: net I_{in} = net I_{out}, where $I = |q| nAuE$

ENERGY CONSERVATION (THE LOOP RULE)

In the steady state, for any round-trip path,

$$\Delta V_1 + \Delta V_2 + \dots = 0$$

We emphasize that the current node rule refers to many electrons moving together, whereas the energy equation is on a per-unit-charge basis.

We will apply these two principles to predicting the behavior of simple circuits, and we will test many of our predictions by measuring the currents in the circuits.

18.9.1 Quantitative measurements of current with a compass

In the following sections we will use our model of circuits to predict the ratio of current in one circuit to current in another circuit, and then we will do experiments to test our predictions. In order to get good experimental results, we need to understand how to use a compass to make accurate measurements of current. There are two important issues to consider.

1) Is magnetic field uniform all along the compass needle?

The compass needle aligns with the net magnetic field at its location. As we have seen, the net magnetic field is the sum of the Earth's magnetic field and the magnetic field made by the current in the wire. In Chapter 17 we found that the magnitude of the magnetic field made by current in a long wire is inversely proportional to distance from the wire:

$$B = \frac{\mu_0}{4\pi} \frac{2I}{r}$$

If the distance from the wire to the center of the needle is significantly different from the distance from the wire to the ends of the needle, the magnetic field at the ends of the needle will be weaker than at the center. In this case the deflection of the needle will be smaller than it should be. If the wire lies on top of the compass, it is about 6 mm away from the center of the needle. If the needle deflects more than $15°$, the ends of the needle are significantly farther away from the wire than the center is, as shown in Figure 18.47, and the deflection will not be proportional to the current.

Elevating the wire

By elevating the wire about 1 cm above the top of the compass, we minimize this effect. Now even if the needle deflects up to $40°$, the ends and the center will still be approximately the same distance from the wire. An easy way to elevate the wire is to support it on two fingers, as shown in Figure 18.48.

2) Is current proportional to deflection angle?

As can be seen from Figure 18.49, the tangent of the deflection angle is proportional to the magnetic field made by the current in the wire. For small

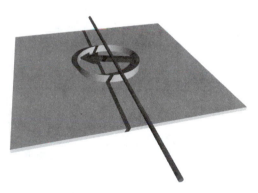

Figure 18.47 If the wire is very close to the compass, with deflection > 15°, the magnetic field at the ends of the needle may be significantly weaker than the magnetic field at the center of the needle.

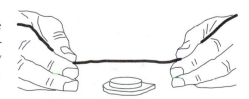

Figure 18.48 One way to elevate the wire by a reproducible distance is to support it on two fingers.

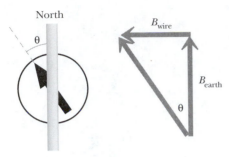

North

θ

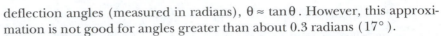

B_{wire}

B_{earth}

θ

Figure 18.49 $B_{wire}/B_{Earth} = \tan\theta$

deflection angles (measured in radians), $\theta \approx \tan\theta$. However, this approximation is not good for angles greater than about 0.3 radians (17°).

Tangent of deflection angle

- For deflection angles less than 15°, we can assume that the magnetic field, and hence the current in the wire, is proportional to deflection angle.
- For deflection angles greater than 15°, we need to use the tangent of the deflection angle to calculate magnetic field (and therefore current).

18.9.2 Application: Twice the length

Figure 18.50 depicts an experiment in which we compare the electron current in a Nichrome wire of length L to the current in a Nichrome wire of length $2L$.

? What does our model for current in a circuit predict for the result of the experiment?

Around the circuit loop we have $emf - EL = 0$, so if we double the length L, we'll get half as big an electric field (since the emf doesn't change). With half the field we get half the drift speed ($\bar{v} = uE$) and half the electron current ($i = nAuE$) and half the conventional current ($I = |q|nAuE$). Therefore we expect half as large a magnetic field and half the compass deflection.

L $2L$

Figure 18.50 How does the current change when we double the length of the Nichrome wire?

Experiment 18.2 Effect of twice the length

Be careful when connecting Nichrome wire. If you run current through a short (< 10 cm) length of the wire it can get hot enough to burn you. (It may even get hot enough to glow.)

Following the directions given in Section 18.9.1, use the compass to compare the amount of current in the two different circuits shown in Figure 18.50:

(a) One battery (to keep the compass deflection small) and half the length of the thin Nichrome wire (about 20 cm). Don't cut the Nichrome wire, just connect clips to a portion half the length of the wire.

(b) One battery and the full length (about 40 cm) of the thin Nichrome wire.

Make careful, quantitative measurements of the angles, and record your measurements.

- Are you able to verify the prediction that the current is halved when you double the length of the Nichrome wire?
- Do you find that the current is unaffected if you bend the wire or add additional connecting wires?

18.9.3 Application: Two identical bulbs in series

Two identical light bulbs in series (Figure 18.51) are essentially the same as one light bulb with twice as long a filament, since the copper connecting wires contribute little energy dissipation. From the preceding experiment with doubling the length of a Nichrome wire, we might expect to get half the current compared with just one bulb.

Figure 18.51 Two identical light bulbs in series.

Experiment 18.3 Current in a two-bulb series circuit

(a) Following the directions given in Section 18.9.1, use a compass to measure the current in the circuit with two round bulbs in series shown in Figure 18.51, and record the observed compass deflections. Compare your observations to your prediction.

(b) Compare the compass deflection you observe in the two-bulb circuit with the compass deflection you observe in a one-bulb circuit. Record your measurements (by removing one bulb from the circuit).

You presumably found that the two-bulb circuit had more than half the current of the one-bulb circuit (if necessary, repeat this measurement carefully). The issue here is that the electron mobility in very hot tungsten (several thousand degrees!) is less than the electron mobility in cooler tungsten. The greater thermal agitation of the atomic cores leads to larger "friction." Note that the bulbs in the two-bulb circuit emit "redder" light than the single bulb, indicating that their filaments are not as hot.

Ex. 18.17 Explain why the current in the two-bulb series circuit is more than half the current in the one-bulb circuit. How is the electric field affected? How is the drift speed affected?

Ex. 18.18 Why didn't we see this effect when we experimented with doubling the length of the Nichrome wire?

18.9.4 Application: Twice the cross-sectional area

Figure 18.52 illustrates an experiment in which we see what happens to the current in a Nichrome wire when we double its cross-sectional area.

? Explain why the electron current in the wire should increase by a factor of two if the cross-sectional area of the wire doubles.

In Figure 18.52 seems plausible that doubling A should double the conventional current $I = |q| nAuE$. It is like having two wires in parallel, doubling the drain on the battery.

We can make this argument quantitative by considering energy. Since we have emf $- EL = 0$ around the loop, the electric field E in the wire does not change when we use a thicker wire, so the current doubles. (This neglects internal resistance in the battery, which has the effect of limiting the amount of current that the battery can supply, so that for large cross-sectional areas and large currents, doubling a large cross-sectional area may not yield twice the current.)

Experiment 18.4 Effect of doubling the cross-sectional area

Use the compass to compare the amount of current in two different circuits (Figure 18.52): one battery (to keep the compass deflection small) and the same lengths of thin and thick Nichrome wires.

Use only one battery. Connect the connecting wires about 30 cm apart on each Nichrome wire (don't cut the Nichrome wire). Following the directions given in Section 18.9.1, make careful, quantitative measurements of the angles.

The thick Nichrome wire has a cross section about twice the cross section of the thin Nichrome wire. Your instructor may give you a more precise ra-

Figure 18.52 How does the current change when we double the cross-sectional area of the Nichrome wire?

tio of the areas. What do you measure for the ratio of the current in the thick wire to the current in the thin wire? Does this agree with predictions?

Your own data are probably consistent with what more precise experiments confirm. These measurements imply that the drift speed of individual electrons is the same in the thick wire and in the thin wire (because the electric field is the same in the two circuits if the length of the wires is the same), and that the many-electron current is distributed uniformly over the cross section.

If the drift speed were not the same everywhere across the cross section of the wire, we couldn't simply write $nA\bar{v}$ for the electron current, using the same $\bar{v}$ for all the electrons passing that point in the circuit. In Section 18.2.4 we proved that the current is indeed uniform across the cross section.

18.9.5 Application: Two bulbs in parallel

Experiment 18.5 A parallel circuit

Connect two round bulbs in "parallel" (that is, one beside the other), as shown in Figure 18.53.

(a) How does the brightness of the round bulbs in this parallel circuit compare to the brightness of two round bulbs in series?

(b) Unscrew one of the bulbs in the parallel circuit. What effect does this have on the brightness of the remaining bulb?

(c) Based on these observations, what relationship would you expect to see between the compass deflections at locations A, B, C, and D, in this circuit? Record your prediction.

(d) Measure the compass deflections at locations A, B, C, and D, and record your measurements. Do your observations agree with your predictions in (c)?

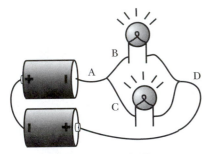

Figure 18.53 Two round bulbs connected in parallel.

In the preceding experiment you should again have seen the current entering a location is equal to the current leaving the same location in the steady state. In particular, the current at A should be equal to the sum of the currents at B and C. If you did not observe this, you should repeat the experiment. We can also think of the two bulbs in parallel as equivalent to increasing the cross-sectional area of one of the bulb filaments.

We can analyze the situation quantitatively. If we follow the path through one of the bulbs, we have $2\text{emf} - EL = 0$, (2emf for two batteries), and L is the length of the bulb filament (neglecting dissipation in the connecting wires). If we follow the path through the other round bulb, we also have $2\text{emf} - EL = 0$. So the electric field $E = 2\text{emf}/L$ is the same in both bulb filaments, and is the same that there would be with only one bulb present.

Figure 18.54 shows an approximate surface-charge distribution on the circuit. Each bulb is flanked by positive and negative charges, which make a large electric field in the bulb filament. Removing one of the bulbs doesn't change the surface-charge distribution very much, and the brightness of the remaining bulb hardly changes. The batteries of course have to deliver more current ($i_B + i_C$) to the two parallel bulbs than they do to one bulb.

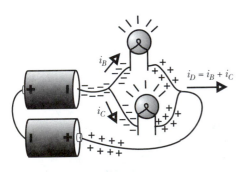

Figure 18.54 Approximate surface charge distribution for two round bulbs in parallel.

18.9.6 Why does current divide between parallel resistors?

When there are parallel branches in a circuit, as in the previous experiment, there is a question of how the current "knows" how to divide between the branches.

In Figure 18.55 we show very approximately the steady state of a circuit containing wide carbon resistors in two parallel branches (Figure 18.55). During the initial transient just after connecting the battery, electrons flowed into both branches, but the surface of the dead-end branch became charged so negatively that no more electrons could enter. In the inset, note the deflection of the electron stream lines due to the lack of charge immediately above the branch point, inside the dead-end branch. The wire is effectively a bit wider as a result of adding the dead-end branch.

Next we complete the parallel connection, which leads to a rearrangement of surface and interface charges (Figure 18.56). The inset shows how the new arrangement of surface charges guides and splits the oncoming electron current, with some of the electrons taking the upper branch and some the lower branch. Note that there is now a larger current through the battery and a larger gradient of surface charge along the wires to drive this larger current.

The steady-state current in each branch is determined by how well each branch conducts current. For example, if the mobility is low in one of the branches, surface charge will build up in such a way as to decrease the fraction of the current that goes down that branch. The extreme case is a dead end, where surface charge builds up until it prevents any current at all from going down the dead-end branch.

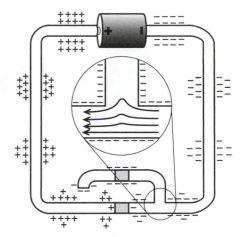

Figure 18.55 A circuit with a (dead-end) parallel branch.

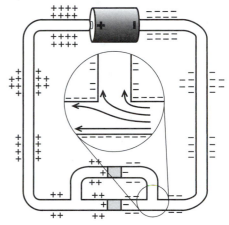

Figure 18.56 Complete the connection: the electron current splits.

18.9.7 Application: Round bulb and long bulb in series

Put a round bulb and a long bulb in series with each other (Figure 18.57). The round bulb doesn't light up! We can explain this puzzling behavior. Consider that:

Both bulb filaments are made of tungsten.

Round bulb has large cross-sectional area A_r, length L_r, electric field E_r, drift speed $\bar{v}_r$.

Long bulb has small cross-sectional area A_b, length L_b, electric field E_b, drift speed $\bar{v}_l$.

Assume for the moment that the electron mobility is the same in both bulbs, even though in fact the glowing long bulb is very hot and the round bulb is apparently cool (since it isn't glowing).

? Using the current node rule, write an equation relating the drift speeds $\bar{v}_r$ and $\bar{v}_l$.

In the steady state, the number of electrons per second passing through each bulb must be the same:

$$nA_r\bar{v}_r = nA_l\bar{v}_l$$

? Based on the equation you just wrote, write an equation relating the electric fields E_r and E_b, and state which is larger, E_r or E_l.

Assuming the same mobility in both filaments, since

$$\bar{v} = uE$$

we know that $nA_ruE_r = nA_luE_l$, so $E_l = \dfrac{A_rE_r}{A_l}$

Since $A_r < A_l$, $E_r > E_l$.

? Applying the loop rule we find that $2\text{emf} = E_rL_r + E_lL_l$, which is approximately equal just to eE_lL_b, since E_r is very small. Explain why the round bulb doesn't glow.

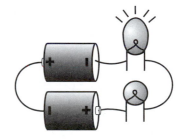

Figure 18.57 A round bulb and a long bulb in series—why doesn't the round bulb light up?

If the round bulb were alone in a circuit, it would be bright, and the electric field in the filament would be:

$$2\text{emf} = E_{\text{bright}}L, \text{ so } E_{\text{bright}} = \frac{2\text{emf}}{L}$$

In the two bulb circuit (if the lengths of the filaments are the same) we have:

$$2\text{emf} = E_r L + E_1 L = L\left(E_r + \frac{A_r E_r}{A_1}\right) \text{ (from previous exercise)}$$

$$= LE_r(1 + A_r/A_1)$$

$$\text{so } E_r = \frac{2\text{emf}}{L(1 + A_r/A_1)} < \frac{2\text{emf}}{L}$$

So the electric field in the round bulb is significantly smaller than it is in the single-bulb circuit. We can estimate how much smaller by noting that we typically measure about 1/4 as much current in a circuit with a single long bulb as in a circuit with a single round bulb. This suggests that $A_r \approx 4A_1$, so the electric field in the round bulb has been reduced by a factor of 5. Evidently this field does not produce sufficient current through the filament to make it glow.

(Moreover, since the round bulb isn't glowing its temperature is low and its mobility is much larger than usual. This further reduces the E_r required to drive the same current that runs through the long bulb.)

18.9.8   Application: Two batteries in series

It should now be clear why two batteries in series can drive more current through a wire than one battery can: the potential difference across two batteries in series is 2emf, and this doubles the electric field everywhere in the circuit. Double the electric field, double the drift speed, double the current (if the mobility doesn't change due to a higher temperature).

In the following experiment we test this prediction with a Nichrome wire rather than a light bulb, because when you add a second battery the bulb filament undergoes a sizable temperature increase that changes the mobility significantly (you can see the color become whiter, which is an indication of the higher filament temperature). With a length of Nichrome wire the temperature stays close to room temperature, so the mobility is nearly the same with one or two batteries.

Experiment 18.6 Effect of two batteries

Use the compass to compare the amount of current in two different circuits: one battery and a length of thin Nichrome wire, and two batteries and the same length of thin Nichrome wire. Following the directions in Section 18.9.1, make careful, quantitative measurements of the angles.

Are you able to verify the prediction that the current doubles when you use two batteries? (If you wish, try three or more batteries with the same length of thin Nichrome wire. But don't use more than two batteries with your light bulbs, because it may burn them out.)

18.10   Observing surface charge in a high voltage circuit

The reason we can't detect the surface charge on the wires of the simple circuits we have been assembling is that the amount of charge on the wires is very small—very little charge is needed to make the small electric fields in the wires and bulb filaments of your circuits. However, it is possible to detect surface charge in a high-voltage circuit (around 10,000 volts is required). The following interesting demonstration of surface charge was suggested by the science education group at the Weizmann Institute in Israel. Perhaps your instructor will be able to show this to you.

Warning: Working with high voltages is dangerous, so don't try this without an instructor present. Keep one hand behind your back when you are near a high voltage circuit. Do not touch the bare wires!

A chain of four identical high resistance resistors is supported in air from Styrofoam columns, so that the resistor chain is far from other objects, including the table, to avoid unwanted polarization effects. The resistors are connected by bare (uninsulated) wires to two high voltage power supplies which maintain one end of the resistor chain at a positive potential and the other end at a negative potential with respect to ground (the round hole in a 3-prong socket). Note that you need to avoid touching the wires when the circuit is connected!

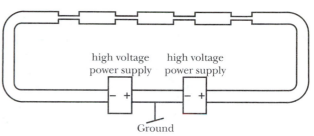

Figure 18.58 A high voltage (low current) circuit for observing surface charge.

We model the resistors as very thin wires. The conventional current everywhere in this series circuit is $I = enA\bar{v} = enAuE$.

? Given the expression for the current, sketch the pattern of electric field around the circuit.

Evidently the electric field must be very much larger inside the thin resistors than inside the thick connecting wires, and the electric field must have the same magnitude inside each resistor. As usual, the pattern of electric field is simple. The feedback mechanism will lead to an arbitrarily complicated distribution of surface charge that produces this simple pattern of field.

? Sketch a plausible approximate distribution of surface charge, consistent with the pattern of electric field.

Very roughly, the electric field inside the left-most resistor (AB) is in large part due to a large negative surface-charge density at A and a medium surface-charge density at B, with a large gradient along the resistor. (This is an approximate analysis, because the field in the resistor is due not just to the nearby charges but to all charges.) The electric field must have the same magnitude in the next resistor (BC) and is largely due to the medium surface-charge density at B and the nearly zero surface-charge density at C. Similar remarks apply to the other two resistors.

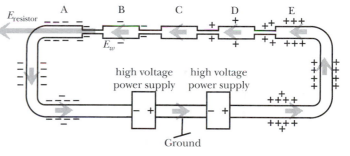

Figure 18.59 Electric field and surface charge for a high voltage circuit.

Inside the connecting wires the electric field is very small, and we indicate a roughly uniform charge distribution along these wires (the details of the actual charge distribution, except for the sign of the charge, are presumably more complicated).

A flexible, thin strip of aluminized plastic (Mylar) is suspended from an insulating rod and brought close to the bare wire near location A.

? What do you expect will happen?

The strip is observed to be attracted to the wire (due to polarization of the strip by the surface charge on the wire) and then to jump away, due to being charged by contact with the bare wire and repelled by the surface charge at that location. The strip is found to be negatively charged, because it is repelled by a plastic pen rubbed through one's hair (a clear plastic pen is known to charge negatively). A similar effect is observed at E, but the strip is observed to be positively charged. A smaller effect is observed at B and D, where the surface-charge density is expected to be smaller. No effect is observed at point C, where by symmetry there is essentially no surface charge. These experimental observations confirm the validity of the approximate surface charge analysis.

The surface-charge density is proportional to the circuit voltage, and only at very high voltages is there enough charge to observe electrostatic repulsion in a mechanical system. It is difficult to observe repulsion by surface charge in a low-voltage circuit, because a charged object is of course attracted to neutral matter, and the attraction between a charged object and a circuit can mask the repulsion, if the surface charge on the circuit is small, as it is in low-voltage circuits.

The object must be very lightweight (thin aluminized Mylar works well) so that one can observe the interaction at a sizable distance, in order to minimize the competition between repulsion and the attraction due to polarization of neutral matter, which falls off with distance much more rapidly than $1/r^2$. Recall that the attraction between a point charge and a small neutral object is proportional to $1/r^5$, since the polarization of the neutral object is proportional to $1/r^2$, and the interaction between a point charge and a permanent dipole is proportional to $1/r^3$.

18.11 Summary

Fundamental Principles

No new fundamental physical principles.

New concepts

- In a uniform wire, a uniform electric field drives a uniform steady-state current.
- Charges on the surface of the wires make electric fields inside the wires.
- There is an initial transient, lasting a minimum of a few nanoseconds, in which feedback establishes the steady-state distribution of surface charges and currents.
- A battery maintains a charge separation and therefore a potential difference. The emf of the battery is the energy input per unit charge required to maintain the charge separation.

Analysis techniques

In analyzing electric circuits we apply two rules:

THE CURRENT NODE RULE

In the steady state, for many electrons
flowing into and out of a region,

electron current: net i_{in} = net i_{out}, where $i = nAuE$

conventional current: net I_{in} = net I_{out}, where $I = |q|nAuE$

ENERGY CONSERVATION (THE LOOP RULE)

In the steady state, for any round-trip path,

$$\Delta V_1 + \Delta V_2 + ... = 0$$

Results

Our model of the microscopic behavior of circuits provides answers to several questions about circuit behavior:

1) Where are the charges that are the sources of the electric field inside the wires? These charges are on the surfaces of the wires, and on and in the battery. The necessary charge distribution is created and maintained by feedback.

2) How can the light come on instantaneously despite the very slow drift speed of the mobile charges? Electric fields propagate at the speed of light. The electrons inside the bulb are affected by the electric field which is quickly (in a few nanoseconds) established in the bulb filament.

3) What does a battery actually do in a circuit? A battery does a fixed amount of work per unit charge (emf), maintaining a charge separation and potential difference across itself, and consequently maintaining a distribution of surface charge all over the circuit.

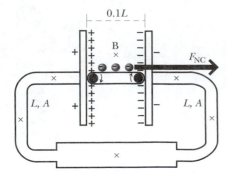

Length 0.6L, cross-sectional area 4A

Figure 18.60 Circuit for example problem.

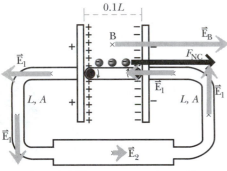

Length 0.6L, cross-sectional area 4A

Figure 18.61 Part (a) electric field at locations in the wire.

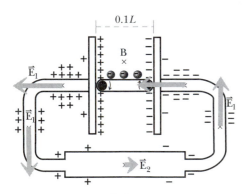

Figure 18.62 Part (b) an approximate surface charge distribution.

18.12 Example problem: A circuit with a wide wire

The circuit in Figure 18.60 contains a low-resistance mechanical battery which exerts a non-Coulomb force F_{NC} to move electrons through the battery (with negligible internal resistance). The end plates of the battery are very large compared to the distance $0.1L$ between the plates (plates not drawn to scale). Two thin Nichrome wires of length L and cross-sectional area A connect the battery to a thick Nichrome wire of length $0.6L$ and cross-sectional area $4A$. The electron mobility of the Nichrome is u, and there are n mobile electrons per cubic meter in the Nichrome.

(a) Show the electric field at the six locations marked with × (including location B between the plates). Pay attention to the relative magnitudes of the six vectors that you draw.

(b) Show the approximate distribution of charge on the surface of the Nichrome wires. Make sure that your distribution is compatible with the electric fields that you drew in part (a).

(c) Calculate the number of electrons that leave the battery every second, in terms of the given quantities L, A, n, u, and F_{NC} (and fundamental constants). Be sure to show all of your work.

(d) Calculate the magnitude of the electric field between the plates of the battery (location B).

Solution

(a) By the current node rule, current in the thin and thick wires will be equal. Since $i = nAuE$, E will be larger in wires with thinner cross-sectional area. This suggests the pattern of electric field shown in the wires in Figure 18.61.

In order for the round trip potential difference to be zero, the electric field between the plates must point to the right and be large compared to the electric field in the wires, as shown in Figure 18.61.

(b) A large gradient of surface charge will produce a large electric field; a small gradient will produce a small field, as shown in Figure 18.62.

(c) Apply basic principles:

Current node rule: current in thin wire (1) = current in thick wire (2)

$$nAuE_1 = n(4A)uE_2.$$

$\Delta V_{\text{round trip}} = 0$ (energy conservation), going counterclockwise from negative plate of mechanical battery:

$$\frac{F_{NC}(0.1L)}{e} - E_1 L - E_2(0.6L) - E_1 L = 0$$

From the current equation:

$$E_2 = \left(\frac{A}{4A}\right)E_1 = \frac{E_1}{4}$$

We now have two equations in two unknowns (E_1 and E_2). Substitute for E_2:

$$\frac{F_{NC}(0.1L)}{e} - 2(E_1 L) - \left(\frac{E_1}{4}\right)0.6L = 0$$

$$E_1 = \frac{F_{NC}}{e(21.5)}$$

Now we can solve for electron current (number of electrons/s leaving the battery):

$$i = nAuE_1 = nAu\left(\frac{F_{NC}}{e(21.5)}\right)$$

(d) Round trip potential difference must be zero, so:

$$E_B(0.1L) - 2E_1L - \left(\frac{E_1}{4}\right)(0.6L) = 0$$

$$E_B = 21.5\left(\frac{F_{NC}}{e(21.5)}\right) = \frac{F_{NC}}{e}$$

The electric field E_B inside the battery is very large compared to the electric field in the wires.

18.13 Review questions

Steady state

RQ 18.1 What is the most important general difference between a system in "steady state" and a system in "static equilibrium"?

RQ 18.2 Describe the following attributes of a *metal wire* in steady state vs. static equilibrium:

	Metal wire in steady state	Metal wire in static equilibrium
Location of excess charge		
Motion of mobile electrons		
E inside the metal wire		

RQ 18.3 How can there be a nonzero electric field inside a wire in a circuit? Isn't the electric field inside a metal always zero?

RQ 18.4 Electron current $i = nA\bar{v} = nAuE$:
   What are the units of electron current?
   What is n? What are its units?
   What is A? What are its units?
   What is $\bar{v}$? What are its units?
   What is u? What are its units?

RQ 18.5 State your own theoretical *and* experimental objections to the following statement: In a circuit with two round bulbs in series, the bulb farther from the negative terminal of the battery will be dimmer, because some of the electron current is used up in the first bulb. Cite relevant experiments.

RQ 18.6 In the parallel circuit shown in Figure 18.63, what are the relationships among i_A, i_B, i_C, and i_D?

Drift speed and mobility

RQ 18.7 Since there is an electric field inside a wire in a circuit, why don't the mobile electrons in the wire accelerate continuously?

RQ 18.8 Why don't all mobile electrons in a metal have exactly the same speed?

RQ 18.9 There are very roughly the same number of iron atoms per m³ as there are copper atoms per m³, but copper is a much better conductor than iron. How does u_{iron} compare with u_{copper}?

Feedback

RQ 18.10 In the few nanoseconds *before* the steady state is established in a circuit consisting of a battery, copper wires, and a single bulb, is the current the same everywhere in the circuit? Explain.

RQ 18.11 During the initial transient leading to the steady state, the electron current going into a bulb may be greater than the electron current leaving the bulb. Explain why and how these two currents come to be equal in the steady state.

Electric field and surface charge distribution

RQ 18.12 A steady-state current flows through the Nichrome wire in the circuit shown in Figure 18.64.

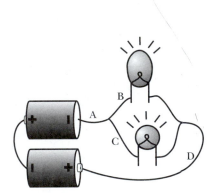

Figure 18.63 A parallel circuit (RQ 18.6).

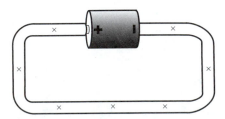

Figure 18.64 A Nichrome wire connected to a single battery (RQ 18.12).

(a) On a copy of this diagram, show the electric field at the locations indicated, paying attention to relative magnitude.

(b) Carefully draw pluses and minuses on your diagram to show the approximate surface charge distribution that produces the electric field you drew. Make your drawing show clearly the differences between regions of high surface charge density and regions of low surface charge density.

RQ 18.13 In the circuit shown in Figure 18.65, all of the wire is made of Nichrome, but one segment has a much smaller cross sectional area.

(a) On a copy of this diagram, *using the same scale for magnitude that you used in RQ 18.12,* show the steady-state electric field at the locations indicated, including in the thinner segment.

(b) Carefully draw pluses and minuses on your diagram to show the approximate surface charge distribution that produces the electric field you drew. Make your drawing show clearly the differences between regions of high surface charge density and regions of low surface charge density.

(c) Use your diagram to explain why there is less current in this circuit than in the circuit shown in RQ 18.12. (Remember to use the same scale for the magnitude of electric field for both circuits.)

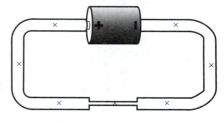

Figure 18.65 A circuit containing two thick Nichrome wires and one thin Nichrome wire (RQ 18.13).

Batteries

RQ 18.14 Criticize the statement below on theoretical and experimental grounds. Be specific and precise. Refer to your own experiments, or describe any new experiments you perform:

> "A flashlight battery always puts out the same amount of current, no matter what is connected to it."

RQ 18.15 What is the difference between emf and electric potential difference?

RQ 18.16 Compare the direction of the average electric field inside a battery to the direction of the electric field in the wires and resistors of a circuit.

Experimental results

RQ 18.17 For each of the following experiments, state what effect you observed (how the current in the circuit was affected) and why, in terms of the relationships:

$$\text{electron current} = nA\bar{v} \text{ and } \bar{v} = uE$$

Experiment	Effect on current (be as quantitative as possible)	Circle parameter(s) that changed and explain briefly
Double the length of a Nichrome wire		$n\ A\ u\ E$
Double the cross-sectional area of a Nichrome wire		$n\ A\ u\ E$
Two identical bulbs in series compared to a single bulb		$n\ A\ u\ E$
Two batteries in series compared to a single battery		$n\ A\ u\ E$

RQ 18.18 In a table like the one at the top of the next page, write an inequality comparing each quantity in the steady state for a narrow resistor and thick connecting wires, which are made of the same material as the resistor.

Electron current in resistor	$<, \leq, =, \geq,$ or $>$	Electron current in thick wires
n_R		n_W
A_R		A_W
u_R		u_W
E_R		E_W
v_R		v_W

Simple Circuits

RQ 18.19 In a circuit with one battery, connecting wires, and a 12 cm length of Nichrome wire, a compass deflection of 6 degrees is observed. What compass deflection would you expect in a circuit containing two batteries in series, connecting wires, and a 36 cm length of thicker Nichrome wire (double the cross-sectional area of the thin piece)? Explain.

RQ 18.20 At a typical drift speed of 5×10^{-5} m/s, it takes an electron about 100 minutes to travel through one of your connecting wires. Why, then, does the bulb light immediately when the connecting wire is attached to the battery?

18.14 Homework problems

Base your analyses in the following problems on the principles discussed in this chapter. If you wish to use formulas derived elsewhere, you must justify them in terms of the microscopic analysis introduced in this chapter.

Problem 18.1 Current in a series circuit

Consider the circuit containing three identical light bulbs shown in a top view in Figure 18.66. North is indicated in the diagram. Compasses are placed under the wires at locations A and B.

(a) The magnitude of the deflection of compass A is 13 degrees away from north. What direction does the needle point? Draw a sketch.

(b) What is the magnitude of the deflection of the needle of compass B? What direction does the needle point? Draw a sketch. Explain your reasoning.

(c) In the steady state 1.5×10^{18} electrons per second enter bulb 1. There are 6.3×10^{28} mobile electrons per cubic meter in tungsten. The cross-sectional area of the tungsten filament in bulb 1 is 1×10^{-8} m^2. The electron mobility in hot tungsten is $1.2 \times 10^{-4} \frac{\text{m/s}}{\text{N/C}}$. Calculate the electric field inside the tungsten filament in bulb 3. Give both the direction and the magnitude of the electric field.

Problem 18.2 Three bulbs

Figure 18.67 is a top view of a circuit containing three identical light bulbs (the rest of the circuit including the batteries is not shown). The connecting wires are made of copper.

(a) The compass is placed on top of the wire, and it deflects 20 degrees away from north as shown (the wire is underneath the compass). What direction are the electrons moving at location P_1? How do you know?

(b) In the steady state, 3×10^{18} electrons pass location P_1 every second. How many electrons pass location P_2 every second? Explain briefly.

(c) Give the relative brightnesses of bulbs B_1, B_2, and B_3. Explain briefly.

(d) There are 6.3×10^{28} mobile electrons per cubic meter in tungsten. The cross-sectional area of the tungsten filament in bulb B_1 is 0.01 mm^2

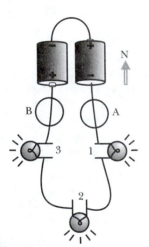

Figure 18.66 Top view of a circuit containing three identical light bulbs in series (Problem 18.1). North is indicated in the diagram. The wires lie on top of the compasses.

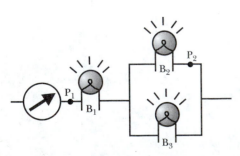

Figure 18.67 Circuit for Problem 18.2.

(which is 10^{-8} m^2). The electron mobility in hot tungsten is 1.2×10^{-4} (m/s)/(N/C). Calculate the electric field inside the tungsten filament in bulb B$_1$. Give both the direction and the magnitude of the electric field.

Problem 18.3 A three-bulb circuit

When a single round bulb of a particular kind and two batteries are connected in series, 3×10^{18} electrons pass through the bulb every second. When a single long bulb of a particular kind and two batteries are connected in series, only 1.5×10^{18} electrons pass through the bulb every second, because the filament has a smaller cross section.

(a) In the circuit shown in Figure 18.68, how many electrons per second flow through the long bulb?

(b) What approximations or simplifying assumptions did you make?

(c) Show approximately the surface charge on a diagram of the circuit.

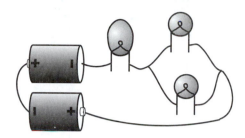

Figure 18.68 Circuit for Problem 18.3.

Problem 18.4 A Nichrome resistor circuit

In the circuit shown in Figure 18.69, the two thick wires and the thin wire are made of Nichrome.

(a) Carefully draw pluses and minuses on your own diagram to show the approximate surface charge distribution in the steady state. Make your drawing show clearly the differences between regions of high surface charge density and regions of low surface charge density.

(b) Show the steady-state electric field at the locations indicated, including in the thin wire.

(c) The emf of the battery is 1.5 volts. In Nichrome there are $n = 9\times10^{28}$ mobile electrons per m^3, and the mobility of mobile electrons is $u = 7\times10^{-5}$ (m/s)/(N/C). Each thick wire has length $L_1 = 20$ cm $= 0.2$ m and cross-sectional area $A_1 = 9\times10^{-8}$ m^2. The thin wire has length $L_2 = 5$ cm $= 0.05$ m and cross-sectional area $A_2 = 1.5\times10^{-8}$ m^2. (The total length of the three wires is 45 cm.) In the steady state, calculate the number of electrons entering the thin wire every second. Do not make any approximations, and do not use Ohm's "law" or series-resistance formulas. State briefly where each of your equations comes from.

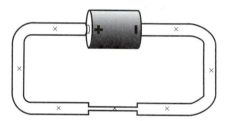

Figure 18.69 Circuit for Problem 18.4.

Problem 18.5 Three bulbs and two bulbs

Three identical round bulbs are in series as shown in Figure 18.70. Thick copper wires connect the bulbs. In the steady state, 3×10^{17} electrons leave the battery at location A every second.

(a) How many electrons enter the second bulb at location D every second? If there is insufficient information to give a numerical answer, state how it compares with 3×10^{17}. Justify your answer carefully.

(b) Next the middle bulb (at DE) is replaced by a wire, as shown in Figure 18.71. Now how many electrons leave the batteries at location A every second? Explain clearly! If you have to make an approximation, state what it is. Do not use ohms or series-resistance formulas in your explanation, unless you can show in detail how these concepts follow from the microscopic analysis introduced in this chapter.

(c) Check your analysis by trying the experiment with a partner. Try to find three round bulbs that glow equally brightly when in series with each other, because bulb construction varies slightly in manufacturing. Remember to arrange the circuit so that the largest compass deflection is no more than 15 degrees. Report the deflections that you observe. Does the experiment agree with your prediction? (If not, can you explain the discrepancy? Be specific. For example, if the current is larger than predicted, explain *why*

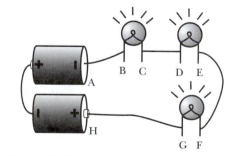

Figure 18.70 Circuit for Problem 18.5a.

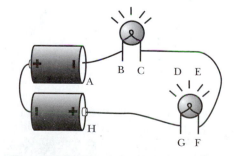

Figure 18.71 Circuit for Problem 18.5b.

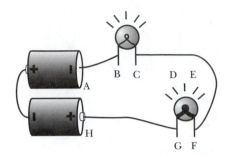

Figure 18.72 Circuit for Problem 18.5d.

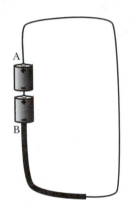

Figure 18.73 Two Nichrome wires (Problem 18.6).

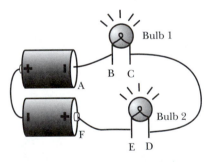

Figure 18.74 Circuit for Problem 18.7.

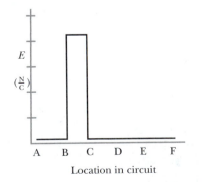

Figure 18.75 Graph of *E* vs. location in circuit, for Problem 18.7d. The magnitude of the electric field in the wires is greatly exaggerated.

it is larger than predicted.)

(d) Finally, the last bulb (at FG) is replaced by a bulb identical in every way except that its filament has twice as large a cross-sectional area, as shown in Figure 18.72. Now how many electrons leave the batteries at location A every second? Explain clearly! If you have to make an approximation, state what it is.

Problem 18.6 Two Nichrome wires
The following questions refer to the circuit shown in Figure 18.73, consisting of two flashlight batteries and two Nichrome wires of different lengths and different thicknesses as shown (corresponding roughly to your own thick and thin Nichrome wires).

The thin wire is 50 cm long, and its diameter is 0.25 mm. The thick wire is 15 cm long, and its diameter is 0.35 mm.

(a) The emf of each flashlight battery is 1.5 volts. Determine the steady-state electric field inside each Nichrome wire. Remember that in the steady state you must satisfy both the current node rule and energy conservation. These two principles give you two equations for the two unknown fields.

(b) The electron mobility in room-temperature Nichrome is about 7×10^{-5} (m/s)/(N/C) . Show that it takes an electron 36 minutes to drift through the two Nichrome wires from location B to location A.

(c) On the other hand, about how long did it take to establish the steady state when the circuit was first assembled? Give an approximate, not precise numerical answer.

(d) There are about 9×10^{28} mobile electrons per cubic meter in Nichrome. How many electrons cross the junction between the two wires every second?

Problem 18.7 Different mobilities
In the circuit shown in Figure 18.74 bulbs 1 and 2 are identical in mechanical construction (the filaments have the same length and the same cross-sectional area), but the filaments are made of different metals. The electron mobility in the metal used in bulb 2 is three times as large as the electron mobility in the metal used in bulb 1, but both metals have the same number of mobile electrons per cubic meter. The two bulbs are connected in series to two batteries with thick copper wires (like your connecting wires).

(a) In bulb 1, the electron current is i_1 and the electric field is E_1. In terms of these quantities, determine the corresponding quantities i_2 and E_2 for bulb 2, and explain your reasoning.

(b) When bulb 2 is replaced by a wire, the electron current through bulb 1 is i_0 and the electric field in bulb 1 is E_0. How big is i_1 in terms of i_0? Explain your answer, including explicit mention of any approximations you must make. Do not use ohms or series-resistance formulas in your explanation, unless you can show in detail how these concepts follow from the microscopic analysis introduced in this chapter.

(c) Explain why the electric field inside the thick copper wires is very small. Also explain why this very small electric field is the same in all of the copper wires, if they all have the same cross-sectional area.

(d) Here is a graph of the magnitude of the electric field at each location around the circuit when bulb 2 is replaced by a wire. Copy this graph and add to it, on the same scale, a graph of the magnitude of the electric field at each location around the circuit when both bulbs are in the circuit. The very small field in the copper wires has been shown much larger than it really is in order to give you room to show how that small field differs in the two circuits.

Problem 18.8 A backwards battery

Some students intended to run a light bulb off two batteries in series in the usual way, but they accidentally hooked up one of the batteries backwards, as shown in Figure 18.76 (the bulb is shown as a thin filament).

(a) Use +'s and —'s to show the approximate steady-state charge distribution along the wires and bulb.

(b) Draw vectors for the electric field at the indicated locations inside the connecting wires and bulb.

(c) Compare the brightness of the bulb in this circuit with the brightness the bulb would have had if one of the batteries hadn't been put in backwards.

(d) Try the experiment to check your analysis. Does the bulb glow about as you predicted?

Problem 18.9 Chemical battery life

Inside a chemical battery it is not actually individual electrons that are transported from the + end to the – end. At the + end of the battery an "acceptor" molecule picks up an electron entering the battery, and at the – end a different "donor" molecule gives up an electron which leaves the battery. Ions rather than electrons move between the two ends to transport the charge inside the battery.

When the supplies of acceptor and donor molecules are used up in a chemical battery, the battery is dead, because it can no longer accept or release electrons. The electron current in electrons per second, times the number of seconds of battery life, is equal to the number of donor (or acceptor) molecules in the battery.

A flashlight battery contains approximately half a mole of donor molecules. The electron current through a round bulb powered by two flashlight batteries in series is about 0.3 ampere. About how many hours will the batteries keep this bulb lit?

Problem 18.10 Two circuits containing long bulbs

Two circuits are assembled using 1.5-volt batteries, thick copper connecting wires, and long bulbs (Figure 18.77). Bulbs A, B, and C are identical long bulbs. Compasses are placed under the wires at the indicated locations.

(a) On sketches of the circuits, draw the directions the compass needles will point, and indicate the approximate magnitude of the compass deflections in degrees. Note that the deflection is given at one location. If you do not have enough information to give a number, then indicate whether it will be greater than, less than, or equal to 5 degrees. (Assume that the compasses are adequately far away from the steel-jacketed batteries.)

(b) Briefly explain your reasoning about the magnitudes of the compass deflections.

(c) On the same sketches of the circuits, show very approximately the distribution of surface charge.

(d) The tungsten filament in each of the bulbs is 4 mm long with a radius of 6×10^{-6} meter. Calculate the electric field inside each of the three bulbs, E_A, E_B, and E_C.

Problem 18.11 Long bulb and round bulb

A circuit is assembled that contains a long bulb and a round bulb as shown in Figure 18.78, with four compasses placed underneath the wires (we're looking down on the circuit). When the long bulb is unscrewed from its socket, the compass on the left (next to the battery) deflects 15 degrees. When the long bulb is screwed back in and the round bulb is unscrewed, the compass on the left deflects 4 degrees.

Figure 18.76 Circuit for Problem 18.8. The bulb is shown as a thin filament.

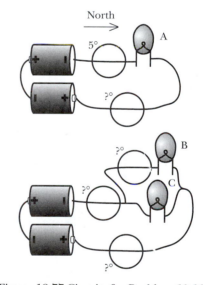

Figure 18.77 Circuits for Problem 18.10.

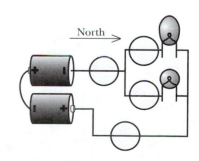

Figure 18.78 Circuit for Problem 18.11.

With both bulbs screwed into their sockets, draw the orientations of the needle on each compass, and write the compass deflection in degrees beside the compass. Explain briefly.

Problem 18.12 Connect two charged spheres together
A solid metal sphere of radius R carries a uniform charge of $+Q$. Another solid metal sphere of radius r carries a uniform charge $-q$. The amount of charge is not enough to cause breakdown in air. The two spheres are very far apart (distance $\gg R$ and distance $\gg r$). At $t = 0$ a very thin wire of length L is connected to the two spheres. The mobility u of free electrons in this wire is very small, and the wire conducts electrons so poorly that it takes about an hour for the system to reach static equilibrium. In a short time Δt (a few seconds) how many electrons leave the sphere of radius r? There are n mobile electrons per cubic meter in the wire, and the wire has a constant cross sectional area A. Explain your work, and any approximations you need to make

Figure 18.79 Two charged spheres are connected by a very thin wire (Problem 18.12).

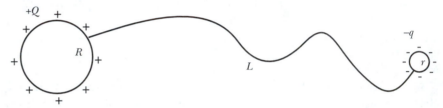

18.15 Answers to exercises

18.1 (page 627) $I_1 + I_4 - I_3 - I_2 = 0$; $I_2 = 3$ A

18.2 (page 627) $I_2 = -2$ A , current is entering node, not leaving

18.3 (page 627): It is not true that current must be the same in every part of a circuit; only that the current entering a node must equal the current leaving a node. Here $i_1 = i_2 + i_3$.

18.4 (page 630) 0.011 N/C

18.5 (page 630) 8×10^{-5} m/s

18.6 (page 631) 1.8×10^{-2} N/C

18.7 (page 631) 1.2×10^{19} electrons/s

18.8 (page 635)

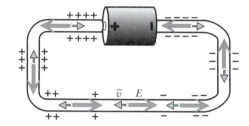

18.9 (page 641)

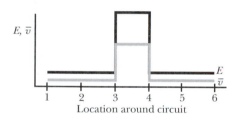

Location around circuit

18.10 (page 642) (Third example on next page.)

No charge gradient in this region, so E would be *very* small.

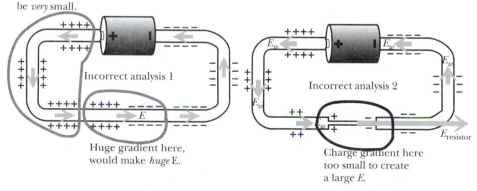

Incorrect analysis 1

Huge gradient here, would make *huge* E.

Incorrect analysis 2

Charge gradient here too small to create a large E.

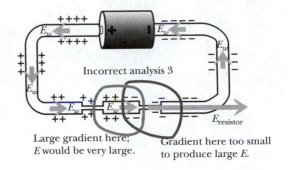

Incorrect analysis 3

Large gradient here; E would be very large.   Gradient here too small to produce large E.

18.11 (page 644)   Treat the mechanical battery as a capacitor. (For a real battery this is a poor approximation, because the endplates are not large compared to its length.) The Coulomb force on the electron is equal to the non-Coulomb force on the electron:

$$Es = \left(\frac{Q}{A\varepsilon_0}\right)s = \text{emf}, \text{ so } Q = \frac{A\varepsilon_0(\text{emf})}{s}$$

18.12 (page 644)

$$Q = \frac{A\varepsilon_0(\text{emf})}{s}$$

$$= \pi(0.01\,\text{m})^2\left(9\times10^{-12}\frac{\text{Nm}^2}{\text{C}^2}\right)\left(\frac{1.5\,\text{volts}}{0.05\,\text{m}}\right)$$

$$= 8.5\times10^{-14}\,C$$

This is much less than the charge on a tape (about 10^{-8} C). One would not expect to observe repulsion, since the repulsive force would probably be smaller than the force attracting the tape to the battery (which had been polarized by the charged tape).

18.13 (page 645)   The electric field in the thick copper connecting wires is very small, so

$$\text{emf} \approx E_{\text{bulb}}L_{\text{bulb}}$$

and the electric field in the bulb filament depends almost entirely on the length of the filament itself.

18.14 (page 646)   +1.5 volts; + end at C

18.15 (page 646)   Approximately 1.5 volts; can't determine I

18.16 (page 646)   5 volts/m; bends don't matter—electric field always parallel to wire; no change—different current but same electric field

18.17 (page 649)   Two bulbs: each bulb is much dimmer than a single bulb would be. Filaments in the dimmer bulbs are cooler, so the electron mobility is higher. Electric field is half as large, but we get more than half the drift speed, hence more than half the current.

18.18 (page 649)   The Nichrome wire did not get hot enough to glow, so apparently the mobility did not change significantly.

Chapter 19

Capacitors, Resistors, and Batteries

Chapter 19

Capacitors, Resistors, and Batteries

In this chapter we will study two remaining aspects of circuits: the role of capacitors in circuits, and a macroscopic analysis that complements and extends the microscopic analysis of the previous chapter.

19.1 Capacitors in circuits

Surface charge rearrangements are extremely rapid, taking place in nanoseconds. In circuits containing only batteries and resistors, this extremely rapid rearrangements of surface charge leads to the establishment of the steady state in just a few nanoseconds. However, there are circuits in which this initial rapid rearrangement is followed by a "quasi-steady state," in which the currents change slowly with time as the circuit slowly comes into static equilibrium. We will experiment with circuits containing capacitors in order to study this time-dependent behavior.

19.2 Charging and discharging a capacitor

Find the capacitor (your instructor will tell you what it looks like).

> **Caution: do not exceed the voltage rating marked on the capacitor—typically a few volts. (If the rating is 2.5 volts, it is normally okay to use two ordinary "1.5 volt" batteries.)**

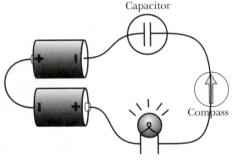

Capacitor

Compass

Figure 19.1 Charging a capacitor.

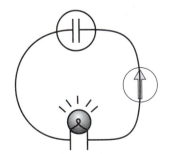

Figure 19.2 Discharging a capacitor.

Experiment 19.1 Charging and discharging a capacitor
- To make sure that the capacitor is initially uncharged, start by connecting a wire across the capacitor for a couple of seconds in order to discharge the capacitor.
- Figure 19.1: Connect the discharged capacitor in series with a round bulb (#14), and run the wire over a compass. The process is called "charging the capacitor." What do you observe?
- After the bulb stopped glowing, was there still current flowing? (You may need to place the wire under the compass in order to see small deflections easily.)
- Figure 19.2: Remove the batteries and connect the bulb directly to the capacitor. The process is called "discharging the capacitor." What do you observe?
- Repeat both stages of the experiment (charging and discharging), but this time use a long bulb (#48) with the two batteries and the capacitor. What difference does the choice of bulb make?
- Sketch graphs of current through the bulb versus time for both kinds of bulbs. On your graphs label the time axis, and note which curve refers to which bulb. Pay special attention to the magnitude of the initial current (you may need to repeat the experiment to observe this).

You see that the time can be varied by varying the resistance. If you like, you can experiment with different lengths or thicknesses of your Nichrome wires, which provide a wide range of charging and discharging times.

19.3 Currents and charges in a capacitor circuit

In this section we will look at the details of current and charge at various places in the circuit. We will do the analyses in terms of electron current, to emphasize a microscopic view of the processes. There are two different time scales for events in a circuit containing a capacitor. The bulb lights instantly when a connection is made (surface charge rearrangement takes only nano-seconds), but the current in the circuit declines slowly over a period of many seconds—the two time scales differ by a factor of 10^{10}.

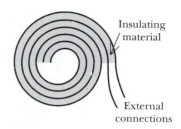

Figure 19.3 Typical capacitor construction—a coiled-up sandwich of metal foils separated by insulating material.

Construction and symbols

A capacitor usually consists of a sandwich of two metal foils separated by insulating material and coiled up into a small package (Figure 19.3). Note that there is no conducting path through the capacitor, even though you are able to light a bulb in series with the capacitor! Charges cannot jump from one plate to the other.

In symbolic circuit diagrams a capacitor is often shown as though we had unrolled the coiled-up capacitor and made external connections to the center of the metal plates (Figure 19.4). The insulating material between the plates is usually not indicated.

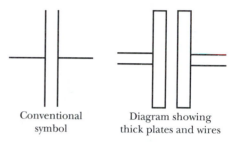

Figure 19.4 Symbols for capacitors.

Ex. 19.1 On the sequence of diagrams in Figure 19.5 at various times during discharging, show the direction and magnitude of the electron current at the locations marked with ×

Ex. 19.2 On the sequence of diagrams in Figure 19.5 at various times during discharging, use +'s and −'s to indicate excess charge on the capacitor plates, and approximate charge on the surface of the wires. Also, draw net electric field vectors on the diagrams at the locations marked ×. The electric field is due both to the charges on the capacitor plates and to surface charges on the wires.

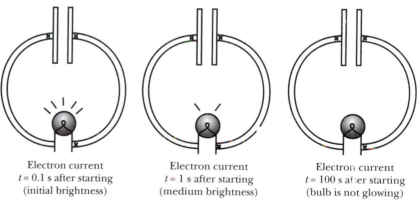

Electron current
$t = 0.1$ s after starting
(initial brightness)

Electron current
$t = 1$ s after starting
(medium brightness)

Electron current
$t = 100$ s after starting
(bulb is not glowing)

Figure 19.5 On copies of this diagram show electron current and electric field vectors at the locations marked with ×. Also show excess charges.

Ex. 19.3 During discharging, why does the bulb get dimmer and dimmer? Explain briefly but carefully.

Ex. 19.4 How does the amount of excess negative charge on one plate of the capacitor compare with the amount of excess positive charge on the other plate of the capacitor?

Ex. 19.5 While the bulb is glowing, how much electron current goes through the gap between the capacitor plates?

Ex. 19.6 You may have noticed that the compass detected some current even after the round bulb had stopped glowing (if not, increase the sensitivity by wrapping several turns around the compass and repeat, or place the wire under the compass in order to be able to see the deflection more easily). How do you explain this?

19.4 Explaining the charging of a capacitor

Explaining the discharging of a capacitor was relatively easy on the basis of charges and fields. The fringe field of the capacitor plus the electric field of the charges on the surfaces of the wires drives current in such a way as to reduce the charge on the capacitor plates (and to reduce the charges on the surfaces of the wires). As a result of these reduced charges, the electric field inside the wires is reduced, and so the current is reduced. The process continues with ever-decreasing current until all the charges have been neutralized. The system reaches static equilibrium, and there is no more current. Explaining the initial charging by the battery is a bit more complicated.

Contribution of charge on capacitor to the field in the wires

We've seen that it takes several seconds for the capacitor to get charged.

> During the first few nanoseconds while the surface charges are arranging themselves, the capacitor makes almost no contribution to the electric field in the wires because there is very little charge on its plates yet!

It is as though the capacitor weren't there—as though there were a continuous wire with no break in it.

? Therefore, how bright will the bulb be initially, compared to a circuit containing just batteries and the bulb?

Evidently the initial current will be the same as it would be in a simple bulb circuit, and the bulb just as bright as usual. But after current has been running in the wires for a second or more, there will be some significant amount of + and – charge on the capacitor plates.

? How will the fringe field of the capacitor affect the net electric field in the neighboring wire—will it increase or decrease the net field? And so what will happen to the current?

The fringe field of the capacitor acts in a direction opposite to the conventional current, thus decreasing the net field. As a result, the current will decrease.

Ex. 19.7 On diagrams like those in Figure 19.6, indicate charge on the capacitor plates and on the surfaces of the wires, and draw the net electric field vectors at the locations marked ×.

19.4.1 Role of the fringe field of the capacitor

Eventually the current stops completely when there is enough charge on the capacitor plates to create a large enough fringe field to counteract the field made by all the other charges. The light stops glowing. We see that for both charging and discharging a key to explaining the qualitative behavior of the circuit is the varying fringe field of the capacitor.

The initial establishment of surface charges takes a few nanoseconds, while it takes several seconds to reach static equilibrium. During the several seconds while the bulb is glowing, the current is not constant, but it changes so slowly that it makes sense to describe the situation as a "quasi-steady state."

? Does the final amount of charge on the capacitor plates depend on which bulb was used during the charging process? Think about this, and we'll discuss this in detail in the next section.

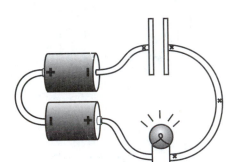

t = 0.1 s after starting
(initial brightness)

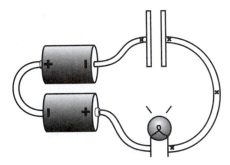

t = 1 s after starting
(medium brightness)

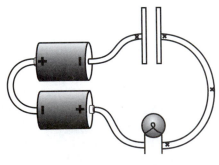

t = 100 s after starting
(bulb is not glowing)

Figure 19.6 On diagrams like these, indicate charges on the capacitor plates and the wires, and the *net* electric field at the locations marked ×.

19.5 The effect of different bulbs

Because charging the capacitor through the long bulb goes on much longer than charging through the round bulb, one could argue that the capacitor gets charged up a lot more when the long bulb is used. Alternatively, one might argue that since the round bulb glows much more brightly, the capacitor would get charged up a lot more when the round bulb is used.

? Which argument do you think is right? Why? (Hint: Think about the fields in the final state after the bulb stopped glowing, as you discussed on the preceding page. Also consider your graph of current vs. time for charging a capacitor in Experiment 19.1 on page 668.)

Experiment 19.2 The effect of different bulbs on final charge
Design an experiment that will show how the final amount of charge on each plate of the capacitor depends on which bulb is involved in charging the capacitor. You might discuss your ideas with other students before carrying out your experiments. (If you wish, you can conduct your study using different lengths or thicknesses of Nichrome wire rather than two kinds of light bulbs.)

Be careful in your experiments. Before charging a capacitor be sure that it doesn't already have some charge on its plates: discharge it fully by connecting a wire across it for a couple seconds. Repeat any numerical measurements several times to make sure you know how reproducible the measurements are: 15.43 seconds and 15.67 seconds may not be significantly different!

Describe your experiments, results, and conclusions.

19.6 How is discharging possible?

Here is a plausible-sounding objection: "When we connect a light bulb to a charged capacitor, the bulb lights up for a while. But I'm really puzzled. I don't see how charge can come off the capacitor and go through the light bulb. The positive and negative charges on the two plates are very close to each other and attract each other strongly. The charge can't leak off!"

? Try to explain the source of this confusion. You might refer to your work on page 669.

You need to look at the electric field in the wire close to but outside the plates. Initially, right next to the negative plate, the electric field in the wire is not zero. The fringe field outside the capacitor points toward the negative plate. Electrons in the wire at this location feel a force and move away from the negative plate. Electrons in the wire near the positive plate likewise move toward the positive plate, under the influence of the electric field in the wire.

19.7 Applications

In this section you are asked to apply the preceding analysis of capacitor circuits to understand the effects on circuit behavior of changes in capacitor design. In analyzing these systems, think about what happens in the first fraction of a second, and how the state of the system at the end of this time interval will affect what happens next.

Ex. 19.8 Consider two capacitors whose only difference is that the plates of capacitor number 2 are larger than those of capacitor number 1 (Figure 19.7). Neither capacitor has an insulating layer

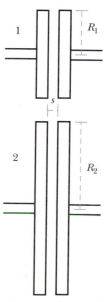

Figure 19.7 What is the effect of increasing the size of the capacitor plates?

between the plates. They are placed in two different circuits having similar batteries and bulbs in series with the capacitor.

Show that in the first fraction of a second the current stays more nearly constant (decreases less rapidly) in the circuit with capacitor number 2. Explain your reasoning in detail. Hint: Show charges on the metal plates, and consider the electric fields they produce in the nearby wires. Remember that the fringe field near a plate outside a circular capacitor is approximately

$$\frac{Q/A}{\varepsilon_0}\left(\frac{s}{2R}\right)$$

A more extensive analysis shows that this trend holds true for the entire charging process: the larger capacitor ends up with more charge on its plates. The capacitor you have been using has a very large plate area, folded into a small package.

Ex. 19.9 Consider two capacitors whose only difference is that the plates of capacitor number 2 are closer together than those of capacitor number 1 (Figure 19.8). Neither capacitor has an insulating layer between the plates. They are placed in two different circuits having similar batteries and bulbs in series with the capacitor.

Show that in the first fraction of a second the current stays more nearly constant (decreases less rapidly) in the circuit with capacitor number 2. Explain your reasoning in detail. Hint: Show charges on the metal plates, and consider the electric fields they produce in the nearby wires. Remember that the fringe field near a plate outside a circular capacitor is approximately

$$\frac{Q/A}{\varepsilon_0}\left(\frac{s}{2R}\right)$$

A more extensive analysis shows that this trend holds true for the entire charging process: the capacitor with the narrower gap ends up with more charge on its plates. The capacitor you have been using has an extremely narrow gap (filled with an insulator).

Ex. 19.10 The insulating layer between the plates of a capacitor not only holds the plates apart to prevent conducting contact but also has a big effect on charging. Consider two capacitors whose only difference is that capacitor number 1 has nothing between the plates, while capacitor number 2 has a layer of plastic in the gap (Figure 19.9). They are placed in two different circuits having similar batteries and bulbs in series with the capacitor.

Show that in the first fraction of a second the current stays more nearly constant (decreases less rapidly) in the circuit with capacitor number 2. Explain your reasoning in detail. Hint: Consider the electric fields produced in the nearby wires by this plastic-filled capacitor.

Suppose the plastic is replaced by a different plastic that polarizes more easily. In the same circuit, would this capacitor keep the current more nearly constant or less so than capacitor 2?

A more extensive analysis shows that this trend holds true for the entire charging process: the capacitor containing an easily polarized insulator ends up with more charge on its plates. The capacitor

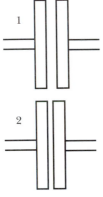

Figure 19.8 What is the effect of making the gap between the capacitor plates smaller?

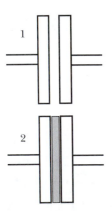

Figure 19.9 What is the effect of putting an insulating layer between the capacitor plates?

you have been using is filled with an insulator that polarizes extremely easily.

Ex. 19.11 Suppose instead of placing an insulating layer between the plates you inserted a metal slab of the same thickness, just barely not touching the plates. In the same circuit, would this capacitor keep the current more nearly constant or less so than capacitor 2 in the preceding exercise?

Note that this is essentially equivalent to making a capacitor with a shorter distance between the plates, and this result is consistent with what we found for a narrower capacitor.

19.7.1   Parallel capacitors

Two capacitors in parallel can be thought of as one capacitor with plates that have twice the area (see Ex. 19.8 on page 671).

Experiment 19.3 Two capacitors in parallel
Work with another group so that you can try charging a round bulb through two capacitors in parallel as shown in Figure 19.10. What do you predict for the charging time compared to a circuit with one capacitor? (Again, consider what happens in the first fraction of a second, and what effect this will have on the process in the next time interval.) What do you observe?

19.7.2   An isolated bulb

In the next experiment you can light a bulb despite the fact that there is no conducting path between the bulb and the battery!

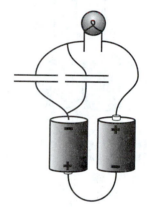

Figure 19.10 Two capacitors in parallel.

Experiment 19.4 Lighting an isolated bulb
Work with another group so that you can make a circuit with two capacitors as shown in Figure 19.11. Note that the long bulb is in a conducting "island" completely isolated from the batteries.

Before running the circuit, predict in detail what you expect will happen, and explain your reasoning. Try to work out agreement among the students in the two groups. Remember that during the few-nanosecond transient, the uncharged capacitors have no effect on the electric field.

Now construct the two-capacitor circuit and observe what happens. Discuss any discrepancies between the prediction you made and what you see. You might use compasses to verify your theory about the currents in the various parts of the circuit.

19.7.3   Reversing a capacitor

Reversing the capacitor in a circuit leads to an interesting effect. Follow the instructions carefully—otherwise you might damage the capacitor or the bulb.

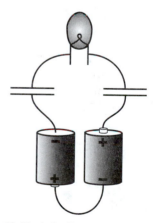

Figure 19.11 A bulb completely isolated from the batteries.

Experiment 19.5 Reversing the capacitor
Charge the capacitor using one battery and a round bulb (use only one battery!). Now that the capacitor is charged, what do you predict would happen if you reversed the connections to the capacitor, leaving all other connections intact, including the battery? Explain your reasoning, including an appropriate diagram.

Now try it. What do you observe? If the observation is different from your prediction, explain what was wrong with the prediction. Why is it important to use only one battery?

19.7.4   What these capacitors are used for

A simple experiment helps understand how these capacitors can be useful.

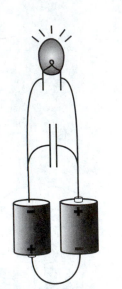

Figure 19.12 What these capacitors are used for.

Experiment 19.6 The usefulness of a capacitor
Connect the capacitor and a long bulb in parallel rather than in series (Figure 19.12). Intermittently break the connection to the batteries for very brief fractions of a second. What happens to the brightness of the bulb during the very brief time that the batteries are disconnected, compared to what happens without a capacitor in the circuit? Explain.

The capacitor in your kit is designed to be used in this way, connected in parallel to a power supply in a computer, to supply the computer circuits with current during momentary power outages. For a given voltage, these capacitors hold almost a million times more charge than ordinary capacitors! This is because the insulator between the plates is exceptionally easy to polarize, and the gap between the plates is exceptionally thin.

More generally, parallel capacitors are used in all kinds of electric circuits to even out rapid changes in voltage. For example, in a power supply, parallel capacitors are used to filter out high-frequency (AC) voltage changes that are riding on top of a constant (DC) voltage. You might say that a capacitor makes a circuit behave in a sluggish manner, being unable to change conditions rapidly.

19.8 Energy considerations

Let's consider briefly the energy aspects of charging and discharging a capacitor through a light bulb.

Ex. 19.12 In the charging phase, from a time just before the circuit is completed to the time when there is essentially no more current, state whether there is energy gain or loss in the battery, in the bulb, in the capacitor, and in the surroundings. (Careful: Think especially carefully about the initial and final states of the bulb.)

Ex. 19.13 You may have noticed that during the discharging phase, the light bulb glows just about as brightly, and for just about as long, as it does during the charging phase. Let E stand for the energy emitted by the light bulb (as light and heat) in the discharging phase, from just before the bulb is connected to the capacitor until the time when there is essentially no more current.

In terms of $+E$ or $-E$, what was the energy change of the battery, capacitor, bulb, and surroundings during the charging phase, and during the discharging phase? One answer is already given in the following table:

	Charging	Discharging
Battery:		
Bulb:		
Capacitor:		
Surroundings:		$+E$

It is somewhat surprising that we can get this much information out of one simple observation.

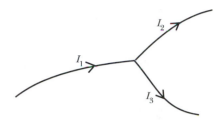

Figure 19.13 In the steady state it must be that $I_1 = I_2 + I_3$.

19.9 The current node rule in a capacitor circuit

The current node rule in circuits results from charge conservation in the steady state. Figure 19.13 shows a simple application of the node rule to a node in a circuit where the incoming current I_1 supplies the outgoing currents I_2 and I_3, so that $\sum(I_{\text{entering}}) = I_1 - I_2 - I_3 = 0$.

Figure 19.14 shows a more complex situation. Consider locations 1 and 2 as nodes. While the capacitor is being charged, there is current running onto one plate of the capacitor without any current coming off that plate (region 1 in Figure 19.14), and as a result the charge on the plate is changing. This is clearly not a steady state. Similarly, current comes off of the other plate (region 2 in Figure 19.14), which changes the charge on that plate.

But if you look at the capacitor as a whole (region 3 in Figure 19.14), just as much current leaves one plate as enters the other plate. Otherwise the two plates together would acquire a nonzero charge that would quickly repel or attract electrons in such a way as to restore the net charge of the capacitor to be zero. So during charging or discharging of a capacitor the current node rule applies to the capacitor as a whole, even though it doesn't apply to just one plate of the capacitor.

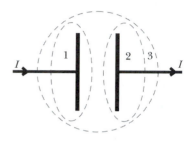

Figure 19.14 In a non-steady-state situation, the current node rule does not apply.

19.10 Capacitance

The capacitors you used in your circuit experiments may have had something like "0.47 F" or "1 F" marked on them, which would mean that they have a "capacitance" of about 0.5 or 1 "farad" (a unit honoring Michael Faraday). Let's see what is meant by this "capacitance" rating.

The more charge on the capacitor plates ($+Q$ and $-Q$), the bigger the electric field in the gap, and the greater the potential difference ΔV across the gap (Figure 19.15). So the absolute magnitude of the potential difference $|\Delta V|$ is proportional to the amount of positive charge Q on the positive plate, and conversely Q is proportional to $|\Delta V|$. We define "capacitance" C as this proportionality constant:

DEFINITION OF CAPACITANCE

$$Q = C|\Delta V|$$

The units of capacitance are coulombs per volt, or "farads."

? By using what you know about E and ΔV, show that $C = (\varepsilon_0 A)/s$ in a parallel-plate capacitor with plate area A and plate separation s.

Since the magnitude of the electric field inside the gap is

$$E = (Q/A)/\varepsilon_0$$

we can calculate the potential difference as

$$\Delta V = [(Q/A)/\varepsilon_0]s,$$

which shows that the potential difference is proportional to the charge Q on one of the plates, as we expect. The capacitance is therefore:

$$C = Q/\Delta V = (\varepsilon_0 A)/s \text{ (parallel plate capacitor)}$$

Note that the capacitance depends on the geometry of the capacitor: the area of the plates and the separation between them.

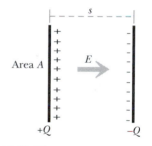

Figure 19.15 The more charge there is on the plates, the bigger the electric field, and the bigger the potential difference.

The concept of capacitance is useful in the detailed analysis of circuits, and we will use it later to study capacitor circuits quantitatively. But bear in mind that capacitance is just a derived quantity relating more fundamental quantities: charge and potential difference (and ultimately charge and field.)

Ex. 19.14 A certain capacitor has rectangular plates 50 cm by 30 cm, and the gap width is 0.25 mm. What is its capacitance?

We see that typical capacitances are very small when measured in farads. A one-farad capacitor is quite extraordinary! Apparently it has a very large area A (all wrapped up in a small package), and a very small gap s.

Ex. 19.15 Moreover, your capacitor has an insulating material between its plates that has the effect of greatly increasing the amount of charge for a given potential difference (see Ex. 19.10 on page 672). In Chapter 16 it was shown that the average electric field in the insulating material is reduced by a factor K, the "dielectric constant." If the insulating material fills the capacitor gap, what is the capacitance $C = Q/\Delta V$?

Ex. 19.16 In your experiments you charged a half-farad or one-farad capacitor in a circuit containing two 1.5-volt batteries, so the final potential difference across the plates was 3 volts. How much charge was on each plate? How many excess electrons were on the negative plate?

19.11 Reflection on capacitors in circuits

We have seen a kind of circuit behavior that may be called a "quasi-steady state." The current in the capacitor circuit is approximately steady over a time of a second or so, but only approximately: the current slowly decreases and finally goes to zero—static equilibrium is attained.

Actually, even an ordinary battery-and-bulb circuit is not in a true steady state, because when the chemicals in the battery are used up, the system comes into static equilibrium (no current). There is an important difference, however. You have seen that in a capacitor circuit the current continually decreases. In a battery-powered circuit, however, the current may stay nearly constant for many hours, then drop to zero in a much shorter time.

We were able to understand the behavior of a capacitor circuit in terms of the fringe field of the capacitor competing with the electric field of the other charges (battery charges and surface charges). The more charge on the capacitor, the larger the fringe field. When charging the capacitor, static equilibrium is finally reached when there is so much charge on the capacitor that the fringe field is large enough to cancel the other contributions to the electric field.

We emphasized a qualitative analysis of capacitor circuits to get at the fundamental issues, and to provide additional practice in reasoning with charge and field. To analyze such circuits quantitatively we need to apply the energy principle in our analysis; later in this chapter we do this, and we will find that we can accurately predict the rate of decrease of the current.

19.12 Macroscopic analysis of circuits

Our study of electric circuits has been based on a microscopic picture of the motion of electrons inside the wires, due to electric fields produced by charges on and in the battery and on the surfaces of the wires. This microscopic picture gives us insight into the fundamental physical mechanisms of circuit behavior. However, it is not easy to measure directly electric field, surface charge, electron drift speed, or mobility.

On the other hand, it is easy to use relatively inexpensive commercially available meters to measure conventional current (rather than electron current), potential difference (rather than electric field), and "resistance" (rather than mobility). For practical purposes, it is useful to describe and analyze circuits in terms of these macroscopic quantities.

We will link what we already know about circuits at the atomic level to a macroscopic description in terms of quantities that can be measured easily with standard meters.

19.13 Resistance

Until now we have focused on a detailed microscopic analysis of the behavior of circuits. To do so we have found it necessary to take into account both properties of particular materials and the geometry of particular circuit elements, such as thin and thick wires. It is sometimes useful to use a single number to summarize the properties of a particular object.

19.13.1 Conductivity: Combining the properties of a material

Conventional current is $I = |q|nA\bar{v} = |q|nAuE$, where $|q|$ is the absolute value of the charge of one of the mobile "charge carriers" (electrons or holes), n is the number density of the charge carriers, A is the cross sectional area of the wire, u is the mobility of the charge carriers, and E is the electric field inside the wire that drives the mobile charges. This formula for conventional current mixes together three quite different kinds of quantities: $|q|$, n, and u are properties of the material, A describes the geometry of the material, and E is the electric field due to charges outside the material (and/or on its surface).

It is helpful to group the material properties together: $I = (|q|nu)AE$. And if we divide both sides of the equation by the area A, we arrive at the following way of describing the situation

$$J = \frac{I}{A} = |q|nuE = \sigma E$$

where J = current density (A/m^2) and σ = conductivity = $(|q|nu)$

By dividing the current by the cross-sectional area, we obtain a "current density" J in amperes per square meter, which is independent of the geometry of the situation. The current density is proportional to the applied electric field, and the proportionality constant σ is called the "conductivity," which lumps together all the relevant properties of the material, $|q|nu$. (σ is a lower-case Greek sigma.)

We can consider $\vec{J}$ to be a vector pointing in the direction of the conventional current flow, which is in the direction of the applied electric field $\vec{E}$ (Figure 19.16):

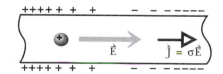

Figure 19.16 $\vec{J}$ is in the direction of the electric field, and in the direction of motion of a fictitious positive charge.

CURRENT DENSITY AND CONDUCTIVITY

$$\vec{J} = \sigma\vec{E}$$

The current density $J = (I/A)$ ampere/m^2
The conductivity $\sigma = |q|nu$ depends on
  • $|q|$, the absolute value of the charge on each carrier;
  • n, the number of charge carriers per m^3; and
  • u, the mobility of the charge carriers.

A higher conductivity means that a given applied electric field $\vec{E}$ gives larger current densities. The conductivity is a property of the material and has nothing to do with what shape the material has, or with how large an electric field is applied to the material.

Ex. 19.17 In copper at room temperature, the mobility of mobile electrons is about 4.5×10^{-3} (m/s)/(volt/meter), and there are about 8×10^{28} mobile electrons per m^3. Calculate the conductivity σ and include the correct units. In actual practice, it is usually easier to measure the conductivity σ and deduce the mobility u from this measurement.

Ex. 19.18 Consider a copper wire with a cross-sectional area of 1 mm^2 (similar to your connecting wires) and carrying 0.3 amperes of current, which is about what you get in a circuit with a round bulb and two batteries in series. Calculate the strength of the very small electric field required to drive this current through the wire.

Ex. 19.19 The conductivity of tungsten at room temperature, $\sigma = 1.8 \times 10^7 (\text{ampere}/m^2) / (\text{volt/meter})$, is significantly smaller than that of copper. At the very high temperature of a glowing lightbulb filament (nearly 3000 degrees Kelvin!), the conductivity of tungsten is 18 times smaller than it is at room temperature. The tungsten filament of one of your round bulbs has a radius of about 0.015 mm. Calculate the electric field required to drive 0.3 amperes of current through the glowing bulb and show that it is huge compared to the field in the connecting copper wires.

19.13.2 Conductivity with two kinds of charge carrier

When electric forces are applied to salt water, both the sodium ions (Na^+) and the chloride ions (Cl^-) move and constitute an electric current. If the electric field is to the right in the water, the force on the Na^+ ions is to the right, but the force on the Cl^- ions is to the left (Figure 19.17).

The flow of Na^+ ions to the right constitutes a conventional current to the right, and the flow of Cl^- ions to the left also constitutes a conventional current to the right, because the motion of these negative ions has the effect of tending to make the region on the right less negative and hence more positive, with an effect similar to the flow of Na^+ ions to the right.

If the charge on the positive ion is q_1 (e in the case of Na^+ ions), the positive-ion contribution to the conventional current in coulombs per second is $q_1 n_1 A v_1$. If the charge on the negative ion is q_2 (the charge on a Cl^- ion is $q_2 = -e$), the negative-ion contribution to the conventional current in coulombs per second is $|q_2| n_2 A v_2$. We have to take the absolute value of the charge because the negative ions contribute conventional current in the same direction as do the positive ions.

Note that the mobilities (and therefore the drift speeds) may differ for the two kinds of ions. Larger ions usually have lower mobilities.

The conventional current in amperes (coulombs per second) is the sum of the conventional current due to the positive charges and the conventional current in the same direction due to the negative charges:

$$I = |q_1| n_1 A v_1 + |q_2| n_2 A v_2$$

Note that both ionic currents contribute to the total current, and that the ionic charge q_1 or q_2 may be a multiple of e. For example, the charge of Zn^{++} is $2e$.

The current density $J = I/A = |q_1| n_1 u_1 E + |q_2| n_2 u_2 E = \sigma E$, so the conductivity in this situation where there are two kinds of charge carrier is

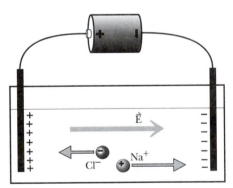

Figure 19.17 Both positive and negative charges move in salt water.

CONDUCTIVITY WITH TWO KINDS OF CHARGE CARRIERS

$$\sigma = |q_1|n_1 u_1 + |q_2|n_2 u_2$$

19.13.3 Resistance combines conductivity and geometry

Often when we have compared two different electric circuits, we have said such things as, "The cross-sectional area goes up, the length decreases, and the mobility increases." It is practical to have a single quantity to describe a resistor that encapsulates both the properties of the material (mobility, mobile-carrier number density) and the geometrical properties of the particular object (cross-sectional area, length). This quantity is called "resistance."

Consider the potential difference ΔV from one end to the other of a wire of constant cross-sectional area A, length L, and uniform composition (Figure 19.18). In the steady state, the electric field inside the wire has the same magnitude everywhere (otherwise the drift speed and current would vary) and is everywhere parallel to the wire (otherwise additional charge would be driven to the surface). Since potential difference is defined as

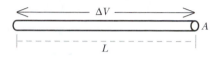

Figure 19.18 A resistor—for example, a Nichrome wire.

$$\Delta V = -\int_i^f \vec{\mathbf{E}} \bullet d\vec{\mathbf{r}},$$

we have $|\Delta V| = EL$. The conventional current density $J = I/A$ is the same everywhere on the cross-sectional area A of the wire, as was proven in the previous chapter (if E varied across the cross section, one could construct a path with a nonzero round-trip potential difference, and uniform E implies uniform v and therefore uniform J).

? For some materials $J = \sigma E$. For such materials, write a formula for I in terms of σ and $|\Delta V|$.

Combining the two equations for $|\Delta V|$ and J, we find a relation between two macroscopic quantities, conventional current I and potential difference ΔV:

$$J = \sigma E$$
$$\frac{I}{A} = \sigma \frac{|\Delta V|}{L}$$

It is useful to rearrange this relation between current and potential difference in the following way:

$$I = \frac{|\Delta V|}{L/(\sigma A)} = \frac{|\Delta V|}{R}$$

where the "resistance" R is measured in volts per ampere or "ohms" (abbreviated Ω and named for Georg Ohm, the physicist who first studied resistance quantitatively in the early 1800s). We have this definition:

DEFINITION OF RESISTANCE

$$R = \frac{L}{\sigma A}$$

Note that resistance depends on
- the properties of the material: $\sigma = |q|nu$
- the geometry of the resistor: L, A

These aspects are lumped into one quantity—resistance. We can make a large-resistance resistor in a variety of ways. The resistor could have a large length L, or a small cross-sectional area A, or a low conductivity σ. Sometimes resistance is expressed in terms of the "resistivity" $\rho = 1/\sigma$. In that case, $R = \rho L/A$.

We can of course rearrange $I = |\Delta V|/R$ and write $|\Delta V| = RI$. But usually it is appropriate to think of the potential difference as causing a current, not of the current causing a potential difference, and the form $I = |\Delta V| / R$ reminds us that the current is the result of applying an electric field, with an associated potential difference.

Note that we haven't developed any new physical principles. We have simply rewritten $I = |q| nA\bar{v} = |q| nA(uE)$ in terms of potential difference, so that we have $I = |\Delta V| / R$.

Reflection: Connecting macroscopic and microscopic viewpoints

We can relate the analysis we've just done to our earlier microscopic analysis of a current in a material with single charge carrier, such as electrons in a metal:

$$\bar{v} = uE$$

$$I = |q| nA\bar{v} = |q| nA(uE)$$

$|\Delta V| = EL$ for a straight wire, so $E = |\Delta V| / L$

$$I = |q| nA\left(u\frac{|\Delta V|}{L} \right) = \frac{|\Delta V|}{\left(\dfrac{L}{|q| nA u} \right)}$$

$$I = \frac{|\Delta V|}{R}, \text{ where } R = \frac{L}{\sigma A}, \text{ with } \sigma = |q| nu$$

You should derive this relationship yourself without referring to notes.

For some kinds of scientific work the microscopic picture in terms of electric field, mobility, and drift speed ($\bar{v} = uE$) is more useful. But when you need to make contact with the technological world of voltmeters and ammeters and ohmmeters, you also need to be able to describe a situation in terms of macroscopic quantities such as potential difference (voltage), resistance, and conventional current.

Here is a side-by-side comparison of the two views:

Microscopic view	Macroscopic view
$\bar{v} = uE$	$J = \sigma E$
$i = nA\bar{v} = nAuE$	$I = |q| nA\bar{v} = \left(\dfrac{1}{R}\right)|\Delta V|$

Warning: don't write $V = IR$! Since "V" refers to the potential at a particular location, this is meaningless.

Ex. 19.20 A carbon resistor is 5 mm long and has a constant cross section of 0.2 mm^2. The conductivity of carbon at room temperature is $\sigma = 3 \times 10^4$ per ohm·m. In a circuit its potential at one end of the resistor is 12 volts relative to ground, and at the other end the potential is 15 volts. Calculate the resistance R and the current I.

Ex. 19.21 A thin copper wire in the same circuit is 5 mm long and has a constant cross section of 0.2 mm^2. The conductivity of copper at room temperature is $\sigma = 6 \times 10^7$ ohm^{-1}m^{-1}. The copper wire is in series with the carbon resistor in the same circuit you just studied, with one end connected to the 15-volt end of the carbon resistor, and the current you calculated runs through the carbon resistor wire. Calculate the resistance R of the copper wire and the potential $V_{\text{at end}}$ at the other end of the wire.

You can see that for most purposes a thick copper wire in a circuit would have practically a uniform potential. This is because the small drift speed in a thick, high-conductivity copper wire requires only a very small electric field, and the integral of this very small field creates a very small potential difference along the wire.

19.13.4 Constant and varying conductivity

We have already seen that the mobility u of the tungsten in a light bulb filament varies with the temperature. For example, the current through one round bulb is less than twice the current through two round bulbs in series, because the mobility is lower at higher temperatures. Like the mobility, the conductivity $\sigma = |q|nu$ and therefore the resistance may change for different amounts of current.

Ohmic resistors

In some materials the conductivity σ is nearly a constant, independent of the amount of current flowing through the resistor. We call such constant-conductivity materials "ohmic." An ohmic material has the property that $I = |\Delta V| / R$, where the resistance $R = L/(\sigma A)$ is constant, independent of the amount of current. Because conductivity depends somewhat on temperature, and higher currents tend to raise the temperature of the material, no material is truly ohmic, but many materials can be considered to be approximately ohmic as long as the temperature doesn't change very much.

Resistors made of ohmic materials are called "ohmic" resistors. Doubling the length L of an ohmic resistor doubles the resistance R and cuts the current in half if the potential difference ΔV across the resistor is kept constant. Doubling the cross-sectional area A cuts the resistance in half and doubles the current if the potential difference ΔV across the resistor is kept constant.

Your Nichrome wires and the carbon resistors used in electronic circuits are nearly ohmic resistors, as long as not enough current runs through them to raise the temperature significantly. In metals the conductivity decreases somewhat at higher temperatures, and metals are only approximately ohmic. You can think of the electrons as having more frequent collisions with atoms that have increased random motion at higher temperatures. You saw experimental evidence for nonohmic behavior in the previous chapter, where the current through two bulbs in series was more than half the current through one bulb.

Ex. 19.22 By actual measurement, the current through a long bulb when connected to two batteries in series (3 volts) is about 100 milliampere (mA); connected to one battery (1.5 volts) the current is about 80 mA; and connected to a small voltage of only 50 millivolts the current is about 6 mA. (Different long bulbs may differ from these values somewhat.) Using the formula $I = |\Delta V| / R$, what is R for each of these cases? Is R a constant? Is a long bulb an ohmic resistor over this whole range of currents?

So from 0 to 3 volts a long bulb is not ohmic, although in the range from 1.5 to 3 volts the resistance changes by only about fifty percent and the bulb can be said to be very roughly ohmic.

Semiconductors

In a metal the current is zero when the potential difference is zero, but the slightest potential difference produces some current, as shown in Figure 19.19. Such behavior implies that at least some of the electrons in a metal are truly free, not bound to the atoms. If they were all bound, we would have

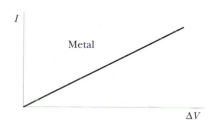

Figure 19.19 Mobile electrons, not bound to atoms—ohmic behavior.

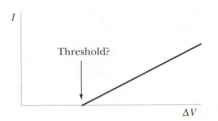

Figure 19.20 If electrons in a metal were bound to the atoms, the current might behave like this.

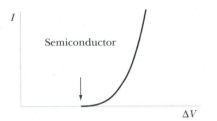

Figure 19.21 Behavior of a pure semiconductor such as silicon or germanium.

to apply some field just to free some up, and then more field to increase the current, as shown in Figure 19.20.

Pure silicon and germanium are nonohmic in a spectacular and spectacularly useful way. In these "semiconductor" materials, the density of mobile electrons n and therefore the conductivity $\sigma = enu$ (where u is the mobile-electron mobility) depend exponentially on the voltage. Applying a slight voltage produces almost no current. After applying enough voltage to get a small current, doubling this voltage may make the current go up by a factor of a hundred rather than by a factor of two! (See Figure 19.21.)

The key difference between a metal and a semiconductor lies in n, the number of available charge carriers per unit volume. In a metal, n is a fixed number (one mobile electron or hole per atomic core). In a semiconductor, at very low temperatures there are no mobile electrons or holes and the material acts nearly like an insulator for small applied fields. However, the electrons are not tightly bound to the atomic cores and it is possible to free up some electrons by applying a large enough electric field, so n is variable and increases with increasing potential difference.

Raising the temperature can also free up some electrons (or holes), and the effect of increasing n more than compensates for the higher rate of collisions with the atomic cores, so that for a semiconductor the resistance decreases with increasing temperature. This is also true for carbon: the resistance of a carbon resistor decreases somewhat with increasing temperature.

Other nonohmic circuit elements

Not all resistors are ohmic, and we have seen that even those called "ohmic" are only approximately so if the temperature changes with current. Moreover, circuit elements other than resistors are definitely not ohmic.

Capacitors aren't ohmic: If you double the current flowing onto and off of the capacitor plates, the capacitor gets charged faster, but it is not the case that $|\Delta V|$ is proportional to I for a capacitor. The voltage is proportional to the charge on one of the plates rather than being proportional to the current: $|\Delta V| = Q/C$. So capacitors are not ohmic devices.

Batteries aren't ohmic: If you double the current through a battery (by changing the circuit elements attached to the battery), the battery voltage hardly changes at all. The battery voltage certainly doesn't double and will actually decrease slightly. It definitely is not the case that $|\Delta V|$ is proportional to I for a battery.

$I = |\Delta V|/R$ HAS LIMITED VALIDITY

It is extremely important to apply $I = |\Delta V|/R$ *only* to resistors. This formula is not a fundamental physics principle with broad validity. It is just an expression of the fact that some materials are (approximately) ohmic.

Conventional symbols

In circuit diagrams we will often use conventional symbols for resistors (jagged lines) and batteries (two parallel lines, the longer one representing the positive end of the battery); see Figure 19.22.

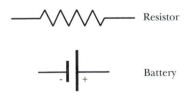

Figure 19.22 Conventional symbols for resistors and batteries.

Ex. 19.23 When a single long bulb is connected to a 1.5-volt battery, the current through the battery is about 80 milliampere. If you add another long bulb in parallel, the battery current of course increases to 160 milliampere. Is the battery ohmic? That is, is the current through the battery proportional to the potential difference across the battery?

19.14 Application: Series and parallel resistance

Series resistance

We can show that several ohmic resistors in series are equivalent to one ohmic resistor with a resistance that is the sum of the individual resistances. Consider the circuit of Figure 19.23, with three ohmic resistors in series with an ideal battery (one with negligible internal resistance).

The current I is the same everywhere in this series circuit.

? Write the energy-conservation loop equation for this circuit, using the fact that $\Delta V = RI$ for an ohmic resistor.

We can rearrange the loop equation, $\text{emf} - R_1 I - R_2 I - R_3 I = 0$, in this form:

$$\text{emf} = (R_1 + R_2 + R_3)I = R_{\text{equivalent}}I$$

Evidently a series of ohmic resistors acts like a single ohmic resistor, with a resistance equal to the sum of all the resistances:

$$R_{\text{equivalent}} = R_1 + R_2 + R_3$$

This is not an entirely obvious result! The proof depends on the following:

- the definition of resistance $R = L/(\sigma A)$;
- the assumption that the conductivity σ is independent of the current for the material in these resistors (ohmic resistors);
- the fact that the magnitude of the sum of all the resistor potential drops is equal to the potential rise across the battery, which is numerically equal to the emf of the battery; and
- the fact that in a series circuit the steady-state current I is the same in every element, due to charge conservation, and an unchanging charge distribution in the steady state.

Because $R = L/(\sigma A) = \rho L/A$, if the various resistors are all made of the same material and have the same cross-sectional area, note that the series-resistance formula is equivalent to just adding the various lengths: $L_{\text{equivalent}} = L_1 + L_2 + L_3$.

? For example, show that two identical resistors in series are like a single resistor that is twice as long.

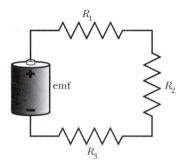

Figure 19.23 Three resistors in series.

Ex. 19.24 A certain ohmic resistor has a resistance of 40 ohms. A second resistor is made of the same material but is three times as long and has half the cross-sectional area. What is the resistance of the second resistor? What is the effective resistance of the two resistors in series?

Parallel resistance

We can show that several ohmic resistors in parallel are equivalent to one ohmic resistor with a resistance that is the reciprocal of the sum of reciprocals of the individual resistances. Consider the circuit of Figure 19.24, with three different ohmic resistors in parallel.

The potential difference is the same across each resistor in this parallel circuit and is numerically equal to the emf of the battery.

? Write the current node equation for this circuit, using the fact that $I = \Delta V/R$ for an ohmic resistor:

We can rearrange the current equation, $I - \dfrac{\text{emf}}{R_1} - \dfrac{\text{emf}}{R_2} - \dfrac{\text{emf}}{R_3} = 0$, in this form:

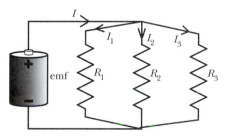

Figure 19.24 Three resistors in parallel.

$$ I = \left(\frac{1}{R_1} + \frac{1}{R_2} + \frac{1}{R_3} \right)(\text{emf}) = \frac{\text{emf}}{R_{\text{equivalent}}} $$

Evidently ohmic resistors in parallel act like a single ohmic resistor, with a resistance whose reciprocal is equal to the sum of the reciprocals of the individual resistances:

$$ \frac{1}{R_{\text{equivalent}}} = \frac{1}{R_1} + \frac{1}{R_2} + \frac{1}{R_3} $$

Because $1/R = \sigma A/L$, if the various resistors are all made of the same material and have the same length, note that the parallel-resistance formula is equivalent to just adding the various cross-sectional areas: $A_{\text{equivalent}} = A_1 + A2 + A_3$.

? For example, show that two identical resistors in parallel are like a single resistor that has twice the cross-sectional area.

This corresponds to what you saw when you had two bulbs in parallel: there was twice as much current through your battery, as though you had one bulb with twice the cross-sectional area.

Ex. 19.25 When glowing, a long bulb has a resistance of about 30 ohms and a round bulb has a resistance of about 10 ohms. If they are in parallel, what is their equivalent resistance? How much current goes through two flashlight batteries in series if a long bulb and a round bulb are connected in parallel to the batteries?

19.15 Work and power in a circuit

If you move a small amount of charge Δq from one location to another, the small amount of electric potential energy change ΔU_e is equal to the charge Δq times the potential difference ΔV, $\Delta U_e = (\Delta q)(\Delta V)$, since electric potential is electric potential energy per unit charge (joules per coulomb, or volts). If this work is done in a time Δt, we can calculate the power, which is energy per unit time:

$$ \text{Power} = \frac{\Delta U_e}{\Delta t} = \frac{(\Delta q)(\Delta V)}{\Delta t} $$

But $\Delta q/\Delta t$ is the amount of charge moved per unit time, which is the current I. So since $\Delta q/\Delta t = I$, we have the following important equation:

POWER FOR ANY KIND OF CIRCUIT COMPONENT

$$ \text{Power} = I\Delta V $$

You should be able to repeat this argument without referring to your notes. This is a very general result, applying to any kind of circuit component. Here are some examples of the application of this equation to batteries, resistors, and capacitors:

Ex. 19.26 Power in a battery. In a certain circuit, a battery with an emf of 1.5 volts generates a current of 3 amperes. What is the output power of the battery? Include appropriate units. (If on the other hand you were charging a rechargeable battery with a charging current of 3 amperes, this would be the input power to the battery.)

Ex. 19.27 Power in a resistor. A resistor R conducts a current I. Show that the power dissipated in the resistor is RI^2. Alternatively, show that the power can be written as $(\Delta V)^2/R$.

Ex. 19.28 Work and energy with a capacitor. A capacitor with capacitance C has an amount of charge q on one of its plates, in which case the potential difference across the plates is $\Delta V = q/C$ (definition of capacitance). The work done to add a small amount of charge dq when charging the capacitor is $dq(\Delta V) = dq(q/C)$. Show by integration that the amount of work required to charge up the capacitor from no charge to a final charge Q is

$$\frac{1}{2}\frac{Q^2}{C}$$

Since this is the amount of work required to charge the capacitor, it is also the amount of energy stored in the capacitor. Substituting $Q = C\Delta V$, we can also express the energy as $\frac{1}{2}C(\Delta V)^2$.

19.16 Application: Real batteries and internal resistance

We have been treating batteries as if they were ideal devices: that is, as if they always managed to maintain the same charge separation regardless of what they are connected to. Real batteries, however, don't quite succeed in maintaining a potential difference equal to the battery emf, due to "internal resistance"—the resistance of the battery itself to the passage of current. We will show that a real battery can be modeled as an ideal, resistance-less battery in series with the internal resistance of the battery.

Consider again our mechanical "battery" consisting of two large charged plates whose charge is continually replenished by transporting electrons on a conveyor belt from the "+" plate to the "−" plate. The electrons on the motor-driven belt are acted on both by the non-Coulomb force exerted by the motor, $\vec{F}_{NC}$, and by the Coulomb force exerted by the charges on the plates, $\vec{F}_C = -e\vec{E}_4$. These two forces oppose each other (Figure 19.25).

In general, the non-Coulomb force F_{NC} has to be somewhat larger than the Coulomb force $F_C = eE_4$ because of finite mobility of the charges moving through the battery. In a chemical battery, this takes the familiar form of collisions, leading to an average drift speed. The drift speed of ions through the battery is proportional to the net force $(F_{NC} - eE_C)$. If σ is the conductivity for moving ions through the battery, and A is the cross-sectional area of the battery, the current density $J = I/A$ is given not by the usual $J = \sigma E$ but by

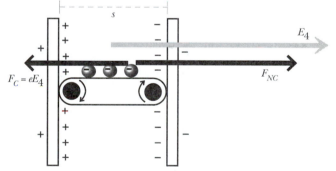

Figure 19.25 A mechanical battery with some friction in the mechanism.

$$J = \frac{I}{A} = \sigma\times(\text{force per unit charge}) = \sigma\left(\frac{F_{NC}}{e} - E_C\right)$$

Rearranging this formula, we have

$$E_C = \frac{F_{NC}}{e} - \frac{I/A}{\sigma}$$

If the (short) length of our mechanical battery is s, the electric field is nearly uniform, and the potential difference across the battery is $\Delta V = E_C s$, so

$$\Delta V = \frac{F_{NC}s}{e} - \frac{s}{\sigma A}I$$

But note that $F_{NC}s/e$ is the work per unit charge or emf of the battery, and $s/(\sigma A)$ has the units of resistance. We call $s/(\sigma A)$ the "internal resistance" r_{int} of the battery, and finally we have

$$\Delta V_{\text{battery}} = \text{emf} - r_{\text{int}}I$$

What this says is this: If there is very little internal resistance r_{int} to movement of ions through the battery (or very little current I), the potential difference ΔV across the battery is numerically nearly equal to the emf of the battery. Up till now we have made this assumption to simplify the discussion. But if the current I through the battery is large enough, the term $r_{\text{int}}I$ may be sizable, and the potential difference across the battery (and therefore across the rest of the circuit) may be significantly less than the battery emf.

Since $\Delta V_{\text{battery}} = \text{emf} - r_{\text{int}}I$, a real (nonideal) battery can be modeled as an ideal battery with the internal resistance in series, so that the potential difference across the battery is less than the emf by an amount $r_{\text{int}}I$, the potential drop across the internal resistance (Figure 19.26).

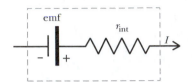

If a resistor R is connected to a real, nonideal battery (Figure 19.27), energy conservation (round-trip potential difference = 0) gives

$$(\text{emf} - r_{\text{int}}I) - RI = 0$$

Solving for the current I, we have

$$I = \frac{\text{emf}}{R + r_{\text{int}}}$$

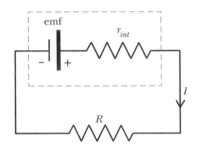

Figure 19.26 A real battery can be modeled as an ideal battery in series with an internal resistance.

Figure 19.27 A nonideal battery connected to a resistor.

Let's see how this works out in practice. A typical 1.5-volt "alkaline" flashlight battery has an internal resistance r_{int} of about 0.25 ohm.

? What current would we get when we connect various resistors to the battery? Calculate the various values to get an idea of the effect:

Ideal battery with zero internal resistance		Real battery with 0.25-ohm internal resistance
100 ohm:	0.015 ampere	100 ohm:
10 ohm:	0.15 ampere	10 ohm:
1 ohm:	1.5 ampere	1 ohm:
0 ohm (short circuit): infinite current!		0 ohm (short circuit):

The currents with the real battery are 0.01496 A, 0.146 A, 1.2 A, and 6 A.

There are several things to notice in this table: With a real battery, with nonzero internal resistance, you don't get 10 times the current through a 1-ohm resistor that you get through a 10-ohm resistor. Also, the internal resistance determines the maximum current that you can get out of a battery—this battery cannot deliver more than 6 amperes, no matter how small a resistor you attach to it. However, if the current is small, the behavior is nearly "ideal."

An example of the role of a battery's internal resistance is seen in the requirements for a car battery, which must deliver a very large current to the starting motor. The battery must be built in such a way as to have a particularly low internal resistance. There are very large metal electrodes inside a car battery (large cross-sectional area A), which helps reduce the internal resistance to the transport of ions.

It should be pointed out that the emf and the internal resistance r_{int} of a chemical battery are only approximately constant. As the battery is used more and more, the emf decreases and the internal resistance increases. However, at any particular moment, the formula $\Delta V_{\text{battery}} = \text{emf} - r_{\text{int}}I$ does relate the potential difference across the battery to the current through the battery (using the values of emf and internal resistance valid at that moment).

Ex. 19.29 What is the potential difference ΔV across a 1.5-volt battery when it is short-circuited, if the internal resistance is 0.25 ohm? What is the average net electric field E inside the battery? For a mechanical "battery," how much charge is on the battery plates?

This again shows the difference between the non-Coulomb emf of the battery, which does not change no matter what you connect to the battery, and the potential difference ΔV across the battery, which depends on how much charge the battery is able to maintain on its ends. Only for an ideal battery with no internal resistance is ΔV exactly equal to the emf. For real batteries that are discharging, ΔV is always somewhat less than the emf (and if you are charging a real battery, you have to apply a ΔV that is somewhat greater than the emf).

Ex. 19.30 Suppose you connect two identical flashlight batteries in series to get a 3-volt emf. Write a loop equation to show that the maximum (short-circuit) current these batteries can deliver is exactly the same as the maximum current from one battery.

Ex. 19.31 Suppose on the other hand you connect two identical flashlight batteries in parallel, which gives you just a 1.5-volt emf. Write loop and node equations to show that the maximum (short-circuit) current these batteries can deliver is double that of one battery.

19.16.1 Experiments dealing with internal resistance

The internal resistance of your batteries is rather small compared with the resistance of a 40-ohm long bulb but not entirely negligible compared with the resistance of a 10-ohm round bulb.

Experiment 19.7 Two bulbs in parallel
Connect up two round bulbs in parallel to two batteries in series. Unscrew one of the two bulbs. When you screw it back in, you may be able to see a slight change in the brightness of the other bulb. Why is this?

Experiment 19.8 Two batteries in parallel
Connect two batteries in parallel to a long bulb. You will see that the bulb has about the same brightness (dimness) as it has with just one battery, because the potential difference is essentially unchanged. What would you expect concerning how long the batteries would last, compared with just one battery? Why?

Experiment 19.9 The short-circuit current
Hold a copper connecting wire high enough above a compass that when the wire is connected by other connecting wires to short-circuit a single battery, you get about a 20-degree compass deflection. Try a different battery, trying not to move the wire. What compass deflections do you get for each battery? For the two batteries in series? For the two batteries in parallel? Are your measurements consistent with your predictions in the exercises above for the maximum current from batteries in series and in parallel?

19.17 Application: Ammeters, voltmeters, and ohmmeters

An important application of circuit theory is to explain the operation of ammeters, voltmeters, and ohmmeters used for measuring conventional currents, potential differences, and resistances. Modern digital "multimeters" combine these functions into one compact instrument, with a switch that lets you choose between measuring current, voltage, or resistance, and with a convenient digital readout, complete with sign.

Using an ammeter

An ammeter must be inserted into a circuit, and the current to be measured is brought into the "+" socket and out the "−" socket. If conventional current flows into the "+" socket, the ammeter indicates a positive current reading (needle deflects to the right, or a digital ammeter displays a positive number). When conventional current flows into the "−" socket, the ammeter indicates a negative current reading (needle deflects to the left, or a digital ammeter displays a negative number).

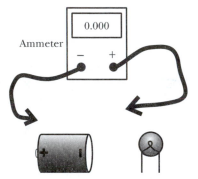

Ammeter

Figure 19.28 How should we connect these devices?

AMMETER SIGN CONVENTION

If conventional current flows into the socket marked "+"
an ammeter indicates a positive current.

(This is backwards from the convention for batteries, since conventional current flows out of the "+" terminal of a battery.)

Ex. 19.32 Here is a digital ammeter with connected leads, a battery, and a bulb (Figure 19.28). What connections would light up the bulb and at the same time measure the current through the bulb in such a way that the ammeter displays a positive number?

A simple ammeter

You have been using a simple ammeter for quite a while, in the form of a compass whose needle is deflected by a passing current. If you could run known amounts of current over the compass, you could calibrate this ammeter by determining the relationship between the known current and the observed deflection angle.

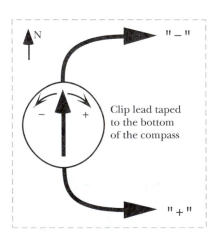

Figure 19.29 A crude ammeter.

Experiment 19.10 Making a simple ammeter

Use your invisible tape to tape one of your connecting wires to the bottom of the compass (to get it out of the way of reading the deflection), aligned with the north-south markings on the compass (Figure 19.29). This combination of compass and attached wire is an ammeter (if you align it north-south, of course).

Label the two connections "+" and "−" in such a way that when conventional current flows into the "+" end there will be a compass needle deflection to the right. If conventional current flows into the socket marked "−" your ammeter needle deflects to the left.

In a circuit consisting of a long #48 bulb connected to two batteries the current was measured with a commercial ammeter to be about 100 milliamperes (different long bulbs may vary somewhat). If you use a round #14 bulb instead of a long bulb, the current is measured to be about 300 milliamperes (different round bulbs may vary somewhat).

Assuming that these currents are correct, calibrate your ammeter using two batteries in series and a long or round bulb:

current, milliamperes	compass deflection, degrees
100 (long bulb)	θ =
300 (round bulb)	θ =

Now you can measure currents in amperes with your ammeter by interpolating in this table. You could even mark the face of the compass with ampere or milliampere values and then be able to read amounts of conventional current directly off the device.

A serious practical problem in using ammeters is that if you want to measure the current somewhere in an existing circuit, you must break the circuit at the place of interest and insert your ammeter.

? For example, if you are trying to measure the current through the bulb in the circuit shown in Figure 19.30, what is wrong with the placement of the ammeter? Try this yourself to verify your analysis: what do you observe?

Connecting your ammeter in parallel to the bulb means that there is a very low-resistance path for the current (a short circuit). Most current will go through the ammeter, with very little current going through the bulb. The bulb no longer glows, and the large current measured by the ammeter has nothing to do with the tiny current that goes through the bulb.

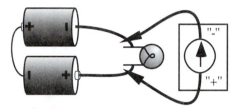

Figure 19.30 This ammeter is placed incorrectly in the circuit.

This may not seem like a problem with your simple circuits, but it can be a big problem if the circuit of interest is built in such a way that you can't break in to insert your ammeter without damaging the circuit. Moreover, a commercial ammeter may blow a fuse or even be permanently damaged if used incorrectly as shown above, because a very large current will run through the very low-resistance ammeter, and most commercial ammeters cannot tolerate large currents.

Sensitivity vs. reality

What if you need to measure very small currents? You could wrap 10 turns of wire around the compass, and then you would get the same compass deflection with a current 10 times smaller than those you measured previously. You then have a more sensitive ammeter. But there is a practical limit to this. The more turns of wire, the more resistance in the ammeter itself. When you insert this more sensitive ammeter into a running circuit, you may be able to measure a smaller current, but you have also changed the circuit by inserting a significant resistance, so you aren't really measuring the current in the undisturbed circuit.

AN AMMETER SHOULD HAVE LOW RESISTANCE

An ammeter should have as small a resistance as possible, so as not to change the current that you are trying to measure.

If on the other hand you make an ammeter very sensitive by using many turns of wire (hopefully of low resistance!), you then have the problem of how to measure large currents. The solution is to place a resistor in parallel with the ammeter to shunt part of the current around the ammeter (the parallel resistor is called a "shunt" resistor).

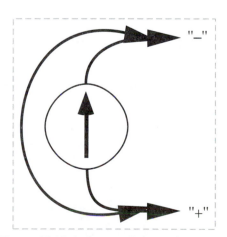

Figure 19.31 A shunt resistor reduces the sensitivity of the ammeter.

Experiment 19.11 Making the ammeter less sensitive
Add another connecting wire (a shunt resistor) in parallel as shown in Figure 19.31. It may seem odd to refer to a copper wire as a resistor, but in this case a copper wire has just as much resistance as the (small) resistance of the ammeter. What deflection angle would you predict when you measure

the current through a round bulb powered by two batteries in series? Try it. What angle do you find? (Be sure to keep the shunt-resistor wire in the plane of the compass as shown in the diagram, to avoid having its current deflect the compass.)

Some commercial ammeters offer different current ranges by means of switch settings that place different shunt resistors in parallel to a sensitive ammeter. Many modern electronic multimeters offer "auto-ranging": they automatically choose an appropriate range to match the measured current. But even these multimeters may have a special socket to use for measuring currents larger than the normal range. When you use this socket, a shunt resistor is connected in parallel with the ammeter.

Voltmeters measure potential difference

If we add a series resistor to an ammeter, we can make a voltmeter to measure potential differences. Suppose we attach an ammeter across a circuit element, through a resistor whose resistance R is very large (Figure 19.32).

? If the measured current through the ammeter is I, what is the potential difference $V_A - V_B$ in this circuit, in terms of R and I?

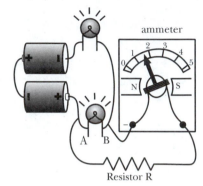

Figure 19.32 An ammeter with a resistor in series forms a voltmeter.

The potential difference along the resistor is $V_A - V_B = RI$. This is the basic principle of a nondigital voltmeter: a small current I running through the resistor is measured by the ammeter, and the potential difference is RI. If the ammeter is relabeled in volts (RI), the device reads potential differences directly. Any voltmeter can be thought of as an ammeter with a series resistance, all in one box (Figure 19.33).

Notice that a voltmeter has two leads and measures potential difference, not potential! Commercial voltmeters have their "+" and "−" sockets arranged as follows:

VOLTMETER SIGN CONVENTION

If the potential is higher at the socket marked "+"
a voltmeter indicates a positive potential difference.

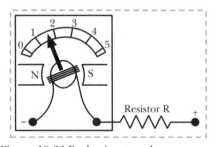

Figure 19.33 Packaging together an ammeter and a resistor makes a voltmeter.

If the "−" socket is at the higher potential, the voltmeter displays a *negative* potential difference.

An ammeter should have very small resistance so that inserting the ammeter into a circuit should disturb the circuit as little as possible. A voltmeter on the other hand should have as large a resistance as possible, so that placing the voltmeter in parallel with a circuit element will disturb the circuit as little as possible, with only a tiny current being deflected through the voltmeter.

A VOLTMETER SHOULD HAVE HIGH RESISTANCE

A voltmeter (which is used in parallel) should have very large resistance.

An ammeter (which is used in series) should have very small resistance.

There is a practical limit to the voltmeter resistance. The larger the resistance of the voltmeter, the more sensitive must be the ammeter part of the voltmeter. Commercial voltmeters often have resistances as high as 10 megohms (10×10^6 ohms), and the ammeter portion of the voltmeter is correspondingly sensitive. Even a commercial voltmeter with a resistance of 10 megohms is useless for measuring potential differences in a circuit consisting of 100-megohm resistors, because placing a 10-megohm voltmeter in parallel with a 100-megohm resistor drastically alters the circuit.

Ex. 19.33 A pair of students measured the current in a series circuit by placing a digital multimeter (set to read current in milliamperes) in series between the two round bulbs (Figure 19.34). Then they wanted to measure the voltage across a bulb, so they simply switched the multimeter to the V setting (volts). They were surprised that the bulbs stopped glowing. What did they do wrong? What did their voltmeter read?

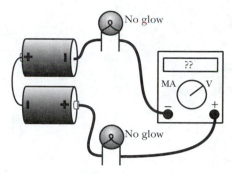

Figure 19.34 This voltmeter is placed incorrectly in the circuit.

Ohmmeters

Commercial multimeters can act not only as ammeters and voltmeters but also as "ohmmeters" to measure the resistance of an element that has been removed from a circuit. The ohmmeter section of the multimeter consists of an ammeter in series with a small voltage source (for example, 50 millivolts). When you connect an unknown resistor to the ohmmeter, the small voltage drives a small current through the resistor and ammeter. The ohmmeter is calibrated to display the applied voltage divided by the observed current, which is the resistance (assuming that the resistance of the ammeter is small compared to the resistance you are trying to measure). You can see that you have to remove the resistor of interest from its circuit before you can use the ohmmeter, because an ohmmeter is an active device with its own voltage source.

How commercial meters are constructed

In older, nondigital ammeters with a swinging pointer, current runs through a multi-turn coil that rotates inside a permanent magnet (Figure 19.35). The magnetic field of the magnet exerts a force on the current-carrying coil (we'll discuss this in detail in the next chapter). The magnetic force is opposed by a spring (not shown in the diagram), so that the deflection angle is proportional to the magnetic force and hence to the current. You may be able to see the coil through the viewing window.

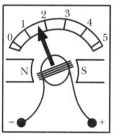

Nondigital ammeter

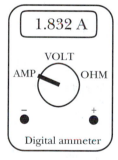

Digital ammeter

Figure 19.35 Two kinds of meters.

A modern digital ammeter has no moving mechanical parts and measures current electronically. Nevertheless, it is similar to your crude ammeter in that it has a very small resistance, and of course you must break a circuit in order to insert the ammeter. Moreover, if you accidentally connect a digital ammeter directly across a battery (thus drawing a very large current), you may blow a fuse in the ammeter or even damage it.

We have seen that the older, nondigital voltmeters are really sensitive ammeters with a large resistor in series. It turns out that modern digital voltmeters read voltage directly, by timing how long it takes a known current to discharge a capacitor! A digital ammeter is basically a voltmeter with a small (shunt) resistance in parallel. A digital ohmmeter drives a known current through the unknown resistor, and measures the resulting potential difference. Although digital meters work differently from nondigital meters, for both kinds of meters it is important to understand that an ammeter has very low resistance, and a voltmeter has very high resistance.

19.18 Application: Quantitative analysis of an RC circuit

We are now in a position to analyze quantitatively a series circuit containing an ideal battery with known emf, a resistor R, and a capacitor C—a so-called "RC" circuit (Figure 19.36). Recall that the potential difference across a capacitor is $\Delta V = Q/C$, where Q is the charge on the positive plate and C is the "capacitance" (page 675; for a parallel-plate capacitor, $C = K\varepsilon_0 A/s$, where A is the area of one of the plates, s is the gap distance between the plates, and K is the dielectric constant of the material filling the gap). The energy equation for the RC circuit in Figure 19.36 is

$$\Delta V_{\text{round trip}} = \text{emf} - RI - Q/C = 0$$

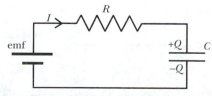

Figure 19.36 An RC circuit.

The initial situation

Recall the observations you made when you connected a battery and round bulb in series to a (discharged) capacitor. Immediately after making the final connection the light bulb was very bright (and then it got dimmer). The initial brightness looks about the same as you get in a circuit containing only a round bulb without a capacitor.

Let's see how we can understand this quantitatively. Rearrange the energy equation to solve for I in terms of the other quantities:

$$I = \frac{\text{emf} - Q/C}{R}$$

? Use this formula for I to explain why the initial brightness is the same as though the capacitor weren't there. (Hint: What is Q initially?)

The capacitor is initially uncharged, so initially $Q = 0$, and plugging this value for Q into the equation for I gives us $I = \text{emf}/R$. This is the same current we would have in a simple circuit without the capacitor. Having understood the initial situation, let's see what happens next.

The rate at which charge Q accumulates on the positive capacitor plate is equal to $I = dQ/dt$. To put it another way, in a time dt the plate receives an additional amount of charge $dQ = Idt$. We rewrite the energy equation:

$$I = \frac{dQ}{dt} = \frac{\text{emf} - Q/C}{R}$$

$$dQ = \left(\frac{\text{emf} - Q/C}{R}\right)dt$$

This is in a form suitable for doing a numerical integration. At each instant, we know Q (initially zero), so we can calculate the change dQ from this equation. Then $Q + dQ$ is the new value of Q, and we can do it again.

Numerical integration

Let's take a couple steps in this numerical integration to see how things go. In the first time interval dt, $dQ = (\text{emf}/R)dt$, since the initial value, Q_0, is zero. The new value, Q_1, is

$$Q_1 = Q_0 + dQ = (\text{emf}/R)dt$$

Now the capacitor has a small nonzero charge Q_1. In the second time interval, we have

$$dQ = \left(\frac{\text{emf} - Q_1/C}{R}\right)dt$$

$$Q_2 = Q_1 + dQ$$

Because Q_1 is nonzero, the increase in charge dQ is smaller this time. As time goes on, and Q gets larger and larger, dQ for each additional time step will be smaller and smaller. Therefore a graph of Q vs. time will look like the upper graph of Figure 19.37. Points are the results of a numerical integration using $R = 10$ ohms (approximately the resistance of a glowing round bulb) and $C = 1.0$ farad, two 1.5 volt batteries in series.

The lower graph of Figure 19.37 is the current flowing into the resistor, calculated as $I = dQ/dt$. As Q increases, I decreases, because the fringe field of the capacitor increasingly opposes the current flow. In terms of potential, as Q increases, the potential difference across the capacitor increases, which means that the potential difference across the resistor must decrease, and therefore $I = \Delta V_{\text{resistor}}/R$ decreases.

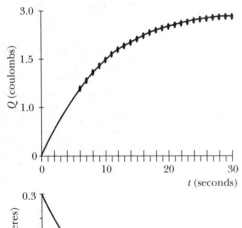

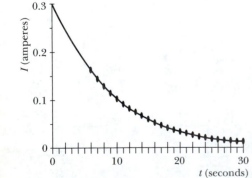

Figure 19.37 Charge and current as a function of time.

Analytical solution

The points on the graphs in Figure 19.37 look like some kind of exponential functions. Let's make a guess that the current is in fact given by

$$I = I_0 e^{-at}$$

where I_0 is the initial current (which we found earlier to be $I_0 = \text{emf}/R$), and a is an unknown constant. Consider this line of reasoning:

$$I = \frac{dQ}{dt} = \frac{\text{emf} - Q/C}{R} \quad \text{(energy equation in terms of } I \text{ and } Q\text{)}$$

$$\frac{dI}{dt} = -\frac{1}{RC}\frac{dQ}{dt} = -\frac{1}{RC}I \quad \text{(since emf}/R \text{ is constant)}$$

$$\frac{dI}{dt} = \frac{d}{dt}(I_0 e^{-at}) = -aI_0 e^{-at} = -aI = -\frac{1}{RC}I \quad \text{(using our guess for } I\text{)}$$

This last equation is true for all time if $a = \dfrac{1}{RC}$, so we have a solution.

CURRENT IN A SERIES RC CIRCUIT

$$I = \left(\frac{\text{emf}}{R}\right)e^{-\frac{t}{RC}}$$

This solution has all the right properties. When $t = 0$, the exponential is equal to 1, and $I = \text{emf}/R$, so the initial brightness is indeed the same as if there were no capacitor. As t becomes very large, the exponential goes to 0, and the current goes to zero, which is what you have observed.

We used a very common and powerful method for finding an analytical solution for a differential equation (an equation involving derivatives): Guess the form of the solution (containing unknown constants such as a), plug this form and its derivatives into the differential equation, and determine what values the constants must have to satisfy the equation ($a = \dfrac{1}{RC}$).

Since $I = dQ/dt$, $Q = \int_0^t I\,dt$, and it is easy to show this:

CHARGE IN A SERIES RC CIRCUIT

$$Q = C(\text{emf})\left[1 - e^{-\frac{t}{RC}}\right]$$

This solution has all the right properties. When $t = 0$, the exponential is equal to 1, and $Q = 0$ (no charge yet on the positive plate of the capacitor). As t becomes very large, the exponential goes to 0, and Q approaches $C(\text{emf})$. This is correct: $\Delta V_{\text{capacitor}} = Q/C = C(\text{emf})/C = \text{emf}$. When the current stops, the potential difference across the capacitor becomes equal to the potential difference across the battery.

? What is the numerical value of the final charge on one plate of the capacitor in your circuit, in terms of coulombs?

We find that $Q_F = C \times \text{emf} = (1.0 \text{ farad})(3 \text{ volts}) = 3.0$ coulombs. That's a huge amount of charge! Compare for example with the amount of charge you found on a strip of invisible tape, which was of the order of 10^{-8} coulomb.

Actually, our analytical solution of the differential equation is physically only an approximation, because as the current changes the resistor changes temperature and changes resistance. This is particularly significant when the resistor is a light bulb, since its resistance drops a great deal as the current decreases and it cools off.

The RC time constant

A rough measure of how long it takes to reach final equilibrium is the "time constant" RC. When the time $t = RC$, the factor $e^{-t/RC}$ has fallen from the value $e^0 = 1$ to the value

$$e^{-\frac{t}{RC}} = e^{-1} = \frac{1}{e} = \frac{1}{2.718} = 0.37$$

? Calculate the "time constant" for a circuit with $R = 10$ ohms (approximately the resistance of a glowing round bulb) and $C = 1.0$ farad. This is the time when the current has decreased from the original 0.30 amperes to 0.11 amperes (0.37 times 0.30 amperes).

The time constant $RC = (10 \text{ ohms})(1.0 \text{ farad}) = 10 \text{ s}$. The bulb will be very dim with this current. This is roughly consistent with what is observed with a round bulb and a one-farad capacitor, which loses most of its brightness in this amount of time.

? Show that the power dissipated in the bulb at $t = RC$ is only 14% of the original power.

The power dissipated in the bulb is $I\Delta V = RI^2$, so a reduction in the current by a factor of 0.37 gives a reduction in the power by a factor of $(0.37)^2 = 0.136$.

? Calculate the "time constant" for a circuit in which you replace the round bulb with a long bulb (whose resistance R when glowing is typically about 30 ohms, which may vary from one bulb to another). Does this time constant agree roughly with what you observed in your experiments?

The time constant $RC = (30 \text{ ohms})(1 \text{ farad}) = 30 \text{ s}$, which agrees roughly with observations with a one-farad capacitor.

Figure 19.38 shows graphs of Q vs. t and I vs. t for the RC circuit with the round bulb, with the 10 second RC time constant indicated.

We now have a complete quantitative description of the behavior of an RC circuit. It should be pointed out that when we did experiments with capacitors we timed how long it took for the current to decrease to a point where the bulb no longer glowed. This is not the same as timing how long it takes for the current to get smaller by a factor of e^{-1}.

19.19 Reflection on the macro-micro connection

In this chapter we made connections between a microscopic and macroscopic view of resistance. In our original microscopic view, individual electrons acquire a drift speed proportional to the electric field, with $\bar{v} = uE$. This electric field is made by all the charges in and on the battery and on the surface of the conducting circuit elements. Electron current i is measured in number of electrons per second.

In the macroscopic view, the electric field made by all the charges produces a conventional current density $J = \sigma E$ and a conventional current $I = JA$ measured in coulombs per second (amperes). The integral of the electric field is potential difference, and any round-trip integral of the electric field due to point charges must equal zero, which represents energy conservation per unit charge. We also used a conventional-current version of the steady-state current node rule.

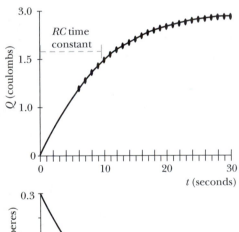

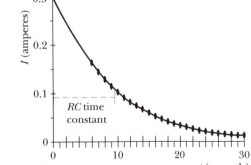

Figure 19.38 Charge and current as a function of time.

19.20 *A complicated resistive circuit

We can use energy conservation (loop equations) and the current node rules to determine how much current would flow in each element of a complicated "multi-loop" circuit made up of batteries and ohmic resistors (resistors for which the current is proportional to the potential difference).

Analyzing complex circuits like the one discussed in this section is more a matter of technology than fundamental physics, so we won't analyze many such circuits in this course. However, it is useful to analyze at least one such circuit as another illustration of the application of energy conservation and the current node rule. We will use the loop and node rules to generate as many independent equations as we have unknown currents, and then solve the system of equations.

Assume that the emf's, battery internal resistances, and R's are known for the circuit shown in Figure 19.39, but the currents are unknown.

We will determine the current in each part of this circuit. The first step in the solution is to assign names and directions to the unknown currents (I_1 through resistor R_1, I_2 through resistor R_2 etc.). We don't know what the actual directions of these unknown conventional currents are, so we guess, and we record our guesses on a diagram, as shown in Figure 19.40. We also draw possible paths to follow for writing loop equations for energy conservation.

We don't have to assign a different current variable to every branch. Consider conventional current I_1. On the circuit diagram we have assumed that it flows from A toward B. This also means that we have assumed the same amount of current I_1 goes from B to C through resistor R_1, and from H to A through resistor R_7. Similar comments apply to the other currents. These simplifications are trivial applications of the current node rule.

First let's use energy conservation for loop 1 (the path *ABCHA*). We can start anywhere around the loop; let's start at location A and walk around the loop, adding up the potential differences we encounter on our walk (Figure 19.41). The potential rises across the battery (the electric field inside the battery is opposite to the direction we're walking) and the potential falls along the resistors, since we are walking in the assumed direction of the conventional current, which runs in the direction of the electric field from higher potential to lower potential.

When we get back to our starting location at A, the net potential change along our walk should be zero, since potential difference in a round trip is zero (and $V_A - V_A = 0$):

$$\text{loop 1: } (\text{emf}_1 - r_1 I_1) + (-R_1 I_1) + (-R_4 I_4) + (-R_7 I_1) = 0$$

Note that in writing this loop rule we implicitly used the current node rule to show that the current is the same I_1 through the battery, r_1, R_1, and R_7. And we used the property of ohmic resistors to be able to relate the current through R_1 to the potential difference along R_1 (as $-R_1 I_1$).

If I_1 turns out to be -1.3 amperes when we solve for the unknown currents, that will merely mean that we guessed the wrong direction and that I_1 actually runs from C to B (and that C is at a higher potential than B). So there is no real penalty for guessing the wrong current directions.

Next consider loop 2 (path *CDEFC*), starting at location C (Figure 19.42). When we get back to our starting location at C, the net potential change along our walk should be zero:

$$\text{loop 2: } (-R_2 I_2) + (-\text{emf}_2 - r_2 I_2) + (-R_6 I_2) + (R_3 I_3) = 0$$

Comments on loop 2: From D to E there is a potential drop (not a rise) across the battery numerically equal to $(-\text{emf}_2 - r_2 I_2)$, because we are traversing the battery from its high-potential $+$ end to its low-potential $-$ end,

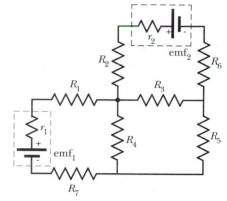

Figure 19.39 A circuit with unknown currents, which we will determine.

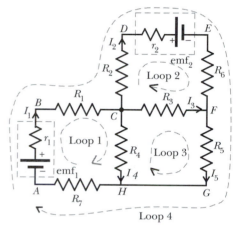

Figure 19.40 Name the unknown currents, draw paths along which to apply energy conservation.

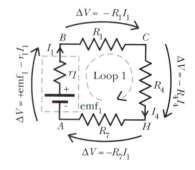

Figure 19.41 Apply energy conservation for a round-trip path, loop 1 (ABCHA).

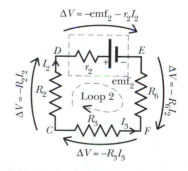

Figure 19.42 Apply energy conservation for a round-trip path, loop 2 (CDEFC).

in the direction of the electric field inside the battery. (If we've guessed right about the direction of I_2, this battery is being charged rather than discharged.) From F to C there is a potential *rise* $+R_3I_3$ (not a drop) because we are heading upstream against the current I_3. That is, the assumed direction of I_3 (from C to F) implies that the potential at C is higher than the potential at F. Again, if we have made the wrong assumption, all that will happen is that the final numerical answer for I_3 will have a negative sign.

? In the same manner, use energy conservation for loop 3 (path $HCFGH$)) and for loop 4 (path $ABCDEFGHA$).

The equation for loop 4 around the outside of the circuit is the algebraic sum of the equations for the other three loops and is not an independent equation. Only three of these four loop equations are actually significant in solving for the unknown currents, although it doesn't matter which three we choose to use in the solution.

Loop 3: $(R_4I_4) + (-R_3I_3) + (-R_5I_5) = 0$

Loop 4: $(\text{emf}_1 - r_1I_1) + (-R_1I_1) + (-R_2I_2) + (-\text{emf}_2 - r_2I_2) +$
$$(-R_6I_2) + (-R_5I_5) + (-R_7I_1) = 0$$

? Now write current node rule equations for nodes C, H, and F, paying attention to signs (+ for incoming current, – for outgoing current).

Adding the equations for node C and node F gives an equation that is equivalent to the equation for node H. So only two of these three node equations are actually independent, and we can use any two of them in solving for the unknown currents.

Node C: $I_1 - I_2 - I_3 - I_4 = 0$ Node F: $I_2 + I_3 - I_5 = 0$

Node H: $I_4 + I_5 - I_1 = 0$

We have a total of five independent equations (three independent loop equations and two independent node equations).

It can be shown that in any resistive circuit there are as many independent loop equations as there are minimal-sized loops (that is, ignoring combined loops such as loop 4). The number of independent node equations can be determined by subtracting the number of independent loop equations from the number of independent currents.

Our five equations are sufficient to be able to solve for the five unknown currents in terms of the known emf's and known resistances. This solution step can involve extremely tedious algebraic manipulation and should usually be turned over to a symbolic manipulation package such as Maple or Mathematica. Of course in some very simple cases it may be easy to solve the equations. But in general the algebra can be quite forbidding.

Once the unknown currents have been determined (typically by using a computer program to solve the equations), we can know essentially everything there is to know about the circuit. For example, we can calculate the potential drop $\Delta V = -RI$ across each resistor, since we know R and I for each resistor. We can therefore also calculate the potential difference between any two locations in the circuit.

Refer to Figure 19.39 on page 695, and suppose $\text{emf}_1 = 12$ volts and $R_1 = 20$ ohms. Suppose after we solved the equations we found that I_1 was 0.4 amperes.

Ex. 19.34 What is the potential difference $V_C - V_B$? How much power is dissipated in R_1? What is the power expenditure of battery 1 (some of which is dissipated in the battery itself due to internal resistance r_1)?

19.21 *Electrons in metals

It is possible to show experimentally that the mobile charged particles in a metal are indeed electrons. For example, magnetic forces can lead to a transverse polarization of a current-carrying wire, and the sign of this polarization is different for positive and negative moving charges (this is called the "Hall effect," and we will study this in the next chapter).

A different kind of experiment was carried out in 1916-1917 by Tolman and Stewart, which showed that the mobile charge carriers in a metal are indeed electrons, which was not known for certain before their experiment. We will describe a simpler experiment which is conceptually equivalent to the experiment they actually carried out.

Suppose that we accelerate a metal bar, which means that we accelerate the atomic cores, which are bound to each other. On the average, the mobile electrons normally feel no force; the mutual repulsion of the mobile electrons is effectively canceled by the attraction to the atomic cores. Therefore as the bar and atomic cores accelerate, the mobile electrons are left behind. As a result, they pile up at the trailing end of the bar (they can't easily leave the bar). This polarizes the bar, with negative charge on the trailing surface and positive charge (deficiency of electrons) on the leading surface, as shown in Figure 19.43.

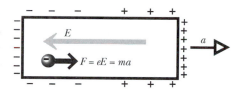

Figure 19.43 In an accelerated metal bar the mobile electrons are initially left behind, polarizing the bar.

The surface charges produce an electric field E to the left, which means that there is a force eE to the right on electrons inside the metal (Figure 19.43). As the polarization increases, E increases until the electric force eE is equal to the mass of the electron times the acceleration of the metal bar, at which point the acceleration of the mobile electrons matches the acceleration of the bar, and then the polarization no longer increases.

The observed direction of the polarization (negative at the left end) led the experimenters to conclude that the mobile particles are indeed negatively charged. You should run through the argument to see that if the movable charges had been positive, the trailing end of the bar would have been positive instead of negative.

Although this experiment does not give separate values for the charge and the mass of the mobile particles, it does provide a measurement of the ratio q/m. There had already been measurements of the ratio e/m for electrons moving in a vacuum, in a device similar to a television tube. The modern value for this ratio for the electron is about

$$\frac{e}{m} = \frac{(1.6 \times 10^{-19}\ \text{C})}{(9 \times 10^{-31}\ \text{kg})} = 1.7 \times 10^{11}\ \text{C/kg}$$

Tolman and Stewart measured the acceleration a, and they used a voltmeter to measure the potential difference ΔV from one end of the metal bar to the other. The potential was lower at the left end of the bar, which shows that the mobile charge carriers are negative. Moreover, the experimental value of q/m had the same value as that obtained for electrons in other experiments. This was very strong evidence that currents in metals consist of moving electrons.

Ex. 19.35 If the length of the bar is L, express q/m in terms of the quantities measured by Tolman and Stewart.

19.22 Summary

Fundamental principles

There were no new fundamental principles.

New concepts

In a capacitor circuit there is a slow transient leading to a final equilibrium state.

Current can run (for awhile) even if there is a gap in a circuit.

The fringe field of the capacitor in the neighboring wires is critical for understanding the behavior of a capacitor circuit in terms of charge and electric field.

Conventional current I:
  direction of motion of positive carriers
  measured in coulombs per second

Conductivity $\sigma = |q| n u$ is a property of a material; $J \equiv \dfrac{I}{A} = \sigma E$

Ohmic and nonohmic resistors

Results

Charging and discharging times are larger with larger plates, smaller gap, or with an insulator in the gap.

Resistance $R = \dfrac{L}{\sigma A}$ incorporates the geometry of a resistor

$I = \dfrac{|\Delta V|}{R}$ is the current through a resistor

$|\Delta V| = RI$ is the potential difference across a resistor

Power $= I \Delta V$ for any circuit element

Potential difference across a real, nonideal battery (Figure 19.44):
$$\Delta V_{\text{battery}} = \text{emf} - r_{\text{int}} I$$

Capacitors in circuits can be analyzed in terms of $\Delta V = Q/C$, where C is the capacitance, and the charge Q and current I vary exponentially with time, with a time constant RC

Capacitance $C = \Delta V / q$ depends on the geometry of the capacitor. For a parallel plate capacitor

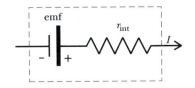

Figure 19.44 Equivalent nonideal battery.

19.23 Example problem: Two different bulbs in series

The circuit in Figure 19.45 is built from components in your electricity kit, consisting of two flashlight batteries, a round bulb, a long bulb, and very low-resistance copper wires.

(a) Calculate the number of electrons that flow out of the battery every second at location A, in the steady state. Assume that in these steady-state conditions the resistance of the round bulb is 10 ohms and the resistance of the long bulb is 40 ohms.

(b) Show the approximate surface charge on the steady-state circuit.

(c) Draw an accurate graph of potential around the circuit, with the x axis running from A to H. Label the y axis with numerical values of the potential.

(d) The tungsten filament in the long bulb is 8 mm long and has a cross-sectional area of 2×10^{-10} m^2. How big is the electric field inside this metal filament?

(e) What is the power output of the battery?

Solution

(a) Loop rule: $2\text{emf} - IR_{\text{round}} - IR_{\text{long}} = 0$

$$I = \frac{2\text{emf}}{(R_{\text{round}} + R_{\text{long}})} = \frac{2(1.5\text{V})}{(10\Omega + 40\Omega)} = 6\times10^{-2}\text{A}$$

$$i = \frac{I}{|q|} = \frac{6\times10^{-2}\text{A}}{1.6\times10^{-19}\text{C}} = 3.8\times10^{17}\text{electrons/s}$$

(b) See Figure 19.46. A bigger electric field is needed in the long bulb (because it has higher resistance), so the surface charge gradient across the long bulb is much bigger than the surface charge gradient across the round bulb. The surface charge gradient across the copper wires is extremely small, since only a very small electric field is needed in the wires. This gradient is too small to show up on this diagram, so the surface charge is shown as approximately uniform on the wires.

(c) See Figure 19.47. Distances on the x axis are not to scale.

(d) The drop in potential across the long bulb is given by $\Delta V_{\text{long}} = R_{\text{long}}I$.

   The electric field in the long bulb filament can be found from the potential difference:

$$E = \frac{\Delta V}{L} = \frac{R_{\text{long}}I}{L} = \frac{(40\Omega)(6\times10^{-2}\text{A})}{(8\times10^{-3}\text{m})} = 300\frac{\text{V}}{\text{m}}$$

(e) For the battery:

$$P = I\Delta V = (6\times10^{-2}\text{A})(3\text{V}) = 0.18\text{W}$$

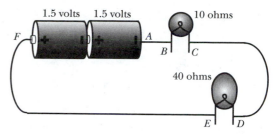

Figure 19.45 A circuit with a long bulb and a round bulb.

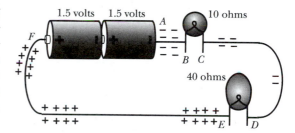

Figure 19.46 Approximate surface charge distribution on circuit in steady state.

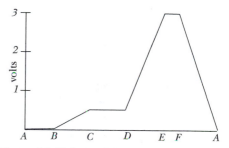

Figure 19.47 Potential (volts) *vs.* location in circuit. Distances (x axis) are not to scale.

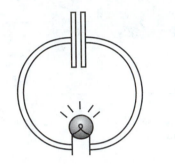

Figure 19.48 Discharging a capacitor.

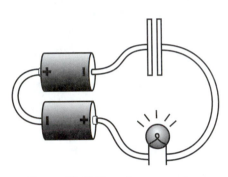

Figure 19.49 Charging a capacitor.

19.24 Review questions

Basic behavior of a capacitor circuit

RQ 19.1 How is the initial current through a bulb affected by putting a capacitor in series in the circuit? Explain briefly.

RQ 19.2 How is the charging time for your capacitor correlated with the initial current? That is, if the initial current is bigger, is the charging time longer, shorter, or the same? Explain briefly.

Explaining the discharging of a capacitor

RQ 19.3 Give a complete but brief explanation for the behavior of the current during the discharging of a capacitor in a circuit consisting of a capacitor and light bulb (Figure 19.48). Explain, don't just describe!

Explaining the charging of a capacitor

RQ 19.4 Give a complete but brief explanation for the behavior of the current during the charging of a capacitor in a circuit consisting of batteries, bulb, and capacitor in series (Figure 19.49). Explain, don't just describe!

The final state

RQ 19.5 How does the final (static-equilibrium) charge on the capacitor plates depend on the kind of bulb or the length of nichrome wire in the circuit during charging? Very briefly state why this is.

Capacitor construction

RQ 19.6 How does the final (static-equilibrium) charge on the capacitor plates depend on the size of the capacitor plates? On the spacing between the capacitor plates? On the presence of a plastic slab between the plates?

RQ 19.7 Researchers have developed an experimental capacitor that takes about eight hours (!) to discharge through one of your long bulbs. They propose using this in electric cars to provide a burst of energy for emergency situations (it would be charged at a slow rate during normal driving). Describe in general terms the key design elements of this extraordinary capacitor.

Conductivity

RQ 19.8 What are the units of conductivity σ, resistivity ρ, resistance R, and current density J?

Ohmic resistors

RQ 19.9 Which of the following are ohmic resistors? For those that aren't, briefly state why they aren't.

> Nichrome wire in your kit
> flashlight bulb in your kit
> plastic rod
> salt water
> silicon (a semiconductor)

RQ 19.10 When a round bulb is connected to one of your flashlight batteries, the current is 0.20 ampere. When you use two of your batteries in series, the current is not 0.40 ampere but only 0.33 ampere. Briefly explain this behavior.

Work and power

RQ 19.11 A desk lamp that plugs into a 120-volt wall socket can use a 60-watt or a 100-watt light bulb. Which bulb has the larger resistance? Explain briefly.

RQ 19.12 If the current through a battery is doubled, by what factor is the battery power increased?

If the current through a resistor is doubled, by what factor is the power dissipation increased?

Explain why these factors are the same or different (depending on what you find).

RQ 19.13 State whether the following statement is true or false, and briefly explain why: "In the two circuits shown in Figure 19.50, the battery output power is greater in circuit 2 because there is an additional resistor R_2 dissipating power."

Internal resistance

RQ 19.14 A certain 6-volt battery delivers 12 amperes when short-circuited. How much current does the battery deliver when a 1-ohm resistor is connected to it?

Series and parallel resistance

RQ 19.15 If the combination of resistors in Figure 19.51 were to be replaced by a single resistor with the equivalent resistance, what should that resistance be?

RQ 19.16 Why can birds perch on the bare wire of a 100,000-volt power line without being electrocuted?

Ammeters and voltmeters

RQ 19.17 Should an ammeter have a low or high resistance? Why? Should a voltmeter have a low or high resistance? Why?

RQ 19.18 You are marooned on a desert island full of all kinds of standard electrical apparatus including a sensitive voltmeter, but you don't have an ammeter. Explain how you could use the voltmeter to measure currents.

Capacitors in circuits

RQ 19.19 A 20-ohm resistor and a 2-farad capacitor are in series with a 9-volt battery. What is the initial current when the circuit is first assembled? What is the current after 50 seconds?

Multi-loop circuits

RQ 19.20 For the following circuit consisting of batteries with known emf and ohmic resistors with known resistance, write the correct number of energy-conservation and current node rule equations that would be adequate to solve for the unknown currents, but do not solve the equations. Label nodes and currents on the diagram, and identify each equation (energy or current, and for which loop or node).

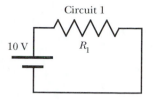

Circuit 1

Circuit 2

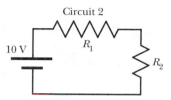

Figure 19.50 Circuits for RQ19.13.

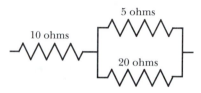

Figure 19.51 Resistors for RQ19.15.

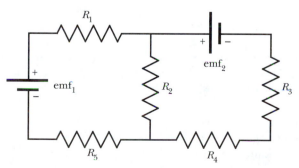

19.25 Homework problems

Problem 19.1 Capacitor circuits with different bulb arrangements

Figure 19.52 shows three circuits labeled A, B, and C. All the long bulbs, capacitors, and batteries are identical, and are like the equipment you used in class. The capacitors are initially uncharged. In each circuit the batteries are connected for a short time T and then disconnected. The time T is only ten percent of the total charging time through a single long bulb, so that the bulb brightness doesn't change much during the time T.

Figure 19.52 Three circuits with long bulbs, a capacitor, and two batteries (Problem 19.1).

(a) In which circuit (A, B, or C) does the capacitor now have the most charge? Explain.

(b) In which circuit (A, B, or C) does the capacitor now have the least charge? Explain.

(c) Design and carry out experiments to check your answers. Describe your experiments and the numerical results. Before each experiment, connect a wire across the capacitor for a few seconds to fully discharge the capacitor. One way to compare the amount of charge stored in the capacitor during the time T is to finish charging it through a single long bulb, and see how much less time is required than when you start with a discharged capacitor.

Problem 19.2 Capacitor circuits with different capacitors

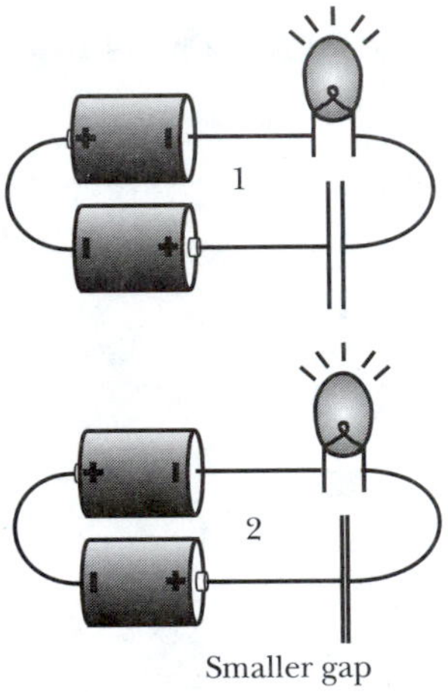

Two circuits (labeled 1 and 2) have different capacitors but the same batteries and long bulbs (Figure 19.53). The capacitors in circuit 1 and circuit 2 are identical except that the capacitor in circuit 2 was constructed with its plates closer together. Both capacitors have air between their plates. The capacitors are initially uncharged. In each circuit the batteries are connected for a time short compared to the time required to reach static equilibrium and then disconnected.

Figure 19.53 Two capacitor circuits with different gaps between the plates (Problem 19.2).

In which circuit (1 or 2) does the capacitor now have more charge? Explain your answer in detail.

Problem 19.3 Predicting and measuring current

Figure 19.54 shows a circuit composed of two batteries, two identical long bulbs, and an initially uncharged capacitor. Three compasses are placed on the desktop underneath the wires, all pointing north before the circuit is closed. Then the gap is closed, and the compass on the left immediately shows about a 20 degree deflection.

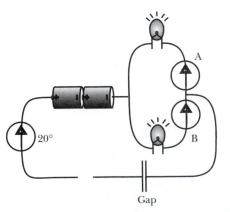

Figure 19.54 Two long bulbs in parallel connected to a capacitor (Problem 19.3).

(a) On a diagram, draw the new needle positions on all three compasses at this time, and write the approximate angle of deflection beside the compasses labelled A and B. Explain carefully.

When this experiment was performed with one of the bulbs removed from its socket, the single bulb glowed for T seconds.

(b) In the two-bulb circuit shown in Figure 19.54, predict how long the two bulbs would glow (in terms of T). Explain carefully.

(c) Carry out the experiment, observing initial compass deflections and length of time of glow with one or two long bulbs. State your numerical results. Do your observations agree with your predictions?

Problem 19.4 Changing the gap after charging

Figure 19.55 shows a circuit consisting of two flashlight batteries, a large air-gap capacitor, and Nichrome wire. The circuit is allowed to run long enough that the capacitor is fully charged with $+Q$ and $-Q$ on the plates.

Next you push the two plates closer together (but the plates don't touch each other). Describe what happens, and explain why in terms of the fundamental concepts of charge and field (do not use "potential" or "capacitance"). Include diagrams showing charge and field.

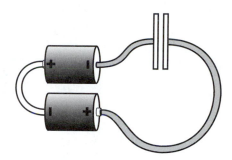

Figure 19.55 Charge the capacitor, then push the capacitor plates closer together (Problem 19.4).

Problem 19.5 Power in a circuit

A battery with negligible internal resistance is connected to a resistor. The power produced in the battery and the power dissipated in the resistor both are P_1.

Another resistor of the same kind is added, so the circuit consists of a battery and two resistors in series.

(a) In terms of P_1, now how much power is dissipated in the first resistor?

(b) In terms of P_1, now how much power is produced in the battery?

(c) The circuit is rearranged so that the two resistors are in parallel rather than in series. In terms of P_1, now how much power is produced in the battery?

Problem 19.6 Fundamentals of a capacitor circuit

An isolated capacitor consisting of two large metal disks separated by a very thin sheet of plastic initially has equal and opposite charges $+Q$ and $-Q$ on the two disks. Then we connect a long thin Nichrome wire to the capacitor as shown below. Describe and explain in detail what happens, including the first few nanoseconds, the approach to the final state, and the final state. Base your explanation on the fundamental concepts of charge and field; do not use the concept of potential in your explanation. To allow room on which to draw, make your diagram like Figure 19.56, with the thickness of the capacitor and the wire greatly exaggerated.

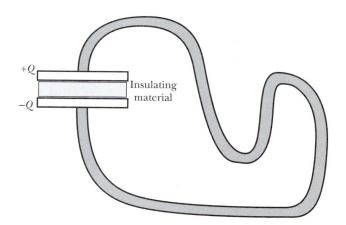

Figure 19.56 A capacitor and a Nichrome wire (Problem 19.6).

Problem 19.7 Capacitor circuits

The charge on an isolated capacitor does not change when a sheet of glass is inserted between the capacitor plates, and we find that the potential difference decreases (because the electric field inside the insulator is reduced by a factor of $1/K$). Suppose instead that the capacitor is connected to a battery, so that the battery tries to maintain a fixed potential difference across the capacitor.

(a) A light bulb and an air-gap capacitor of capacitance C are connected in series to a battery with known emf. What is the final charge Q on the positive plate of the capacitor?

(b) After fully charging the capacitor, a sheet of plastic whose dielectric constant is K is inserted into the capacitor and fills the gap. Does any current run through the light bulb? Why? What is the final charge on the positive plate of the capacitor?

Problem 19.8 Remove plastic slab from capacitor

A capacitor with a slab of glass between the plates is connected to a battery by Nichrome wires and allowed to charge completely. Then the slab of glass is removed. Describe and explain what happens. Include diagrams. If you give a direction for a current, state whether you are describing electron current or conventional current.

Problem 19.9 Insert plastic into a capacitor

A resistor with resistance R and an air-gap capacitor of capacitance C are connected in series to a battery (whose strength is "emf").

(a) What is the final charge on the positive plate of the capacitor?

(b) After fully charging the capacitor (so there is no current), a sheet of plastic whose dielectric constant is K is inserted into the capacitor and fills the gap. Explain why a current starts running in the circuit. You can base your explanation either on electric field or on electric potential, whichever you prefer.

(c) What is the initial current through the resistor just after inserting the sheet of plastic?

(d) What is the final charge on the positive plate of the capacitor after inserting the plastic?

Problem 19.10 A battery with internal resistance

(a) You short-circuit a 9-volt battery by connecting a short wire from one end of the battery to the other end. If the current in the short circuit is measured to be 18 amperes, what is the internal resistance of the battery?

(b) What is the power generated by the battery?

(c) How much energy is dissipated in the internal resistance every second? (Remember that one watt is one joule per second.)

(d) This same battery is now connected to a 10-ohm resistor. How much current flows through this resistor?

(e) How much power is dissipated in the 10-ohm resistor?

(f) The leads to a voltmeter are placed at the two ends of the battery of this circuit containing the 10-ohm resistor. What does the meter read?

Problem 19.11 Establishing a potential difference

The deflection plates in an oscilloscope are 10 cm by 2 cm with a gap distance of 1 mm. A 100-volt potential difference is suddenly applied to the initially uncharged plates through a 1000-ohm resistor in series with the deflection plates. How long does it take for the potential difference between the deflection plates to reach 95 volts?

Problem 19.12 Three bulbs

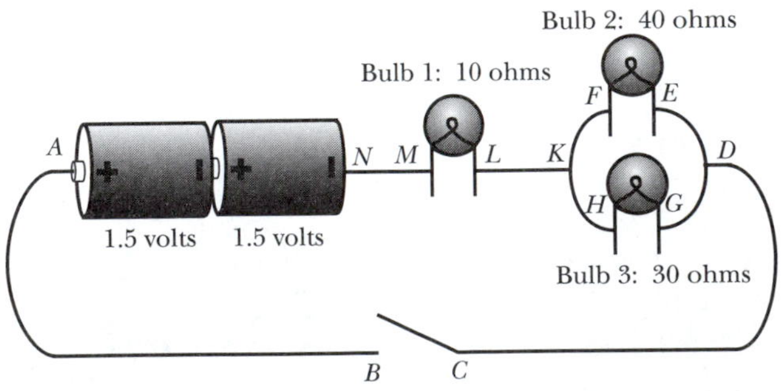

A circuit is made of two 1.5-volt batteries and three light bulbs as shown in Figure 19.57. When the switch is closed and the bulbs are glowing, bulb 1 has a resistance of 10 ohms, bulb 2 has a resistance of 40 ohms, bulb 3 has a resistance of 30 ohms, and the copper connecting wires have negligible resistance. You can also neglect the internal resistance of the batteries.

(a) With the switch open, indicate the approximate surface charge with +'s and −'s on the circuit diagram.

(b) With the switch open, find the potential differences $V_B - V_C$ and $V_D - V_K$.

(c) After the switch is closed and the steady state is established, the currents through bulbs 1, 2, and 3 are I_1, I_2, and I_3 respectively. Write loop and node equations that could be solved to determine these three unknown currents, but do not solve the equations. Label on the diagram what current directions, loops, and nodes you are using, and explain which equation is which. In order to learn about the general approach, do not use formulas

Figure 19.57 Three bulbs (Problem 19.12).

for series and parallel resistance in this problem. (You can use these formulas to check your work.)

(d) In terms of the unknown currents I_1, I_2, and I_3, what is the potential difference $V_C - V_F$ (with the switch closed)?

(e) In terms of the unknown currents I_1, I_2, and I_3, how much power is delivered by the batteries (with the switch closed)?

(f) Now, to see what's involved in a full solution, show how to solve your equations and show that, to the nearest milliampere, $I_1 = 0.111$ A, $I_2 = 0.047$ A, and $I_3 = 0.063$ A.

(g) How many electrons leave the battery at location N every second?

(h) What is the numerical value of $V_C - V_F$?

(i) What is the numerical value of the power delivered by the batteries?

(j) The tungsten filament in the 40-ohm bulb is 8 mm long and has a cross-sectional area of 2×10^{-10} m^2. How big is the electric field inside this metal filament?

Problem 19.13 Multiple loops

Figure 19.58 shows a circuit consisting of two batteries (with negligible internal resistance), six ohmic resistors, and connecting wires that have negligible resistance. Unknown currents I_1, I_2, I_3, I_4, I_5, and I_6 have their directions marked on the circuit diagram.

(a) Write down a set of equations that could be solved for the six unknown currents. Be sure to explain fully how you get these equations!

When a correct set of equations is solved, the currents are as follows (to the nearest milliampere), and you should use these values of the currents to get numerical results in the rest of this problem:

$I_1 = 0.440$ ampere $I_2 = 0.332$ ampere $I_3 = 0.005$ ampere
$I_4 = 0.109$ ampere $I_5 = 0.326$ ampere $I_6 = 0.435$ ampere

(b) Show by actual substitution that these values of the currents satisfy your equations.

(c) Suppose you connect the negative lead of a voltmeter to location G and the positive lead of the voltmeter to location C. What does the voltmeter read, including both magnitude and sign?

(d) What is the power output of the 5-volt battery?

(e) The 12-ohm resistor is made of a very thin metal wire that is 3 mm long, with a diameter of 0.1 mm. What is the electric field inside this metal resistor?

Problem 19.14 A multiple loop circuit

Figure 19.59 shows a circuit containing two batteries (with negligible internal resistance) and five ohmic resistors. The connecting wires have negligible resistance. The letters A through H are shown to make it possible to refer to specific parts of the circuit.

(a) Write all the equations necessary to solve for the unknown currents I_1, I_2, I_3, I_4, and I_5, whose directions are indicated on the circuit diagram. Do not solve the equations, but do explain very clearly what your equations are based on, and to what they refer.

Assume that a computer program has solved your equations in terms of the known battery voltages and known resistances, so that the currents I_1, I_2, I_3, I_4, and I_5 are known.

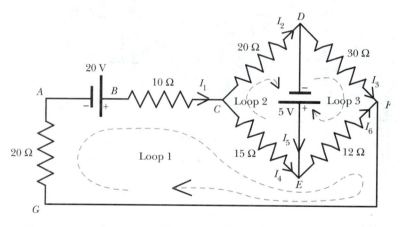

Figure 19.58 Multiple loops (Problem 19.13).

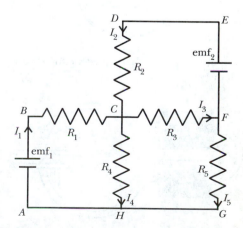

Figure 19.59 A multiple loop circuit (Problem 19.14).

(b) In terms of the known quantities, calculate $V_D - V_A$, and check that your sign makes sense.

(c) In terms of the known quantities, calculate the power produced in battery number 2.

Problem 19.15 Two resistors

In circuit 1 (Figure 19.60), ohmic resistor R_1 dissipates 5 watts; in circuit 2, ohmic resistor R_2 dissipates 20 watts. The wires and batteries have negligible resistance. The circuits contain 10V batteries.

(a) What is the resistance of R_1 and of R_2?

(b) Resistor R_1 is made of a very thin metal wire that is 3 mm long, with a diameter of 0.1 mm. What is the electric field inside this metal resistor?

(c) The same resistors are used to construct circuit 3, using the same 10V battery as before. Make a complete accurate graph of potential for circuit 3. Label the y axis numerically. Be sure to explain the significant features of your graph.

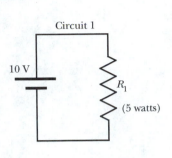

Circuit 1

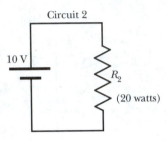

Circuit 2

Figure 19.60 (Two circuits (Problem 19.15).

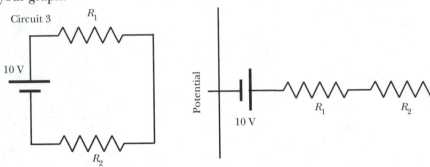

Circuit 3

Figure 19.61 A two-resistor circuit (Problem 19.15c).

(d) On a copy of the diagram of circuit 3, place +'s and −'s to indicate the distribution of surface charge everywhere on the circuit. (No explanation is needed for this part d.)

(e) In circuit 3, calculate the number of electrons entering R_1 every second, and the number of electrons entering R_2 every second.

(f) What is the power output of the battery in circuit 3?

Problem 19.16 Capacitance of a spherical capacitor

Figure 19.62 shows a spherical metal shell of radius r_1 which has a charge Q (on its outer surface) and which is surrounded by a concentric spherical metal shell of radius r_2 which has a charge $-Q$ (on its inner surface).

(a) Use the definition of capacitance: $Q = C|\Delta V|$ to find the capacitance of this spherical capacitor.

(b) If the radii of the spherical shells r_1 and r_2 are large and nearly equal to each other, show that C can be written as $\varepsilon_0 A/s$ (which is also the formula for the capacitance of a parallel-plate capacitor) where $A = 4\pi r^2$ is the surface area of one of the spheres, and s is the small gap distance between them $(r_2 = r_1 + s)$.

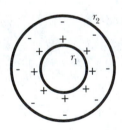

Figure 19.62 A spherical capacitor (Problem 19.16).

Problem 19.17 Pull capacitor plates apart

A capacitor is connected to batteries by Nichrome wires and allowed to charge completely (Figure 19.63). Then the plates are suddenly moved farther apart. Describe what happens and explain in detail why it happens, based on fundamental physical principles. If you give a direction for a current, state whether you are describing electron current or conventional current. Include appropriate diagrams to support your explanation.

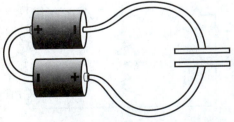

Figure 19.63 Pull plates apart (Problem 19.17).

Problem 19.18 Discharging a capacitor

Two of your 1.5-V batteries in series were connected to a capacitor consisting of two very large metal disks placed very close to each other, so that the disks became charged (left disk positive). The batteries were then removed. The capacitor has properties similar to those of the capacitor you used in class.

Next, thick copper wires and a round bulb were attached to the charged capacitor as shown in Figure 19.64. Two magnetic compasses lying on top of the copper wires initially pointed North. (When this round bulb was connected directly to the two batteries in series we got a compass deflection of 15°.)

(a) Briefly describe in detail what you would observe about the bulb and both compasses during the next minute. You do not need to explain yet why this happens-just describe in detail what you would see.

(b) At a time that is 0.01 second after assembling this circuit, draw and label the electric field at the 8 locations marked × on the diagram. Pay attention to the relative magnitudes of your vectors. Make sure you have clearly labeled your electric field vectors (for example, E_{bulb}, etc.).

(c) Sketch roughly a possible charge distribution on the diagram. If necessary for clarity, add comments on the diagram about the charge distribution.

(d) The length of the left copper wire is L, the length of the right copper wire is also L, the length of the bulb filament is L_{bulb}, and the gap between the capacitor plates is s. The diameter of the copper wires is d, and the diameter of the bulb filament is d_{bulb}. The mobility in the copper is u, and the mobility in the hot tungsten is u_{bulb}. The number of mobile electrons per unit volume is approximately the same in the copper and the tungsten. At a time that is 0.01 second after assembling this circuit, write two valid physics equations involving the magnitudes of the electric fields, using the labels that you gave for the electric fields in part (b). What are the principles underlying these physics equations?

(e) In part (a) you described what you would see happening with the bulb and compasses over the course of a minute. Now explain qualitatively but in detail why this happens. Be brief but be complete in your explanation.

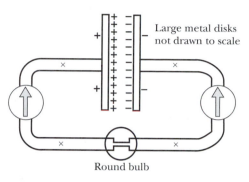

Figure 19.64 Discharging a capacitor (Problem 19.18).

Problem 19.19 A hot slab

A long iron slab of width w and height h emerges from a furnace, as shown in Figure 19.65. Because the end of the slab near the furnace is hot, and the other end is cold, the electron mobility increases significantly with the distance x.

The electron mobility is $u = u_0 + kx$, where u_0 is the mobility of the iron at the hot end of the slab. There are n iron atoms per cubic meter, and each atom contributes one electron to the sea of mobile electrons (we can neglect the small thermal expansion of the iron). A steady-state conventional current runs through the slab from the hot end toward the cold end, and an ammeter (not shown) measures the current to have a magnitude I in amperes. A voltmeter is connected to two locations a distance d apart, as shown.

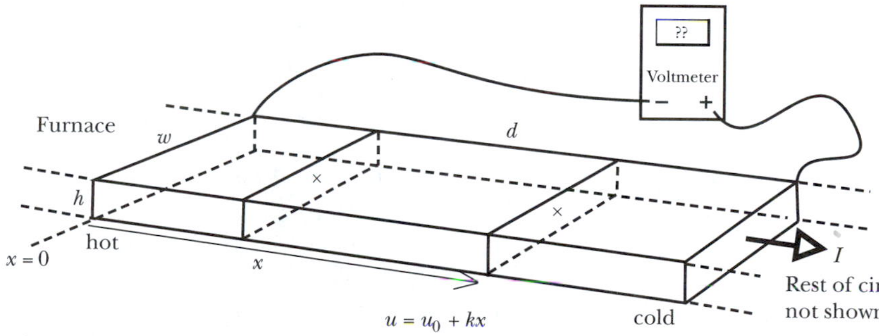

(a) Show the electric field inside the slab at the two locations marked with ×. Pay attention to the relative magnitudes of the two vectors that you draw.

(b) Explain why the magnitude of the electric field is different at these two locations.

(c) At a distance x from the left voltmeter connection, what is the magni-

Figure 19.65 A hot slab emerges from a furnace (Problem 19.19).

tude of the electric field in terms of x and the given quantities w, h, d, u_0, k, n, and I (and fundamental constants)?

(d) What is the sign of the potential difference displayed on the voltmeter? Explain briefly.

(e) In terms of the given quantities w, h, d, u_0, k, n, and I (and fundamental constants), what is the magnitude of the voltmeter reading? Check your work.

(f) What is the resistance of this length of the iron slab?

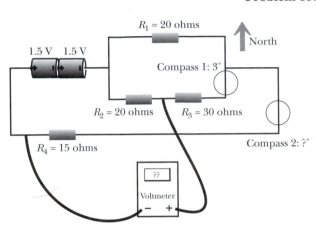

Figure 19.66 A multiresistor circuit (Problem 19.20).

Problem 19.20 A multiresistor circuit

Two flashlight batteries in series power a circuit consisting of four ohmic resistors as shown in Figure 19.66, connected by copper wires with negligible resistance. The internal resistance of the batteries is negligible. Two magnetic compasses are placed underneath the wires, initially pointing northward before closing the circuit. A very-high-resistance voltmeter is connected as shown.

(a) Starting from fundamental principles, write equations that could be solved to determine the conventional current through each resistor, and show the directions of these currents on the diagram. Explain where each of your equations comes from. Do not use formulas for parallel and series resistances (but you are free to use such formulas to check your work if you like). If these equations are solved, we find that the current through R_1 is 0.073 ampere and the current through R_4 is 0.102 ampere.

(b) Sketch the positions of the compass needles. Compass 1 deflects by 3 degrees. What is the deflection angle of compass 2?

(c) The resistors are made of a material which has 8×10^{28} free electrons per cubic meter and a mobility of 3×10^{-5} (m/s)/(N/C). Find E_2, the magnitude of the electric field in resistor R_2, which is in the form of a short wire with a constant cross-sectional area of 6×10^{-10} m^2.

(d) Is the field E_1 in resistor R_1 larger, smaller, or the same as E_2?

(e) What is the length of resistor R_2?

(f) What does the voltmeter read, including sign?

(g) How much energy in joules is expended by the batteries in moving a singly-charged ion from one end of one of the batteries to the other end of that same battery?

(h) How much power output is there from one of the batteries?

Problem 19.21 Circuit with meter

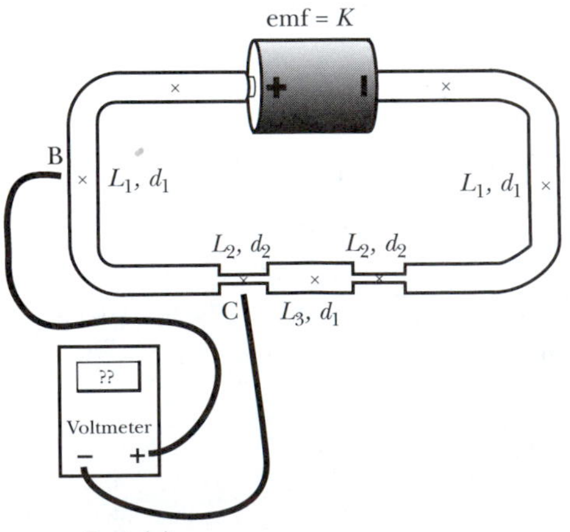

Figure 19.67 shows a circuit consisting of a battery, whose emf is K, and five Nichrome wires, three thick and two thin as shown. The thicknesses of the wires have been exaggerated in order to give you room to draw inside the wires. The internal resistance of the battery is negligible compared to the resistance of the wires. The voltmeter is not attached until part (e) of the problem.)

(a) Draw and label appropriately the electric field at the locations marked $\times$ inside the wires, paying attention to appropriate relative magnitudes of the vectors that you draw.

(b) Show the approximate distribution of charges for this circuit. Make the important aspects of the charge distribution very clear in your drawing, supplementing your diagram if necessary with very brief written descriptions on the diagram. Make sure that parts (a) and (b) of this problem are consistent with each other.

(c) Assume that you know the mobile-electron density n and the electron

emf = K

L_1, d_1 L_1, d_1
L_2, d_2 L_2, d_2
C L_3, d_1
B

??
Voltmeter
−   +

Part (e)

Figure 19.67 A circuit with Nichrome wires of different thicknesses and a voltmeter (Problem 19.21).

mobility u at room temperature for Nichrome. The lengths (L_1, L_2, L_3) and diameters (d_1, d_2) of the wires are given on the diagram. Calculate accurately the number of electrons that leave the negative end of the battery every second. Assume that no part of the circuit gets very hot. Express your result in terms of the given quantities $(K, L_1, L_2, L_3, d_1, d_2, n,$ and $u)$. Explain your work and identify what principles you are using.

(d) In the case that $d_2 \ll d_1$, what is the *approximate* number of electrons that leave the negative end of the battery every second?

(e) A voltmeter is attached to the circuit with its + lead connected to location B (halfway along the left-most thick wire) and its − lead connected to location C (halfway along the left-most thin wire). In the case that $d_2 \ll d_1$, what is the approximate voltage shown on the voltmeter, including sign? Express your result in terms of the given quantities $(K, L_1, L_2, L_3, d_1, d_2, n,$ and $u)$.

Problem 19.22 Capacitor and voltmeter

A capacitor consists of two rectangular metal plates 3 meters by 4 meters, placed a distance 2.5 mm apart in air, as shown in Figure 19.68. The capacitor is connected to a 9-volt battery long enough to charge the capacitor fully, and then the battery is removed.

(a) Prove that there will not be a spark in the air between the plates.

(b) How much charge is on the positive plate of the capacitor?

With the battery still disconnected, you insert a slab of plastic 3 meters by 4 meters by 1 mm between the plates, next to the positive plate. This plastic has a dielectric constant of 5.

(c) After inserting the plastic, you connect a voltmeter to the capacitor. What is the initial reading of the voltmeter?

(d) The voltmeter has a resistance of 100 megohms (10^8 ohms). What does the voltmeter read 3 seconds after being connected?

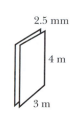

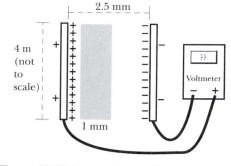

Figure 19.68 A capacitor and a voltmeter (Problem 19.22).

Problem 19.23 A bent bar

A 40-cm-long high-resistance wire with rectangular cross section 7 mm by 3 mm is connected to a 12-volt battery through an ammeter, as shown in Figure 19.69. The resistance of the wire is 50 ohms. The resistance of the ammeter and the internal resistance of the battery can be considered to be negligibly small compared to the resistance of the wire.

Leads to a high-resistance voltmeter are connected as shown, with the − lead connected to the inner edge of the wire, at the top (location A), and the + lead connected to the outer edge of the wire, at the bottom (location C). The distance along the wire between voltmeter connections is 5 cm.

(a) On a copy of the diagram, show the approximate distribution of charge.

(b) On a copy of the diagram, draw the electric field inside the wire at the 3 locations marked ×.

(c) What is the magnitude of the electric field at location B?

(d) What does the voltmeter read, both magnitude and sign?

(e) What does the ammeter read, both magnitude and sign?

(f) In a 60-second period, how many electrons are released from the − end of the battery?

(g) There are 1.5×10^{26} free electrons per cubic meter in the wire. What is the drift speed v of the electrons in the wire?

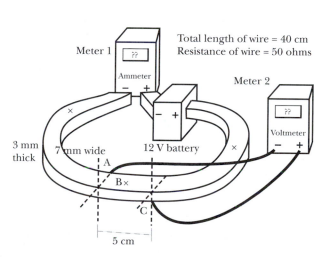

Figure 19.69 A bent bar (Problem 19.23).

(h) What is the mobility u of the material that the wire is made of?

(i) Switch meter 1 from being an ammeter to being a voltmeter. Now what do the two meters read?

(j) The 12-volt battery is removed from the circuit and both the ammeter and voltmeter are connected in parallel to the battery. The voltmeter reads 1.8 volts, and the ammeter reads 20.4 amperes. What is the internal resistance of the battery?

19.26 Answers to exercises

19.1 (page 669)

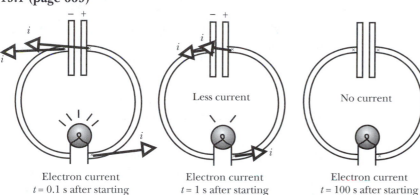

Electron current
$t = 0.1$ s after starting
(initial brightness)

Electron current
$t = 1$ s after starting
(medium brightness)

Electron current
$t = 100$ s after starting
(bulb is not glowing)

19.2 (page 669)

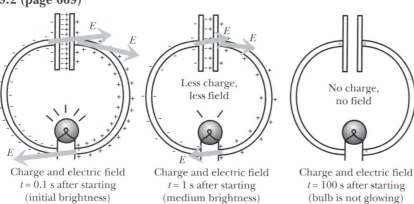

Charge and electric field
$t = 0.1$ s after starting
(initial brightness)

Charge and electric field
$t = 1$ s after starting
(medium brightness)

Charge and electric field
$t = 100$ s after starting
(bulb is not glowing)

19.3 (page 669) The current runs in a direction that reduces the charge on the plates. As the charge on the plates decreases, the magnitude of the electric field in the wires decreases. This produces a slower drift speed for the electrons, and hence the current is decreased. Less current gives less brightness.

19.4 (page 669) The current is the same at all locations in the circuit, so the number of electrons flowing onto one plate during charging must equal the number of electrons flowing off the other plate. So at any moment the amount of negative charge on the negative plate equals the amount of positive charge on the positive plate.

19.5 (page 669) No current. Charges cannot jump from one plate to the other.

19.6 (page 669) The compass was deflected (a little bit) for several seconds after the bulb stopped glowing. Apparently current is still flowing, but it is too little to light the bulb.

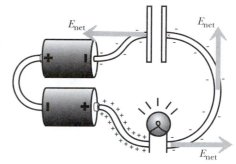

$t = 0.1$ s after starting
(initial brightness)

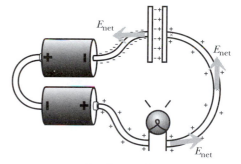

$t = 1$ s after starting
(medium brightness)

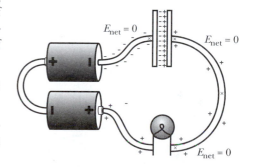

$t = 100$ s after starting
(bulb is not glowing)

Figure 19.70 Three stages in charging the capacitor through a bulb.

19.7 (page 670)   At $t = 0.1$ s, there is hardly any charge on the capacitor, and the net electric field is due mainly to the battery and surface charges (Figure 19.70).

At $t = 1$ s, there is some charge on the capacitor. The net field is smaller now due to the opposing fringe field of the capacitor, so the current is smaller.

Finally there is so much charge on the capacitor plates that the fringe field cancels the fields due to the battery and surface charges. The net field is zero, and there is no current.

19.8 (page 671)   Consider the very first short time interval—say 0.01 second. At the beginning of this interval both capacitors are have very little charge. The electric field in the wires is due almost entirely to the battery and surface charge on the wires, and is the same in both cases. Thus the number of electrons flowing onto the left plate and off of the right plate is approximately the same in the first 0.01 second.

At the end of this time interval, both capacitors have nearly the same small amount of charge, Q. This is the situation illustrated in Figure 19.71.

The fringe field of the smaller capacitor is this:

$$E_{\text{fringe},1} \approx \frac{Q/A}{\varepsilon_0}\left(\frac{s}{2R_1}\right) = \frac{1}{\varepsilon_0}\frac{Q}{\pi R_1^2}\left(\frac{s}{2R_1}\right)$$

The fringe field of the larger capacitor is smaller, since $R_2 > R_1$:

$$\left[E_{\text{fringe},2} = \frac{1}{\varepsilon_0}\frac{Q}{\pi R_2^2}\left(\frac{s}{2R_2}\right)\right] < E_{\text{fringe},1}$$

The smaller fringe field of the large capacitor means a larger *net* field, and larger drift speed. So the current flowing onto the large capacitor has decreased less than the current flowing onto the small capacitor.

Comment on the final charge: As we will see in section 19.18, it turns out that the electron current in a capacitor circuit decreases exponentially with time as shown in the accompanying graph (Figure 19.72). The area under a curve of electron current vs. time is the number of electrons deposited on the negative plate of the capacitor. You can see that a slow decrease of current is associated with a larger final charge being deposited on the plate, because the area under the curve is larger. Note that the two curves refer to *different* capacitors but the *same* bulb, so that the initial current is the same in both cases. This is not the same as your experiments with the same capacitor and different bulbs, where the initial current differs.

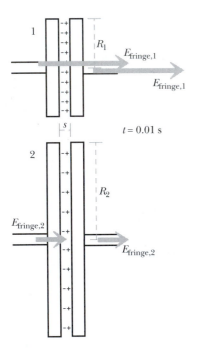

Figure 19.71 At $t = 0.01$ s both capacitors have similar charge but different E_{fringe}.

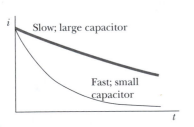

Figure 19.72 Electron current vs. time for two different capacitors.

19.9 (page 672) The analysis here is the same as for a capacitor with bigger plates.

After 0.01 s, both capacitors have nearly the same small amount of charge, Q. This is the situation illustrated in Figure 19.73.

Now the fringe field of the thinner capacitor is smaller, because $s_2 < s_1$.

The smaller fringe field of the thinner capacitor means a larger *net* field, and larger drift speed. So the current flowing onto the thinner capacitor has decreased less than the current flowing onto the thicker capacitor.

19.10 (page 672) Again we have to look at the electric field in the wire next to a plate of the capacitor.

After 0.01 s, both capacitors have nearly the same small amount of charge, Q. This is the situation illustrated in Figure 19.74.

The field of the polarized plastic is opposite in direction to E_{fringe}, so the field due to the battery and surface charges is reduced by a smaller amount. The net field in the wire is larger than it is without the plastic. So the current flowing onto the capacitor containing the plastic has decreased less than the current flowing onto the capacitor with the air gap.

19.11 (page 673) A metal is more easily polarized than any other material, because the mobile electrons are quite free to move within the metal. So the current would change least quickly with metal between the plates.

19.12 (page 674) The battery loses energy, the bulb neither loses nor gains (it wasn't glowing originally, and it isn't glowing afterwards either), the capacitor gains energy (enough to power a bulb later), and the surroundings gain energy (light and heat transfer from the bulb).

19.13 (page 674) Since the bulb glows the same in both charging and discharging, it must be that the surroundings get an increase $+E$ in both cases. The energy of the bulb stays the same in both cases, because it is dark to begin with and dark at the end. The battery isn't in the discharging circuit, so it must be that the energy change in the capacitor during discharging must be $-E$, to supply $+E$ to the surroundings. So the capacitor must have gained $+E$ energy during charging, and the battery had to lose $2E$, half of which was stored in the capacitor, and half went to the surroundings. Here is the summary:

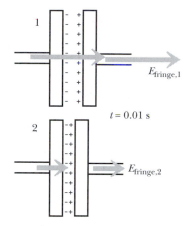

Figure 19.73 At $t = 0.01$ s both capacitors have similar charge but different E_{fringe}.

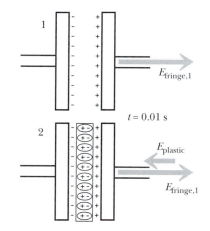

Figure 19.74 The field due to the polarized insulator opposes the fringe field.

	Charging	Discharging
Battery:	$-2E$	0
Bulb:	0	0
Capacitor:	$+E$	$-E$
Surroundings:	$+E$	$+E$

19.14 (page 676): 5.4×10^{-9} farad

19.15 (page 676): $C = K(\varepsilon_0 A)/s$

19.16 (page 676):   0.5 F: 1.5 coulomb; 9.4×10^{18} electrons

                          1 F: 3 coulomb; 1.9×10^{19} electrons

19.17 (page 678): $\sigma = 6 \times 10^7 (\text{ampere/m}^2) / (\text{volt/m})$

19.18 (page 678):   0.005 volt/m

19.19 (page 678):   424 volts/m

19.20 (page 680):   0.83 Ω; 3.6 amperes

19.21 (page 680): 4.3×10^{-4} Ω; 15.0015 volts

19.22 (page 681):   30 Ω; 19 Ω; 8 Ω; clearly not ohmic

19.23 (page 682):   No; double the current with the same voltage

19.24 (page 683)   240 ohms; 280 ohms

19.25 (page 684)   7.5 ohms; 0.4 ampere

19.26 (page 684):   4.5 watts

19.27 (page 684): $\text{Power} = I\Delta V = I(RI) = RI^2 = \left(\dfrac{\Delta V}{R}\right)\Delta V = \dfrac{(\Delta V)^2}{R}$

19.29 (page 687):   0; 0; 0

19.30 (page 687): $(\text{emf} - rI) + (\text{emf} - rI) = 0$, so $I = \dfrac{2\,\text{emf}}{2r} = \dfrac{\text{emf}}{r}$

19.31 (page 687): $\text{emf} - rI_1 = 0$; $\text{emf} - rI_2 = 0$; $I = I_1 + I_2$

$$I = \frac{2\,\text{emf}}{r}$$

19.32 (page 688):   Figure 19.75

19.33 (page 691):   Voltmeter has very high resistance, so current extremely small; reads 3 volts (+3–3 = 0 round trip potential diff.)

19.34 (page 696) $V_C - V_B = -8$ volts

                          Power dissipated in R_1 = 3.2 watts

                          Battery output power = 4.8 watts

19.35 (page 697): $\dfrac{q}{m} = \dfrac{aL}{\Delta V}$

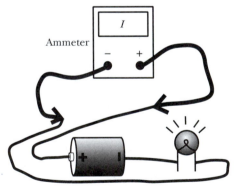

Ammeter

Figure 19.75 How to connect the circuit to get a + reading on the ammeter (Ex. 19.32).

Chapter 20

Magnetic Force

Chapter 20

Magnetic Force

The study of magnetic forces—the effect of magnetic fields on moving charges—brings together many of the ideas we have encountered in the previous three chapters. In our study of the Hall effect we will see that magnetic forces can allow us to determine experimentally the sign of the mobile charges in a particular material. We will be able to determine the effect of a magnetic field on a current-carrying wire simply by adding up the effects on the individual moving charges that make up the current. Once we understand the magnetic force on a wire, we will be able to understand the torque that a magnetic field applies to a coil of wire (which is closely related to the torque that a magnetic field applies to a magnet such as a compass needle), and also to understand the behavior of atoms in magnetic fields. Examining magnetic forces in different reference frames will allow us to learn more about the connection between electric and magnetic fields.

Review: magnetic field of a moving charge

Recall from Chapter 17 that the magnetic field associated with a moving charge can be calculated from the Biot-Savart law:

$$\vec{B} = \frac{\mu_0}{4\pi} \frac{q\vec{v} \times \hat{r}}{r^2}$$

From this law and the definition of current we derived an expression for the magnetic field contributed by a small current element:

$$\Delta\vec{B} = \frac{\mu_0}{4\pi} \frac{I\Delta\vec{l} \times \hat{r}}{r^2}$$

and we were able to integrate this form of the Biot-Savart law to predict the magnetic field of various configurations of currents, such as straight wires, loops, and solenoids.

In this chapter we will study the effects of magnetic fields. Often our analysis will require two steps: first, find the field at a location made by a moving charge or a current, and second, find the effect of this field on a different moving charge. Just as the electric field made by a charge does not affect the source charge itself, so the magnetic field made by a moving charge affects other moving charges, not the source charge.

20.1 Magnetic force on a moving charge

In oscilloscopes used for precision scientific work, the electrons are accelerated by a horizontal electric field (Problem 16.3 on page 574), and the moving electrons are deflected vertically by a vertical electric field (Problem 16.14 on page 576). In the picture tubes found in home television sets and CRT computer monitors the electrons are accelerated in the same way, but the moving electrons are deflected vertically by a magnetic field. This magnetic field is horizontal and is produced by currents in coils positioned as shown in Figure 20.1 (not shown are other coils whose magnetic field deflects the electrons from side to side).

There is a rather complicated geometrical relationship between the direction of the magnetic force on a moving charge and

- the direction of the applied magnetic field,
- the direction of the velocity of the moving charge, and
- the sign of the moving charge.

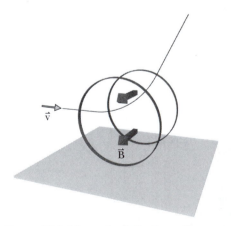

Figure 20.1 Magnetic deflection of a moving electron.

Experiments in which these parameters are systematically varied show that the magnetic force on a moving charge is given by the following important formula, involving a cross product:

MAGNETIC FORCE

$$\vec{F}_{magnetic} = q\vec{v} \times \vec{B}$$

q is the charge (including sign) of the moving charge.
$\vec{v}$ is the velocity of the moving charge.
$\vec{B}$ is the applied magnetic field (in tesla).

Since $\vec{F}_{magnetic} = q\vec{v} \times \vec{B}$, we can see that a tesla has the units of F/qv:

$$T = \frac{N}{C\cdot m/s} = \frac{N}{A\cdot m}$$

Use the right-hand rule for cross products to verify that the formula for magnetic force gives the observed direction of magnetic force in Figure 20.1, and in the side view shown in Figure 20.2. Remember that the electron charge $q = -e$, which means that the magnetic force on a moving electron is in the opposite direction to the direction of the cross product $\vec{v} \times \vec{B}$.

The formula $\vec{F} = q\vec{v} \times \vec{B}$ summarizes the experimental observations of the effect of a magnetic field on a moving charge.

? What do these observations say about the effect that a magnetic field has on a charge that is not moving? How does this compare with the effect that an electric field has on a charge that is not moving?

If the charge is not moving, the magnetic field has no effect, whereas electric fields affect charges even if they are at rest.

Magnetic force cannot change a particle's speed

Since the magnetic force on a moving charged particle acts at a right angle to the velocity of the particle, it is always true that the quantity $\vec{F} \cdot d\vec{l} = 0$. Thus a magnetic force cannot change the kinetic energy of a charged particle—it can change the direction of the particle's momentum, but not the magnitude of its momentum (or velocity).

Magnitude of the magnetic force

According to the formula $\vec{F}_{magnetic} = q\vec{v} \times \vec{B}$, there should be a magnetic force on a charged piece of invisible tape when you move it near a magnet. However, the magnitude of this magnetic force is extremely small compared to the familiar electric force acting between the tape and the magnet (due to the charged tape polarizing the metal).

The magnetic force on a single moving electron in a television tube is significant because the particle's acceleration is proportional to the ratio q/m:

$$\left|\frac{d\vec{p}}{dt}\right| = qvB\sin\theta$$

$$\left|\frac{d\vec{v}}{dt}\right| = \frac{q}{m}vB\sin\theta \quad (\text{for } v \ll c)$$

The q/m ratio is enormous for a single electron:

$$(1.6\times10^{-19}\text{ C})/(9\times10^{-31}\text{ kg}) \approx 10^{11}\text{ C/kg}$$

But q/m is very small for a charged tape:

$$(10^{-8}\text{ C})/(10^{-4}\text{ kg}) \approx 10^{-4}\text{ C/kg}$$

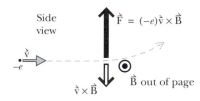

Figure 20.2 Side view of the magnetic deflection of the moving electron.

Magnetic forces on moving objects that have been charged by rubbing will always be tiny, because only an extremely small fraction of the atoms in the object have lost or gained an electronic charge. Moreover, the magnetic force is proportional to the speed of the moving particle, and the speed of electrons in a television tube is very high (nearly 10^8 m/s in Problem 16.3 on page 574).

Ex. 20.1 In Chapter 1 you found that a charged invisible tape has a net charge of about 10^{-8} C. You could move it past a magnet with a speed of about 10 m/sec. You were able to estimate the magnetic field near your bar magnet to be on the order of 0.1 tesla or so (Experiment 17.8, page 604: Dependence on distance for a bar magnet). Compare the approximate magnetic force on a moving invisible tape with the approximate electric force that two charged tapes exert on each other when they are 10 cm apart.

20.1.1 Circular motion in a magnetic field

In a television tube the region of magnetic field that deflects the electrons is mainly confined to the region of the deflection coils, and the trajectory of an electron is nearly straight before entering the magnetic field region and after leaving it (Figure 20.3).

Figure 20.3 Electron trajectory through a limited region of magnetic field.

However, if we have a large enough region of uniform magnetic field, a charge can go around and around in a circle, and the curved portion of the electron trajectory in a television tube is a circular arc (if the magnetic field is fairly uniform). There are many other examples of circular orbits of charged particles in magnetic fields, including particle trajectories in particle accelerators.

Consider a proton moving horizontally with speed v in a large region of uniform vertically upward magnetic field (neglect the much smaller effects of gravity). The magnetic force $\vec{F} = q\vec{v} \times \vec{B}$ is perpendicular to the velocity $\vec{v}$, which means that the magnetic force will change the direction of the velocity without changing the magnitude of the velocity (the speed v), because no work is done by the magnetic force.

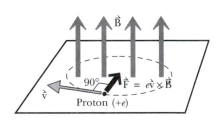

Figure 20.4 Circular motion of a proton in a region of uniform magnetic field.

Circular motion at any speed

This situation is similar to that of a rock going around in a circle at the end of a rope, or the Moon orbiting the Earth in a circular orbit. Circular motion can be most easily analyzed by recalling that for any rotating vector $\vec{X}$ (whose magnitude doesn't change), we have

$$\left|\frac{d\vec{X}}{dt}\right| = \omega|\vec{X}| \quad \text{(rotating vector; } \omega \text{ is angular speed)}$$

The momentum $\vec{p}$ of the particle is a rotating vector, so the momentum principle is this, where ω is the angular speed in radians per second:

$$\left|\frac{d\vec{p}}{dt}\right| = \omega p = \left|q\vec{v} \times \vec{B}\right| = |q|vB\sin 90°$$

$$\omega\frac{mv}{\sqrt{1 - v^2/c^2}} = |q|vB$$

$$\omega = \frac{|q|B}{m}\sqrt{1 - v^2/c^2}$$

Circular motion at low speeds

An important special case is motion at speeds small compared to the speed of light c, in which case the angular speed ω is (approximately) independent of the speed v:

$$\omega \approx \frac{|q|B}{m} \text{ if } v \ll c$$

An alternative derivation of this result starts from the non-relativistic form of the momentum principle:

$$\frac{d\vec{p}}{dt} \approx m\frac{d\vec{v}}{dt} = \vec{F}_{net}$$

The acceleration of a particle in circular motion is $\dfrac{v^2}{R}$, so

$$\frac{mv^2}{R} = |q|vB\sin 90°, \text{ and since } \omega = \frac{v}{R}$$

$$m\omega = |q|B, \text{ or } \omega = \frac{|q|B}{m}$$

Time required for a circular orbit

Let $T =$ the time required for a charged particle to go around once in a complete circular orbit.

? Show that $T = 2\pi\dfrac{m}{|q|B}$ for nonrelativistic speeds ($v \ll c$).

Since by definition the angular speed ω is equal to 2π radians per the amount of time T required to go around once, $\omega = 2\pi / T$, and $\omega \approx |q|B/m$ if $v \ll c$, the result follows. This is an unusual result, because it is independent of the speed v and radius r. It says that the particle will take the same amount of time to go around in a circle in a uniform magnetic field, no matter how fast it is going (as long as $v \ll c$). The faster it goes, the bigger the radius of the circle, but the longer distance around the bigger circle is exactly compensated by the higher speed. This independence of the orbit time is the basis for the cyclotron (see Section 20.1.4 on page 722).

Determining the momentum of a particle

The position vector $\vec{r}$ is a rotating vector, and we have

$$\left|\frac{d\vec{r}}{dt}\right| = v = \omega r, \text{ so } \omega = \frac{v}{r}$$

Putting this value of ω into the momentum principle, we have this:

$$\left|\frac{d\vec{p}}{dt}\right| = \omega p = \left(\frac{v}{r}\right)p = |q|vB$$

Solving for the magnitude of the momentum p, we have this important result:

$$p = |q|Br, \text{ valid even for relativistic speeds}$$

This result is used to measure momentum in high-energy particle experiments. Particles with known charge q pass through a region of known magnetic field B, and the radius r of the circular trajectory is measured. The momentum of the particle is then known to be $p = |q|Br$, even if the speed of the particle is relativistic (approaching the speed of light).

Ex. 20.2 Suppose a proton has a component of velocity parallel to the magnetic field as well as perpendicular to it (Figure 20.5). What is the effect of the magnetic field on this parallel component of the velocity? What will the trajectory of the proton look like?

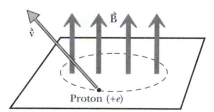

Figure 20.5 What kind of trajectory will the proton follow (Ex. 20.2)?

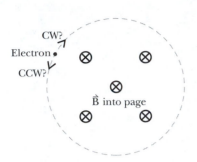

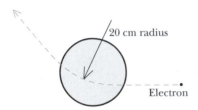

Figure 20.6 Which direction will the electron go in this uniform magnetic field (Ex. 20.3)?

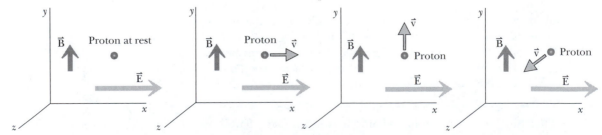

Figure 20.7 How large a magnetic field is needed to get a radius of curvature of 20 cm (Ex. 20.4)? What is the direction of the field?

Ex. 20.3 In Figure 20.6 there is a magnetic field going into the page, produced by a large current-carrying coil (not shown). Would an electron go in a clockwise or counter-clockwise circle? Explain briefly. (Hint: try both directions, and see which direction is consistent with the magnetic force.) Note that we're not asking what magnetic field is made by the moving electron, but what effect the coil's magnetic field has *on* the moving electron.

Ex. 20.4 To become familiar with the order of magnitude of magnetic effects, consider the situation in a typical home television set. In order to bend the electron trajectory so that the electron hits the top of the screen rather than going straight through to the center of the screen, you need a radius of curvature in the magnetic field of about 20 cm (Figure 20.7). If the electrons are accelerated through a 15000-volt potential difference, they have a speed of 0.7×10^8 m/s (Problem 16.3 on page 574). Calculate the magnitude of the magnetic field required to make the electrons hit the top of the screen. Is the magnetic field into or out of the page?

20.1.2 Combination of electric and magnetic forces: The Lorentz force

If a charged particle is moving in a magnetic field, it is subject to a magnetic force $\vec{F}_m = q\vec{v} \times \vec{B}$. If there is an electric field in the region, the particle is also subject to an electric force $\vec{F}_e = q\vec{E}$. The combined effect of electric and magnetic fields is called the "Lorentz force."

THE LORENTZ FORCE

$$\vec{F} = q\vec{E} + q\vec{v} \times \vec{B}$$

In one sense, the terms in this formula are a kind of definition of electric and magnetic fields in terms of the observable effects that these fields have on charges.

Coulomb's law for the electric field of a point particle, and the Biot-Savart law for the magnetic field of a point particle, are adjusted to predict fields that yield the observed forces on other charged particles according to the Lorentz force formula. The constants $1/(4\pi\varepsilon_0)$ and $\mu_0/(4\pi)$ in the equations for electric and magnetic field, together with the definition of the unit of charge, the coulomb, make the field and force equations consistent with each other.

Ex. 20.5 Consider the following situation, in which there is a uniform electric field in the *x* direction and a uniform magnetic field in the *y* direction. For each example of a proton at rest or moving in the *x*, *y*, or *z* direction, what is the direction of the Lorentz force on the proton at this instant?

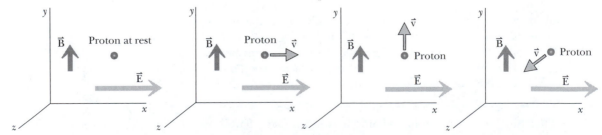

Ex. 20.6 In each of the following diagrams, what is the direction of the Lorentz force on an electron at this instant?

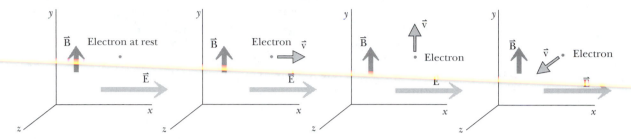

20.1.3 Application: A velocity selector

It is possible to arrange electric and magnetic fields so that the combined electric and magnetic force $\vec{F} = q\vec{E} + q\vec{v} \times \vec{B}$ acting on a positive charge $+q$ moving with velocity $\vec{v}$ is zero.

? With the velocity $\vec{v}$ in the x direction (Figure 20.8), try to sketch a three-dimensional diagram of an arrangement of $\vec{E}$ and $\vec{B}$ that could lead to a net Lorentz force of zero, in which case the particle will move with constant velocity through the region.

Several arrangements are possible, one of which is shown in Figure 20.9.

? If the magnetic field in this situation has a magnitude B, what must be the magnitude E of the electric field?

Since the magnetic and electric forces must be equal and opposite to produce a zero net force,

$$eE = evB\sin 90°$$
$$E = vB$$

? Now suppose a particle with the same charge q comes through this region, traveling in the same direction but at a slower speed. Explain briefly why it will *not* move in a straight line.

Now $qE > qvB$, so the net force isn't zero.

? What if a particle with the correct velocity but with a *negative* charge $-q$ comes through. Will it pass straight through or not? (Make a diagram.)

Yes. Both the electric force and the magnetic force are flipped 180°, and the net force is still zero.

So a region containing crossed electric and magnetic fields (that is, perpendicular to each other) forms a kind of velocity selector, allowing particles to pass straight through only if they have the chosen speed and direction. You might notice that the cross product $\vec{E} \times \vec{B}$ points in the direction of the velocity that can pass straight through.

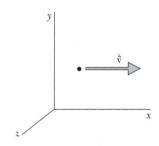

Figure 20.8 Sketch electric and magnetic fields in three-dimensional space so that the Lorentz force on the particle is zero.

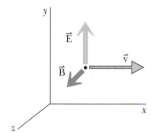

Figure 20.9 One configuration of fields which can allow a charged particle to travel in a straight line.

Ex. 20.7 If an electron travelling at a velocity of <0,0.9*c*,0> m/s passes through a region where the magnetic field is <1.0,0,0> T, what electric field would be necessary to keep the electron travelling in a straight line? Give $\vec{E}$ as a vector.

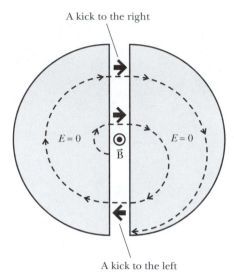

A kick to the right

$E = 0$ $E = 0$

$\vec{B}$

A kick to the left

Figure 20.10 A cyclotron. Magnetic field is uniform in the $+y$ direction; the proton orbits in the x-z plane.

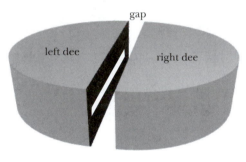

gap

left dee   right dee

Figure 20.11 3D view of a cyclotron.

20.1.4 Application: A cyclotron

One of the principal methods for studying the nucleus of an atom is to shoot high-velocity protons at the nucleus and observe reactions that take place. In order to drive a proton into a nucleus, we have to accelerate the proton through a potential difference of a million volts or more. For example, there is a million-volt potential difference from very far away ("infinity") to the surface of a carbon nucleus ($Q = 6e$), whose radius is about 10^{-14} meters. A compact device for creating such large potential differences is the "cyclotron." Ernest Lawrence invented the idea for a cyclotron in 1929 (see Richard Rhodes, *The Making of the Atomic Bomb*, Simon & Schuster 1986, pp. 145-148; this book contains an excellent history of early 20th-century physics).

A cyclotron consists of two D-shaped metal boxes (called "dees") placed in a strong magnetic field (Figure 20.10 and Figure 20.11). Protons can enter the dees through slits on the flat faces. Near the center of the cyclotron is a source of low-velocity ionized hydrogen (protons). Inside the metal dees there is no electric field (because polarization of the metal box cancels any fields of external charges—see Chapter 21), and a proton follows a circular path in the magnetic field. As a proton goes from the left dee to the right dee, it receives a kick to the right due to a potential difference across the two dees. This increases the proton's momentum, so its path in the right dee has a larger radius of curvature.

While the proton is inside the right dee (which has no electric field inside), we reverse the potential difference across the dees so that when the proton goes from the right dee to the left dee at the bottom of the drawing, it again receives a kick, this time to the left, resulting in a higher momentum and a larger radius of curvature in the left dee. By repeatedly kicking the proton through a modest potential difference, we can accelerate the proton up to very high momentum and energy that would otherwise require a practically unattainable potential difference of millions of volts.

As long as $v \ll c$, the period of a proton's orbit does not change as its speed increases (see page 719), so it is possible to apply a sinusoidally oscillating potential difference $\Delta V = \Delta V_0 \cos(\omega t)$ across the dees to accelerate the proton.

However, if the equivalent accelerating potential is more than about 100 million volts, as it is in modern accelerators, the speed v of a proton becomes a sizable fraction of the speed of light c, and relativistic effects are significant. The proton's momentum is no longer simply $mv = eBR$ but is

$$\frac{mv}{\sqrt{1 - v^2/c^2}} = eBR$$

The period of the motion is no longer independent of the speed, and either the frequency of the accelerating voltage or the strength of the magnetic field must be adjusted during the acceleration to compensate.

A modern example of a circular accelerator is Fermilab, outside Chicago. In the final stage of acceleration in the Fermilab accelerator the protons travel in a circle in a donut-shaped tube having a radius of a kilometer! A smaller accelerator injects protons into this ring with a speed very close to the speed of light, and their speed gets even closer as they are accelerated, varying only slightly from $0.995c$ to $0.9999995c$. The magnetic field is increased during the acceleration to keep the protons at the same radius despite a large change in the factor $1/\sqrt{1 - v^2/c^2}$, which increases from about 10 to about 1000, at which point $B = (mv/eR)/\sqrt{1 - v^2/c^2}$ = about 3 tesla, which is provided by electromagnets containing superconducting coils. The time to go around once is nearly constant as in a cyclotron, but for a completely different reason (nearly constant speed approximately equal to c, at a constant radius of one kilometer).

Homework problems 20.4 and 20.5 deal with the operation of a cyclotron.

20.2 The Hall effect

Mobile electrons moving inside a metal wire can be deflected to the surface of the wire by a magnetic field that is perpendicular to the wire. In Figure 20.12, large coils (not shown) apply a uniform magnetic field of magnitude B throughout the region of the circuit. Verify for yourself that the moving electrons will be deflected toward the bottom surface of the wire (shown with a rectangular cross-section for simplicity). This will make the bottom surface negatively charged and the top surface positively charged (due to a deficiency of electrons). These charges are in addition to the usual surface charges, which for clarity we don't show on the diagram.

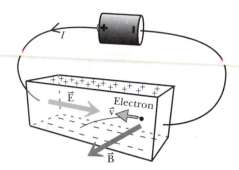

Figure 20.12 Magnetic deflection of electrons in a wire.

As time goes on, more and more electrons will pile up on the bottom of the wire. However, this piling-up won't continue forever. When enough electrons have piled up on the bottom (with a corresponding depletion of electrons on the top surface), these extra surface charges will produce a new electric field that applies a vertical electric force to the moving electrons inside the wire.

? What is the direction of this new electric field?

Inside the metal wire, the new electric field points down, pointing away from the positive charge on the top and towards the negative charge on the bottom. We'll call this the "transverse" electric field, $\vec{E}_\perp$.

? When no more electrons pile up on the bottom, what is the magnitude $E_\perp$ of the transverse electric field, in terms of the drift speed $\bar{v}$ of the free electrons and the magnetic field B? (Hint: If no more electrons are piling up, what do you know about the electric and magnetic forces on the moving electrons?)

Pile up will cease when the electric force grows to be as large as the magnetic force, at which time the net vertical force becomes zero, and the electrons move horizontally through the wire, with no vertical deflection. When this happens we have $eE_\perp = e\bar{v}B\sin 90°$, so $E_\perp = \bar{v}B$. This analysis is very similar to the analysis of the velocity selector on page 721.

? Suppose you attach the leads of a digital voltmeter to the top and bottom surfaces as shown in Figure 20.13. Determine the magnitude and sign of the voltage displayed on the voltmeter (the wire has height h, depth d, and length L; there is a uniform magnetic field of magnitude B).

Since the electric field points down inside the wire, the electric potential is higher on top. The electric field has the uniform value $E_\perp = \bar{v}B$ throughout the wire, so the voltmeter will read a positive value $+E_\perp h = +\bar{v}Bh$.

? How big is this voltage if there is no magnetic field?

Since the potential difference is $\bar{v}Bh$, the voltage will be zero if $B = 0$.

The appearance of a sideways or transverse potential difference across a current-carrying wire in the presence of a magnetic field is called the "Hall effect," named for its discoverer.

The Hall effect and the sign of the charge carriers

The Hall effect offers a very special insight into the nature of the charge carriers responsible for a current. To see why, consider what would happen if *positive* particles (holes in the case of aluminum or zinc) were moving to the *right* in the wire as in Figure 20.14, instead of electrons moving to the left.

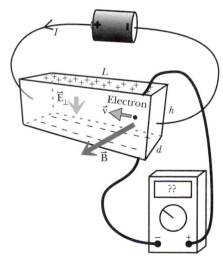

Figure 20.13 Measuring the transverse potential difference. The voltmeter leads are directly across from each other on the top and bottom surfaces.

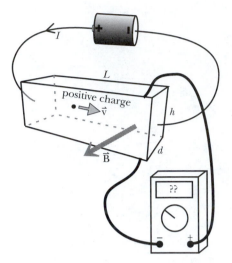

Figure 20.14 What happens if the charge carriers are positive? The voltmeter leads are directly across from each other on the top and bottom surfaces.

? On a diagram like Figure 20.14, show the charge pile-up on the top and bottom surfaces and the electric field due to that pile-up. What would be the sign of the voltage displayed on the voltmeter?

Now it is the bottom of the wire that has additional positive surface charge, rather than the top being positive as it was with negative charge carriers. The transverse electric field now points up, and the voltmeter reads a negative voltage.

Most electrical measurements come out the same whether electrons are moving inside the wire or positive particles (holes in a metal) are moving in the opposite direction. But the Hall effect differs in the two cases.

HALL EFFECT AND THE SIGN OF CHARGE CARRIERS

By measuring the Hall effect for a particular material,
we can determine the sign of the moving particles
that make up the current.

The measured Hall effect for many metals is consistent with the moving particles being negative (electrons). In some metals including aluminum and zinc, and in "p-type" semiconductors, it is positive holes in the electron sea that carry the current.

Let's see how big the Hall effect is in a metal. First we'll do the analysis with algebraic symbols. Then we'll evaluate the result numerically for a particular situation. Consider a bar of metal of length L, height h, and depth d, carrying a conventional current $I = |q| n A \bar{v}$ to the right in the presence of a perpendicular magnetic field B, as shown in Figure 20.13 or Figure 20.14.

? Predict the Hall-effect potential difference ΔV_{Hall} across the bar in terms of $J = I/A$, B, $|q|$, n, and the dimensions of the wire.

The result is $|\Delta V_{Hall}| = |[J/(|q|n)]Bh|$. If we measure the current density $J = I/A$ and the Hall-effect voltage ΔV_{Hall}, and if we know from other measurements the charge q (which is $-e$ for most metals), we can determine n, the number of charge carriers per cubic meter, since n is the only unknown quantity in our expression for ΔV_{Hall}.

The Hall-effect voltage for some monovalent metals such as sodium and copper shows n to be the same as the number of atoms per cubic meter, which shows that each atom gives up one electron into the free-electron sea. In other metals there may be more than one electron contributed per atom.

Ex. 20.8 Calculate the Hall voltage ΔV_{Hall} for the case of a ribbon of copper 5 mm high and 0.1 mm deep, carrying a current of 20 amperes in a magnetic field of 1 tesla. For copper, $q = -e$, and $n = 8.4 \times 10^{28}$ free electrons per m^3 (one per atom).

This is a very small voltage, but it can be measured with a sensitive voltmeter. In semiconductors the density of charge carriers n may be quite small compared to a metal, in which case the Hall-effect voltage can be quite large, since it is inversely proportional to n.

Ex. 20.9 It is instructive to compare this transverse Hall voltage in copper with the voltage you would measure *along* the copper ribbon. Calculate the potential difference you would measure along 5 mm of the ribbon; the conductivity of copper is:

$$\sigma = 6 \times 10^7 \frac{A/m^2}{volt/m}$$

Since this voltage is much larger than the Hall voltage, experimenters must take special precautions to place their voltmeter leads exactly across from each other, to avoid a spurious reading due to the potential difference along the bar. One way to check for this problem is to remove the magnetic field, in which case the Hall voltage should go to zero.

Summary of Hall effect

In the steady state, the Hall effect is an example of the straight line motion of charged particles through crossed electric and magnetic fields. Measuring the Hall effect for a material gives us valuable information about the sign of the charge carriers and about the number of them per unit volume. To give a striking example of its use, the Hall effect has confirmed that in some metals and semiconductors, it is positive holes in the electron sea rather than the electrons themselves that are the dominant charge carriers.

A very practical use of the Hall effect turns the relationship around. There are commercial devices that measure the magnitude and direction of an unknown magnetic field by observing the Hall voltage in a material in which the sign and number density of the mobile charges is known.

Remember that the surface charges which accumulate due to the Hall effect are not the only charges on the wires; there are also the usual surface charges responsible for the longitudinal electric field in the wires. The distinction is that the ordinary surface charges produce electric fields inside the wire in the direction of the conventional current, whereas the Hall-effect surface charges produce a transverse electric field $\vec{E}_{\perp}$, perpendicular to the current.

20.3 Magnetic force on a current-carrying wire

Often we are interested in the magnetic force exerted on a large number of charges moving through a wire in a circuit. According to the superposition principle, we can simply add up the magnetic force contributions on the individual charges. Because the net force on a system (the wire) is the sum of all the forces acting on its individual components, adding up these forces will give the net force on the entire wire.

Let's calculate the magnetic force on a bunch of moving positive charges contained in a small volume with length Δl and cross-sectional area A (Figure 20.15). Their average drift velocity is $\vec{v}$. If there are n moving charges per unit volume, there are $nA\Delta l$ moving charges in this small volume.

n charges/m^3

Figure 20.15 Calculate the magnetic force on a short current-carrying wire.

? Show that the net magnetic force on all the moving charges in this short length of wire is

$$\Delta\vec{F}_{\text{magnetic}} = I\Delta\vec{l} \times \vec{B}$$

where $\Delta\vec{l}$ is a vector with magnitude Δl, pointing in the direction of the conventional current I.

Since there are $nA\Delta l$ moving charges affected by the magnetic field, the total force on this piece of wire is

$$(nA\Delta l)(q\vec{v} \times \vec{B}) = (qnA\bar{v})(\Delta\vec{l} \times \vec{B})$$

and the conventional current is $I = |q|nA\bar{v}$

so we get $\Delta\vec{F}_{\text{magnetic}} = I\Delta\vec{l} \times \vec{B}$

? In a metal wire the actual moving charges are negative electrons traveling in the direction. Explain briefly why the formula is valid even if the actual charge carriers are negative.

The electrons drift in a direction opposite to the conventional current, and their charge is negative. Therefore the quantity $q\vec{v}$ has two canceling minus signs, and the formula in terms of conventional current works fine, even if the actual charge carriers are negative electrons. Note the contrast with the Hall effect, where the actual sign of the moving charges makes a difference.

So we have an alternative form for the magnetic force on a short length of current-carrying wire:

FORCE ON A SHORT LENGTH OF CURRENT-CARRYING WIRE

$$\Delta\vec{F}_{\text{magnetic}} = I\Delta\vec{l} \times \vec{B}$$

I is the conventional current
$\Delta\vec{l}$ is the length of this segment of the wire and points
     in the direction of the conventional-current flow
$\vec{B}$ is the applied magnetic field

This is not a new fundamental equation. It is merely the magnetic force added up for all the individual moving charges in a piece of wire. One way to keep this straight is to memorize the derivation. The key point is that there are $nA\Delta l$ electrons in a short length of wire, each moving with average speed $\bar{v}$, so that the sum of all the $|q|\bar{v}$'s is $(nA\Delta l)|q|\bar{v} = (|q|nA\bar{v})\Delta l = I\Delta l$.

? Figure 20.16 shows a circuit consisting of a battery and a long wire, part of which is in a uniform magnetic field pointing out of the page. What is the direction of the force on those parts of the wire that are in the magnetic field?

Along each wire, the magnetic force points outward from the loop (toward the left along the left wire, down along the bottom wire, and toward the right along the right wire). Note that you get this result whether you consider positive charges drifting in the direction of the conventional current, or electrons drifting in the opposite direction.

? Given those force directions, what is the net force on the wire in terms of the magnetic field B, current I, height h, and length L? (Note: add up the contributions of lots of $\Delta\vec{l}$'s.)

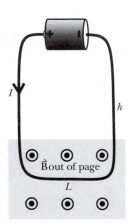

Figure 20.16 Magnetic forces on a loop of wire carrying conventional current I.

The outward forces on the left and right wires cancel (though they tend to stretch the loop), and the net magnetic force is down. The magnitude is $(IL)B\sin 90° = IBL$.

? To get an idea of the numerical order of magnitude, suppose $B = 1$ tesla, $I = 10$ amperes, $h = 50$ cm, and $L = 10$ cm. What is the magnitude of the force?

The magnetic force depends only on L, not h, and $IBL = 1\text{ N}$, a rather small force (1 N is approximately the gravitational force on a small apple).

> Keep in mind that there is really only one magnetic force expression, not two. Always remember that $\Delta\vec{F}_{\text{magnetic}} = I\Delta\vec{l} \times \vec{B}$ is simply the result of adding up the effects on many moving charges in a short length of wire, each of which experiences a magnetic force $\vec{F} = q\vec{v} \times \vec{B}$.

Experiment 20.1 Observing magnetic force

Observe for yourself the force that a magnetic field exerts on a current-carrying wire. Short-circuit one or more batteries and observe how a current-carrying wire deflects near your bar magnet (make and break the circuit to see the direction of the deflection). Use your 2-meter-long wire and let a portion of it hang as freely as possible over the edge of the desk with a weight on the top part, to keep the lower part of the wire from moving when you connect and disconnect the circuit (Figure 20.17). When you connect the wire you can see the wire jerk sideways slightly. The force is small, so observe closely.

Try different directions of current and magnetic field (using the N and S markings on your magnet obtained previously, or by checking with your compass), and take turns with a partner in predicting the direction of the deflection. Make sketches of the various situations and use the formula to explain what you observe.

Experiment 20.2 Bar magnet and coil
Wrap your long wire into a coil, then hang the coil by its connecting wires and bring your bar magnet near the hanging coil.

(a) Under what conditions does the bar magnet repel the coil? Under what conditions does the bar magnet attract the coil?

(b) Sketch a situation where the coil twists under the influence of the bar magnet. Explain briefly in terms of what forces act on what parts of the coil.

Experiment 20.3 Series and parallel battery connections
(a) If you assume that all of your batteries have the same internal resistance, what would you predict would happen to the magnetic force if you had more batteries in series? In parallel? One of your batteries has a resistance of about a quarter or a half an ohm; the resistance of your wire is much smaller than that. (See your calculation of short-circuit current for series and parallel batteries on page 687.)

(b) Using various battery combinations (including combining batteries with your partner), observe the deflection of a wire or a coil by your bar magnet. What do you observe, and how do your observations compare with your predictions? (Hint on technique: you can compare the effects of different numbers of batteries in parallel by connecting some of them in and out of the circuit while observing the deflection of the wire.)

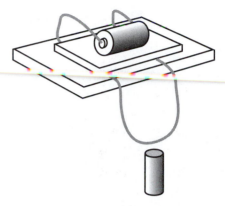

Figure 20.17 Observe a force on a current-carrying wire due to your bar magnet.

20.3.1 Why does the wire move?

There is a subtle point about the magnetic force that a magnetic field exerts on a current-carrying wire. The magnetic force only acts on moving particles, which in the case of a copper wire are the drifting electrons. The stationary positive atomic cores do not experience a magnetic force. So why does the entire wire move?

The answer to this puzzle is closely related to the Hall effect. In Figure 20.18, consider again a current-carrying metal wire in a magnetic field, with a growing number of electrons piling up on the bottom surface (and a deficiency of electrons accumulating on the top surface). Again, for clarity we don't show the usual surface charges along the wire; we just show the extra surface charges due to the magnetic deflection of the electrons.

The extra surface charges exert an electric force $eE_\perp$ upward on the moving electrons which in the steady state just balances the magnetic force evB. The extra surface charges also exert an electric force $eE_\perp$ on the positive atomic cores, and this is an unbalanced force, because the atomic cores initially are not moving and hence are not subject to a magnetic force.

? What is the direction of this electric force exerted by the surface charges on the atomic cores?

Evidently the positively charged atomic cores are pushed downward. The electric force due to the Hall-effect surface charges forces the wire as a whole to move in the direction of $\Delta\vec{F}_{magnetic} = I\Delta\vec{l} \times \vec{B}$, and the magnitude of the electric force is equal to this force. The overall effect is that in Figure

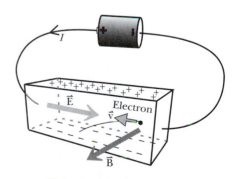

Figure 20.18 Deflected electrons drag the wire along too.

20.18 the atomic cores are dragged downwards as the electrons are dragged downward. The motion of the wire is an electric side effect of the magnetic force on the moving electrons.

Another way of thinking about this is to say that the electrons are bound electrically to the metal wire—they can't simply fall out of the wire. So when the electrons are pushed down by the magnetic force, they drag the wire along with them.

20.3.2 Application: Forces between parallel wires

An important application of the theory is the calculation of the magnetic force that parallel wires exert on each other, because the ampere is defined as the amount of conventional current in two very long parallel wires, a meter apart, which gives a force of one wire on the other of 2×10^{-7} newton per meter of length, from which it follows that $\mu_0/4\pi$ is *exactly* equal to 10^{-7} tesla·m/ampere.

Figure 20.19 shows two very long parallel straight wires of length L, a distance d apart. The upper wire carries a conventional current I_1 and the lower wire carries a conventional current I_2 in the same direction. There are no magnets present.

Magnetic field due to upper wire

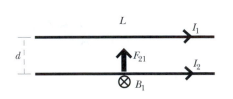

Figure 20.19 Parallel current-carrying wires exert magnetic forces on each other.

In order to calculate the force that the upper wire exerts on the lower wire, first we need to determine the magnitude and direction of the magnetic field $\vec{B}_1$ produced by the upper wire (carrying I_1) at the location of the lower wire. In Chapter 17 we calculated the magnetic field near a long straight wire. Using that result we have $B_1 \approx [\mu_0/(4\pi)][2I_1/d]$. The direction is into the page, using the right-hand rule.

Magnetic force on lower wire

Now we can calculate the magnitude of the magnetic force $\vec{F}_{21}$ exerted on the lower wire by the field produced by the upper wire:

$$F_{21} = I_2 L B_1 \sin 90° = I_2 L \left(\frac{\mu_0}{4\pi} \frac{2I_1}{d} \right)$$

Using the right hand rule with $I\Delta\vec{l} \times \vec{B}$, the direction of the force is up, so that the lower wire is attracted to the upper wire.

Force on upper wire

Find the magnitude and direction of the magnetic force $\vec{F}_{12}$ exerted on the upper wire. This requires two steps:

? First, what are the magnitude and direction of the magnetic field produced by the lower wire (carrying I_2) at the location of the upper wire?

We have $B_2 = [\mu_0/(4\pi)][2I_2/d]$. The direction is out of the page.

? What is the magnitude and direction of the magnetic force $\vec{F}_{12}$ exerted on the upper wire by the field produced by the lower wire? Is Newton's third law obeyed (reciprocity of forces)? That is, is it the case that $\vec{F}_{12} = -\vec{F}_{21}$?

The magnitude of the force on the upper wire is this:

$$F_{12} = I_1 L B_2 \sin 90° = I_1 L \left(\frac{\mu_0}{4\pi} \frac{2I_2}{d} \right)$$

Using the right hand rule with $I\Delta\vec{\mathbf{l}} \times \vec{\mathbf{B}}$, the direction of the force is down, so that the upper wire is attracted to the lower wire. The forces on the two wires have the same magnitude and opposite directions, so in this case the reciprocity of the magnetic forces holds. (But this need not always be the case: see Problem 20.2.)

Currents in opposite directions

? If the two currents run in the same direction, we see that the magnetic interaction is an attraction. If the two currents run in opposite directions, is the magnetic interaction an attraction or a repulsion? Prove your assertion by showing the direction of a magnetic field and the force it exerts.

In this case you should find that two currents running in opposite directions repel each other. For electric forces we say that "likes repel, unlikes attract," but for the magnetic forces between two parallel current-carrying wires, we can say "likes attract, unlikes repel."

Ex. 20.10 To get a feel for the size of the effects, calculate the magnitude of the force per meter if $I_1 = I_2 = 10$ amperes, and the distance between the wires is $d = 1$ centimeter.

20.4 Motional emf: Currents due to magnetic forces

A magnetic field exerts a force on a current-carrying wire. This effect can be reversed: moving a wire through a magnetic field makes a current run in the wire, which provides a way to generate electricity from mechanical work.

Polarization by magnetic forces

Consider a metal bar of length L that is moving through a region of uniform magnetic field, into the page, with magnitude B (Figure 20.20). Because a mobile electron inside the moving metal bar is moving to the right, it experiences a magnetic force.

? If the bar is moving at a speed v, what is the magnitude and direction of the magnetic force on this electron?

The Lorentz force $\vec{\mathbf{F}} = (-e)\vec{\mathbf{v}} \times \vec{\mathbf{B}}$ points downward, and the magnitude of the force is $evB\sin(90°) = evB$. As a result of this magnetic force, the mobile-electron sea will shift downward.

? What will the resulting approximate charge distribution on the metal bar be? What is the direction of the electric field $\vec{\mathbf{E}}$ produced by this charge distribution inside the metal?

The magnetic force on the moving bar polarizes the bar so that it becomes negative at the bottom and positive at the top. As a result, there is an electric field $\vec{\mathbf{E}}$ inside the metal that points downward (Figure 20.21). A mobile electron inside the metal is subject to an electric force due to this electric field as well as to the magnetic force.

? If the bar is kept moving at a constant speed v, how big will E be compared to B, in the steady state? Why? (Hint: why doesn't the polarization keep increasing as more and more electrons pile up at one end of the bar?)

At first the only force on a mobile electron is the downward magnetic force, but the more the mobile-electron sea shifts downward, the larger the upward electric force $-e\vec{\mathbf{E}}$ on an electron. Polarization continues until the

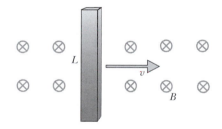

Figure 20.20 A metal bar moving at constant velocity through a region of uniform magnetic field into the page.

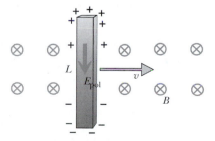

Figure 20.21 The metal bar is polarized as shown, with a downward Coulomb electric field inside the metal.

electric field grows so large that the net downward force ($evB - eE$) is zero. Therefore in the steady state we have $E = vB$.

This situation is a bit odd. Here we have a nonzero electric field inside a metal in static equilibrium (bar moving at constant speed, no accelerations). There is no current, because the electric force is balanced by a magnetic force: the net force on a mobile electron inside the metal is zero.

? What is the potential difference ΔV from one end of the bar to the other? Which end is at the higher potential?

Since the electric field $E = vB$, and v and B are the same throughout the bar, the electric field E is uniform, and the potential difference is simply $\Delta V = EL = vBL$, with the upper (positive) end of the bar at the higher potential (moving upward in the bar, we go against the direction of the electric field, which means we're moving toward higher potential).

? Once the bar is completely polarized, how much force must we apply to the bar to keep it moving at a constant speed v?

In the steady state there is no current running in the bar, so there is no net force on the bar due to the magnetic field, and the polarized bar will coast along with no force required to keep it going.

20.4.1 Non-Coulomb work

? Does this polarized bar remind you of any object we've studied previously? What object or device have we seen before that is most similar to the polarized bar?

This looks like a battery, because a charge separation is maintained from one end to the other by a non-Coulomb force, in this case a magnetic force. We can exploit this effect to generate an electric current.

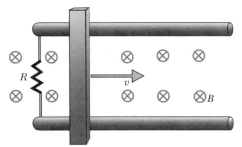

Figure 20.22 Pull the metal bar at constant speed along frictionless metal rails. A resistor is connected between the rails.

Suppose you pull the metal bar at a constant speed v along frictionless metal rails through the uniform magnetic field (Figure 20.22). The moving metal bar and the metal rails have negligible resistance, but a resistor with resistance R is connected between the rails. A current will run in this circuit, driven by a "battery" whose role is played by the moving bar. Electrons are driven the "wrong" way through the moving bar, moving toward the negative end of the bar. We show an approximate surface-charge distribution on the circuit (Figure 20.23).

? Why do the electrons move the "wrong" way through the bar?

The electron current through the resistor is continually depleting the charges on the ends of the moving bar, with the result that the electric field inside the bar is always slightly less than is needed to balance the magnetic force. The upward electric force eE is slightly less than the downward magnetic force evB, so the electrons move toward the negative end of the bar.

The magnetic force evB is a non-Coulomb force (F_{NC} in our earlier discussions of batteries). The work done by this force in moving an electron from one end of the bar to the other is

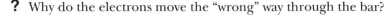

$$F_{NC}L = evBL$$

? The emf is the work done per unit charge in moving a charge from one end of a "battery" to the other, so what is the emf of this "battery"?

The non-Coulomb work per unit charge is $(evBL)/e = vBL$. If the resistance of the bar is negligible, $\Delta V = \text{emf} = vBL$. If the bar has some resistance r_{int}, this resistance is like the internal resistance of any other kind of battery, and $\Delta V = \text{emf} - r_{int}I$ (see the discussion starting on page 685).

The emf that is created by moving a wire in a magnetic field is called "motional emf." This effect is reminiscent of the Hall effect, because when you

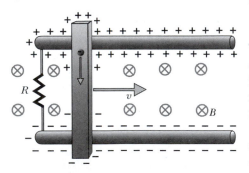

Figure 20.23 Approximate surface-charge distribution on the circuit.

move a wire through a magnetic field, the magnetic force creates a polarization (and electric field) in a direction perpendicular to the motion.

? Does it look like the potential difference $\Delta V = -\oint \vec{E} \bullet d\vec{l}$ for a round trip around the circuit in Figure 20.24 could be equal to zero, as should be the case for the path integral of electric fields due to point charges? (In Figure 20.24, pay attention to the directions of the electric field in the moving bar, in the rails, and in the resistor.)

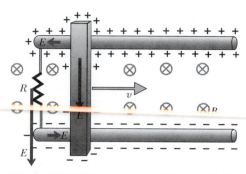

Figure 20.24 The pattern of electric field in the circuit driven by the moving bar.

Going around the loop counter-clockwise from the top end of the moving bar, we get a small potential drop in the rail, a large potential drop in the resistor, a small drop in the rail, and a large potential increase in the moving bar. So it does look plausible that the net ΔV for the round trip could be zero, which we know should be true.

This situation is quite similar to a circuit with a "mechanical battery" in it, where charges move through the mechanical battery in the "wrong" direction, due to non-Coulomb forces, while the round-trip potential difference is zero. Here the non-Coulomb force is the magnetic force.

Ex. 20.11 Can you light a flashlight bulb by moving a connecting wire rapidly past your bar magnet? Let's put in the numbers. You found that near your bar magnet the magnetic field was about 0.1 tesla, and the length of wire affected by the magnetic field might be about a centimeter. You might be able to move the wire at a speed of 10 meters per second past the magnet. What is the emf you can produce? How does this compare with the emf of a flashlight battery?

Small size of motional emf

From the preceding exercise dealing with your flashlight bulb you can see that it is hard to see the effects of motional emf with your equipment. To get a sizable emf you need larger magnetic field B, extending over larger regions (L), with very fast motion v of the wires through the magnetic field, and many turns of wire, each contributing to the net emf.

Also, if you move a wire rapidly past your magnet, the emf doesn't last very long. And finally, your light bulb (or compass) isn't a very sensitive detector of small currents. Consequently, to study this motional emf you need better lab equipment than is in your electricity kit. Your instructor may arrange a demonstration of these effects, or a formal laboratory exercise where you can observe the phenomena for yourself.

20.4.2 Forces on the moving bar and energy conservation

If the metal bar were to slide effortlessly along the rails at a constant speed v, and the resistor were to continually warm the surroundings, we would be getting something for nothing. So there must be an energy input into this system. We must exert a force on the bar to keep it moving at constant speed (Figure 20.25), because there is a magnetic force on the moving bar, due to the conventional current I in the bar.

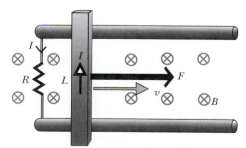

Figure 20.25 You have to keep pulling on the bar in order to keep current running in this circuit.

? What is the direction of this magnetic force? What is the direction and magnitude of the force that we have to exert on the bar to keep it going at a constant speed?

Since there is a (conventional) current upward in the moving metal bar, there is a magnetic force $\left| I\Delta \vec{l} \times \vec{B} \right| = ILB\sin(90°) = ILB$ to the left on the bar. In order for the bar to keep moving at constant speed, the net force on the bar must be zero, so we have to exert a force of magnitude ILB to the right.

In a short time Δt we move the bar a distance Δx. We do an amount of work $F\Delta x$ in a time Δt, so we have to supply energy at a rate of $F\Delta x/\Delta t = Fv$.

? Is this power input Fv numerically equal to the power dissipated in the resistor?

We have power $= (ILB)v = I(LBv) = I(\text{emf})$. But $I(\text{emf})$ is equal to the power output of the "battery" that is input to the resistor. Moreover, $I = \text{emf}/R$, so the power dissipated in the resistor is $\text{emf}^2/R = RI^2$, as expected.

This is the basic principle of electric generators: mechanical power is converted into electric power. Commercially, the mechanical energy is supplied by falling water at power dams, or by expanding steam in a turbine (where the steam is heated by chemical or nuclear reactions—burning coal or oil, or fissioning uranium). A practical generator has a rotating coil rather than a sliding bar, and we'll discuss practical generators later in this chapter.

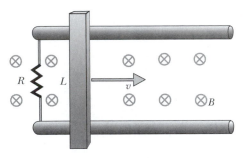

Figure 20.26 Pull the metal bar at constant speed along frictionless metal rails. A resistor R is connected between the rails.

Ex. 20.12 Here is a review of the basic issues: A metal rod of length L is dragged horizontally at constant speed v on frictionless conducting rails through a region of uniform upward magnetic field of magnitude B (Figure 20.26). There is a resistance R. What is the magnitude of the induced emf? What is the direction of the conventional current I? What is the magnitude of the current I? What is the magnitude and direction of the force you have to apply to keep the rod moving at a constant speed v?

20.5 Magnetic forces in moving reference frames

The magnetic field produced by a moving charge depends on the velocity of the charge, and this leads to the curious result that a stationary Jack may observe a different magnetic field than a moving Jill does (see page 591). Suppose that Jack holds a stationary charge, which of course produces a pure electric field and no magnetic field. If Jill runs past Jack, she sees a mixture of electric and magnetic fields produced by what is for her a moving charge.

The magnetic force experienced by a moving charge also depends on the velocity of the charge, which is another reason why observations of magnetic effects may differ in different moving reference frames. If Jack holds two charged tapes that repel each other, he describes this by saying that one tape produces a purely electric field, and the other tape is affected by that electric field. Jill on the other hand as she moves past Jack says that one (moving) tape produces a mixture of electric and magnetic fields, and the other (moving) tape feels both electric and magnetic forces.

Two moving protons

We are now in a position to do a detailed calculation of these odd effects. We'll consider two protons initially traveling parallel to each other with the same speed v, a distance r apart, and we will calculate the magnitudes and directions of the electric and magnetic forces that proton 1 exerts on proton 2 at this moment (Figure 20.27). The electric force is easy. Proton 1 makes an electric field $\vec{E}_1$ at the location of proton 2, and proton 2 experiences an electric force $\vec{F}_{21,e} = q_2\vec{E}_1$:

$$F_{21,e} = \frac{1}{4\pi\varepsilon_0}\frac{e^2}{r^2}, \text{ downward}$$

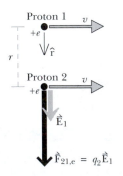

Figure 20.27 Proton 1 exerts an electric force on proton 2.

We're making an approximation when we use Coulomb's law to calculate the electric force. Due to relativistic effects, the electric field $\vec{E}_1$ made by the moving charge (proton 1) is somewhat different from the electric field made by a stationary charge. We'll say a bit more about this approximation later.

The magnetic force involves cross products both in the field and in the force. To determine the magnetic force $\vec{F}_{21,m}$ acting on proton 2, first we need to calculate the magnetic field $\vec{B}_1$ made by proton 1 at the location of proton 2. It helps enormously to draw the unit vector on the diagram (from source to observation point; that is, from proton 1 to proton 2).

? Give the direction and the magnitude of $\vec{B}_1 = \dfrac{\mu_0}{4\pi}\dfrac{q_1\vec{v}_1 \times \hat{r}}{r^2}$.

The magnetic field $\vec{B}_1$ is into the page at the location of proton 2, and its magnitude is $B_1 = [\mu_0/(4\pi)](ev/r^2)\sin 90°$.

? Now that you know the direction and magnitude of the magnetic field $\vec{B}_1$, determine the magnitude and direction of the magnetic force exerted on proton 2, $\vec{F}_{21,m} = q_2\vec{v}_2 \times \vec{B}_1$.

The direction is upward, and the magnitude is $F_{21,m} = \dfrac{\mu_0}{4\pi}\dfrac{e^2 v^2}{r^2}$.

The ratio of the magnetic and electric forces

Figure 20.28 summarizes the situation. The magnetic force on proton 2 is opposite to the electric force, and the net force on proton 2 is reduced from what it would be for a stationary proton.

? Show that the ratio of the magnetic force to the electric force on proton 2,

$$\frac{F_{21,m}}{F_{21,e}} = \left(\frac{\mu_0}{4\pi}\frac{e^2 v^2}{r^2}\right)\Big/\left(\frac{1}{4\pi\varepsilon_0}\frac{e^2}{r^2}\right) = \frac{v^2}{c^2}$$

where c is the speed of light (3×10^8 m/s). Note that

$$\frac{1}{\mu_0\varepsilon_0} = \left(\frac{1}{4\pi\varepsilon_0}\right)\Big/\left(\frac{\mu_0}{4\pi}\right) = \left(9\times10^9\,\frac{\text{N}\cdot\text{m}^2}{\text{C}^2}\right)\Big/\left(10^{-7}\,\frac{\text{T}}{\text{A}\cdot\text{m}}\right) = (3\times10^8\,\text{m/s})^2$$

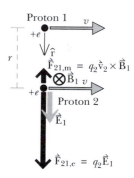

Proton 1

Figure 20.28 The magnetic force on proton 2 is opposite to the electric force.

There is something peculiar here: Why should the speed of light show up in a comparison between electric and magnetic forces? What does the speed of light have to do with this? There are some deep issues here.

This result means that at ordinary speeds ($v \ll c$) the magnetic interaction between two charged particles is much weaker than the electric interaction. However, if v approaches the speed of light c, the magnetic force may be comparable to the electric force. Even though at low speeds magnetic forces are inherently weak compared to electric forces, magnetic forces are more important in many industrial applications, because the magnetic force on a current-carrying wire is typically much larger than an electric force on the wire, since the net charge of the wire is nearly zero (neglecting the small amount of surface charge).

? Now determine the full Lorentz force, $\vec{F} = q\vec{E} + q\vec{v} \times \vec{B}$, of proton 1 on proton 2 (magnitude and direction).

The electric force is downward and the magnetic force is upward, so the Lorentz force on proton 2 is

$$\frac{1}{4\pi\varepsilon_0}\frac{e^2}{r^2} - \frac{\mu_0}{4\pi}\frac{e^2 v^2}{r^2} = \frac{1}{4\pi\varepsilon_0}\frac{e^2}{r^2}\left[1 - \frac{v^2}{c^2}\right],\ \text{downward}$$

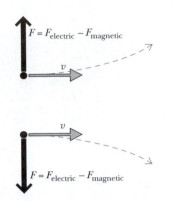

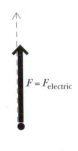

Figure 20.29 Jack is sitting still, and this is what he sees.

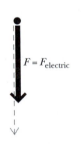

Figure 20.30 Jill runs along beside the protons, at speed v; this is what she sees.

? Describe the subsequent trajectories of the protons.

The protons repel each other, so their trajectories are curving lines moving apart (Figure 20.29).

Ex. 20.13 As a numerical example, suppose $v = 1500$ m/s, a very high speed typical of a molecule of hydrogen gas at room temperature: What is the numerical ratio of the magnetic force to the electric force for protons moving at this speed?

20.5.1 Jack and Jill and Einstein

The fact that the net force is smaller on the moving protons may seem innocuous, but it has startling consequences. Suppose Jack is sitting still and sees the protons go by. The protons repel each other along curving paths, with a force somewhat weakened by the magnetic effects (Figure 20.29).

Suppose Jill is running along beside the protons, with the same speed v that the protons have initially. She sees the two protons repel each other straight away from each other (Figure 20.30). The repulsion is purely electric, with no weakening of this electric repulsion (the magnetic field of one proton is zero at the location of the other proton, because $\hat{v} \times \hat{r} = 0$ along the line of motion).

Let both Jack and Jill use their wristwatches to time how long it takes for the protons to hit the floor and ceiling of the room, starting from halfway between floor and ceiling.

? Given the difference in the force that accelerates the protons vertically, which observer will measure a shorter time for the protons to reach the floor and ceiling?

Jill sees a pure electric repulsion, while Jack sees the electric repulsion partially counteracted by an attractive magnetic force. To take a concrete example, suppose Jack's watch advances by 20 nanoseconds during this process. Jill's watch might advance by only 15 nanoseconds (if v is a large fraction of the speed of light), corresponding to her observation that the protons get to the floor and ceiling in less time due to the greater force of repulsion.

But that's crazy! Surely Jack and Jill wouldn't measure different amounts of time for the same process? It was Einstein who thought very deeply about this and related paradoxes and concluded that, much as it violates common sense, it has to be the case that Jill's wristwatch seems to run slower for Jack than his own wristwatch!

This sounds utterly absurd, yet a great variety of measurements of this kind all agree with Einstein's radical proposal that time runs at different rates for different observers. One of the reasons that Einstein was driven to these strange notions was that by the time that he proposed his special theory of relativity (1905), it was clear that the existing theories of electricity and magnetism seemed to lead to paradoxes, as we have seen in the observations of Jack and Jill. In fact, the title of Einstein's 1905 paper was "On the electrodynamics of moving bodies."

The net force (electric minus magnetic) observed by Jack is always smaller than the purely electric force observed by Jill, and Jack and Jill disagree on how long it takes the protons to move apart, and on whether their watches are marking time correctly. Yet both Jack and Jill can correctly predict how much their own watches will advance during the time it takes for the protons to hit the floor and ceiling.

The details of our calculations are not correct, because relativistic aspects of the fields have not been taken into account. Both the electric field and the magnetic field of a fast-moving charge are different from the fields given by Coulomb's law and the Biot-Savart law for a slow-moving charge. However, both fields are modified by the same factor $1/\sqrt{1 - v^2/c^2}$, so the ratio of the magnetic force to the electric force is equal to v^2/c^2 for all speeds. And Jack and Jill will indeed observe different times for the same process.

An asymmetry in the measurements

You may wonder about the asymmetry in the measurements that Jack and Jill make. If Jack thinks that Jill's watch runs slow, why doesn't Jill think that Jack's watch runs slow? After all, Jill should be able to take the point of view that she is standing still while Jack moves past her.

In reality, Jack cannot make his measurements unaided. Rather, Jack and his friend Fred must station themselves an appropriate distance apart in the room. Jack records the time (on Jack's watch) when the protons pass by him, and he also records the time that he sees on Jill's watch at that moment. Fred records the later time (on Fred's watch) when the protons reach him and hit the floor and ceiling, and he also records the time that he sees on Jill's watch at that moment. Two people, Jack and Fred, are needed to make these measurements. Otherwise Jack would have to apply corrections to his measurements, due to the time it takes for light from Jill's watch to reach him from the distant location.

After Fred has made his measurements, Jack and Fred compare notes and find that the times that they wrote down differ by 20 nanoseconds (they had previously synchronized their watches). Their two views of Jill's watch indicate that her watch advanced only 15 nanoseconds. Jack and Fred describe Jill's watch as running slow. The situation is different for Jill, because she looks at two *different* watches, Jack's and Fred's. The asymmetry lies in the fact that there are two observers sitting in the room (Jack and Fred), but only one observer moving through the room (Jill).

On the other hand, suppose Jill sits at the front of a space ship moving past Jack, and her friend Sally sits further back in the space ship. Jill notes the time on her own watch and on Jack's watch when she passes Jack, and Sally does the same when she passes Jack. When Jill and Sally compare notes, they will describe Jack's watch as seeming to run slow.

20.6 Relativistic field transformations

Jack and Jill observed different electric and magnetic fields in their different reference frames, though they both could correctly predict the motion of the protons in their own reference frames, in terms of the fields they observe and the Lorentz force.

The theory of relativity predicts how electric and magnetic fields transform when you change from one reference frame to another. We state these transforms without proof. In the following, the "unprimed" quantities (E_x, etc.) are the fields that are observed in the "lab" frame (fixed to our laboratory). The "primed" quantities (E_x', etc.) are the fields that are observed in a reference frame that is moving with speed v in the +x direction relative to the lab frame (Figure 20.31).

$$E_x' = E_x \qquad E_y' = \frac{(E_y - vB_z)}{\sqrt{1 - v^2/c^2}} \qquad E_z' = \frac{(E_z + vB_y)}{\sqrt{1 - v^2/c^2}}$$

$$B_x' = B_x \qquad B_y' = \frac{\left(B_y + \dfrac{v}{c^2}E_z\right)}{\sqrt{1 - v^2/c^2}} \qquad B_z' = \frac{\left(B_z - \dfrac{v}{c^2}E_y\right)}{\sqrt{1 - v^2/c^2}}$$

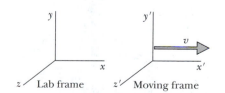

Figure 20.31 The moving reference frame moves with speed v in the +x direction relative to the lab frame.

20.6.1 Magnetic field of a moving charged particle

Let's apply these transforms to a positively charged particle that is at rest in the lab frame. It of course makes only an electric field, yet if you run past the particle, you observe it in motion, and you observe not only an electric field but also a magnetic field.

In the lab frame the electric field of the stationary charge is given by:

$$\vec{E} = \frac{1}{4\pi\varepsilon_0}\frac{q}{r^2}\hat{r}$$

We would like to determine the fields made by this charge in the moving reference frame. In the moving reference frame the particle is seen to be moving with speed v in the $-x$ direction (Figure 20.32).

In the lab frame there is no magnetic field. But straight above the moving particle, E_y is positive, so we have this in the moving reference frame:

$$B_z' = \frac{\left(B_z - \frac{v}{c^2}E_y\right)}{\sqrt{1-v^2/c^2}} = \frac{-\frac{v}{c^2}E_y}{\sqrt{1-v^2/c^2}}$$

At low speeds, this is approximately

$$B_z' = -\frac{v}{c^2}\left(\frac{1}{4\pi\varepsilon_0}\frac{q}{r^2}\right)\ (v \ll c)$$

The magnetic field above the positively charged particle is in the $-z$ direction (Figure 20.32). This is consistent with the prediction of the Biot-Savart law for the direction of magnetic field above the charge. In an exercise at the end of this section you will show that the transforms predict the usual curly magnetic field all the way around the moving charge.

Let's calculate the magnitude of the magnetic field in the moving frame. In analyzing the electric and magnetic forces between two moving protons (the Jack and Jill episode), we found that

$$\frac{1}{\mu_0\varepsilon_0} = c^2 \text{ so } \frac{1}{4\pi\varepsilon_0} = c^2\frac{\mu_0}{4\pi}$$

Therefore we can write

$$B_z' = -\frac{v}{c^2}\left(\frac{1}{4\pi\varepsilon_0}\frac{q}{r^2}\right) = -\frac{\mu_0}{4\pi}\frac{qv}{r^2}$$

This is exactly the magnitude of the magnetic field of the moving charge that is predicted by the Biot-Savart law, which is valid for $v \ll c$. This is partial verification that the field transforms are consistent with what you already know about electric and magnetic fields.

ELECTRIC AND MAGNETIC FIELDS ARE INTERRELATED

Electric fields and magnetic fields are closely related to each other, since a mere change of reference frame mixes the two together. One can say that magnetic fields are a relativistic consequence of electric fields.

Ex. 20.14 Use the field transforms to find the direction of the magnetic field in front of, behind, and below the positively charged particle in the moving reference frame, and therefore show that there is a curly magnetic field all the way around the moving charge, as expected (Figure 20.33).

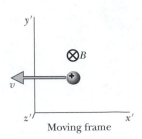

Figure 20.32 In the moving reference frame the particle moves with speed v in the $-x$ direction. A magnetic field appears in this frame.

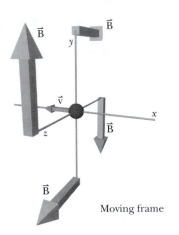

Figure 20.33 Use the field transforms to show that in the moving frame there is a curly magnetic field all the way around the moving positive charge.

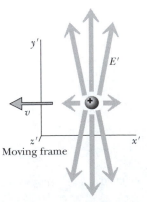

Figure 20.34 The electric field of the particle in the moving frame, where the particle is moving rapidly.

20.6.2 Electric field of a rapidly moving charged particle

If the speed of the moving reference frame is very large (very near c), the electric field of the particle is altered significantly in the moving frame. The magnitude of the electric field in front of and behind the particle in the moving frame is unchanged: $E_x' = E_x$. But straight off to the side of the particle the electric field is much larger at high speeds, since we have

$$E_y' = \frac{(E_y + vB_z)}{\sqrt{1 - v^2/c^2}} = \frac{E_y}{\sqrt{1 - v^2/c^2}}$$

which is much greater than E_y if v approaches c. Figure 20.34 on the preceding page shows the general pattern of electric field in the moving frame, near the charge that is rapidly moving in this reference frame. The region of high field strength becomes a thinner and thinner pancake as the speed of the particle approaches the speed of light.

20.6.3 Moving through a region of uniform magnetic field

The previous examples involved a situation where there was no magnetic field in the rest frame of the particle but there were both electric and magnetic fields in a frame where the particle was moving. Next we'll consider a region of uniform magnetic field in the lab frame, but no electric field, and see what fields are present in the moving reference frame. In the lab frame the magnetic field is $< 0, 0, -B >$, so $B_z = -B$ (Figure 20.35).

? Suppose you move with speed v in the $+x$ direction through this region. Using the field transforms given above, what electric field would you observe?

The transforms show that in the moving reference frame, you will observe an upward electric field of magnitude $E = vB$ ($+y$ direction; see Figure 20.36), though there was no electric field in the lab frame:

$$E_y' = \frac{0 - vB_z}{\sqrt{1 - v^2/c^2}} = \frac{0 - v(-B)}{\sqrt{1 - v^2/c^2}} \approx vB \text{ if } v << c$$

Suppose we are in this moving frame, and we observe a metal bar that is at rest in this frame. The metal bar polarizes because there is an upward electric field of magnitude $E = vB$ everywhere in this region (Figure 20.37). The bar polarizes until there is sufficient charge buildup to produce a downward electric field of magnitude $E_{pol} = vB$, at which point the net electric field inside the metal is zero, and there is static equilibrium.

In the lab frame, the metal bar moves at constant speed v in the $+x$ direction through a region of uniform magnetic field and zero electric field (Figure 20.38). We discussed this situation earlier in the chapter in connection with motional emf. In the lab frame the moving metal bar polarizes due to the magnetic force qvB on the moving mobile charges. The polarization charges produce a downward electric field E_{pol} ($-y$ direction). At equilibrium it must be that $qE_{pol} = qvB$, so $E_{pol} = vB$.

The amount of charge buildup is the same in either reference frame, but the explanation that observers in the two frames give for the cause of the charge buildup is different. In the moving reference frame, where the bar is at rest, there is an upward electric field of magnitude $E = vB$ that polarizes the bar until this electric field becomes balanced by the downward electric field $E_{pol} = vB$ of the polarization charges. In the lab frame, where the bar is in motion, there is a magnetic force that polarizes the bar until the magnetic force becomes balanced by the electric force of the polarization charges; the downward electric field has magnitude $E_{pol} = vB$.

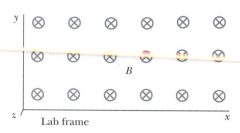

Figure 20.35 A region in the lab frame of uniform magnetic field in the $-z$ direction.

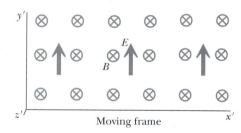

Figure 20.36 In the moving frame there is an electric field in the $+y$ direction of magnitude $E = vB$.

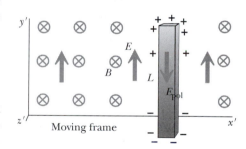

Figure 20.37 A metal bar at rest in the moving frame is polarized by the upward applied electric field. The polarization charges produce a downward electric field $E_{pol} = vB$.

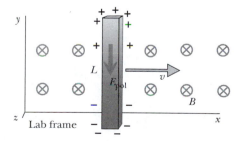

Figure 20.38 In the lab frame the moving metal bar is polarized by the magnetic force on the moving charges. The polarization charges produce a downward electric field $E_{pol} = vB$.

THE PRINCIPLE OF RELATIVITY

There may be different mechanisms for different observers in different reference frames, but all observers can correctly predict what will happen in their own frames, using the same relativistically correct physical laws.

20.6.4 Moving through a velocity selector

As another example, consider the velocity selector discussed earlier and shown in Figure 20.39. In the lab frame a charged particle traveling with speed v moves in the x direction through a region where there is an electric field of magnitude E in the $+y$ direction and a magnetic field of magnitude B in the $+z$ direction. If $E = vB$, the charged particle passes through with no deflection, because the electric force of magnitude qE is equal and opposite to the magnetic force of magnitude qvB.

? Suppose you move with speed v in the $+x$ direction through this region, without the charged particle present. Using the field transforms given above, what electric and magnetic fields would you observe?

The only fields you would observe in the moving frame are these:

$$E_y{}' = \frac{(E_y - vB_z)}{\sqrt{1 - v^2/c^2}} = \frac{E - vB}{\sqrt{1 - v^2/c^2}} = \frac{vB - vB}{\sqrt{1 - v^2/c^2}} = 0$$

$$B_z{}' = \frac{\left(B_z - \dfrac{v}{c^2}E_y\right)}{\sqrt{1 - v^2/c^2}} = \frac{B - \dfrac{v}{c^2}E}{\sqrt{1 - v^2/c^2}} = \frac{B - \dfrac{v}{c^2}(vB)}{\sqrt{1 - v^2/c^2}} = \frac{(1 - v^2/c^2)}{\sqrt{1 - v^2/c^2}}B$$

? Given that you observe these fields, what will happen to a charged particle initially at rest in your reference frame?

A charge at rest is unaffected by the magnetic field, and there is no electric field, so there is no force on the charged particle. This makes sense, because a particle at rest in your reference frame moves with constant velocity in the lab frame, straight through the velocity selector.

Note again that observers in the two reference frames differ in their analyses of the situation, but both correctly predict what happens to the charged particle. You predict that a charged particle at rest will remain at rest, because there is no electric field, and the Lorentz force is zero. An observer in the lab frame calculates an electric force and a magnetic force, determines that they are equal and opposite to each other, and predicts that the particle will have a uniform velocity since the net force on the particle is zero.

20.6.5 An alternative form of the transforms

You may have noticed terms that look like a cross product in the transforms for the y and z components of the fields—the components perpendicular to the motion. We can write the transforms for the fields parallel to the motion and perpendicular to the motion in the following way:

$$E_{\parallel}{}' = E_{\parallel} \qquad\qquad\qquad B_{\parallel}{}' = B_{\parallel}$$

$$E_{\perp}{}' = \frac{(\vec{E} + \vec{v} \times \vec{B})_{\perp}}{\sqrt{1 - v^2/c^2}} \qquad\qquad B_{\perp}{}' = \frac{\left(\vec{B} - \dfrac{\vec{v} \times \vec{E}}{c^2}\right)_{\perp}}{\sqrt{1 - v^2/c^2}}$$

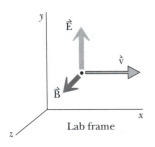

Figure 20.39 A velocity selector. If $E = vB$, the charged particle passes through with no deflection.

Ex. 20.15 Try using these versions of the transforms to reanalyze one or more of the phenomena treated earlier: fields of a moving charge, the metal bar moving through a magnetic field, or the velocity selector.

If you would like to know more about relativity, there are many fine books on the subject, or you might like to take a course on relativity. Here are two excellent introductory books:

R. Resnick, *Introduction to Special Relativity* (Wiley, 1968)
E. Taylor and J. Wheeler, *Spacetime Physics* (Freeman, 1992)

20.7 Magnetic torque on a magnetic dipole moment

A magnetic field can cause a current-carrying coil of wire to twist. You may have seen this for yourself, if you were able to suspend your coil of wire and see it twist in the magnetic field of your bar magnet (or you may have floated your coil and seen it twist in the magnetic field of the Earth). Moreover, we have seen that a compass needle can be thought of as a collection of atomic current loops, and you have certainly seen a compass needle twist in a magnetic field.

Consider a rectangular current-carrying loop of wire in a uniform magnetic field (that is, the magnetic field has the same direction and magnitude throughout this region). The loop is free to rotate on an insulating horizontal axle (Figure 20.40).

? In Figure 20.40, what is the direction of the magnetic force on each of the four sides of the loop?

On the sides of length h the magnetic force is horizontal and points outward, tending to stretch the loop. On the other sides, of length w, the magnetic forces are horizontal and tend to make the loop twist on the axle. This may be easiest to see in the side view shown in Figure 20.40.

? In the side view, will the loop rotate clockwise ("to the right") or counter-clockwise ("to the left")?

The magnetic forces act to twist the loop counter-clockwise.

? When the plane of the loop is perpendicular to the magnetic field, the magnetic forces don't exert any twist. But one of these orientations is unstable: if you nudge it slightly, the loop will flip over. Which of the two orientations shown in side view in Figure 20.41 is stable, and which is unstable?

The situation on the top is stable: the magnetic forces twist the loop back toward the horizontal plane. The situation on the bottom is unstable: a small displacement away from the horizontal plane leads to magnetic forces that rotate it even farther out of the plane.

It is often easier to talk about the twist on a current-carrying loop of wire in terms of the "magnetic dipole moment" (see page 601). The magnetic dipole moment $\vec{\mu}$ of a current-carrying loop of wire is defined as a vector pointing in the direction of the magnetic field that the loop makes along its axis, which is given by a right-hand rule (this magnetic field made by the loop is of course different from the applied magnetic field that makes the loop twist). The magnitude of the magnetic dipole moment is $\mu = IA$.

Our rectangular loop has an area $A = wh$, so its magnetic dipole moment is $\mu = IA = Ihw$. In Figure 20.42, note that the coil tends to twist in a direction to make $\vec{\mu}$ line up with the applied magnetic field $\vec{B}$.

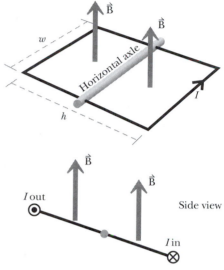

Figure 20.40 A current-carrying loop connected to an insulating horizontal axle is free to rotate.

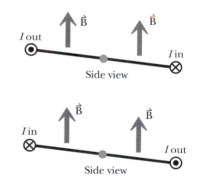

Figure 20.41 Which of these orientations is stable?

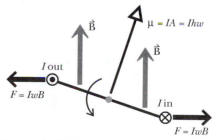

Figure 20.42 The magnetic dipole moment of the loop is a vector that tends to line up with the applied magnetic field.

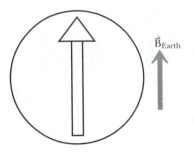

Figure 20.43 What is the direction of the atomic magnetic dipole moments inside this compass needle, which points North?

Ex. 20.16 Inside the magnetized compass needle shown in Figure 20.43, which points north, what is the direction of a typical atomic magnetic dipole moment $\vec{\mu}$?

Ex. 20.17 If the needle is moved away from north and then released, what will happen? Explain briefly.

At last we have an explanation in terms of the fundamental principles of magnetism for the behavior of a compass needle, and why it always lines up with the applied magnetic field. (A practical compass has some friction to damp out the oscillations: without friction the needle would swing back and forth forever, like a frictionless pendulum. Many compasses are filled with liquid to damp out the oscillations.)

20.7.1 Quantitative torque

The fact that a magnetic dipole moment $\vec{\mu}$ twists in a magnetic field is very important. The torque provided by each of the magnetic forces around the axle is equal to the distance from the axle (the "lever arm") times the component of the force perpendicular to the lever arm. The twist applied to this rectangular loop is due to the magnetic forces on the *w*-lengths of the loop, as is seen in Figure 20.44.

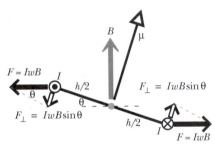

Figure 20.44 The torque due to magnetic forces on a rectangular current-carrying loop in a magnetic field.

The perpendicular component of each force is $F_\perp = IwB\sin\theta$, and the lever arm is $h/2$. (Note that the other component of F, and the forces on the *h*-lengths of the loop, just tend to stretch the loop without causing a twist.) Each of the forces exerts a torque of $(F_\perp)(h/2)$, so the total torque is $2IwB\sin\theta(h/2) = (Iwh)B\sin\theta$.

Since $\mu = Iwh$, we can express the torque in the form $\vec{\mu} \times \vec{B}$, because this cross product has a magnitude $|\vec{\mu}||\vec{B}|\sin\theta = \mu B\sin\theta$. The direction of this torque vector is along the axle around which the loop rotates. If the thumb of your right hand points in the direction of $\vec{\mu} \times \vec{B}$, your fingers curl around in the direction that the coil will twist. In Figure 20.44, the torque vector is out of the page, and the coil rotates counter-clockwise.

We showed that the torque is $\mu B\sin\theta$ for a rectangular loop, but a loop of any shape can be approximated by a set of current-carrying rectangles along whose adjacent sides the current cancels (Figure 20.45). This shows that the torque is $\mu B\sin\theta$ for any kind of loop.

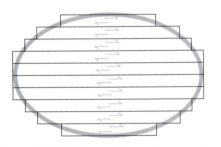

Figure 20.45 A loop of any shape can be approximated by a set of rectangular loops.

TORQUE ON A MAGNETIC DIPOLE MOMENT

$$\vec{\tau} = \vec{\mu} \times \vec{B}$$

20.8 Potential energy for a magnetic dipole moment

If allowed to pivot freely, a magnetic dipole moment turns toward alignment with the applied magnetic field. That means that an aligned magnetic dipole moment is associated with lower potential energy in the magnetic field, and the system tries to go to the lower potential-energy configuration. If we calculate how much work it takes to move the magnetic dipole moment out of alignment, we will have a measure of the increased potential energy associated with being out of alignment.

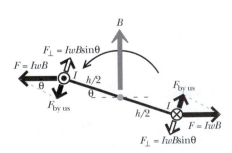

Figure 20.46 Calculating how much work is required to turn a current loop in a magnetic field.

We consider the rectangular loop shown in Figure 20.46, and we calculate how much work we have to do to move it from an initial angle $\theta = \theta_i$ to a final angle $\theta = \theta_f$. We move slowly, without changing the kinetic energy of the loop. The loop is always very nearly in static equilibrium.

As we rotate the loop through a small angle $d\theta$, both of the forces that we exert act through a distance $(h/2)d\theta$, because arc length is radius $(h/2)$ times angle ($d\theta$ measured in radians). If we force the loop to rotate from the initial angle $\theta = \theta_i$ to a final angle $\theta = \theta_f$, the amount of work we do by exerting our two forces must be calculated by an integral, because F isn't constant. The work that we do goes into changing the magnetic potential energy U_m of the system consisting of the magnetic dipole moment in the magnetic field:

$$\text{work} = \Delta U_m = \int_{\theta_i}^{\theta_f} 2IwB\sin\theta\left(\frac{h}{2}d\theta\right) = IwhB\int_{\theta_i}^{\theta_f}\sin\theta\,d\theta$$

$$\Delta U_m = IwhB[-\cos\theta]_{\theta_i}^{\theta_f} = -\mu IwhB[\cos\theta_f - \cos\theta_i]$$

$$\Delta U_m = \Delta(-\mu B\cos\theta) \text{ since } \mu = IA = Iwh$$

It is customary to define the potential energy of a magnetic dipole moment in a magnetic field by choosing the zero of potential energy in such a way that one writes

$$U_m = -\mu B\cos\theta$$

This can also be written as a dot product:

THE POTENTIAL ENERGY FOR A MAGNETIC DIPOLE MOMENT

$$U_m = -\vec{\mu} \cdot \vec{B}$$

This convention corresponds to calling the potential energy zero when $\theta = 90$ degrees, since $\cos(90°) = 0$.

? Given that $U_m = -\vec{\mu} \cdot \vec{B}$, which of the alignments in Figure 20.47 corresponds to the lowest possible potential energy? To the highest possible potential energy?

The lowest energy corresponds to $\vec{\mu}$ and $\vec{B}$ pointing the same direction, so that $U_m = -\mu B\cos 0 = -\mu B$. This is also the most stable configuration; if you twist the magnetic dipole moment away from this position, it will be twisted back by the magnetic field. The highest energy corresponds to $\vec{\mu}$ and $\vec{B}$ pointing in opposite directions, so that $U_m = -\mu B\cos 180° = +\mu B$. This is an unstable configuration; a slight displacement will lead to a large twist. The other two configurations correspond to medium energy: $U_m = 0$ (because $\cos 90° = 0$).

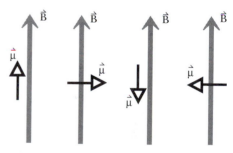

Figure 20.47 Which alignment represents the lowest potential energy? The highest?

This picture of the potential energy of a magnetic dipole moment in a magnetic field is very important in atomic and nuclear physics. In a magnetic field, atoms or nuclei can make transitions between higher and lower energy states that correspond to different orientations of the magnetic dipole moments of the atoms or nuclei in the magnetic field.

For example, in the magnetic imaging equipment now used in hospitals, the patient lies inside a large solenoid. In the presence of the magnetic field of the solenoid, nuclei in the patient's body have slightly different energies depending on whether the nuclear magnetic dipole moment of the nucleus is in the direction of or opposite to the applied field. Sensitive equipment detects transitions between these two slightly different energy levels.

Ex. 20.18 What is the energy difference between the highest and lowest states?

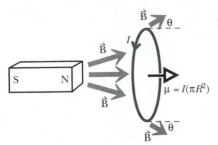

Figure 20.48 A bar magnet exerts a force on a current-carrying loop.

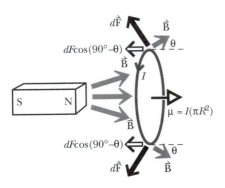

Figure 20.49 Force on a current loop by a bar magnet.

20.8.1 Force on a magnetic dipole moment

We have successfully explained the twist of a magnetic dipole moment in a magnetic field. There remains the question of how to explain the net force on a magnetic dipole moment in a magnetic field. For example, how can we explain the attraction or repulsion between two magnets?

We can show that if a magnetic dipole moment is placed in a nonuniform magnetic field, it experiences a force. Figure 20.48 shows a circular current loop placed near the north pole of a bar magnet, along the axis of the magnet. The magnetic field of the bar magnet diverges with an angle θ as shown and so is non-uniform.

? Show that the net force on the current loop is to the left, with magnitude $F = I(2\pi R)B\sin\theta = (2\mu B/R)\sin\theta$, where R is the radius of the circular loop.

As shown in Figure 20.49, each segment of the ring experiences a magnetic force $d\vec{F} = Id\vec{l} \times \vec{B}$. The horizontal component of the force is $dF\cos(90° - \theta) = dF\sin\theta$. The vertical components of the forces on two opposite pieces of the ring cancel, so the net force is in the $-x$ direction (toward the bar magnet).

$$F_{\text{net}} = IB\sin\theta \int dl = IB(2\pi R)\sin\theta$$

$$\text{Since } \mu = IA = I(\pi R^2), F_{\text{net}} = \frac{\mu(2B\sin\theta)}{R}$$

If we knew how the magnetic field spread out in angle (that is, how big θ is at a distance R off-axis), we could calculate the net force F (see Problem 21.5). Without that information, we can nevertheless determine F by using an argument based on potential energy, which we will in the next section.

Ex. 20.19 In what direction would the net force be if the bar magnet were turned around, or the current in the loop reversed?

Ex. 20.20 What can you say about the net force if the field were uniform ($\theta = 0$)?

20.8.2 Calculating the force on a magnetic dipole moment

Using an energy argument, we can calculate the force on a magnetic dipole moment due to a nonuniform magnetic field. For example, we can calculate the force on one magnet due to another magnet.

If we move a magnetic dipole moment from a region of low magnetic field to a region of high magnetic field, the magnetic potential energy $U_m = -\vec{\mu} \bullet \vec{B}$ will change. If released from rest, the object will spontaneously move to the place where the potential energy is lower, picking up kinetic energy as it goes. If we prevent the kinetic energy from changing, by moving the object at a slow constant speed, we have to exert a force and do some work. By calculating how much work we have to do, we can determine how big a force we have to apply, and this is numerically equal to the magnetic force on the object. This will let us calculate how big a force one magnet exerts on another magnet.

Consider the situation shown in Figure 20.50. A magnetic dipole moment is aligned with the magnetic field made by a bar magnet, and the magnet exerts a force F_{mag} to the left on the magnetic dipole moment, as we saw in the previous section. We exert a force $F_{\text{by us}}$ to the right, infinitesimally larger than F_{mag}, and we move the magnetic dipole moment a short distance Δx to the right.

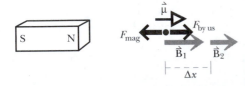

Figure 20.50 A bar magnet acts on a magnetic dipole moment.

The magnet makes a non-uniform magnetic field (proportional to $1/x^3$ if we are far from the magnet), and we are moving the magnetic dipole moment from a larger magnetic field B_1 to a smaller magnetic field B_2. The magnetic potential energy increases from $-\mu B_1$ to the somewhat less negative $-\mu B_2$. Therefore we have to do an amount of work

$$F_{\text{by us}} \Delta x = \Delta U_m - (-\mu B_2) - (-\mu B_1) = -\mu(B_2 - B_1) = -\mu \Delta B$$

This is greater than zero because $B_2 < B_1$. The magnitude of the force we have to exert is this:

$$F_{\text{by us}} = \frac{\Delta U_m}{\Delta x} = -\mu \frac{\Delta B}{\Delta x} \rightarrow -\mu \frac{dB}{dx} \text{ (a positive value; } dB/dx < 0)$$

The force that the magnet exerts to the left on the magnetic dipole moment is numerically equal to the force that we exert. This magnitude is also correct in the case of repulsion. If the magnetic dipole moment is at some angle to the magnetic field, there is both a net force and a twist.

This is a special case of the general result that the force associated with a potential energy is the negative gradient of that potential energy:

$$F_x = -\frac{dU}{dx} \text{ (general result)}$$

$$F_x = -\frac{d(-\vec{\mu} \cdot \vec{B})}{dx} = \mu \frac{dB}{dx} \text{ (our specific case)}$$

In this specific case, the quantity $\vec{\mu} \cdot \vec{B}$ is decreasing in the $+x$ direction, so the formula predicts that the x component of the force on the magnetic dipole moment is in the $-x$ direction, which is correct.

Note that there is no force if the field is uniform ($dB/dx = 0$), in agreement with our results on the previous page, though there is a twist. This is the situation with a compass needle affected by the magnetic field of the Earth, which hardly varies over short distances and therefore applies an extremely weak net force, but there is a twist due to equal and opposite forces that align the needle to point north.

We can now calculate the force that one bar magnet exerts on another. The magnetic field along the axis of one magnet, far from that magnet, is

$$B_1 \approx \frac{\mu_0}{4\pi} \frac{2\mu_1}{x^3}$$

where μ_1 is the total magnetic dipole moment of all the atomic magnetic dipole moments in the magnet.

? Show that the attractive force that this magnet would exert on a magnetic dipole moment μ_2 lying along the axis of μ_1 (Figure 20.51) is

$$F \approx 3\mu_2 \left(\frac{\mu_0}{4\pi} \frac{2\mu_1}{x^4} \right)$$

Figure 20.51 The magnetic dipole moment μ_2 of the second magnet lies along the axis of μ_1 of the first magnet.

Just take the negative gradient of the magnetic potential energy to obtain this result. This says that two magnets should attract or repel each other with a force proportional to $1/x^4$.

Moreover, using our measurements of the magnetic dipole moments of our bar magnets in the previous chapter, we should be able to calculate how close one of our magnets must be to another one in order to pick up the magnet.

Experiment 20.4 A bar magnet picks up another bar magnet
Prediction: On page 604 you measured the magnetic dipole moment μ_1 of your magnet, which you should record with this experiment. Also record the magnetic dipole moment μ_2 of your partner's magnet. Predict the numerical value of the distance x at which the magnetic attraction between these two magnets just equals the weight Mg of one magnet (you presumably measured the mass M in Problem 17.2 Predicting the magnetic dipole moment of your magnet, page 608).

Measurement: Place a magnet upright on the desk and try to lift it with the other magnet. Measure the distance x *between their centers* at the point where the lower magnet is lifted, and compare with your theoretical prediction. There is a ruler on the inside back cover of this textbook. In using the center-to-center distance we are averaging over distance, for a force that varies like $1/x^4$. This works out better than one might expect.

Problem 20.1 Picking up a paper clip
Now you are in a position to be able to explain why your magnet can pick up an unmagnetized paper clip. Put together what you know about ferromagnetic materials and about the forces on magnetic moments to explain qualitatively why the magnet picks up the paper clip.

20.8.3 *The Stern-Gerlach experiment

An important milestone in our understanding of atoms was the Stern-Gerlach experiment (1922). Silver was vaporized in an oven, and a beam of neutral silver atoms was defined by a series of slits (Figure 20.52). On theoretical grounds, each silver atom was expected to have angular momentum, and a magnetic dipole moment proportional to the angular momentum. The beam of neutral atoms passed through a strongly non-uniform magnetic field, and then struck a cold glass plate onto which the atoms condensed, leaving a visible silvery trace on the glass. (The entire apparatus was in a vacuum to avoid collisions with air molecules.)

If the atoms have a magnetic dipole moment, they should experience a vertical force, up or down in varying degrees depending on the orientation of their magnetic dipole moments. One would expect a random orientation of the magnetic dipole moments, so that there should be a vertical band of silver deposited on the glass. For example, if the magnetic dipole moment happens to be horizontal, the atom goes straight through without any deflection. (The motion is actually more complicated because the magnetic dipole moment oscillates back and forth in the magnetic field, like the motion around north of an undamped compass needle, unless it happens to point in the direction of the magnetic field. This does not affect the general conclusions, however.)

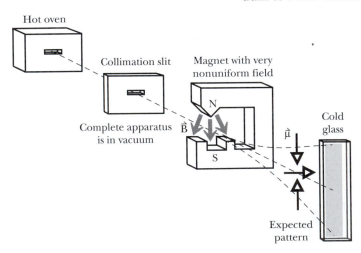

Figure 20.52 Classically one expects a continuous pattern of silver deposited on the glass.

Instead of being spread out uniformly as one would expect, what Stern and Gerlach actually observed was that the atoms always hit either at the top or at the bottom of the expected region (Figure 20.53).

This was interpreted as indicating that the orientation of the angular momentum of the atom is "quantized"—that whenever the atom is observed in this way we find that a component of angular momentum can have only certain quantized values. This peculiar quantization effect is typical of the behavior of atoms and subatomic particles.

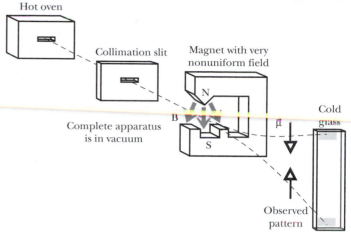

Figure 20.53 What one observes is a quantized pattern of silver deposited on the glass.

20.9 *Motors and generators

Electric motors are the most important everyday application of the torque that a magnetic field exerts on a current-carrying coil. Our daily lives are greatly affected by electric motors in refrigerator compressors, water pumps, computer disk drives, VCRs, elevators, automobile starter motors, clocks, etc.

We have seen that a current-carrying coil twists to its most stable position in a magnetic field. How can we get the coil to turn continuously? The key is to make electrical connections to the coil in such a way that just as it is coming to its stable position, we reverse the direction of the current. Assuming that it is rotating fast enough at that moment, the coil will continue on around in the same direction to the new stable position rather than simply going backwards. Then as the coil approaches the new stable position we again switch the direction of the current.

A simple way to achieve continuous rotational motion of the coil is with a "split-ring commutator" that automatically changes the direction of the current through the coil at just the right moment (Figure 20.54). Springy metal tabs or brushes make contact between the battery and the commutator. While actual motors have a wide variety of designs, this simple single-loop motor illustrates the basic principle involved in motors driven by "direct current" (DC, as opposed to alternating current, AC).

20.9.1 *Electric generators

It would be inconvenient to build a generator in the form of a bar riding on rails, such as we considered earlier in this chapter. It is mechanically awkward to move a bar or loop back and forth in a linear motion, and it is difficult to avoid having a lot of friction between a bar and supporting rails. Commercial generators rotate a loop of wire in a magnetic field rather than moving a bar back and forth. This has the same effect of achieving a charge separation along part of the rotating loop, but rotary motion is much easier to arrange mechanically, and with relatively low friction on the axle.

20.9.2 *Rotating loop in an external magnetic field

Suppose we turn a crank to rotate a loop of dimensions $w \times h$ at a constant angular speed $\omega = d\theta/dt$ (in radians per second) inside a uniform magnetic field made by some large magnet or coils (Figure 20.55). The tangential speed of the left or right wire is $v = \omega(h/2)$, since the wire is a distance $h/2$ from the axle.

? Look at the magnetic forces on conventional positive charges inside the wires. At the moment shown in Figure 20.55, in which direction will conventional current run in this loop, clockwise or counter-clockwise?

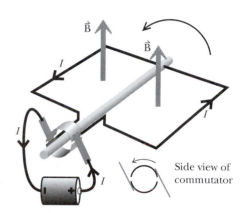

Figure 20.54 A "split-ring commutator" rotates with the coil, and as the coil reaches its most stable position the current in the coil reverses direction.

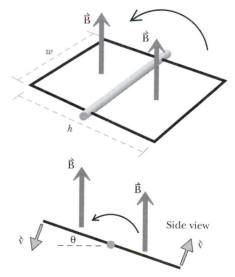

Figure 20.55 Rotating a loop in a magnetic field produces a current.

The magnetic forces drive current out of the page on the right wire and into the page on the left wire. Therefore there will be a clockwise conventional current, as seen from above (Figure 20.56). (As far as the front and back wires are concerned, the magnetic forces merely polarize them, perpendicular to the wires.)

? At a moment when the loop is at an angle θ to the horizontal, what is the magnitude of the magnetic force that acts on a charge carrier q in the right wire?

The magnetic force on a charge carrier q in the wire is $\vec{F} = q\vec{v} \times \vec{B}$, where $\vec{v}$ is the velocity of the wire. The magnitude of this force is $F = qvB\sin\theta$.

? At a moment when the loop is at an angle θ to the horizontal, what is the motional emf created along the right wire in Figure 20.55?

The magnetic force is uniform throughout the right wire at this instant, so the motional emf, non-Coulomb work per unit charge, is given by the magnetic force times the length w, divided by q:

$$\text{emf along left wire} = \frac{(qvB\sin\theta)w}{q} = vBw\sin\theta$$

The left wire contributes the same emf, acting in the same direction around the loop, so the emf around the loop is this:

$$\text{emf around loop} = 2vBw\sin\theta$$

Since $v = \omega(h/2)$, and $\sin\theta = \sin(\omega t)$, the emf can be written as

$$\text{emf around loop} = \omega B(hw)\sin(\omega t)$$

where hw is the area of the loop.

We can take current out of the loop through metal tabs pressed against rotating rings connected to the loop, as shown in Figure 20.56.

? If a light bulb with resistance R is connected as shown in Figure 20.56, what is the magnitude of the current I in the loop at this instant, assuming that the loop itself has negligible resistance?

The current will be $I = \omega B(hw)\sin(\omega t)/R$. We see that if we rotate a loop at constant speed inside a magnetic field, we generate an emf and can drive current through a light bulb. This is the basic principle of the electric generators used to power our cities, farms, and factories. The only significant difference from this simple generator is that a commercial generator has many turns in the loop to get a bigger emf. Having N turns is equivalent to having N single-loop generators in series, so we get N times the single-loop emf.

The emf produced in this rotating loop is proportional to $\sin(\omega t)$, so it will drive current through a bulb first one way, then the other way, with a sinusoidal variation. Such currents are called "alternating currents" or AC, and this is what is supplied at wall sockets. This is in contrast to the steady one-directional current maintained by a battery (or a bar sliding on rails through a uniform magnetic field), which is called "direct current" or DC.

If the frequency of the alternating current is high enough, a bulb filament doesn't have time to cool off very much between peaks in the cycle, so that the light seems steady to the eye (and there is persistence of vision in the eye, as well). This is the situation with incandescent bulbs in reading lamps, where the frequency f is 60 hertz in some countries and 50 hertz in others (with the angular frequency $\omega = 2\pi f$ in radians per second).

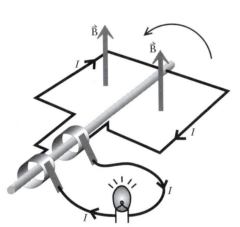

Figure 20.56 If you rotate the loop you can light the bulb, whose resistance is R.

Ex. 20.21 For what value of the angle $\theta = \omega t$ is the instantaneous emf a maximum? A minimum?

Ex. 20.22 Rotate a 40-turn rectangular loop that is 0.1 m by 0.2 m in a uniform magnetic field of 2 tesla, turning at a rate of 30 revolutions (2π radians) per second. What is the ~~maximum~~ emf?

20.9.3 *Power required to turn a generator

Presumably we have to exert ourselves to rotate the loop in the magnetic field—we're not going to light the bulb for nothing! Let's see how much power we have to input to make the loop rotate. In the previous section we calculated the current I through an attached light bulb (as a function of the angle $\theta = \omega t$ of the loop to the horizontal). In order to turn the loop in Figure 20.57, we have to exert a varying force $F_{\text{by us}} = IwB\sin\theta$ at both ends of the loop, perpendicular to the loop, to balance the perpendicular component $F_{\perp} = IwB\sin\theta$ of the magnetic force on the loop.

When we rotate the loop through a small angle $\Delta\theta$, we move each end a short distance $(h/2)\Delta\theta$ and do a small amount of work

$$\Delta W = 2[IwB\sin\theta]\left[\frac{h}{2}\Delta\theta\right]$$

Divide this result by Δt and take the limit as Δt approaches zero to calculate the mechanical power dW/dt we supply:

$$\frac{dW}{dt} = 2[IwB\sin\theta]\left[\frac{h}{2}\frac{d\theta}{dt}\right] = I[Bwh\omega\sin\theta]$$

Given your calculation in the preceding section that $I = Bwh\omega\sin(\theta)/R$ when a bulb of resistance R is attached (where $\theta = \omega t$), we can write this:

$$\frac{dW}{dt} = I[RI] = RI^2$$

This is the power dissipated in the resistor at this instant. So we don't get something for nothing; the output power of the resistor that heats the room is equal to the input power that we supply. An electric "generator" might better be called a "converter," because it doesn't create electric power for free, it merely converts mechanical power into electric power.

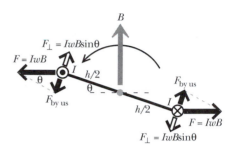

Figure 20.57 Calculating how much work is required to turn a current loop in a magnetic field.

20.10 Summary

Fundamental principles

The Lorentz force

$$\vec{F} = q\vec{E} + q\vec{v} \times \vec{B}$$

New concepts

The pattern of electric and magnetic fields is different in stationary and moving reference frames. A phenomenon such as the polarization of a metal bar may have different explanations in different reference frames. The following transforms relate the fields $\vec{E}'$ and $\vec{B}'$ in a reference frame moving at speed v to the fields $\vec{E}$ and $\vec{B}$ in a stationary reference frame.

$$E_x' = E_x \qquad E_y' = \frac{(E_y - vB_z)}{\sqrt{1 - v^2/c^2}} \qquad E_z' = \frac{(E_z + vB_y)}{\sqrt{1 - v^2/c^2}}$$

$$B_x' = B_x \qquad B_y' = \frac{\left(B_y + \dfrac{v}{c^2}E_z\right)}{\sqrt{1 - v^2/c^2}} \qquad B_z' = \frac{\left(B_z - \dfrac{v}{c^2}E_y\right)}{\sqrt{1 - v^2/c^2}}$$

Results

Force on a short length of current-carrying wire

$$\Delta\vec{F}_{\text{magnetic}} = I\Delta\vec{l} \times \vec{B}$$

Circular motion in a uniform magnetic field

$$\omega = \frac{|q|B}{m}\sqrt{1 - \frac{v^2}{c^2}} \approx \frac{|q|B}{m} \text{ if } v << c$$

Momentum $p = |q|Br$, valid even for relativistic speeds.

The *Hall effect*: a sideways electric field in a current-carrying wire in a magnetic field due to the magnetic force on moving particles. The Hall effect allows us to determine the sign and number density of the mobile charges in a particular material.

Motional emf is the integral of the magnetic force that acts on mobile charges in a wire moving through a magnetic field.

A current-carrying coil or magnetic dipole moment experiences a torque in a magnetic field, and twists to align with the applied magnetic field.

$$\vec{\tau} = \vec{\mu} \times \vec{B} \text{ magnetic torque on a magnetic dipole moment}$$

Potential energy associated with a magnetic dipole moment is $U_m = -\vec{\mu} \bullet \vec{B}$

The force on a magnetic dipole moment is $F_x = -\dfrac{dU_m}{dx} = \dfrac{d(\vec{\mu} \bullet \vec{B})}{dx}$

20.11 Example problems

20.11.1 Example 1: Electron and current loop

A 2-turn circular loop of wire of radius R is fed a large conventional current I from long straight wires as shown. A charge of magnitude $+Q$ is a distance $3R$ to the right of the center of the loop. An electron traveling upward with speed v passes through the center of the loop, and the net force on this electron at this instant is zero. The Earth's magnetic field is negligible in this region. What are the magnitude and direction of the current I? Assume all other quantities are known.

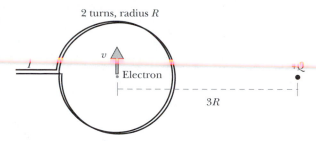

Figure 20.58 An electron inside two turns of current-carrying wire.

Solution

The key idea here is that $\vec{F}_{mag} + \vec{F}_{el} = 0$ at this instant.

Direction of current

At the location of the electron, the electric field due to the positive charge is in the $-x$ direction, so the electric force on the electron is in the $+x$ direction, as shown in Figure 20.59. Since at the instant specified the net force on the electron is zero, the magnetic force on the electron must be equal and opposite to the electric force. $\vec{F}_{mag}$ is in the $-x$ direction, but the electron's charge is negative, so $\hat{v} \times \vec{B}$ must be in the $+x$ direction, as shown in Figure 20.60. By trial and error with the right-hand rule (thumb in direction of $\hat{v} \times \vec{B}$, fingers initially pointing up), we find that $\vec{B}$ must be in the $+z$ direction. Applying the right hand rule for a current loop, we find that conventional current must flow counter-clockwise around the loops to make a magnetic field in this direction.

Figure 20.59 The directions of the electric and magnetic forces on the electron.

Magnitude of current

Since the net force on the electron is zero, we know that:

$$\left|\vec{F}_{mag}\right| = \left|\vec{F}_{el}\right|$$

$$\frac{1}{4\pi\varepsilon_0}\frac{eQ}{(3R)^2} = evB$$

$$B = \frac{\frac{1}{4\pi\varepsilon_0}\frac{Q}{9R^2}}{v}$$

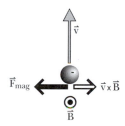

Figure 20.60 Finding the direction of the magnetic field.

We found in Chapter 17 that the magnetic field of a loop was given by:

$$B_{loop} = \frac{\mu_0}{4\pi}\frac{2\pi R^2 I}{(z^2 + R^2)^{3/2}}$$

At the center of the loops $z = 0$, so $B_{loop} = \frac{\mu_0}{4\pi}\frac{2\pi I}{R}$, and for 2 loops:

$$\frac{\frac{1}{4\pi\varepsilon_0}\frac{Q}{9R^2}}{v} = 2\frac{\mu_0}{4\pi}\frac{2\pi I}{R}$$

$$I = \frac{1}{\varepsilon_0\mu_0}\frac{Q}{36\pi Rv} = \frac{Qc^2}{36\pi Rv} \text{ since } \frac{1}{\varepsilon_0\mu_0} = c^2$$

This quantity has units of C/s, which is correct for conventional current.

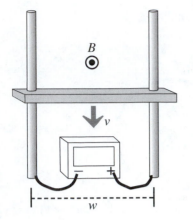

Figure 20.61 As the metal bar moves downward, what does the voltmeter read?

Figure 20.62 Magnetic force on a positive charge; resulting polarization of the bar; and electric field due to polarization.

20.11.2 Example 2: A metal bar moving downward

In Figure 20.61 a horizontal metal bar is moving downward in the plane of the page with electrical contact along two metal vertical bars that are a distance w apart. Throughout this region there is a uniform horizontal magnetic field with magnitude B out of the plane of the page, made by large coils that are not shown. At the instant that the moving bar has a speed v, what is the reading on the voltmeter? Give magnitude and sign, and explain briefly.

Solution

The charged particles in the bar are moving through a magnetic field, and therefore experience a magnetic force. It does not matter what sign we assume for the mobile charges, so we will assume they are positive because it simplifies keeping track of signs.

$\vec{v} \times \vec{B}$ is to the left, so the bar polarizes as shown in Figure 20.62.

The polarization takes an extremely short time (less than nanoseconds), after which for every mobile charge in the bar:

$$\vec{F}_{mag} + \vec{F}_{el} = 0$$
$$qvB\sin(90°) = qE$$
$$E = vB$$

Since E is uniform throughout the bar,

$$|\Delta V| = \left| -\int \vec{E} \cdot d\vec{l} \right| = vBw$$

The left end of the bar is at a higher potential than the right end. Since the high potential end is connected to the negative terminal of the voltmeter, the reading on the voltmeter will be negative. The reading on the voltmeter is therefore $-vBw$.

20.11.3 Example 3: Determining electrical properties

An experiment was carried out to determine the electrical properties of a new conducting material. A bar was made out of the material, 9 cm long with a rectangular cross section 4 cm wide and 2 cm deep. The bar was inserted into a circuit with a 1.5-volt battery and carried a constant current of 0.7 ampere (Figure 20.63). The resistance of the copper connecting wires and the ammeter, and the internal resistance of the battery, were all negligible compared to the resistance of the bar.

(a) On the diagram, show the approximate distribution of charge.

(b) What is the resistance in ohms of the bar?

(c) How much power is supplied by the battery?

Using large coils not shown in Figure 20.64, a uniform magnetic field of 1.7 tesla was applied perpendicular to the bar (to the left, as shown in Figure 20.64). A voltmeter was connected across the bar, with the connections across the bar carefully placed directly across from each other as shown. The voltmeter read −0.29 millivolts. Assume that there is only one kind of mobile charge in the bar material.

(d) What is the sign of the mobile charges in the bar, and which way do they move? Explain carefully, using diagrams to support your explanation.

(e) How long does it take for a mobile charge to go from one end of the bar to the other?

(f) What is the mobility u of the mobile charges?

(g) If each mobile charge is singly charged ($|q| = e$), how many mobile charges are there in 1 m^3 of this material?

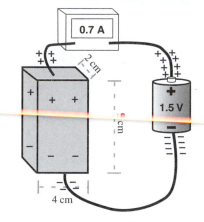

Figure 20.63 Measuring electrical properties of a new material (parts a-c).

Solution

(a) See surface charges shown on Figure 20.63.

(b) Loop rule: $(1.5 \text{ V}) + (-R(0.7 \text{ A})) = 0$, so $R = \dfrac{1.5 \text{ V}}{0.7 \text{ A}} = 2.1 \text{ ohms}$.

(c) Power = $(1.5 \text{ V})(0.7 \text{ A}) = 1.05 \text{ watt}$

(d) Let's try assuming the charge carriers are negative, in which case they drift upward through the bar as shown on Figure 20.64 and experience a magnetic force toward the back. Negative charge builds up on the back, positive on the front. Since the positive terminal of the voltmeter is connected to the lower potential surface, the voltmeter should read negative, and it does. (If the assumption of negative charges had given the wrong prediction for the sign of the voltmeter reading, we would have tried positive charges.)

(e) In the steady state, the transverse magnetic and electric forces balance, so $eE_\perp = evB$. We have $E_\perp = (\Delta V_\perp)/d = vB$, where $d = 2 \text{ cm}$, so

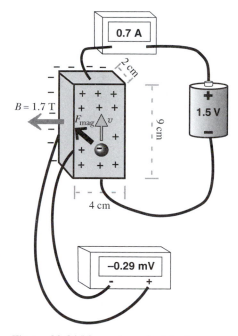

Figure 20.64 Measuring electrical properties of a new material (parts d-g).

$$v = \frac{\Delta V_\perp}{Bd} = \frac{0.29 \times 10^{-3} \text{ volt}}{(1.7 \text{ T})(0.02 \text{ m})} = 8.5 \times 10^{-3} \text{ m/s}$$

$$\Delta t = \frac{0.09 \text{ m}}{8.5 \times 10^{-3} \text{ m/s}} = 10.6 \text{ s}$$

(f) $v = uE$, where $E = (1.5 \text{ V})/(0.09 \text{ m}) = 16.7 \text{ V/m}$ is the longitudinal electric field in the direction of the current:

$$u = v/E = (8.5 \times 10^{-3} \text{ m/s})/(16.7 \text{ V/m}) = 5 \times 10^{-4} \text{ (m/s)/(V/m)}$$

(g) $I = enAv$, so $n = I/(eAv)$, so we have this:

$$n = \frac{(0.7 \text{ A})}{(1.6 \times 10^{-19} \text{ C})(0.02 \text{ m})(0.04 \text{ m})(8.5 \times 10^{-3} \text{ m/s})} = 6.4 \times 10^{23} \text{ m}^{-3}$$

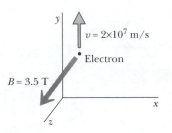

Figure 20.65 Magnetic force on a moving electron (RQ 20.1).

20.12 Review questions

Magnetic force on a moving charge

RQ 20.1 An electron is moving in the y direction at a speed $v = 2 \times 10^7$ m/s at a point where there is a magnetic field $B = 3.5$ tesla in the z direction (Figure 20.65). What is the magnitude and direction of the magnetic force on the moving electron? Draw the force vector on a diagram of the situation.

Electric and magnetic force

RQ 20.2 An object of charge q is moving in the x direction with speed v. Throughout the region there is a uniform electric field in the y direction of magnitude E (Figure 20.66). Determine a direction and magnitude B for a uniform magnetic field such that the net electric and magnetic force on the moving charge is zero. Draw $\vec{B}$ on a diagram of the situation.

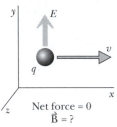

Figure 20.66 What is the magnetic field direction and magnitude so that the net force on the electron is zero (RQ 20.2)?

Circular motion

RQ 20.3 An alpha particle (consisting of two protons and two neutrons) is moving at constant speed in a circle, perpendicular to a uniform magnetic field applied by some current-carrying coils, making one clockwise revolution every 80 nanoseconds (Figure 20.67). If the speed is small compared to the speed of light, what is the numerical magnitude B of the magnetic field made by the coils? What is the direction of this magnetic field?

Figure 20.67 Some coils (not shown) make a magnetic field in which an alpha particle moves in a circle (RQ 20.3).

RQ 20.4 An electron is moving with a speed v in the plane of the page, and there is a uniform magnetic field B into the page throughout this region; the magnetic field is produced by some large coils which are not shown in Figure 20.68. Draw the trajectory of the electron, and explain qualitatively.

Figure 20.68 Trajectory of an electron (RQ 20.4).

The Hall effect

RQ 20.5 A copper wire with square cross section carries a conventional current I to the left. There is a magnetic field B perpendicular to the wire (Figure 20.69). Describe the direction of $E_\perp$, the transverse electric field inside the wire due to the Hall effect, and explain briefly.

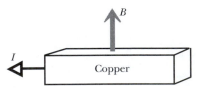

Figure 20.69 What is the direction of the transverse Hall-effect electric field (RQ 20.5)?

RQ 20.6 You hold your magnet perpendicular to a bar of copper which is connected to a battery (Figure 20.70). Describe the directions of the components of the electric field at the center of the copper bar, both parallel to the bar and perpendicular to the bar. Explain carefully.

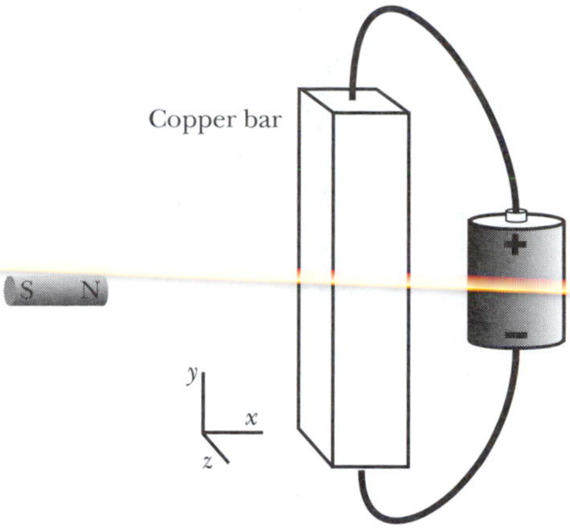

Two straight wires

RQ 20.7 A 1.5-volt battery is connected to a resistive wire whose resistance is 5 ohms (Figure 20.71). The wire is laid out in the form of a rectangle 50 cm by 3 centimeters. What are the approximate magnitude and direction of the magnetic force that the upper part of the wire exerts on the lower part of the wire in the diagram? Explain briefly.

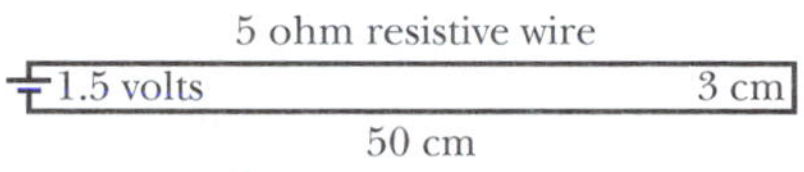

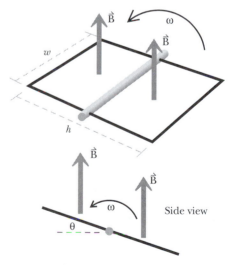

Figure 20.70 Effect of a magnet on a copper current-carrying wire (RQ 20.6).

5 ohm resistive wire

1.5 volts　　　　　　　　　3 cm

50 cm

Figure 20.71 What is the force that the upper wire exerts on the lower wire (RQ 20.7)?

Magnetic force on a current-carrying wire

RQ 20.8 If your flashlight battery has an internal resistance of about one-quarter ohm, and your bar magnet produces a magnetic field of about 0.1 tesla near one end of the magnet, what is the approximate magnitude of the magnetic force on 5 centimeters of a wire that short-circuits the battery (Figure 20.72)? Indicate the direction of this force on a diagram. Check your direction experimentally.

Motional emf

RQ 20.9 Two metal wires of length w are attached to two metal wires of length h to form a rectangle of dimensions w by h (Figure 20.73). The rectangle rotates counter-clockwise on a frictionless axle (due to external forces) with constant angular velocity ω in a uniform upward magnetic field B.

　(a) At the moment when the rectangle is at an angle θ from the horizontal, what is the magnitude of the emf around the circuit and the direction of the current I?

　(b) Why are external forces required to rotate the loop, even though the rectangle rotates at a constant angular speed on a frictionless axle? Show the direction of the external forces on a diagram.

RQ 20.10 A metal rod of length L slides horizontally at constant speed v on frictionless insulating rails through a region of uniform upward magnetic field of magnitude B (Figure 20.74). On a diagram, show the polarization of the rod, and the direction of the Coulomb electric field inside the rod. Explain briefly. What is the magnitude E_C of the Coulomb electric field inside the rod? What is the potential difference across the rod? What is the emf across the rod? What is the magnitude and direction of the force you have to apply to keep the rod moving at a constant speed v?

Figure 20.72 Your bar magnet exerts a force on a current-carrying wire (RQ 20.8).

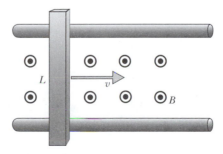

Figure 20.73 A loop is rotated in a magnetic field and develops an emf (RQ 20.9).

Figure 20.74 A metal rod slides on insulating rails through a region of uniform magnetic field (RQ 20.10).

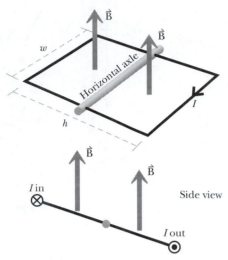

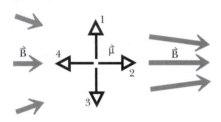

Figure 20.75 A rectangular current-carrying loop of wire rotates in a magnetic field (RQ 20.11).

Figure 20.77 What is the initial force on a magnetic dipole moment in one of these four orientations (RQ 20.13)?

Torque on a current loop

RQ 20.11 A current-carrying rectangular loop of wire mounted on a pivot, held at an angle to a uniform magnetic field (Figure 20.75). If the loop is released, will it turn clockwise or counter-clockwise? Explain briefly, and use the diagram in your explanation.

RQ 20.12 A bar magnet is mounted on a pivot (Figure 20.76) and released in the presence of a magnetic field $\vec{B}$ made by coils that are not shown. Does the magnet initially turn clockwise or counterclockwise? Explain qualitatively in detail why it turns, and why it turns in the predicted direction.

Figure 20.76 Effect on a pivoted bar magnet of an applied magnetic field (RQ 20.12).

Force on a magnetic dipole moment

RQ 20.13 Figure 20.77 shows a region of non-uniform magnetic field. The field is stronger to the right. What is the initial direction of the magnetic force on a magnetic dipole moment $\vec{\mu}$ in each of the four orientations shown in the diagram? (Hint: a magnetic dipole moment if free to move goes to a region of lower potential energy. Or check with what you know about the interaction of two magnets.)

Different reference frames

RQ 20.14 If you travel at speed v along with the moving bar in Figure 20.74, in your reference frame the charged particles in the bar are not moving. Explain why, in your moving reference frame, the bar still polarizes.

RQ 20.15 A proton moves with speed v in the $+x$ direction. In the lab (stationary) reference frame, what electric and magnetic fields would you observe a distance r above the proton (in the $+y$ direction)? In a reference frame moving with the proton, what electric and magnetic fields would you observe at the same location?

20.13 Homework problems

20.13.1 In-line problem from this chapter

Problem 20.1 Picking up a paper clip, page 744
Now you are in a position to be able to explain why your magnet can pick up an unmagnetized paper clip. Put together what you know about ferro-magnetic materials and about the forces on magnetic moments to explain qualitatively why the magnet picks up the paper clip.

20.13.2 Additional problems

Problem 20.2 Moving proton and moving electron
An electron and a proton are both in motion near each other as shown in Figure 20.78; at this instant they are a distance r apart. The proton is moving with speed v_p at an angle θ to the horizontal, and the electron is moving straight up with speed v_e.

(a) Calculate the x and y components of the force that the proton exerts on the electron.

> First calculate the electric field and magnetic field that the proton produces at the location of the electron, then calculate the forces that these fields exert on the electron.

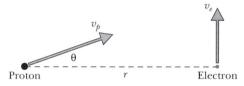

Figure 20.78 A proton and an electron exert forces on each other (Problem 20.2).

(b) Calculate the x and y components of the force that the electron exerts on the proton.

> First calculate the electric field and magnetic field that the electron produces at the location of the proton, then calculate the forces that these fields exert on the proton.

(c) There is momentum associated with the electric and magnetic fields of these particles, as we will see in a later chapter. If we don't account for this "field momentum" we can't be sure without an explicit calculation whether the principle of reciprocity (Newton's third law) holds true for the proton and electron or not. Are your results from parts (a) and (b) consistent or inconsistent with the principle of reciprocity? That is, check whether the x and y components of the forces are equal and opposite.

(d) Will the total momentum of the two particles remain constant? (In a completely correct calculation we would have to use the relativistically correct fields. However, we would still reach the same conclusion.)

Problem 20.3 Mass spectrometer
A mass spectrometer is a tool used to determine accurately the mass of individual ionized atoms or molecules, or to separate atoms or molecules that have similar but slightly different masses. For example, you can deduce the age of a small sample of cloth from an ancient tomb, by using a mass spectrometer to determine the relative abundances of carbon-14 (whose nucleus contains 6 protons and 8 neutrons) and carbon-12 (the most common isotope, whose nucleus contains 6 protons and 6 neutrons).

Background: In organic material the ratio of ^{14}C to ^{12}C depends on how old the material is, which is the basis for "carbon-14 dating." ^{14}C is continually produced in the upper atmosphere by nuclear reactions caused by "cosmic rays" (high-energy charged particles from outer space, mainly protons), and ^{14}C is radioactive with a half-life of 5700 years. When a cotton plant is growing, some of the CO_2 it extracts from the air to build tissue contains ^{14}C which has diffused down from the upper atmosphere. But after the cotton has been harvested there is no further intake of ^{14}C from the air, and the cosmic rays that create ^{14}C in the upper atmosphere can't penetrate the atmosphere and reach the cloth. So the amount of ^{14}C in cotton cloth continually decreases with time, while the amount of non-radioactive ^{12}C remains constant.

Here is a particular kind of mass spectrometer (Figure 20.79). Carbon from the sample is ionized in the ion source at the left. The resulting singly ionized ^{12}C$^+$ and ^{14}C$^+$ ions have negligibly small initial velocities (and can be considered to be at rest). They are accelerated through the potential difference ΔV_1. They then enter a region where the magnetic field has a fixed magnitude $B = 0.2$ tesla. The ions pass through electric deflection plates that are 1 cm apart and have a potential difference ΔV_2 that is adjusted so that the electric deflection and the magnetic deflection cancel each other for a particular isotope: one isotope goes straight through, and the other isotope is deflected and misses the entrance to the next section of the spectrometer. The distance from the entrance to the fixed ion detector is a distance of 20 cm.

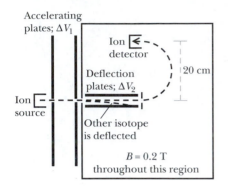

Figure 20.79 A mass spectrometer (Problem 20.3).

There are controls that let you vary the accelerating potential ΔV_1 and the deflection potential ΔV_2 in order that only ^{12}C$^+$ or ^{14}C$^+$ ions go all the way through the system and reach the detector. You count each kind of ion for fixed times and thus determine the relative abundances. The various deflections insure that you count only the desired type of ion for a particular setting of the two voltages.

Carry out the following calculations, and give brief explanations of your work:

(a) Determine which accelerating plate is positive (left or right), which deflection plate is positive (upper or lower), and the direction of the magnetic field.

(b) Determine the appropriate numerical values of ΔV_1 and ΔV_2 for ^{12}C. Carry out your intermediate calculations algebraically, so that you can use the algebraic results in the next part.

(c) Determine the appropriate numerical values of ΔV_1 and ΔV_2 for ^{14}C.

Problem 20.4 Cyclotron

The design and operation of a cyclotron is discussed in Section 20.1.4.

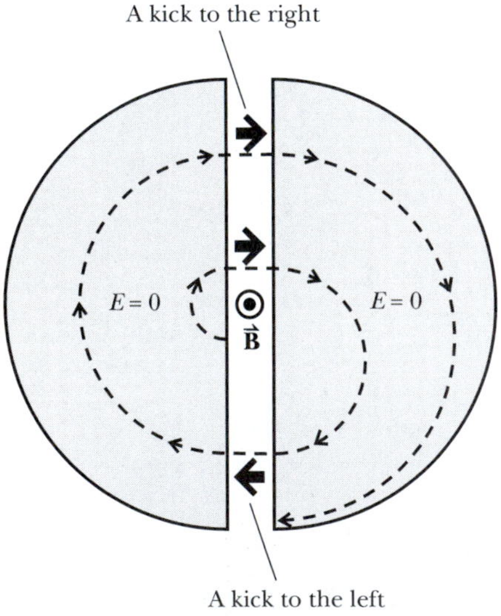

(a) Show that the "period" of the motion, the time between one kick to the right and the next kick in the same direction, does not depend on the current speed of the proton (at speeds small compared to the speed of light). As a result, we can place across the dees a simple sinusoidal potential difference having this period and achieve continual acceleration out to the maximum radius of the cyclotron. See Figure 20.80.

(b) One of Ernest Lawrence's first cyclotrons, built in 1932, had a diameter of only about 30 cm and was placed in a magnetic field of about 1 tesla. What was the frequency (= 1/period, in hertz = cycles per second) of the sinusoidal potential difference placed across the dees to accelerate the protons?

(c) Show that the equivalent accelerating potential of this little cyclotron was about a million volts! That is, the kinetic energy gain from the center to the outermost radius was $\Delta K = e\Delta V_{eq}$, with $\Delta V_{eq} = 10^6$ volts. (Hint: Calculate the final speed at the outermost radius.) (continued on next page)

(d) If the sinusoidal potential difference applied to the dees had an amplitude of 500 volts (that is, it varied between +500 and −500 volts), show that it took about 65 microseconds for a proton to move from the center to the outer radius.

Figure 20.80 A cyclotron (Problem 20.4).

Problem 20.5 Computer model of a cyclotron

In this problem you will write a computer program to compute and display the motion of a proton inside a cyclotron (see Section 20.1.4). The first cyclotrons were small desktop devices; we will model a small device in order to understand the basic physical phenomena. Consider a cyclotron whose radius is 5 cm, with a gap of 0.5 cm between the dees (Figure 20.81). External magnets create a uniform magnetic field of 1.0 tesla in the $+y$ direction throughout the device. A proton starts essentially from rest at the center of the cyclotron. The maximum potential difference between the dees is 5000 V. Note that the electric field inside the dees is 0 at all times.

(a) What is the angular frequency ω of the sinusoidally oscillating potential difference you will need to apply?

(b) Compute and display the trajectory of a proton in the cyclotron. You may assume that the proton does not reach a speed sufficiently close to the speed of light to require a relativistically correct calculation, but do check this assumption. To get your program going, you can use a large value of dt (such as 1E–10 seconds), but you will need to decrease this value later.

(c) Plot the proton's kinetic energy K (in eV) *vs.* time. Explain the shape of the graph you observe.

(d) What happens if you use a different angular frequency for the potential difference? For example, try an angular frequency of 0.6ω.

(e) If the proton gains a significant amount of kinetic energy while inside the dees, your dt is too large. Adjust your dt appropriately. What value of dt gives reasonable results?

(f) How much time does the proton take to reach the outer edge of the cyclotron?

(g) What is K (in eV) when the proton reaches the outer edge of the cyclotron?

(h) Given your answer to (g), how many orbits must the proton have made before reaching the outer edge of the cyclotron?

(i) What is the final speed of the proton? Were you justified in assuming that $v \ll c$?

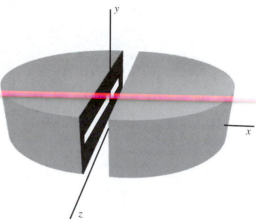

Figure 20.81 Geometry of a cyclotron (Problem 20.5).

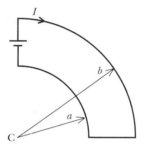

Figure 20.82 A circuit produces magnetic field and affects a moving electron (Problem 20.6).

Problem 20.6 Force on an electron due to a circuit

A Nichrome wire has the shape of two quarter-circles of radius a and b, connected by straight sections (Figure 20.82). The wire carries conventional current I in the direction shown.

(a) Calculate the direction and magnitude of the magnetic field at point C, the center of the quarter-circles. Briefly explain your work, including a diagram.

(b) At a particular instant, an electron is at point C, traveling to the right with speed v. Calculate the direction and magnitude of the magnetic force on the electron. Briefly explain your work, including a diagram.

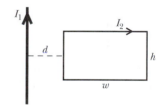

Figure 20.83 A current-carrying wire interacts with a current-carrying loop (Problem 20.7).

Problem 20.7 Force on a loop due to a straight wire

A long wire carries a current I_1 upward, and a rectangular loop of height h and width w carries a current I clockwise (Figure 20.83). The loop is a distance d away from the long wire. The long wire and the rectangular loop are in the same plane. Find the magnitude and direction of the net force exerted by the long wire on the rectangular loop. Briefly explain your work, including a diagram.

Problem 20.8 Electric and magnetic forces on an electron

Two long straight wires carrying conventional current I are connected by a three-quarter-circular arc of radius R (Figure 20.84). (The rest of the circuit is far enough away that it contributes little magnetic field in this region.)

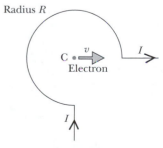

Figure 20.84 A portion of a circuit—the rest of the circuit is far away and has little effect (Problem 20.8).

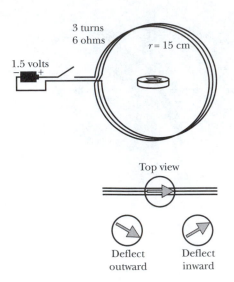

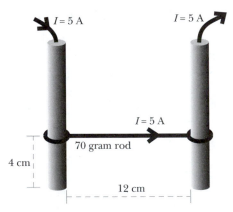

Figure 20.85 A coil produces a magnetic field that affects a compass or an electron (Problem 20.9).

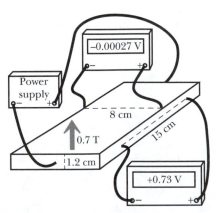

Figure 20.86 Magnetically lifting a current-carrying metal rod (Problem 20.10).

Figure 20.87 A slab of unknown material is connected to a power supply (Problem 20.11).

Electric fields are also present in this region, due to charges not shown on the drawing. An electron is moving to the right with speed v as it passes through the center C of the arc, and at that instant the net electric and magnetic force on the electron is zero.

(a) What is the direction of the magnetic field at the center C, due to the current I?

(b) What is the direction of the electric field at the center C?

(c) What is the magnitude of the magnetic field at the center C, due to the current I?

(d) What is the magnitude of the electric field at the center C?

Problem 20.9 Force on an electron due to a coil

A thin circular coil of radius $r = 15$ cm contains $N = 3$ turns of Nichrome wire. The coil has a resistance of 6 ohms. A small compass is placed at the center of the coil. With the battery disconnected, the compass needle points to the right, in the plane of the coil (Figure 20.85). The apparatus is located near the equator, where the Earth's magnetic field B_{Earth} is horizontal, with a magnitude of about 4×10^{-5} tesla.

(a) When the 1.5-volt battery is connected, predict the deflection of the compass needle. If you have to make any approximations, state what they are. Is the deflection outward or inward as seen from above? What is the magnitude of the deflection?

(b) The compass is removed. The current in the coil continues to run. An electron is at the center of the coil and is moving with speed $v = 5 \times 10^6$ m/s into the page (perpendicular to the plane of the coil). In addition to the magnetic fields in the region due to the Earth and the coil, there is an electric field at the center of the coil (due to charges not shown in the diagram), and this electric field points upward and has a magnitude $E = 250$ volts/meter. What is the magnitude and direction of the force on the electron at this moment? (You can neglect the gravitational force, which can easily be shown to be negligible.)

Problem 20.10 Magnetic levitation

A metal rod of length 12 cm and mass 70 grams has metal loops at both ends, which go around two metal poles (Figure 20.86). The rod is in good electrical contact with the poles but can slide freely up and down. The metal poles are connected by wires to a battery, and a 5 ampere current flows through the rod. A magnet supplies a large uniform magnetic field B in the region of the rod, large enough that you can neglect the magnetic fields due to the 5-ampere current. The rod sits at rest 4 cm above the table. What is the magnitude and direction of the magnetic field in the region of the rod? Briefly explain your work, including a diagram.

Problem 20.11 Measuring the properties of a slab of material

A slab made of unknown material is connected to a power supply as shown in Figure 20.87. There is a uniform magnetic field of 0.7 tesla pointing upward throughout this region (perpendicular to the horizontal slab). Two voltmeters are connected to the slab and read steady voltages as shown. The connections across the slab are carefully placed directly across from each other. Assume that there is only one kind of mobile charges in this material, but we don't know whether they are positive or negative.

(a) Determine the (previously unknown) sign of the mobile charges, and state which way these charges move inside the slab. Explain carefully, using diagrams to support your explanation.

(b) What is the drift speed $\bar{v}$ of the mobile charges?

(c) What is the mobility u of the mobile charges?

(continued on next page)

(d) The current running through the slab was measured to be 0.3 ampere. If each mobile charge is singly charged ($|q| = e$), how many mobile charges are there in 1 m^3 of this material?

(e) What is the resistance in ohms of a 15-cm length of this slab?

Problem 20.12 A falling metal bar

A metal bar of mass M and length L slides with negligible friction but with good electrical contact down between two vertical metal posts (Figure 20.88). The bar falls at a constant speed v (zero acceleration). The falling bar and the vertical metal posts have negligible electrical resistance, but the bottom rod is a resistor with resistance R. Throughout the entire region there is a uniform magnetic field of magnitude B coming straight out of the page. Explain every step clearly and in detail.

(a) Calculate the amount of current I running through the resistor.

(b) On a diagram, clearly show the surface-charge distribution all the way around the circuit, and the direction of the conventional current I. Explain.

(c) Calculate the constant speed v of the falling bar.

(d) Show that the rate of change of the gravitational energy of the Universe is equal to the rate of energy dissipated in the resistor.

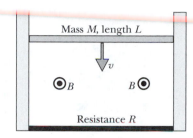

Figure 20.88 A metal bar falls through a region of magnetic field (Problem 20.12).

Problem 20.13 Dragging a loop through a magnetic field

A rectangular loop of wire, with width w, height h, and resistance R, is dragged at constant speed v to the right through a region where there is a uniform magnetic field of magnitude B out of the page (Figure 20.89). The magnetic field is negligibly small outside that region. For each stage of the motion, determine the current I in the loop and the force $\vec{F}$ required to maintain the constant speed. There are five stages:

1) before entering the magnetic-field region;
2) while the loop is part-way into the magnetic-field region (this is the stage that is shown in Figure 20.89);
3) while the loop is entirely inside the magnetic-field region;
4) while the loop is part-way out of the magnetic-field region;
5) after leaving the magnetic-field region.

Explain briefly in each case. In addition to determining the magnitude of $\vec{F}$ and I for each stage, draw a diagram for each of the 5 stages and show the direction of $\vec{F}$ and I on the diagram. Also show the approximate surface-charge distribution on the loop.

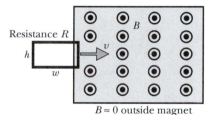

Figure 20.89 A loop of wire is dragged through a region of magnetic field (Problem 20.13).

Problem 20.14 Dragging a loop through two regions of magnetic field

In Figure 20.90, on the left is a region of uniform magnetic field B_1 into the page, and adjacent on the right is a region of uniform magnetic field B_2 also into the page. The magnetic field B_2 is smaller than B_1 ($B_2 < B_1$). You pull a rectangular loop of wire of length L, height h, and resistance R from the first region into the second region, on a frictionless surface. While you do this you apply a constant force F to the right, and you notice that the loop doesn't accelerate but moves with a constant speed.

Calculate this constant speed v in terms of the known quantities B_1, B_2, L, h, R, and F, and explain your calculation carefully. Also show the approximate surface-charge distribution on the loop.

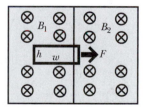

Figure 20.90 A loop of wire is dragged through two magnets (Problem 20.14).

Problem 20.15 An electron moving through a coil

Two long straight wires carrying a large conventional current I are connected by one-and-a-quarter turns of wire of radius R (Figure 20.91). An electron is moving to the right with speed v at the instant that it passes through the center of the arc. You apply an electric field $\vec{E}$ at the center of the arc

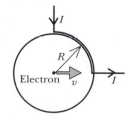

Figure 20.91 An electron moves to the right at the center of the arc (Problem 20.15).

in such a way that the net force on the electron at this instant is zero. (You can neglect the gravitational force on the electron, which is easily shown to be negligible, and the magnetic field of the coil is much larger than the magnetic field of the Earth.)

Determine the direction and magnitude of the electric field $\vec{E}$. Be sure to explain your work fully; draw and label any vectors you use.

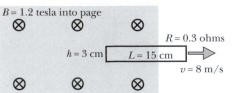

Figure 20.92 A rectangular loop emerges from a region of uniform magnetic field (Problem 20.16).

Problem 20.16 Pull a rectangular loop through a magnetic field

In Figure 20.92 a rectangular loop of wire $L = 15$ cm long by $h = 3$ cm high, with a resistance of $R = 0.3$ ohms, moves with constant speed $v = 8$ m/s to the right, emerging from a rectangular region where there is a uniform magnetic field into a region where the magnetic field is negligibly small. In the region of magnetic field, the magnetic field points into the page, and its magnitude is $B = 1.2$ tesla.

(a) What are the direction and magnitude of conventional current in the loop? Explain briefly.

(b) What are the direction and magnitude of the force that you have to apply in order to make the loop move at its constant speed?

Problem 20.17 A unipolar generator

A "unipolar" generator consists of a copper disk of radius R rotating in a uniform, steady magnetic field B perpendicular to the disk, out of the page (Figure 20.93). (The magnetic field is produced by large coils carrying constant current, not shown in the diagram.) Sliding contacts are made at the center (on the axle) and at the rim of the disk, and the wires are connected to a voltmeter. If the outer rim travels counter-clockwise at a speed v, what does the voltmeter read (magnitude and sign)?

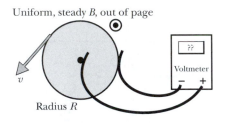

Figure 20.93 A "unipolar" generator (Problem 20.17).

Problem 20.18 Measuring properties of some material

An experiment was carried out to determine the electrical properties of a new conducting material. A bar was made out of the material, 18 cm long with a rectangular cross section 3 cm high and 0.8 cm deep. The bar was part of a circuit and carried a steady current (Figure 20.94 shows only part of the circuit). A uniform magnetic field of 1.5 tesla was applied perpendicular to the bar, coming out of the page (using some coils that are not shown).

Two voltmeters were connected along and across the bar and read steady voltages as shown (mV = millivolt = 10^{-3} volt). The connections across the bar were carefully placed directly across from each other to eliminate false readings corresponding to the much larger voltage along the bar. Assume that there is only one kind of mobile charge in this material.

(a) What is the sign of the mobile charges, and which way do they move? Explain carefully, using diagrams to support your explanation.

(b) What is the drift speed v of the mobile charges? Explain your reasoning.

(c) What is the mobility u of the mobile charges?

(d) The current running through the bar was measured to be 0.6 ampere. If each mobile charge is singly charged ($|q| = e$), how many mobile charges are there in 1 m^3 of this material?

(e) What is the resistance in ohms of this length of bar?

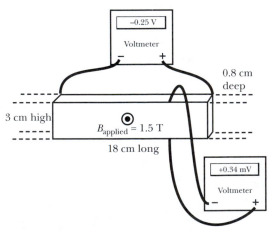

Figure 20.94 Measuring the properties of a new material (Problem 20.18).

Problem 20.19 Measuring a current indirectly

In Figure 20.95, a bar of length L, width w, and thickness d is positioned a distance h underneath a long straight wire that carries a large but unknown steady current I_{wire} (w is small compared to h, and h is small compared to the length of the long straight wire). The material that the bar is made of is known to have n positive mobile charge carriers ("holes") per cubic meter, each of charge e, and these are the only mobile charge carriers in the bar. An ammeter measures the small current I passing through the bar.

(a) Draw connecting wires from the voltmeter to the bar, showing clearly how the ends of the connecting wires must be positioned on the bar in order to permit determining the amount of current in the long straight wire I_{wire}. If the voltmeter reading ΔV is positive, does the current in the long straight wire run to the left or to the right? Explain briefly.

(b) If the voltmeter reading is ΔV, what is the current in the long straight wire, I_{wire}? Express your answer for the large current I_{wire} only in terms of the known quantities: h, L, w, d, n, $I_{ammeter}$, ΔV, and known physical constants such as e.

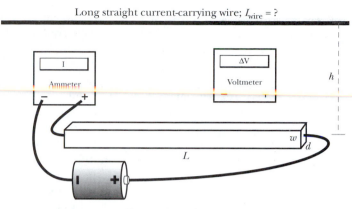

Figure 20.95 Determine the current in the long straight wire (Problem 20.19).

Problem 20.20 A mass spectrometer

In the simple mass spectrometer shown in Figure 20.96, positive ions are generated in the ion source. They are released, traveling at very low speed, into the region between two accelerating plates between which there is a potential difference ΔV. In the gray region there is a uniform magnetic field $\vec{B}$; outside this region there is negligible magnetic field. The semicircle traces the path of one singly charged positive ion of mass M, which travels through the accelerating plates into the magnetic field region, and hits the ion detector as shown.

Determine the appropriate magnitude and direction of the magnetic field $\vec{B}$, in terms of the known quantities shown on the diagram. Explain all steps in your reasoning.

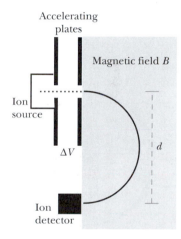

Figure 20.96 A mass spectrometer (Problem 20.20).

Problem 20.21 Two arcs and two straight wires

In Figure 20.97 two straight wires carrying conventional current I are connected by a three-quarter-circular arc of radius R_1 and a one-quarter-circular arc of radius R_2. Electric fields are also present in this region, due to charges not shown on the drawing.

An electron is moving down with speed v as it passes through the center C of the arc, and at that instant the net electric and magnetic force on the electron is zero. (The Earth's magnetic field is negligible here.)

(a) What is the direction of the magnetic field at the center C, due to the current I? Explain briefly.

(b) What is the direction of the electric field at the center C? Explain your reasoning clearly.

(c) What is the magnitude of the magnetic field at the center C, due to the current I?

(d) What is the magnitude of the electric field at the center C? Explain briefly.

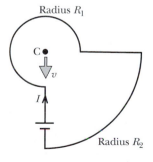

Figure 20.97 Two arcs and two straight wires (Problem 20.21).

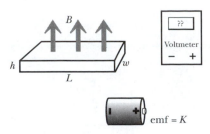

Figure 20.98 Find the magnitude B of the magnetic field (Problem 20.22).

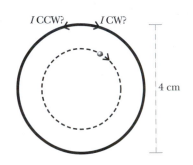

Figure 20.99 An orbiting nucleus (Problem 20.23).

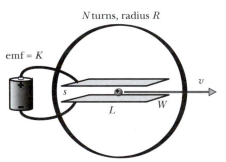

Figure 20.100 Plates inside a coil (Problem 20.24).

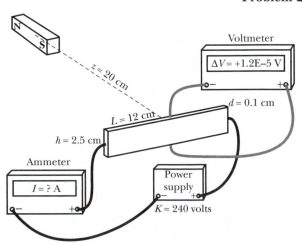

Figure 20.101 Determining a magnetic dipole moment (Problem 20.25).

Problem 20.22 Measuring magnetic field

You have a battery with known emf = K, a sensitive voltmeter, and a slab of high-resistance metal in which the mobile charges are electrons with mobility u; there are n electrons per m^3 (Figure 20.98). The metal slab has length L, height h, and width w. You also have plenty of very low-resistance wires you can use to connect these components together. There is a uniform magnetic field pointing up throughout the region where the slab is located; this magnetic field is produced by a large magnet which is not shown. The goal is to figure out the magnitude B of this magnetic field, using the available equipment.

(a) Draw wires on the diagram to show how you would connect the equipment.

(b) Explain in detail, including appropriate diagrams, how you will determine B, and give an algebraic expression for B involving the reading ΔV on the voltmeter and the known quantities (K, u, n, L, h, w).

Problem 20.23 An orbiting nucleus

A long solenoid with diameter 4 cm is in a vacuum, and a lithium nucleus (4 neutrons and 3 protons) is in a clockwise circular orbit inside the solenoid (Figure 20.99). It takes 50 ns (50×10^{-9} s) for the lithium nucleus to complete one orbit.

(a) Does the current in the solenoid run clockwise or counter-clockwise? Explain, including physics diagrams.

(b) What is the magnitude of the magnetic field made by the solenoid?

Problem 20.24 Plates inside a coil

A battery with known emf = K is connected to two large parallel metal plates (Figure 20.100). Each plate has a length L and width W, and the plates are a very short distance s apart. The plates are surrounded by a vertical thin circular coil of radius R containing N turns through which runs a steady conventional current I. The center of the coil is at the center of the gap between the plates. At a certain instant, a proton (charge +e, mass M) travels through the center of the coil to the right with speed v, and the net force on the proton at this instant is zero (neglecting the very weak gravitational force). What is the magnitude and direction of conventional current in the coil? Explain clearly.

Problem 20.25 Determining a magnetic dipole moment

The center of a a large bar magnet is 20 cm from a thin plate of high-resistance material 12 cm long, 2.5 cm high, and 0.1 cm thick that is connected to a 240 volt power supply whose internal resistance is negligible (Figure 20.101). The bar magnet is perpendicular to the plate. The mobility of charge carriers in the thin plate is 3.5×10^{-4} (m/s)/(N/C), and the number density of the singly-charged charge carriers is 4×10^{23} m^{-3}. A voltmeter is connected precisely vertically across the thin plate and reads +1.2×10^{-5} volt . A low-resistance ammeter is in series with the rest of the circuit.

(a) Are the charge carriers electrons or holes? Explain, including a physics diagram.

(b) What is the magnetic dipole moment of the bar magnet? Include units.

(c) What does the ammeter read, including sign?

Problem 20.26 A falling metal rod

A neutral metal rod of length 60 cm falls toward the Earth. The rod is horizontal and oriented east-west.

(a) Which end of the rod, east or west, has excess electrons? Explain, using physics diagrams.

(b,) At a moment when the rod's speed is 4 m/s, approximately how many excess electrons are at the negative end of the rod?

Problem 20.27 A conducting bar

In Figure 20.102 there is a bar 11 cm long with a rectangular cross section 3 cm high and 2 cm deep, connected to a 1.2-volt battery and an ammeter. The resistance of the copper connecting wires and the ammeter, and the internal resistance of the battery, are all negligible compared to the resistance of the bar.

Using large coils not shown on the following diagram, a uniform magnetic field of 1.8 tesla was applied perpendicular to the bar (out of the page, as shown). A voltmeter was connected across the bar, with the connections across the bar carefully placed directly across from each other.

The mobile charges in the bar have charge $+e$, their density is 7×10^{23} per cubic meter, and their mobility is 3×10^{-5} (m/s)/(V/m) .

Predict the reading of the voltmeter, including sign. Explain carefully, using diagrams to support your explanation. Remember that a voltmeter reads positive if the "+" terminal is connected to higher potential.

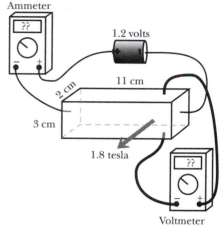

Figure 20.102 A conducting bar (Problem 20.27).

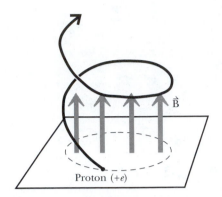

Figure 20.103 A general trajectory in a uniform magnetic field is a a helix.

20.14 Answers to exercises

20.1 (page 718) $F_{magnetic} \approx 1 \times 10^{-8}$ N ; $F_{electric} \approx 9 \times 10^{-5}$ N , much larger

20.2 (page 719)   For the component of velocity parallel to the magnetic field, $\vec{v} \times \vec{B} = 0$, so only the perpendicular component of the velocity is affected. The trajectory is a circular helix, as shown in Figure 20.103.

20.3 (page 720)   Clockwise; $q\vec{v} \times \vec{B} = -e\vec{v} \times \vec{B}$ points inward and turns the electron's momentum.

20.4 (page 720) $B = 2 \times 10^{-3}$ T , into the page.

20.5 (page 720)

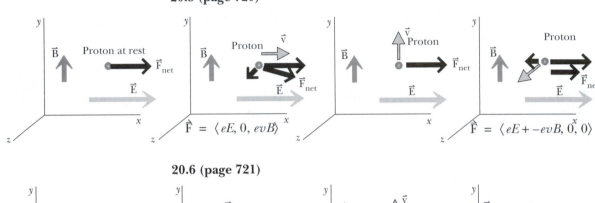

20.6 (page 721)

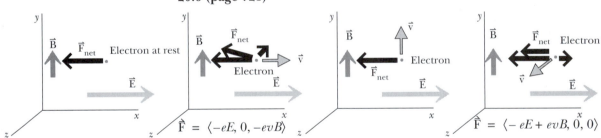

20.7 (page 721) $\vec{E} = \langle 0, 0, 2.7 \times 10^{8} \rangle$ V/m . Note that a particle travelling that fast would presumably be in an evacuated chamber, so an electric field of this magnitude would not cause a spark.

20.8 (page 724) $\Delta V_{Hall} = 1.5 \times 10^{-5}$ volts

20.9 (page 724) $\Delta V_{wire} = 3.3 \times 10^{-3}$ volts $\gg \Delta V_{Hall}$

20.10 (page 729) $F = 2 \times 10^{-3}$ N per meter of wire

20.11 (page 731)   0.01 volts ($\ll$ 1.5 volts)

20.12 (page 732)   emf $= vBL$; I counterclockwise; $I = vBL/R$; $F = vB^2L^2/R$ to the right

20.13 (page 734) $F_{magnetic}/F_{electric} = 2.5 \times 10^{-11}$

20.14 (page 736)   Below the charge, $E_y < 0$ and $E_z = 0$, so

$$B_y' = \frac{\left(B_y + \frac{v}{c^2}E_z\right)}{\sqrt{1 - v^2/c^2}} = 0$$

$$B_z' = \frac{\left(B_z - \frac{v}{c^2}E_y\right)}{\sqrt{1 - v^2/c^2}} = \frac{-\frac{v}{c^2}E_y}{\sqrt{1 - v^2/c^2}} > 0$$

Therefore $B_y' = 0$ and $B_z' > 0$, as expected.

In front of the charge, $E_y = 0$ and $E_z > 0$, so

$$B_y' = \frac{\left(B_y + \frac{v}{c^2}E_z\right)}{\sqrt{1 - v^2/c^2}} = \frac{\frac{v}{c^2}E_z}{\sqrt{1 - v^2/c^2}} > 0$$

$$B_z' = \frac{\left(B_z - \frac{v}{c^2}E_y\right)}{\sqrt{1 - v^2/c^2}} = 0$$

Therefore $B_y' > 0$ and $B_z' = 0$, as expected.

Behind the charge, $E_y = 0$ and $E_z < 0$, so

$$B_y' = \frac{\left(B_y + \frac{v}{c^2}E_z\right)}{\sqrt{1 - v^2/c^2}} = \frac{\frac{v}{c^2}E_z}{\sqrt{1 - v^2/c^2}} < 0$$

$$B_z' = \frac{\left(B_z - \frac{v}{c^2}E_y\right)}{\sqrt{1 - v^2/c^2}} = 0$$

Therefore $B_y' < 0$ and $B_z' = 0$, as expected.

20.16 (page 740)   Atomic magnetic dipole moments point north.

20.17 (page 740)   Swings toward north; magnetic dipole moments twist to align with the magnetic field.

20.18 (page 741) $2\mu B$

20.19 (page 742)   Net force would reverse and be repulsive.

20.20 (page 742)   Net force would be zero.

20.21 (page 747)   emf maximum when $\theta = 90°$ (velocity at right angle to magnetic field), minimum when $\theta = 0$ (velocity in the direction of magnetic field, so cross product zero).

20.22 (page 747)   302 volts

Chapter 21

Patterns of Field in Space

Chapter 21

Patterns of Field in Space

Electric charges produce electric and magnetic fields, and we know how to calculate or estimate the magnitude and direction of the field due to a particular distribution of charges (including moving charges). But sometimes it is useful to reason in the other direction: from an observed three-dimensional pattern of electric or magnetic field it may be possible to infer what charges are responsible for this pattern. Gauss's law relates patterns of electric field over a closed surface to the amount of charge inside that closed surface. Ampere's law relates patterns of magnetic field around a closed path to the amount of current passing through that closed path.

21.1 Patterns of electric field in space: Gauss's law

We begin our study of patterns of field in space with patterns of electric field. Figure 21.1 shows a box that contains some charge. Suppose you can't look inside the box, but you can measure the electric field all over the surface of the box.

? Can you conclude that there is some charge in the box? Positive or negative charge?

The fact that the electric field points outward all over the box implies that there must be some positive charge inside the box.

Figure 21.2 shows another box (for clarity, the electric field on the surface is displayed with the head rather than the tail of the vector at the observation location). Again, suppose you can't look inside the box, but you can measure the electric field all over the surface of the box.

? Can you conclude that there is some charge in the box? Positive or negative charge?

Since the electric field points into the box, all over its surface, there must be some negative charge inside the box.

Suppose you find a box on whose surface the electric field has a pattern like that in Figure 21.3.

? Can you conclude that there is some charge in the box? Positive or negative charge? Can you think of a charge distribution that could produce this pattern of electric field?

The electric field points out of the box on part of the surface, and into the box on other parts of the surface. A possible charge distribution that would produce such a pattern of electric field is a very large positively charged plate to the left of the box, or a very large negatively charged plate to the right of the box, or plates of both kinds, in which case the box lies in the gap between the plates of a capacitor.

It appears that if we find this kind of distribution of electric field over the surface of the box, we can conclude that there is no net charge inside the box. Although at first glance it might appear that there could be an electric dipole inside the box, a dipole would produce the observed pattern of electric field on the ends of the box ($\vec{E}$ to the right) but not on the sides of the box ($\vec{E}$ to the left), so this is not a possibility.

Suppose you find a pattern of electric field on a box as shown in Figure 21.4.

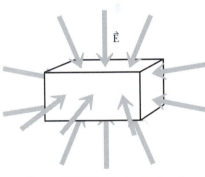

Figure 21.1 What's in this box?

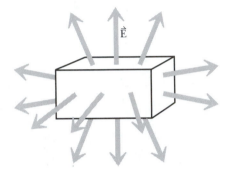

Figure 21.2 What's in this box?

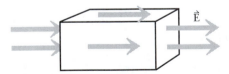

Figure 21.3 What's in this box?

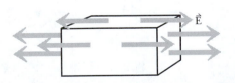

Figure 21.4 What's in this box?

? Can you think of a distribution of charge that could produce such a pattern of electric field on the surface of the box? (Hint: There may be some charge outside the box in addition to the charge inside the box.)

A charge distribution that could produce this pattern of field over the surface of the box is a very large vertical positively charged plate that cuts through the middle of the box, making electric field to the left and right of the plate. In this case there is positive charge inside the box (and outside, too).

On the other hand, if your measurements of the electric field don't surround a region, you may not be able to conclude much. Figure 21.5 shows a pattern of electric field on just one face of a box.

? Give three different charge configurations that could be responsible for the field pattern observed in Figure 21.5.

Figure 21.5 What can you conclude?

Possibilities include a very large positively charge plate to the left of the surface, a very large negatively charged plate to the right of the surface, or both. We see an important difference between an "open" surface such as this one and a "closed" surface such as a box or a spherical surface, which surrounds a region in space. Qualitatively, it seems clear that measurements of electric field all over a closed surface can tell us quite a bit about the charge inside that closed surface. But measurements of electric field on an open surface can't even tell us whether nearby charges are positive or negative.

Gauss's law is
- a quantitative relationship between
- measurements of electric field on a closed surface and
- the amount and sign of the charge inside that closed surface.

We will study Gauss's law for the additional insights it provides about the relationship between charge and field, and for certain important results that can be obtained using Gauss's law that are almost impossible to obtain using Coulomb's law alone, even though Gauss's law and Coulomb's law are essentially equivalent to each other (if the charges aren't moving).

Moreover, the concept of "flux" that arises in the context of Gauss's law plays an important role in our later study of the effects of time-varying magnetic field (Faraday's law).

21.2 Definition of "electric flux"

It is clear that there is some connection between the pattern of electric field on a closed surface and the amount of charge inside that surface, but how can we turn this into a quantitative relationship? We need to define a quantitative measure of the amount and direction of electric field over an entire surface. This measure is called "electric flux." We will develop a list of properties that a quantitative definition of electric flux should have.

21.2.1 Property 1: Direction of electric field

Compare the electric field patterns on the surfaces of the three boxes shown in Figure 21.6, which contain positive, negative, and zero charge.

Evidently one property of "electric flux" is that it should be positive where the electric field points out of the box, negative where the electric field points into the box, and zero where the electric field does not pierce the surface. In that way, the total flux for the third box in Figure 21.6 would consist of +1 unit on the right face, −1 unit on the left face, and 0 on the other faces, for a total of zero, which corresponds to the zero charge that is inside the box.

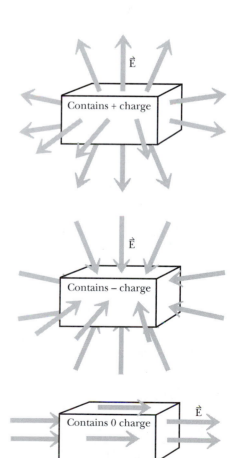

Figure 21.6 Compare patterns of electric field on these three boxes.

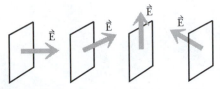

Figure 21.7 The electric field at different angles to a surface.

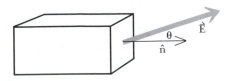

Figure 21.8 A unit vector pointing outward.

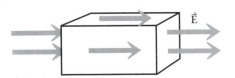

Figure 21.9 The total electric flux on the surface of this box is zero.

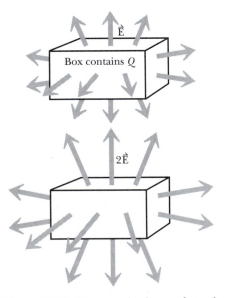

Figure 21.10 How much charge does the lower box contain?

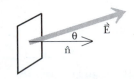

Figure 21.11 $E\cos\theta = \vec{E} \bullet \hat{n} = E|\hat{n}|\cos\theta$

We want the sign of electric flux to be related to the angle the electric field makes with the surface. This angle factor varies between +1 (outward and perpendicular to the surface) and –1 (inward and perpendicular to the surface), and is zero if the electric field is parallel to the surface. These values of +1, –1, and 0 are the extremes, and we need some way to express varying degrees of "outward-goingness" in the range from –1 to +1.

Figure 21.7 shows a series of open surfaces where this angle factor varies smoothly from +1 down to 0, and on to negative values. This suggests parametrizing the angle factor as the cosine of an angle between the electric field vector and the "outward-going normal," a unit vector $\hat{n}$ perpendicular to the surface and pointing outward from the closed surface (Figure 21.8).

? Show that the cosine of this angle θ has the appropriate properties we're looking for: +1 if $\vec{E}$ points straight outward, –1 if $\vec{E}$ points straight inward, 0 if $\vec{E}$ is parallel to the surface, and an intermediate value between +1 and –1 in other cases.

Bear in mind that the electric flux on a closed surface is the sum of the electric flux on the open surfaces that make up that closed surface. For example, on the box in Figure 21.9 we see negative flux on the left face, positive flux on the right face, and zero flux on the other faces. The total flux on this closed surface is the sum of the flux on each face, and the total flux adds up to zero, corresponding to there being no charge inside this box.

21.2.2 Property 2: Magnitude of electric field

How should the electric flux depend on the magnitude of the electric field? Consider the electric field patterns on the surfaces of the two boxes shown in Figure 21.10.

? The upper box contains a charge Q. If the electric field vectors are twice as long on the surface of the lower box, how much charge is inside the lower box?

Doubling an amount of charge doubles the electric field made by that charge. Evidently the lower box contains an amount of charge $2Q$. So the definition of electric flux should have the property that twice the field gives twice the flux. Therefore electric flux ought to be proportional to E as well as being proportional to $\cos\theta$. Apparently the definition of electric flux should contain the product $E\cos\theta$.

We can write $E\cos\theta$ as $\vec{E} \bullet \hat{n} = E|\hat{n}|\cos\theta$, because the dot product involves the cosine of the angle between the two vectors, and the magnitude of a (dimensionless) unit vector $|\hat{n}|$ is 1 (Figure 21.11).

? Show that the dot product $\vec{E} \bullet \hat{n}$ is $+E$ if the electric field points straight outward (perpendicular to the surface), $-E$ if the electric field points straight inward, and 0 if $\vec{E}$ is parallel to the surface.

Outward: $\vec{E} \bullet \hat{n} = <+E,0,0> \bullet <1,0,0> = +E+0+0 = +E$
Inward: $\vec{E} \bullet \hat{n} = <-E,0,0> \bullet <1,0,0> = -E+0+0 = -E$
Parallel: $\vec{E} \bullet \hat{n} = <0,E,0> \bullet <1,0,0> = +0+0+0 = 0$

21.2.3 Property 3: Surface area

Finally, consider the two boxes shown in Figure 21.12. The small box contains a charge Q, and there are no charges outside the small box. If we enclose the same charge Q in a larger box, the electric field vectors on the surface of the larger box are smaller, because the surfaces are farther from the enclosed charge.

We need to define the electric flux in such a way that the flux is the same for the large electric fields on the surface of the small box or for the small electric fields on the surface of the large box. This suggests that the definition of electric flux ought to involve the area A of a surface, not just $\vec{E} \bullet \hat{n}$, in order to compensate for how big we draw the enclosing surface. Maybe electric flux should be defined as $\vec{E} \bullet \hat{n}A$. In fact, if you double all the distances, the surface area increases by a factor of four, while the magnitude of the electric field decreases by a factor of four. So including A in the definition ought to compensate exactly for changing the size of the box.

The magnitude and direction of $\vec{E}$ may vary from location to location on a large surface, so we should define electric flux on a small area ΔA over which there is little variation in the electric field.

Taking all these factors into account, perhaps a useful definition of electric flux over a small area $\Delta A = \Delta x \Delta y$ would be $\vec{E} \bullet \hat{n}\Delta A$, and we would add up all such contributions over an extended surface to get the electric flux over that entire surface (Figure 21.13).

DEFINITION OF ELECTRIC FLUX ON A SURFACE

$$\sum_{\text{surface}} \vec{E} \bullet \hat{n}\Delta A$$

21.2.4 Perpendicular field or perpendicular area

We define electric flux as the sum of $\vec{E} \bullet \hat{n}\Delta A$ over a surface. This definition takes into account the direction of the electric field, the magnitude of the electric field, and the surface area. One of the ways to calculate the flux is to think of $E\cos\theta$ as the component of electric field perpendicular to the surface, $E_\perp = E\cos\theta$, so that the flux on a small area is $E_\perp \Delta A$ (Figure 21.14).

Alternatively, we can calculate the flux on a small area as the product of the electric field and the projection of the surface area ΔA perpendicular to the direction of the electric field (Figure 21.15). If we call this perpendicular projection of the area $\Delta A_\perp$, the flux is $E\Delta A_\perp$.

So we have three equivalent ways of writing the flux: $\vec{E} \bullet \hat{n}\Delta A$, $E_\perp \Delta A$, or $E\Delta A_\perp$. In many books you will find yet another notation for electric flux: $\vec{E} \bullet \Delta \vec{A}$. What this notation means is that $\Delta \vec{A}$ is considered to be a vector pointing in the direction of the outward-going normal, but with a magnitude ΔA: $\Delta \vec{A} = \hat{n}\Delta A$.

Ex. 21.1 If the electric field in Figure 21.14 is 1000 volts/meter, and the field is at an angle of 30 degrees to the outward-going normal, what is $E_\perp$? What is the flux $E_\perp \Delta A$ on a small area 1 mm by 2 mm?

Ex. 21.2 Suppose that the small area ΔA in Figure 21.15 measures 1 mm by 2 mm, and it is oriented in such a way that its outward-going normal is at an angle of 30 degrees to the direction of the electric field. What is $\Delta A_\perp$? If the electric field is 1000 volts/meter, what is the flux $E\Delta A_\perp$?

21.2.5 Adding up the flux

The electric flux on a small area ΔA is $\vec{E} \bullet \hat{n}\Delta A$. To calculate the electric flux over an extended surface we need to divide the surface into little areas, calculate the flux on each little area, and then add up all these contributions to the flux. Symbolically, we have

electric flux on a surface = $\sum \vec{E} \bullet \hat{n}\Delta A$

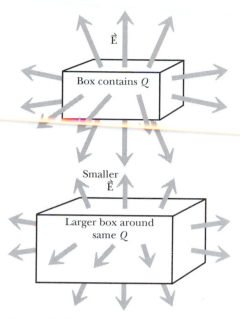

Figure 21.12 A larger box containing the same charge has smaller electric field on its surface.

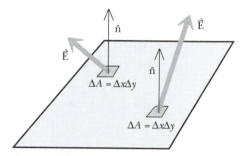

Figure 21.13 Electric flux on a surface is calculated by adding up the contributions $\vec{E} \bullet \hat{n}\Delta A$ on each little piece of area $\Delta A = \Delta x \Delta y$.

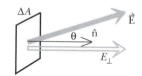

Figure 21.14 $\vec{E} \bullet \hat{n}\Delta A = E_\perp \Delta A$.

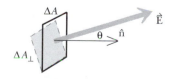

Figure 21.15 $\vec{E} \bullet \hat{n}\Delta A = E\Delta A_\perp$.

If we divide the surface into an infinite number of infinitesimal areas, this summation becomes an integral. This is similar to what we did in Chapter 15 to calculate the electric field due to charges distributed over a ring or a disk:

$$\text{electric flux on a surface} = \int \vec{\mathbf{E}} \bullet \hat{\mathbf{n}}\, dA$$

This "surface integral" can also be written in the following equivalent forms:

$$\text{electric flux on a surface} = \int E_\perp\, dA = \int E\, dA_\perp = \int \vec{\mathbf{E}} \bullet d\vec{\mathbf{A}}$$

We have seen that it is the electric flux on closed surfaces, such as a box, that tells us something about what charge is inside that closed surface. Whenever we need to add up the electric flux all over a closed surface, we can remind ourselves of this with the following notation:

$$\text{electric flux on a closed surface} = \oint \vec{\mathbf{E}} \bullet \hat{\mathbf{n}}\, dA = \sum_{\text{surface}} \vec{\mathbf{E}} \bullet \hat{\mathbf{n}} \Delta A$$

A circle superimposed on an integral sign is a standard notation for "completeness," in this case, that the flux should be integrated over an entire closed surface.

21.3 Gauss's law

We can now state Gauss's law, which we will then prove:

GAUSS'S LAW

$$\sum_{\text{surface}} \vec{\mathbf{E}} \bullet \hat{\mathbf{n}} \Delta A = \frac{\sum q_{\text{inside}}}{\varepsilon_0}, \text{ or}$$

$$\oint \vec{\mathbf{E}} \bullet \hat{\mathbf{n}}\, dA = \frac{\sum q_{\text{inside}}}{\varepsilon_0}$$

When we add up the electric flux on the entire area of a closed surface, this quantitative measure of the pattern of electric field does turn out to be proportional to the amount of charge inside that closed surface, as we had guessed from the patterns we saw at the beginning of this chapter.

Proving Gauss's law

What would we need to do in order to prove that Gauss's law is correct, that the electric flux over a closed surface is indeed proportional to the charge inside the surface?

- Show that the proportionality constant is indeed $1/\varepsilon_0$.
- Show that the size and shape of the surface don't matter.
- Show that it is true for any number of point charges inside the surface.
- Show that charges outside the surface contribute zero net flux.

The following sections give a formal proof of the correctness of Gauss's law, by addressing each of these issues. The proof is a geometry-based proof rather than an algebra-based proof. By now you should be convinced that there is indeed some kind of relationship between the charge inside a closed surface and the total electric flux on that closed surface. The formal proof nails this down, quantitatively.

21.3.1 Determining the proportionality constant

We start by determining the proportionality constant connecting charge inside a closed surface and net electric flux on the surface. To do this, we calculate the net electric flux for a surface for which we can make the calculation easily: a closed spherical surface surrounding a positive point charge $+Q$ at the center of the sphere (Figure 21.16). It is extremely important to understand that this spherical surface is of our own choosing and is purely mathematical in nature: it has nothing to do with any actual spherical shell made of some material.

Note that because it is hard to draw and interpret three-dimensional diagrams, we will usually draw two-dimensional diagrams that are cross-sectional slices through the 3-D surface. You must, however, keep in mind that the spherical surface we have drawn around the charge is a closed 3-D surface that encloses the charge Q.

? We need to divide the spherical surface into small pieces ΔA and evaluate $\vec{E} \bullet \hat{n} \Delta A$ for each piece. Because the closed spherical surface is centered on the point charge, what can you say about the angle θ in $\vec{E} \bullet \hat{n} = E\cos\theta$, at all locations on the surface? (See Figure 21.17.)

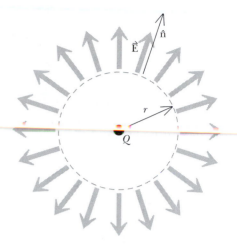

Figure 21.16 A point charge Q surrounded by a mathematical spherical surface of our choosing, of radius r.

Since both the electric field $\vec{E}$ and the outward-going normal $\hat{n}$ are perpendicular to our chosen spherical surface at all locations on the surface, the angle between the two vectors is zero, and $\cos\theta = +1$.

? We need to add up $E\cos\theta \Delta A$ over the entire surface. What can you say about the magnitude of E at all locations on the surface, if the radius of the surface is r?

Because we chose our mathematical surface to be centered on the charge Q, the magnitude of the electric field at all locations on the sphere is simply

$$E = \frac{1}{4\pi\varepsilon_0}\frac{Q}{r^2}$$

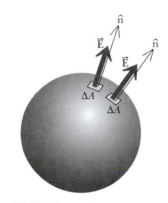

Figure 21.17 We need to calculate the flux on every little piece ΔA of the surface.

? The total surface area of a sphere is $4\pi r^2$. If you add up all the pieces $E\cos\theta \Delta A$ over the entire surface, what do you get, in terms of the charge Q at the center of the sphere?

$E\cos\theta \Delta A$ is simply equal to $E\Delta A$ (since $\cos\theta = +1$), and E is constant, so the net flux over the entire surface is given by

$$\sum_{\text{surface}} \vec{E} \bullet \hat{n} \Delta A = \left(\frac{1}{4\pi\varepsilon_0}\frac{Q}{r^2}\right)(+1)(4\pi r^2) = \frac{Q}{\varepsilon_0}$$

Our hope is that the total sum of electric flux on our chosen closed surface should be proportional to the charge inside the closed surface:

$$\sum_{\text{surface}} \vec{E} \bullet \hat{n} \Delta A = kq_{\text{inside}}$$

? Is this the case? What is the proportionality factor k?

Evidently for the spherical surface we get $k = 1/\varepsilon_0$, since the charge inside is simply equal to Q. This looks promising. We've got a quantitative measure of the electric field pattern on a closed surface (the total electric flux), and at least in this one case of a spherical surface centered on a point charge the total electric flux does turn out to be proportional to the charge inside the surface.

? What if the charge at the center of our chosen sphere were negative $(-Q)$? Would the relationship still hold? Check that it does, by considering what $\cos\theta$ is when the inside charge is negative.

In this case the flux would be negative (due to the cosine factor being -1), and proportional to the negative charge inside $(-Q)$. This is additional evidence that the $\cos\theta$ term is an appropriate part of the definition of flux.

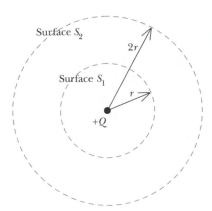

Figure 21.18 Consider the electric flux on two different spherical surfaces, chosen by us (and shown in cross section).

21.3.2 The size of the surface doesn't matter

Note that our result $k = 1/\varepsilon_0$ didn't depend on the radius r of our chosen spherical surface. Why is this? Suppose we consider another spherical surface with twice the radius. We draw cross sections of the two spherical surfaces S_1 and S_2 (Figure 21.18).

? How does E_2 on surface S_2 differ from E_1 on surface S_1? How does the total surface area differ (the surface area of a sphere is $4\pi r^2$)? How does the total electric flux differ? Is $\sum_{\text{surface}} \vec{E} \bullet \hat{n}\Delta A = q_{\text{inside}}/\varepsilon_0$ valid for both surfaces?

The magnitude of the electric field drops by a factor of 4, since it is proportional to $1/r^2$. The surface area increases by the same factor of 4, since the surface area is proportional to r^2. Therefore the electric flux on the outer surface is exactly the same as on the inner surface, and the same proportionality factor $k = 1/\varepsilon_0$ holds.

This example confirms that it was appropriate to include the area in the definition of electric flux. Note that this result depends critically on the fact that the electric field of a point charge is proportional to $1/r^2$. If it were proportional to $1/r^{2.001}$, say, the electric flux would be smaller on the outer surface than on the inner surface, and the concept of electric flux wouldn't be very useful.

So the relationship

$$\sum_{\text{surface}} \vec{E} \bullet \hat{n}\Delta A = \frac{q_{\text{inside}}}{\varepsilon_0}$$

is true no matter what the sign of q_{inside} is, or how large we draw the enclosing spherical surface.

21.3.3 The shape of the surface doesn't matter

Next consider an arbitrary closed surface surrounding the point charge, with a small spherical surface of radius r_1 for comparison (Figure 21.19). Remember that only cross sections of the surfaces are shown; the two dashed curves really represent closed surfaces that enclose the charge Q. We draw a cone that includes a small area of both surfaces.

The quantity $\Delta A_\perp = \Delta A \cos\theta$ is the projection of the actual area ΔA onto a spherical surface drawn through that location, centered on the point charge (Figure 21.20). The radius R of this circular projected area is proportional to the distance r_2 from the charge Q: $R = r_2 \tan\phi$. Therefore the area $\Delta A_\perp$ is proportional to r_2^2.

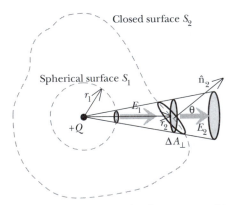

Figure 21.19 Compare the flux on an arbitrary closed surface with the flux on a small spherical surface.

? Explain why the electric flux on the portion of the outer surface enclosed by the cone is equal to the electric flux on the corresponding portion of the inner spherical surface. That is, why is $E_2 \Delta A_{2\perp}$ equal to $E_1 \Delta A_{1\perp}$?

We have $E_2/E_1 = r_1^2/r_2^2$, since the magnitude of the electric field is proportional to $1/r^2$. We also have $\Delta A_{2\perp}/\Delta A_{1\perp} = r_2^2/r_1^2$, since the circular projected area is proportional to r^2. Therefore $(E_2 \Delta A_{2\perp})/(E_1 \Delta A_{1\perp})$ is equal to 1.

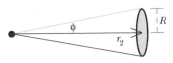

Figure 21.20 The perpendicular area $\Delta A_\perp = \pi R^2$, which is proportional to r_2^2.

? But if in every direction we find that the electric flux on any portion of the outer surface is the same as the electric flux on the projection of that area onto the small sphere, how does the total flux on the outer surface compare to the total flux on the inner sphere?

Evidently the total electric flux on the arbitrarily shaped outer surface is exactly equal to the total electric flux on the inner sphere. This is really encouraging!

We've shown that

$$\sum_{\text{surface}} \vec{E} \bullet \hat{n} \Delta A = \frac{q_{\text{inside}}}{\varepsilon_0}$$

is valid not only for an enclosing spherical surface but that it holds true for any enclosing surface whatsoever. Again, the key reason why this works is the $1/r^2$ property of Coulomb's law: the electric field on a larger enclosing surface is exactly enough smaller to make the product of field and area come out to be exactly the same. If electric field didn't fall off exactly like $1/r^2$, electric flux wouldn't be a useful concept, because the total flux on differently shaped enclosing surfaces would differ.

21.3.4 Charges outside the surface contribute zero net flux

For the formula to be valid in all cases, it should be true that the total flux is zero if there is no charge inside the closed surface, even if there are other charges *outside* the closed surface. Consider the situation in Figure 21.21 where a point charge $+Q$ is outside a closed surface (which is shown in cross section).

? Draw a cone from the charge, cutting through the near and far sides of the closed surface. Again note that $\Delta A_{\perp} = \Delta A \cos\theta$ is the projection of the actual area ΔA at the two locations where the cone pierces the surface. Show that the magnitude of the electric flux on the inner surface and on the outer surface is the same, but that they differ in sign.

We can use the same argument we made before about the flux on two surfaces enclosed within a cone. However, in this case the flux on the first surface is negative (electric field pointing into the volume), whereas on the second surface the flux is positive (electric field pointing out of the volume). Therefore the fluxes add up to zero inside the cone.

? Since this is true for any pair of inner and outer sections of the closed surface, what is the total electric flux over the entire surface, due to the point charge outside the closed surface?

This cancellation holds true for any cone we draw, so the net flux added up over the entire surface is zero.

Point charges outside a closed surface certainly do produce electric field and electric flux on the surface, but when we add up all the contributions we find the total electric flux due to these outside charges is zero:

$$\sum_{\text{surface}} \vec{E} \bullet \hat{n} \Delta A = q_{\text{inside}}/\varepsilon_0$$

is valid even when q_{inside} happens to be zero.

21.3.5 Superposition and Gauss's law

Consider three point charges, two inside and one outside of a mathematical closed surface (Figure 21.22). For each individual charge we have

$$\sum_{\text{closed}} \vec{E}_1 \bullet \hat{n} \Delta A = Q_1/\varepsilon_0$$

$$\sum_{\text{closed}} \vec{E}_2 \bullet \hat{n} \Delta A = Q_2/\varepsilon_0$$

$$\sum_{\text{closed}} \vec{E}_3 \bullet \hat{n} \Delta A = 0 \ (\text{since } Q_3 \text{ is outside the surface})$$

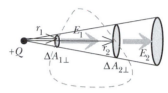

Figure 21.21 A point charge outside contributes zero net flux.

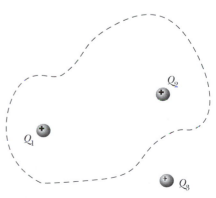

Figure 21.22 The charge outside the surface contributes zero to the net electric flux.

If we add these up, we have

$$\sum_{\text{closed}} (\vec{E}_1 + \vec{E}_2 + \vec{E}_3) \bullet \hat{n} \Delta A = \frac{Q_1 + Q_2}{\varepsilon_0}$$

But $\vec{E}_1 + \vec{E}_2 + \vec{E}_3$ is the net electric field due to all the charges, both inside and outside the surface, and it is this net field that we would measure on the surface. Generalizing to any number of point charges inside and outside a closed surface, we have proved what is called Gauss's law:

GAUSS'S LAW

$$\sum_{\text{surface}} \vec{E} \bullet \hat{n} \Delta A = \frac{\sum q_{\text{inside}}}{\varepsilon_0} \quad \text{or}$$

$$\oint \vec{E} \bullet \hat{n} \, dA = \frac{\sum q_{\text{inside}}}{\varepsilon_0}$$

This is Gauss's law—the quantitative form of what we saw qualitatively: the pattern of electric field on a closed surface tells us something about the charges that are inside that closed surface. Now we see that by expressing the pattern of electric field on a surface quantitatively in terms of electric flux, we know exactly how much charge is inside our chosen surface, even though the electric field we measure may be partly due to charges outside the surface.

21.3.6 Law or theorem?

Since we derived Gauss's law from Coulomb's law, it might make sense to refer to "Gauss's theorem." However, it is possible in a similar way to derive Coulomb's law from Gauss's law, so the two laws seem completely equivalent, despite their very different appearance. Although Coulomb's law and Gauss's laws are essentially the same law, some situations are more easily analyzed with Coulomb's law, and other situations are more easily analyzed with Gauss's law.

However, Gauss's law and Coulomb's law are not equivalent if the charges are moving, and the two laws differ drastically in their predictions if the charges are moving at speeds that are comparable to the speed of light. The issue is retardation. The electric field at a location in space due to a fast-moving charge moving in a general way cannot be calculated from Coulomb's law but involves a complicated calculation that takes into account the past history of the charge's motion: the electric field at the observation location *now* depends on the location of the source charge *then*.

Unlike Coulomb's law, it turns out that Gauss's law is correct in all cases. Suppose a large number of observers with synchronized watches are deployed all over a closed surface. By prior agreement, at 9:35 AM they all measure the electric flux on their own small piece of the surface and then report their measurements.

The sum of all these electric flux measurements is equal to the sum of all the charges (divided by ε_0) that were inside the surface at 9:35 AM, even if some of the charges were moving at speeds comparable to the speed of light. So Gauss's law is consistent with the special theory of relativity, even though Coulomb's law isn't. (There is a practical limitation to this measurement: Since the observers can't transmit their reports faster to us than the speed of light, by the time we add up all the flux contributions and thereby determine how much charge was inside the surface at 9:35 AM, the time is already quite a bit later!)

21.4 Reasoning from Gauss's law

We've used the total electric flux on a closed surface to find out how much charge is inside that closed surface. This means that we are using knowledge of the electric field to find out something about charges. We can in some cases use Gauss's law in the other direction, to derive the magnitude of the electric field due to a charge distribution. We'll study several such examples. To apply Gauss's law you

- choose a (mathematical) closed surface passing through locations where you know $\vec{E}$ or want to find $\vec{E}$, and
- which encloses a region in which you know what charges there are or want to find out what charges there are.

21.4.1 The electric field of a large plate

Suppose we don't know or don't remember the formula for the electric field near the center of a large uniformly charged plate. All we know is that near the center of the plate the electric field must be perpendicular to the plate, as shown in Figure 21.23, where we have also drawn a mathematical box (a "Gaussian surface") that is pierced by the large charged plate.

Let the area of the end face of the Gaussian surface be A_{box}. Assume that we don't know the magnitude of either E_{right} or E_{left}, but by symmetry we can assume that they have the same magnitude $E_{\text{right}} = E_{\text{left}} = E$.

? Calculate the total flux in terms of this unknown field E, and use Gauss's law to determine the magnitude of this unknown field. Note that the charge enclosed in the box is $(Q/A)A_{\text{box}}$, where Q is the total charge on the plate, and A is the total area of the plate.

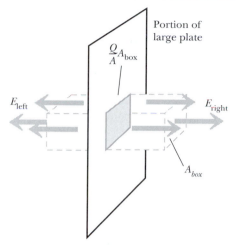

Figure 21.23 A portion of a large charged plate, with a Gaussian surface in the form of a box whose end area is A_{box}

If we draw the unit normal vector $\hat{n}$ on each face of the box, we see that there is nonzero electric flux only on the ends of the box, so the total electric flux over the entire box is the flux on the two ends, $2E(\cos 0)A_{\text{box}}$. Gauss's law yields

$$2EA_{\text{box}} = \frac{(Q/A)A_{\text{box}}}{\varepsilon_0}$$

from which we obtain the correct result for the electric field near a large uniformly charged plate:

$$E = \frac{(Q/A)}{2\varepsilon_0}$$

Contrast this simple calculation of the electric field near a large plate with the lengthy calculation for a disk that you did in Chapter 15!

Here you see the power of Gauss's law. In situations where we have some partial knowledge about the electric field (here it was the direction and symmetry of the field), it may be possible to determine the electric field due to a complex distribution of charges without carrying out a complicated integral.

Note that the area of the end face of our chosen mathematical Gaussian surface drops out of the calculation. This must happen, because the physical result for the electric field cannot depend on the size of some mathematical surface that we happen to choose for our own convenience.

21.4.2 The electric field of a uniform spherical shell

As a second example of using partial knowledge about the electric field pattern to deduce the magnitude of the electric field, consider a uniformly charged spherical shell. Recall from Chapter 15 that integrating Coulomb's law to get the electric field of this charge distribution is quite difficult. Suppose that we don't know how large the field is on two Gaussian surfaces, in-

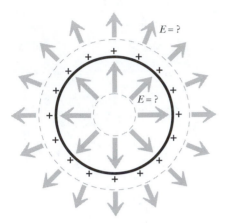

Figure 21.24 Two spherical Gaussian surfaces drawn inside and outside of a uniformly charged spherical surface (cross sections).

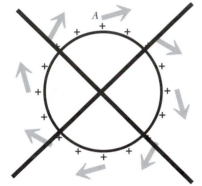

Figure 21.25 If the electric field of the charged sphere were not radial, it would have to look like this, which is impossible.

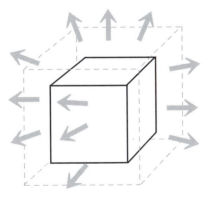

Figure 21.26 Trying to use Gauss's law to determine the electric field of a charged cube.

side and outside of the charged spherical shell (Figure 21.24). We can, however, conclude something about the direction and magnitude of the electric field—by symmetry it must be radial and uniform on a spherical surface centered on the spherical shell. If the electric field weren't radial, by symmetry the pattern would have to look something like the one shown in Figure 21.25, but in that case the round-trip integral of electric field, $V_A - V_A$, wouldn't be zero; we can therefore rule out this field pattern.

? First consider the outer mathematical sphere of radius r, shown in cross section in Figure 21.24. Use Gauss's law to show that the magnitude of the electric field is indeed

$$E = \frac{1}{4\pi\varepsilon_0}\frac{Q}{r^2}$$

? Next consider the inner mathematical surface in Figure 21.24. Use Gauss's law to show that the magnitude of the electric field is indeed $E = 0$.

The electric flux on the inner surface is $E(\cos 0)(4\pi r^2)$, but the charge inside the inner surface is zero. Hence E must be zero. So Gauss's law lets us determine the magnitude of the electric field inside and outside of a uniformly charged spherical shell, with very simple calculations. In contrast, determining the electric field of a uniformly charged spherical shell by integrating Coulomb's law over the charge distribution involves rather difficult trigonometry and integral calculus (see Chapter 15). The calculation using potential in Chapter 16 was also much more difficult than using Gauss's law.

21.4.3 The electric field of a uniform cube

Next let's try to use Gauss's law to determine the electric field near a cube that is uniformly charged positive. This cube must be made of an insulator, because on a metal cube the charge would not be distributed uniformly. It seems plausible to choose a cubical box as the Gaussian surface surrounding the charged cube (Figure 21.26).

However, the electric field on a face of the cubical Gaussian surface is not uniform in direction or magnitude as it was in the case of the large plate or the sphere.

? Is Gauss's law valid for this cubical Gaussian surface?

Gauss's law is always valid, no matter what the situation. It is definitely the case that the total electric flux over the entire Gaussian surface is numerically equal to the charge on the cube, divided by ε_0.

? Explain why we can't use Gauss's law to determine the electric field near a cube.

Although Gauss's law is valid, it doesn't help. The magnitude of the electric field varies all over the surface, and so does the angle between the electric field and the outward-going normal. There's no symmetry to simplify the integral, and neither E nor $\cos\theta$ can be factored out of the flux calculation. Therefore we can't carry out the flux calculation in terms of a single unknown quantity E.

21.4.4 Proving some important properties of metals

We have repeatedly stated that there cannot be any excess charge in the interior of a metal: any excess charge must be on a surface. Using Gauss's law, we can now prove this for the case of a metal in static equilibrium. We will use "proof by contradiction," by assuming that there is some excess charge inside the metal and using Gauss's law to demonstrate a contradiction.

Consider a piece of metal that has been polarized by external charges and is in static equilibrium. We assume (incorrectly) that there is some excess charge in the interior, as shown, and we draw a Gaussian surface (closed mathematical surface) that encloses that excess charge (Figure 21.27). The Gaussian surface is drawn in such a way that all elements of the surface are inside the metal. (Remember that this surface is just a mathematical construct, so it's fine to put it inside a solid object.)

? Since the metal is in static equilibrium, what is the magnitude of the electric field everywhere on the Gaussian surface? What is the total electric flux on that surface? Using Gauss's law, what can you conclude about the charge inside the Gaussian surface?

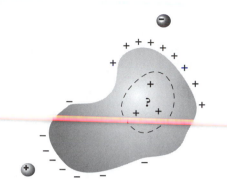

Figure 21.27 Can there be excess charge in the interior of a metal?

Everywhere inside a metal in static equilibrium, the electric field is zero, because if it were nonzero, electrons would flow and increase the polarization of the metal, until eventually in static equilibrium the net field in the interior will become zero. Since the field is zero on our mathematical surface, which is inside the metal, the total electric flux on that surface must be zero. Gauss's law tells us that the charge inside the surface must be zero.

So we reach a contradiction, and our initial assumption must be wrong: in static equilibrium there cannot be any excess charge in the interior of a metal—any excess charge therefore must be on the surface. Note that in this analysis we knew something about the electric field ($E = 0$), and from that we could deduce something about the charges.

21.4.5 A hole inside a metal object

Suppose there is a hole inside the metal (Figure 21.28).

Charge on the inner surface?

? Using a Gaussian surface drawn in the metal but close to the boundary of the hole, what can you conclude about the net charge on the surface of the hole?

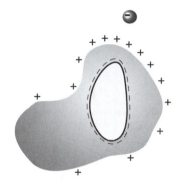

Figure 21.28 A hole inside a charged, polarized, piece of metal.

Again, all over a Gaussian surface inside a metal in static equilibrium the electric field is zero, so the total electric flux is zero, and charge enclosed inside the Gaussian surface must be zero.

Electric field inside the hole?

Is the electric field inside the hole zero or nonzero? It might conceivably be nonzero if there were "+" and "−" charges on this interior surface, as long as they add up to zero. However, we can show that E must be zero inside the hole. Consider the potential difference calculated along two different paths: A to C to B, and A to D to B (Figure 21.29).

? Along the path A to D to B, through the metal, what is the potential difference ΔV? Since potential difference is independent of path, what must be the potential difference along the path A to C to B?

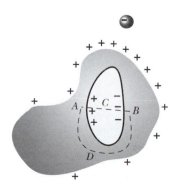

Figure 21.29 Could there be a nonzero electric field inside the hole, as long as the net charge on the surface of the hole is equal to zero?

All locations along the path A to D to B pass through the metal, where the electric field is zero, so the potential difference (path integral of the electric field) must be zero. There would be a potential difference along the path A to C to B if there were an electric field inside the hole made by charges on the surface of the hole, but the potential difference $V_B - V_A$ must be independent of path. Therefore we have reached a contradiction, and our assumption that there are charges on the surface of the hole must be false.

We have proved that in static equilibrium:

• there are no charges at all on the surface of an empty hole in a metal
• the electric field is zero everywhere inside an empty hole in a metal.

Reasoning less formally

Having gone through formal reasoning to show that an empty hole in a metal in static equilibrium contains no charges or electric field, it may be helpful to think about this important result in another way. If the metal were solid, the electric field in static equilibrium would be zero throughout, and there would be no excess charge anywhere in the interior, as we concluded earlier. Imagine somehow removing lots of neutral, unpolarized atoms from the interior, as though they could be eaten away with acid. These neutral, unpolarized atoms contributed nothing to the electric field, so removing them will not affect the distribution of charge or field. Hence we conclude that the hole that we've made inside the metal will contain neither field nor charge.

Charges inside a hole

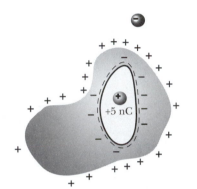

Figure 21.30 Now the hole is not empty but contains a +5 nC charge.

On the other hand, suppose that there is a charge of +5 nC floating inside the hole (Figure 21.30).

? Using Gauss's law with the indicated Gaussian surface, what can you conclude about the charge on the inner surface of the hole?

Since the flux on the chosen surface must be zero (because the electric field is zero all over this surface in the metal), the net charge inside the surface must be zero. Therefore there must be –5 nC of charge distributed on the inner surface of the hole. So in this case, where we have deliberately placed extra charge in the hole, there are charges and fields inside the hole.

21.4.6 "Screening"

We used Gauss's law (and potential theory) to show that in static equilibrium the electric field is zero not only in the interior of a metal but even in an empty hole inside the metal. This has important practical consequences. A metal box "screens" its contents from the effects of charges outside the box (Figure 21.31).

For many purposes the box doesn't even have to have solid metal sides. A box with some holes in the sides, or a box with sides made of wire mesh or screen, can be quite effective in reducing the field inside the box.

Figure 21.31 $E = 0$ inside an empty metal box, even though the box has a net charge and is polarized by an external charge.

Ex. 21.3 How does this "screening" really work? Are the electric fields produced by the outside charges blocked from getting inside the box? What about the principle that electric fields pass right through matter? Can you explain the seeming paradox?

21.4.7 Gauss's law applied to circuits

All of the examples so far have dealt with static equilibrium. But Gauss's law expresses a completely general relationship between charges and electric field, so it applies to electric circuits, too. Figure 21.32 shows a portion of a current-carrying wire of constant cross section A and uniform composition.

? In the steady state, excess charges arrange themselves on the surface of such a wire so that the electric field E has the same magnitude at different locations along the wire, and is everywhere parallel to the wire. In Figure 21.32 the dashed tube-shaped Gaussian

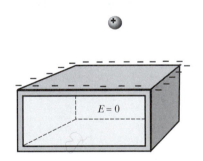

Figure 21.32 A Gaussian surface just inside a current-carrying wire in a circuit.

surface nearly fills the wire. Use Gauss's law to prove that there is no excess charge in the interior of the wire.

Along the sides of the tube the electric field is parallel to the surface and contributes no flux. The right end of the tube contributes an amount of flux *EA*, but the left end contributes –*EA*, so the total electric flux on the Gaussian surface is zero, which means that there must not be any net charge inside the Gaussian surface, all excess charge is on the outer surface of the wire.

See the example problem on page 796, which applies Gauss's law to a junction between two wires of different materials.

21.5 Gauss's law for magnetism

There is a dramatic difference between electric and magnetic dipoles. The individual positive and negative electric charges ("monopoles") making up an electric dipole can be separated from each other, and these point charges make outward-going or inward-going electric fields (which vary like $1/r^2$). One might expect a magnetic dipole such as a bar magnet to be made of positive and negative magnetic monopoles, but although scientists have searched carefully, no one has ever found an individual magnetic monopole. Such a magnetic monopole would presumably make an outward-going or inward-going magnetic field (which would vary like $1/r^2$), but such a pattern of magnetic field has never been observed (Figure 21.33).

The non-observation of such a field pattern can be expressed as a "Gauss's law" for magnetism, that the net flux of the magnetic field $\vec{B}$ over a closed surface is equal to the (non-existent) "magnetic charge" inside:

GAUSS'S LAW FOR MAGNETISM

$$\oint \vec{B} \bullet \hat{n}\, dA = 0$$

If you cut a magnet in two, you don't get two magnetic monopoles—you just get two magnets!

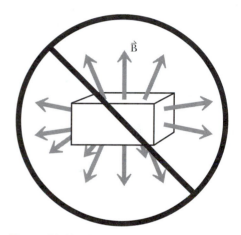

Figure 21.33 This pattern of magnetic field has never been observed.

21.6 Patterns of magnetic field in space: Ampere's law

There is a relationship called Ampere's law that is an alternative version of the Biot-Savart law, just as Gauss's law is an alternative version of Coulomb's law. Ampere's law refers to a closed path rather than a closed surface.

We will study Ampere's law for the additional insights it provides about the relationship between current and magnetic field, and for certain important results that are most easily obtained using Ampere's law. Moreover, Ampere's law is relativistically correct, as is Gauss's law. Even more importantly, in a later chapter we will study Maxwell's extension of Ampere's law that makes a connection between time-varying electric fields and magnetic fields, which leads to understanding the nature of light.

Patterns of magnetic field

Moving charges produce magnetic fields, and we know how to calculate or estimate the magnitude and direction of the magnetic field due to a particular distribution of moving charges or currents. But sometimes it is useful to reason in the other direction: from an observed pattern of magnetic field it may be possible to infer what currents are responsible for this pattern.

Suppose you can't look inside the circular region shown in Figure 21.34, but you can measure the magnetic field all along the boundary of the region.

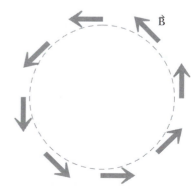

Figure 21.34 Is there current passing through this region? Does conventional current run into or out of the page?

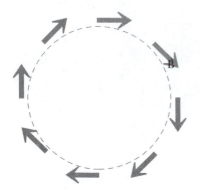

Figure 21.35 Is there current passing through this region? Does conventional current run into or out of the page?

? Can you conclude that there is some current inside the region? Does conventional current go into the page or out of the page?

A likely possibility is that a current-carrying wire is located at the center of the circle, with conventional current headed out of the page. This would produce the observed pattern of magnetic field.

Again, suppose you can't look inside the circular region shown in Figure 21.35, but you can measure the magnetic field all along the boundary of the region.

? Can you conclude that there is some current inside the region? Does conventional current go into the page or out of the page?

A wire carrying conventional current into the page at the center of the circle would produce the observed pattern of magnetic field.

Ampere's law is a quantitative relationship between measurements of magnetic field along a closed path and the amount and direction of the currents passing through that boundary. Note the close parallel with Gauss's law, which relates measurements of electric field on a closed surface to the amount of charge inside that closed surface.

Quantifying the magnetic field pattern

It is clear that there is some connection between the pattern of magnetic field along a closed path and the amount of current passing through the region bounded by that path, but how can we turn this into a quantitative relationship? We need to define a quantitative measure of the amount and direction of magnetic field along an entire closed path. Let's consider the magnetic field of a very long current-carrying wire a distance r from the wire (Figure 21.36), which we calculated using the Biot-Savart law in Chapter 17:

$$B_{\text{wire}} \approx \frac{\mu_0}{4\pi}\frac{2I}{r} \quad \text{(a distance } r \text{ from a very long wire)}$$

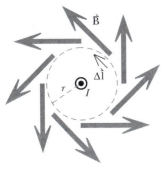

Figure 21.36 The magnetic field a distance r from a very long straight wire.

Given the curly character of the magnetic field around a wire, it is probably useful to integrate $\int \vec{B} \bullet d\vec{l}$ along a round-trip circular path around a wire and see what we get:

$$\oint \vec{B} \bullet d\vec{l} = \left(\frac{\mu_0}{4\pi}\frac{2I}{r}\right)\oint dl, \text{ because } \vec{B}\| d\vec{l} \text{ all along the path, and } r = \text{constant}$$

$$\oint \vec{B} \bullet d\vec{l} = \left(\frac{\mu_0}{4\pi}\frac{2I}{r}\right)(2\pi r) = \mu_0 I$$

This is Ampere's law: the path integral of the magnetic field is equal to μ_0 times the current passing through the region enclosed by the path. (This is similar to Gauss's law, in which the surface integral of the electric field is equal to $1/\varepsilon_0$ times the charge inside the closed surface.)

In order to obtain a general proof of this quantitative relationship between current and magnetic field, we need to show that we can follow any path that surrounds the current, and that currents outside the path contribute zero to the path integral.

A circular path with twice the radius

It is easy to see that our result is independent of the radius r of the path we choose. For example, suppose we choose a path along a larger circle of radius $2r$ (Figure 21.37).

? Why is the path integral the same along this longer path?

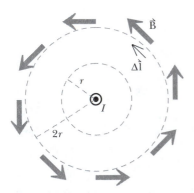

Figure 21.37 The magnetic field a distance $2r$ from a very long straight wire.

We know that the magnetic field near a very long wire is proportional to $1/r$, where r is the perpendicular distance to the wire. So on a circle of twice the radius (and twice the circumference) the magnetic field is half as big, making the path integral be the same as before. The path integral of the magnetic field is independent of how big the circle is. This is similar to Gauss's law, where if you double the radius of a Gaussian sphere surrounding a charge, the surface area of the sphere goes up by a factor of four, but the electric field drops by a factor of four.

A noncircular path

Suppose we evaluate the path integral along a noncircular path. Focus on the contribution $\vec{B}_1 \bullet \Delta\vec{l}_1$ along a short arc of the circle and the contribution $\vec{B}_2 \bullet \Delta\vec{l}_2$ along the corresponding section of the noncircular path (Figure 21.38).

In Figure 21.38 the parallel projection $\Delta l_{2\parallel}$ of $\Delta\vec{l}_2$ along the direction of $\vec{B}_2$ is equal to $\Delta l_2\cos\theta$, and by similar triangles it is also true that

$$\frac{\Delta l_{2\parallel}}{r_2} = \frac{\Delta l_1}{r_1}.$$

Therefore

$$\vec{B}_2 \bullet \Delta\vec{l}_2 = B_2\Delta l_2\cos\theta = B_2\Delta l_{2\parallel} = B_2(r_2/r_1)\Delta l_1$$

But since the magnetic field near a long straight wire is inversely proportional to the distance from the wire, we have $B_2/B_1 = r_1/r_2$, so

$$\vec{B}_2 \bullet \Delta\vec{l}_2 = B_1\Delta l_1 = \vec{B}_1 \bullet \Delta\vec{l}_1$$

Since this is true for any arc, it must be that along the noncircular path $\oint \vec{B} \bullet d\vec{l}$ has the same value it has along the circular path, which is $\mu_0 I$. We've almost got enough to prove Ampere's law. All that remains is to show that any current-carrying wires outside the closed path don't contribute.

Effect of currents outside the path

Consider Figure 21.39, in which there is a current outside the path.

? Briefly explain why $\vec{B}_2 \bullet \Delta\vec{l}_2 = -\vec{B}_1 \bullet \Delta\vec{l}_1$ as we go around counterclockwise, and why $\oint \vec{B} \bullet d\vec{l} = 0$ in this case.

If $\Delta\vec{l}_1$ and $\Delta\vec{l}_2$ are short enough that they approximate arc segments on the two circles, then the length $\Delta l_{2\perp} = \Delta l_2\cos\theta$ goes up by r, but B goes down by $1/r$, so the magnitude of the product $B\Delta l_\perp$ is the same at both locations. As we go around the path counter-clockwise, $\vec{B}_1 \bullet \Delta\vec{l}_1$ is negative but $\vec{B}_2 \bullet \Delta\vec{l}_2$ is positive, so $\vec{B}_1 \bullet \Delta\vec{l}_1 + \vec{B}_2 \bullet \Delta\vec{l}_2 = 0$. Adding up all such pairs, we find that a current that is outside the chosen closed path adds zero to the path integral $\oint \vec{B} \bullet d\vec{l}$.

Let's consider a specific example of three current-carrying wires (Figure 21.40). We draw a mathematical boundary that encloses two of the wires, and we walk counter-clockwise along this path. For each individual current we have

$$\oint \vec{B}_1 \bullet d\vec{l} = \mu_0 I_1 \text{ (since } I_1 \text{ heads out of the page)}$$

$$\oint \vec{B}_2 \bullet d\vec{l} = -\mu_0 I_2 \text{ (since } I_2 \text{ heads into the page)}$$

$$\oint \vec{B}_3 \bullet d\vec{l} = 0 \text{ (since } I_3 \text{ is outside the boundary)}$$

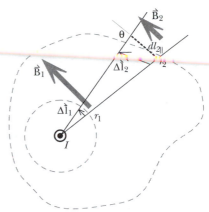

Figure 21.38 Comparing the path integral of magnetic field on a noncircular path to the path integral on a circular path.

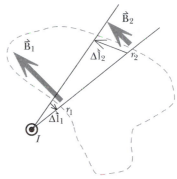

Figure 21.39 A current outside the path makes a zero net contribution to the path integral of the magnetic field.

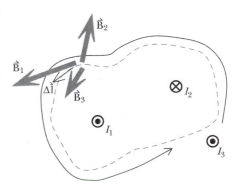

Figure 21.40 Add up all the $\vec{B} \bullet \Delta\vec{l}$ contributions along this counter-clockwise path.

If we add these up, we have

$$\oint (\vec{B}_1 + \vec{B}_2 + \vec{B}_3) \bullet d\vec{l} = \mu_0 (I_1 - I_2)$$

But $(\vec{B}_1 + \vec{B}_2 + \vec{B}_3)$ is the net magnetic field $\vec{B}$ due to *all* the currents, both inside and outside the boundary, and it is this net field that we would measure along the boundary. Note that we count a current positive or negative depending on whether it makes a magnetic field in the direction of our counter-clockwise walk along the path, or opposite to our walk, and this is equivalent to specifying the sign of the current depending on whether the current comes out of the page or goes into the page.

Putting it all together, we see that the path integral of magnetic field is equal to μ_0 times the sum of the enclosed currents (plus or minus), and currents outside the path contribute zero to the sum. This relationship is called Ampere's law:

AMPERE'S LAW

$$\oint \vec{B} \bullet d\vec{l} = \mu_0 \sum I_{\text{inside path}}$$

All the currents in the universe contribute by superposition to the magnetic field, yet Ampere's law refers only to those currents inside the chosen path because the outside currents contribute zero to the path integral. There is a strong parallel with Gauss's law.

Despite its very different form, Ampere's law is essentially equivalent to the Biot-Savart law from which it was derived. However, in some cases it is much easier to use Ampere's law than the Biot-Savart law. Moreover, Ampere's law is relativistically correct, whereas the Biot-Savart law is not, since it doesn't include retardation. This is just like the relationship between Gauss's law and Coulomb's law.

Note that Ampere's law involves the component of $\vec{B}$ parallel to a path, while Gauss's law involves the component of $\vec{E}$ perpendicular to a surface.

Ex. 21.4 Along the path shown in Figure 21.41 the magnetic field is measured and is found to be uniform in magnitude and always tangent to the circular path. If the radius of the path is 3 cm and B along the path is 1.3×10^{-6} T, what are the magnitude and direction of the current enclosed by the path?

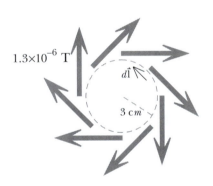

Figure 21.41 A magnetic field of 1.3×10^{-6} T is measured everywhere on a circular path of radius 3 cm.

21.6.1 Which currents are "inside" the path?

Before we apply Ampere's law to some important situations, we should pause to be more precise about how to decide what currents are "inside" a closed boundary. In the simple case of a single wire it is easy to say, but in more complicated cases this can be a bit tricky. We offer a procedure that is easy to follow and which moreover leads in a later chapter to an important extension of Ampere's law, made by Maxwell.

Just as every application of Gauss's law must start with choosing a closed surface to which to apply the law, so every application of Ampere's law must start with choosing a closed path that is the boundary of the region of interest.

After choosing a closed path, imagine stretching a soap film over your chosen path. You calculate the net current through the boundary by counting what current-carrying wires pierce your soap film. You walk counter-clockwise around the path, adding up $\oint \vec{B} \bullet d\vec{l}$ as you go. You count as positive any currents that pierce the soap film coming up through the film. To be more precise, let the fingers of your right hand curl counter-clockwise around the path, and count a current as positive if it goes in the direction of your thumb. Any currents that pierce the film going in the opposite direction are counted as negative.

In Figure 21.42 we have $\oint \vec{B} \bullet d\vec{l} = \mu_0 \sum I_{\text{inside path}} = \mu_0(I_1 - I_2)$, since I_1 pierces the imaginary soap film in the upward direction, I_2 in the downward direction, and I_3 doesn't pierce the film.

Though we will not carry out the proof (which is based on dividing up the soap film into little areas and applying Ampere's law to the boundary of each area, then adding up the results), it can be shown that Ampere's law is still valid even if the closed boundary path does not lie in a plane, or even if the soap film that is anchored on the boundary is deformed into a non-flat shape. Certainly in the case of the straight current-carrying wires shown above, deforming the soap film upward or downward would make no difference in counting the piercings.

For a more interesting case, consider a wire with an S-shaped bend in it, with a closed boundary drawn around the wire as shown in Figure 21.43. How much current is "inside" the boundary? If we count the current by counting the piercings of the soap film, we can show that we get the same (correct) answer for any and all deformations of the soap film, as long as the film stays anchored at its edges to the boundary path.

In Figure 21.43, there is just one piercing of the imaginary soap film, and $\sum I = +I$. In Figure 21.44, there are three piercings of the imaginary soap film, but adding up all the currents we again get $\sum I = (+I) + (-I) + (+I) = +I$. So deforming the imaginary soap film does not affect the accounting for the amount of current enclosed by the boundary of that film (our chosen closed path).

21.6.2 Applications of Ampere's law

We'll apply Ampere's law to some situations where there are lots of currents distributed over a region. Here is a summary of the key steps in applying Ampere's law:

1) Choose a (mathematical) closed path as a boundary.
2) Stretch an imaginary soap film over the boundary.
3) Walk around the boundary counter-clockwise, integrating $\oint \vec{B} \bullet d\vec{l}$.
4) Add up the positive and negative currents that pierce the soap film; this is $\sum I_{\text{inside path}}$. Count as positive those currents that pierce the imaginary soap film coming out of the film (that is, in the direction of your right thumb with the fingers of your right hand curling around in the direction of your walk); currents that pierce the film going into the film are counted as negative.
5) Apply Ampere's law, equating the two sums: $\oint \vec{B} \bullet d\vec{l} = \mu_0 \sum I_{\text{inside path}}$

We'll apply this procedure to find the magnetic field due to a long thick wire, a solenoid, and a toroid.

A long thick wire

Consider the problem of determining the magnetic field near a long *thick* wire (Figure 21.45). If we use the Biot-Savart law, we face the difficult problem of setting up and evaluating an integral not only along the length of the wire but also across the thickness of the wire, with a varying distance to those current elements. However, the result is very easy to get using Ampere's law.

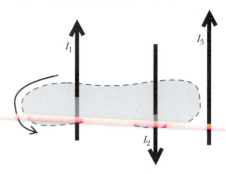

Figure 21.42 Curl the fingers of your right hand in the direction of the counter-clockwise path. Piercings in the direction of your thumb (I_1) count positive; piercings in the opposite direction count negative (I_2). Don't count I_3, because it doesn't pierce the soap film.

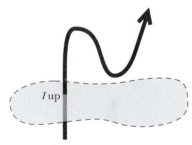

Figure 21.43 Just one piercing of the (flat) soap film.

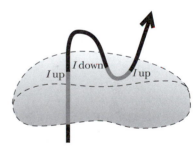

Figure 21.44 Here the dome-shaped soap film is pierced three times, two up and one down.

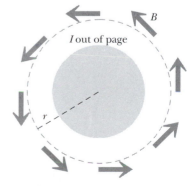

Figure 21.45 Using Ampere's law to find the magnetic field of a thick wire.

Draw a circular path of radius r around the thick circular wire, assuming by symmetry that the magnetic field B is tangent to the path and constant in magnitude. Stretch an imaginary soap film over this circular path.

? Using Ampere's law, calculate the magnitude of the magnetic field at a distance r from the center of a long thick wire.

The path integral of the magnetic field is simply $B(2\pi r)$, since B is by symmetry constant in magnitude and is always parallel (tangential) to the circular path. The amount of current piercing the imaginary soap film is $\mu_0 I$. So we have this:

$$B(2\pi r) = \mu_0 I$$

$$B = \frac{\mu_0 I}{2\pi r} = \frac{\mu_0}{4\pi}\frac{2I}{r} \quad \text{(thick wire)}$$

This simple result (equivalent to the magnetic field of a thin wire) would have been very difficult to obtain using the Biot-Savart law directly. So even though we derived Ampere's law from the Biot-Savart law, there are symmetrical situations where it is much easier to use Ampere's law.

A solenoid

A long coil of wire with a small diameter is called a "solenoid." The electromagnet that you made by winding a wire on a nail was an iron-filled solenoid. We can calculate the magnetic field along the axis of a solenoid consisting of lots of circular loops. It turns out that the magnetic field is nearly the same in magnitude and direction throughout the interior of the solenoid, as long as you don't get too near the ends. One of the uses of solenoids is to make a uniform magnetic field. Solenoids also play an important role in some kinds of electronic circuits.

Figure 21.46 shows a portion of a very long solenoid. We assume that if we are far from the ends, the magnetic field inside the solenoid is fairly uniform and the magnetic field outside is very small due to near cancellation of the contributions of the near and far sides of the loops. (See analysis using the Biot-Savart law, at the end of Chapter 17.)

We choose a rectangular path as shown in Figure 21.46, and we stretch an imaginary soap film over the rectangular frame. If there are N loops of wire along the entire length L of the solenoid, there are $(N/L)d$ wires that pierce the soap film, coming out of the page.

? Use Ampere's law to show that the magnetic field inside the solenoid is $B \approx \mu_0 NI/L$, which agrees with the much more difficult analysis at the end of Chapter 17 based on the Biot-Savart law.

The path integral is just Bd, because along the two ends of the rectangular path the magnetic field is perpendicular to the path, and along the outside portion of the path the magnetic field is assumed to be very small. The current piercing the soap film is the number of piercings times I, or $[(N/L)d]I$. Ampere's law says that

$$Bd = \mu_0[(N/L)d]I$$

from which we easily conclude this:

$$B = \frac{\mu_0 NI}{L} \quad \text{(solenoid)}$$

The length d does not appear in the result, because the magnetic field cannot depend on the length of the mathematical path that we chose.

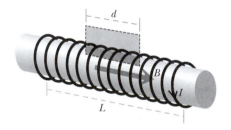

Figure 21.46 A solenoid, with a rectangular path chosen in such a way that the surface bounded by the path is pierced by some of the current-carrying wires.

No matter where we choose to place that part of the path that is inside the solenoid, we would get this same result. That means that we expect the magnetic field to be quite uniform inside the solenoid, as long as we are far from the ends. This is confirmed by a numerical computation using the Biot-Savart law, as shown below. If you look closely, you should see some tiny dots outside the central regions of the solenoid. These dots are actually arrows representing the magnetic field at these locations. Clearly, our assumption is justified that the field is small outside the solenoid, far from the ends.

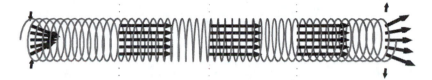

A toroid

Figure 21.47 shows a toroid, a "doughnut" wrapped with wire that is essentially a solenoid bent around to make the ends meet. By symmetry, the magnetic field on a circular path inside the toroid is tangent to the path and constant in magnitude, as shown in a top view in Figure 21.48.

We draw a closed path in the form of a circle of radius r inside the windings. If there are N loops of wire wound on the solenoid, there are N wires that pierce a soap film stretched over that closed path.

> **?** Use Ampere's law to show that the magnetic field inside the toroid, a distance r from the center, is $B = (\mu_0 NI)/(2\pi r)$.

The path integral of the magnetic field is simply $B(2\pi r)$, since $\vec{B}$ is constant in magnitude and always tangential (parallel to $d\vec{l}$). The current piercing the soap film is NI, so Ampere's law is this:

$$B(2\pi r) = \mu_0 NI$$

From this we find the magnetic field:

$$B = \frac{\mu_0 NI}{2\pi r} \quad \text{(toroid)}$$

Note that B is proportional to $1/r$, and the magnetic field varies across the cross section of the toroid. We could not have easily used the Biot-Savart law to obtain this result, because the integration over all the current elements would have been extremely difficult. Yet Ampere's law gives us the result right away, with little calculation.

21.7 Maxwell's equations

In this chapter we have developed three new equations: Gauss's law for electricity, Gauss's law for magnetism, and Ampere's law for magnetism. In a way, we already have an "Ampere's law for electricity." The round-trip path integral of the electric field produced by point charges is zero:

$$\oint \vec{E} \bullet d\vec{l} = 0 \text{ since round-trip potential difference is zero}$$

The four equations taken together are the basis for the four equations known collectively as "Maxwell's equations," though two of the equations are incomplete, as we will see in later chapters. What we have just called "Ampere's law for electricity" when completed is called "Faraday's law."

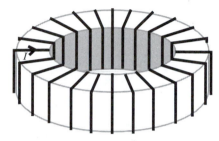

Figure 21.47 A toroid, which is essentially a solenoid bent around to make the ends meet.

Figure 21.48 Top view of the toroid.

What we have called "Ampere's law" when completed is called the "Ampere-Maxwell law." In summary, here are the four equations:

MAXWELL'S EQUATIONS (INCOMPLETE)

$$\oint \vec{E} \bullet \hat{n}\, dA = \frac{\sum q_{\text{inside}}}{\varepsilon_0} \qquad \text{Gauss's law for electricity}$$

$$\oint \vec{B} \bullet \hat{n}\, dA = 0 \qquad \text{Gauss's law for magnetism}$$

$$\oint \vec{E} \bullet d\vec{l} = 0 \qquad \text{Incomplete version of Faraday's law}$$

$$\oint \vec{B} \bullet d\vec{l} = \mu_0 \sum I_{\text{inside path}} \qquad \begin{array}{l}\text{Ampere's law} \\ \text{(incomplete version of Ampere-Maxwell law)}\end{array}$$

Note the structure of these laws. Two of the equations involve integrals over a surface (for electric field or magnetic field), and two involve integrals along a path (for electric field or magnetic field). The incompleteness in the last two laws comes from effects we have not yet studied. The complete forms of these laws include terms involving the time derivatives of electric and magnetic fields.

21.8 *The differential form of Gauss's law

This section deals with an advanced topic, which is not required as background for later work in this course. It includes the important issue of the relativistic correctness of Gauss's law.

We have pointed out several times that unless charges are motionless or moving very slowly compared to the speed of light, there is a flaw in an attempt to calculate electric field at some location by using Coulomb's law. The problem is relativistic retardation. The electric field depends on where fast-moving particles were at some time in the past, not where they are now. So Coulomb's law is not consistent with the theory of special relativity. Gauss's law on the other hand *is* consistent with relativity. In this section we will see why.

One way we could avoid problems with retardation would be to come up with some physical law that would relate a property of the electric field at some location and time (x, y, z, t) to source charges at the *same* location and time (x, y, z, t). If we could write our law in this form, relating source and fields at the same location and time, retardation wouldn't be a problem, and our law could be consistent with special relativity, unlike Coulomb's law.

In brief, we're looking for
- a property of the electric field $\vec{E}$ at some (x, y, z, t),
- that is related to source charges at the *same* (x, y, z, t).

There is a difficulty in this scheme with point charges, since the electric field is infinite at the location of a point charge. After we discuss the general idea, we'll explain how point charges are dealt with.

21.8.1 *Divergence

We will now introduce an appropriate property of the electric field, called the "divergence," which has the characteristics we want: we can relate the divergence to the charges at the same location and time.

Consider a very small Gaussian surface around a location in space, enclosing a very small volume ΔV. (Here V is volume, not potential!) In this region there is a charge density ρ measured in coulombs per cubic meter (ρ is lowercase Greek "rho"). The charge inside the tiny volume is $\rho \Delta V$ (multiply coulombs per cubic meter times volume in cubic meters). From Gauss's law, we know that the net flux on the surface of the tiny volume is:

$$\oint \vec{E} \bullet \hat{n}\, dA = \frac{\rho \Delta V}{\varepsilon_0}$$

Divide by the tiny volume ΔV:

$$\frac{\oint E \bullet \hat{n}\, dA}{\Delta V} = \frac{\rho}{\varepsilon_0}$$

On the left side of this equation we have the net flux per unit volume. In the limit as the volume becomes arbitrarily small, $\Delta V \rightarrow 0$, this ratio is called the "divergence" (div) of the electric field at this location:

$$\mathrm{div}(\vec{E}) \equiv \lim_{\Delta V \rightarrow 0} \frac{\oint \vec{E} \bullet \hat{n}\, dA}{\Delta V} \quad \text{(Note that this is a scalar, not a vector.)}$$

This is the property we are looking for, because we now have an equation in a form that doesn't have retardation problems:

$$\mathrm{div}(\vec{E}) = \frac{\rho}{\varepsilon_0}$$

Why is this property of the electric field called "divergence"? Look again at the pictures at the beginning of this chapter, and at Figure 21.49. If there is positive charge inside a (vanishingly small) volume, the electric field directions "diverge," the net electric flux is positive, and $\mathrm{div}(\vec{E})$ is positive. If there is negative charge inside the tiny volume, the electric field directions converge ("anti-diverge"), the net electric flux is negative, and $\mathrm{div}(\vec{E})$ is negative. If there is no charge inside the volume, the electric field looks as though it simply flows through the volume at this location, with neither net divergence nor net convergence, the net electric flux is zero, and $\mathrm{div}(\vec{E})$ is zero. In this introduction to the topic, we write the divergence as $\mathrm{div}(\vec{E})$, an older notation that reminds us of the definition. Later we will introduce a different notation, $\vec{\nabla} \bullet \vec{E}$, that is normally used for this property.

Divergence is relativistically correct

The divergence (net electric flux per unit volume, in the limit as the volume goes to zero) is a measure of an important property of the electric field at some location and time, and this is proportional to the charge density at that *same* location and time. This is called a "local" relationship between charge and field, because the charge and the divergence property are evaluated at the same location and time, rather than at different locations and times. This avoids the problem of relativistic retardation and forms the basis for what is called a "local field theory" that is consistent with special relativity.

   In the beginning of this chapter we developed what is called the "integral form" of Gauss's law, involving an integral of electric flux over a (possibly large) closed surface. What we have just developed is the "differential form" of Gauss's law:

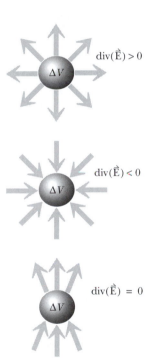

Figure 21.49 The divergence of the electric field at a location enclosed by a tiny volume $\Delta V \rightarrow 0$.

DIFFERENTIAL FORM OF GAUSS'S LAW

$$\mathrm{div}(\vec{E}) = \frac{\rho}{\varepsilon_0}, \text{ where the divergence is defined as}$$

$$\mathrm{div}(\vec{E}) \equiv \lim_{\Delta V \rightarrow 0} \frac{\oint \vec{E} \bullet \hat{n}\, dA}{\Delta V}, \text{ which is a scalar quantity}$$

Ex. 21.5 Based on its definition as electric flux per unit volume, what are the units of the divergence of the electric field? Show that these units are the same as the units of ρ/ε_0.

Ex. 21.6 A lead nucleus is spherical with a radius of about 7×10^{-15} m. The nucleus contains 82 protons (and typically 126 neutrons). Because of their motions the protons can be considered on average to be uniformly distributed throughout the nucleus. Based on the net flux at the surface of the nucleus, calculate the divergence of the electric field as electric flux per unit volume. Repeat the calculation at a radius of 3.5×10^{-15} m (you can use Gauss's law to determine the magnitude of the electric field at this radius). Also calculate the quantity ρ/ε_0 inside the nucleus.

21.8.2 *Why the integral form of Gauss's law is relativistically correct

In the beginning of this chapter we derived the integral form of Gauss's law by starting from Coulomb's law for stationary charges. The differential form was in turn derived from the integral form, yet the differential form of Gauss's law has the important property of being consistent with the theory of relativity, so it is actually more general than the Coulomb's law that we started from. Now we start from the differential form of Gauss's law, which is relativistically correct, and show that the integral form is also relativistically correct.

There is a purely mathematical theorem, the "divergence theorem," that shows that the volume integral of the divergence of a vector field,

$$\int \text{div}(\vec{E})\, dV$$

throughout a large volume at a particular time is equal to a familiar surface integral of the perpendicular component of that vector field,

$$\oint \vec{E} \bullet \hat{n}\, dA$$

on the closed bounding surface of the volume. In other words, the mathematical "divergence theorem" gives this result:

$$\int \text{div}(\vec{E})\, dV = \oint \vec{E} \bullet \hat{n}\, dA$$

But $\text{div}(\vec{E}) = \rho/\varepsilon_0$, so

$$\int \text{div}(\vec{E})\, dV = \int \frac{\rho}{\varepsilon_0}\, dV = \frac{Q_{\text{inside}}}{\varepsilon_0}$$

and the volume integral of the charge density ρ is the charge contained inside the bounding surface at this moment in time. You can think of this as simultaneously calculating the divergence of every tiny volume ΔV inside the surface to get the amount of charge in that tiny volume, then adding up all these charges to get the total charge inside the surface.

Combining these results we obtain the integral form of Gauss's law, starting from the differential form, and using only a purely mathematical theorem in the derivation:

$$\oint \vec{E} \bullet \hat{n}\, dA = \frac{Q_{\text{inside}}}{\varepsilon_0}$$

At a fixed moment in time we have integrated the differential form of Gauss's law, which is relativistically valid, and we have recovered the integral form of Gauss's law. This is why even the integral form of Gauss's law, unlike Coulomb's law, is consistent with the special theory of relativity (see earlier discussion in section 21.3.6 on page 776).

21.8.3 *Dealing with point charges

When calculating the divergence by taking the limit as the volume goes to zero, $\Delta V \rightarrow 0$, if ΔV shrinks down onto the exact location of a point charge such as an electron, the limiting process doesn't work—the flux per unit volume becomes infinite. (At a location arbitrarily close to the point charge, there is no difficulty, and the charge density at that location is zero.)

The mathematical difficulty with point charges is handled formally through the use of the "delta function." The delta function is defined to be zero everywhere except at one point, where it is infinite, and its integral is exactly 1. In other words, the delta function is an infinite spike with zero width, constructed in such a way that the product of the infinite height and the zero width is finite (Figure 21.50). This peculiar entity is not a true function but is an example of a class of mathematical objects called "generalized functions" or "distributions."

An equivalent way of dealing with the mathematical problem is to approximate the point charge as a sphere of charge with a very small radius and a very large charge density. In that case the divergence limit is well defined throughout the tiny sphere of charge, as it was in your calculations of the divergence inside a lead nucleus (Exercise 21.6).

In the practical case of the interior of ordinary atomic matter, we can stop shrinking the volume at a size that, though tiny, is still large enough to include lots of atoms. In this case we are averaging the divergence and the charge density over a tiny volume containing many protons and electrons. This is similar to what was done at the end of Chapter 16, where we found an average electric field inside an insulator, averaged over many atoms.

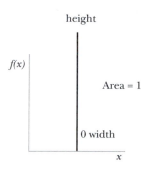

Figure 21.50 The one-dimensional delta function.

21.8.4 *Another way to calculate the divergence

It is possible to calculate the divergence directly from the definition, as a limit as $\Delta V \rightarrow 0$. In this section we present another way to calculate the divergence. Consider the tiny box-shaped surface shown in Figure 21.51. In this tiny region the electric field is in the x direction, but varying in magnitude. Let's use the definition of divergence to calculate this property at a location inside the surface, in the limit that the volume goes to zero:

$$\text{div}(\vec{E}) \equiv \lim_{\Delta V \rightarrow 0} \frac{\oint \vec{E} \bullet \hat{n}\, dA}{\Delta V} = \lim_{\Delta V \rightarrow 0} \frac{(E_2 - E_1)\Delta y \Delta z}{\Delta x \Delta y \Delta z}$$

$$\text{div}(\vec{E}) = \lim_{\Delta x \rightarrow 0} \frac{(E_2 - E_1)}{\Delta x}$$

But this limit is the definition of the derivative of the electric field with respect to x:

$$\text{div}(\vec{E}) = \frac{dE}{dx}$$

More precisely, the divergence in this case is equal to the "partial derivative" of E_x with respect to x, where the special "partial derivative" notation indicates that we hold y and z constant while taking the derivative:

$$\text{div}(\vec{E}) = \frac{\partial E_x}{\partial x}$$

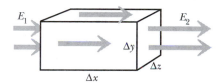

Figure 21.51 Calculating the divergence of the electric field on the closed surface of a tiny volume.

More generally, there may be contributions to the divergence involving E_y or E_z, which we would calculate in the same way. Adding together all these contributions to the divergence, we find the following:

DIFFERENTIAL FORM OF GAUSS'S LAW

$$\text{div}(\vec{E}) = \frac{\partial E_x}{\partial x} + \frac{\partial E_y}{\partial y} + \frac{\partial E_z}{\partial z} = \frac{\rho}{\varepsilon_0}$$

21.8.5 *Electric potential and Gauss's law

Remember that electric field can be expressed as the negative gradient of the electric potential (just as force can be expressed as the negative gradient of potential energy). Writing V for electric potential (not volume!), we have this:

$$E_x = -\frac{\partial V}{\partial x}, \; E_y = -\frac{\partial V}{\partial y}, \; E_z = -\frac{\partial V}{\partial z}$$

Putting these gradient forms into the differential form of Gauss's law gives the following important equation:

$$\frac{\partial^2 V}{\partial x^2} + \frac{\partial^2 V}{\partial y^2} + \frac{\partial^2 V}{\partial z^2} = -\frac{\rho}{\varepsilon_0}$$

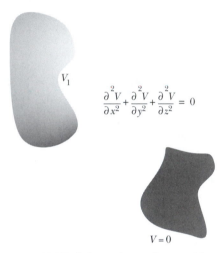

$$\frac{\partial^2 V}{\partial x^2} + \frac{\partial^2 V}{\partial y^2} + \frac{\partial^2 V}{\partial z^2} = 0$$

V_1

$V = 0$

Here we have a differential equation relating the electric potential at some location to the charge density at that same location. A frequent use of this equation is in determining the electric potential in empty space (where the charge density ρ is zero), near some charged pieces of metal that are maintained at known potentials (Figure 21.52). If we knew the charge distribution on these pieces of metal, we could calculate the electric potential and electric field anywhere, through superposition. But often we know only the fixed potentials of the pieces of metal, not the charge distribution.

In such cases we can use the following differential equation, called "Laplace's equation":

$$\frac{\partial^2 V}{\partial x^2} + \frac{\partial^2 V}{\partial y^2} + \frac{\partial^2 V}{\partial z^2} = 0 \quad \text{(in empty space)}$$

Figure 21.52 A boundary value problem. One piece of metal is maintained at a fixed potential V_1 relative to a potential defined to be $V = 0$ for the other piece. In the empty space Gauss's law gives a differential equation to be solved for $V(x, y, z)$.

Throughout the empty space between the pieces of metal, the electric potential must be a solution of this differential equation. There exist both analytical and numerical methods for solving Laplace's equation. At locations very near the pieces of metal, the solution of Laplace's equation for V in the empty space must equal the known potentials. This is an example of what is called a "boundary value problem," which is an important class of problem in science and engineering.

Ex. 21.7 In a certain region of space, the electric potential depends on position in the following way, where V_0 and a are constants: $V = V_0 + a(x^2 + y^2 + z^2)$. What is the charge density in this region?

Ex. 21.8 In a certain region of space, the electric potential depends on position in the following way, where V_0 and b are constants: $V = V_0 + bxyz$. What is the charge density in this region?

21.8.6 *Standard notation

To help introduce the topic, we have written the divergence as $\text{div}(\vec{E})$. The standard notation normally used for divergence is this:

$$\vec{\nabla} \bullet \vec{E} = \frac{\partial E_x}{\partial x} + \frac{\partial E_y}{\partial y} + \frac{\partial E_z}{\partial z} = \frac{\rho}{\varepsilon_0}$$

The $\vec{\nabla}$ (or "del") operator is expressed in the following form:

$$\vec{\nabla} = \frac{\partial}{\partial x}\hat{\imath} + \frac{\partial}{\partial y}\hat{\jmath} + \frac{\partial}{\partial z}\hat{k}$$

An "operator" operates on some quantity to its right. In this case the operation involves partial differentiation. The usual rules of calculating the dot product give the following for the quantity $\vec{\nabla} \cdot \vec{E}$.

$$\left(\frac{\partial}{\partial x}\hat{\imath} + \frac{\partial}{\partial y}\hat{\jmath} + \frac{\partial}{\partial z}\hat{k}\right) \cdot (E_x\hat{\imath} + E_y\hat{\jmath} + E_z\hat{k}) = \frac{\partial E_x}{\partial x} + \frac{\partial E_y}{\partial y} + \frac{\partial E_z}{\partial z}$$

Also, note that the del operator can be used to express the concept of gradient:

$$\vec{E} = -\left(\frac{\partial V}{\partial x}\hat{\imath} + \frac{\partial V}{\partial y}\hat{\jmath} + \frac{\partial V}{\partial z}\hat{k}\right) = -\vec{\nabla} V$$

We can rewrite the differential form of Gauss's law like this:

$$\vec{\nabla} \cdot (\nabla V) = \frac{\partial^2 V}{\partial x^2} + \frac{\partial^2 V}{\partial y^2} + \frac{\partial^2 V}{\partial z^2} = -\frac{\rho}{\varepsilon_0}$$

This combination, the divergence of a gradient, is called the "Laplacian" and is written like this:

$$\nabla^2 V = \frac{\partial^2 V}{\partial x^2} + \frac{\partial^2 V}{\partial y^2} + \frac{\partial^2 V}{\partial z^2} = -\frac{\rho}{\varepsilon_0}$$

In regions where there are no charges, we have Laplace's equation:

$$\nabla^2 V = \frac{\partial^2 V}{\partial x^2} + \frac{\partial^2 V}{\partial y^2} + \frac{\partial^2 V}{\partial z^2} = 0 \quad \text{(in empty space)}$$

21.9 *The differential form of Ampere's law

We have pointed out several times that unless charges are moving very slowly compared to the speed of light, there is a flaw in trying to calculate magnetic field at some location by using the Biot-Savart law. The problem is relativistic retardation. The magnetic field depends on where fast-moving particles were at some time in the past, not where they are now. So the Biot-Savart law is not consistent with the theory of special relativity. Ampere's law on the other hand is consistent with relativity. In this section we will see why.

The argument is very similar to the one presented in connection with Gauss's law. We look for

- a property of the magnetic field $\vec{B}$ at some (x, y, z, t),
- that is related to moving source charges at the *same* (x, y, z, t).

21.9.1 *Curl

We introduce an appropriate property of the magnetic field, called the "curl." Consider a very small Amperian path around a location in space, enclosing a very small area ΔA. In this region there is a current density $\vec{J}$ measured in amperes per square meter. The direction of the vector $\vec{J}$ is the direction of the conventional current at that location. The current passing through the tiny area is $\vec{J} \cdot \hat{n}\Delta A$ (multiply amperes per square meter times area in square meters), where $\hat{n}$ is a "normal" unit vector perpendicular to the area. From Ampere's law, we know that the path integral of magnetic field around the perimeter of the tiny area is this:

$$\oint \vec{B} \cdot d\vec{l} = \mu_0 \vec{J} \cdot \hat{n}\Delta A$$

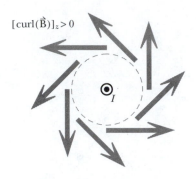

$[\mathrm{curl}(\vec{\mathrm{B}})]_z > 0$

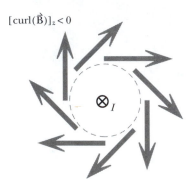

$[\mathrm{curl}(\vec{\mathrm{B}})]_z < 0$

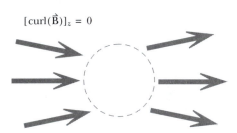

$[\mathrm{curl}(\vec{\mathrm{B}})]_z = 0$

Figure 21.53 The curl of the magnetic field at a location surrounded by a tiny $\Delta A \to 0$.

Divide by the tiny area ΔA:

$$\frac{\oint \vec{\mathrm{B}} \bullet d\vec{l}}{\Delta A} = \mu_0 \vec{\mathrm{J}} \bullet \hat{\mathrm{n}}$$

On the left side of this equation we have the path integral of magnetic field per unit area. In the limit as the area becomes arbitrarily small, $\Delta A \to 0$, this ratio is called the component in the $\hat{\mathrm{n}}$ direction of the "curl" of the magnetic field at this location:

$$[\mathrm{curl}(\vec{\mathrm{B}})] \bullet \hat{\mathrm{n}} \equiv \lim_{\Delta A \to 0} \frac{\oint \vec{\mathrm{B}} \bullet d\vec{l}}{\Delta A}$$

This is the property we are looking for. We obtain an equation in a form that doesn't have retardation problems:

$$\mathrm{curl}(\vec{\mathrm{B}}) = \mu_0 \vec{\mathrm{J}} \text{ (note that this is a } vector \text{ equation)}$$

Why is this property of the magnetic field called the "curl" of the magnetic field? Look at Figure 21.53. If there is a current crossing through a (vanishingly small) area, the magnetic field "curls" around the current.

Curl is relativistically correct

The curl (net path integral of magnetic field per unit area, in the limit as the area goes to zero) is a measure of an important property of the magnetic field at some location and time, and this is proportional to the current density at that *same* location and time. This is called a "local" relationship between current and field, because the current and the curl property are evaluated at the same location and time, rather than at different locations and times. This avoids the problem of relativistic retardation and forms the basis for what is called a "local field theory" that is consistent with special relativity.

Earlier in this chapter we developed what is called the "integral form" of Ampere's law, involving a path integral of magnetic field over a (possibly long) closed path. What we have just developed is the "differential form" of Ampere's law:

DIFFERENTIAL FORM OF AMPERE'S LAW

$$\mathrm{curl}(\vec{\mathrm{B}}) = \mu_0 \vec{\mathrm{J}}$$

where the component of the curl in the $\hat{\mathrm{n}}$ direction, perpendicular to a tiny area ΔA, is defined as

$$[\mathrm{curl}(\vec{\mathrm{B}})] \bullet \hat{\mathrm{n}} \equiv \lim_{\Delta A \to 0} \frac{\oint \vec{\mathrm{B}} \bullet d\vec{l}}{\Delta A}$$

Using the "del" operator, Ampere's law is written like this:

$$\vec{\nabla} \times \vec{\mathrm{B}} = \mu_0 \vec{\mathrm{J}}$$

The differential form of Ampere's law is correct in linking current density to magnetic field at the same location and time. However, we will see later that Ampere's law is incomplete because current is not the only source of magnetic field.

Ex. 21.9 Based on its definition as path integral of magnetic field per unit area, what are the units of the curl of the magnetic field? Show that these units are the same as the units of $\mu_0 J$.

21.10 Summary

Fundamental principles

GAUSS'S LAW

$$\oint \vec{E} \bullet \hat{n} \, dA = \frac{\sum q_{inside}}{\varepsilon_0}$$

GAUSS'S LAW FOR MAGNETISM

$$\oint \vec{B} \bullet \hat{n} \, dA = 0$$

AMPERE'S LAW

$$\oint \vec{B} \bullet d\vec{l} = \mu_0 \sum I_{inside \; path}$$

Alternative notations for Gauss's law:

$$\sum_{surface} \vec{E} \bullet \hat{n} \Delta A = \frac{\sum q_{inside}}{\varepsilon_0} \quad \text{(a finite sum of electric flux)}$$

$$\oint E_\perp \, dA = \frac{\sum q_{inside}}{\varepsilon_0} \quad \text{(where } E_\perp = E \text{ perpendicular to surface)}$$

$$\oint E \, dA_\perp = \frac{\sum q_{inside}}{\varepsilon_0} \quad \text{(where } A_\perp = A \text{ perpendicular to electric field)}$$

$$\oint \vec{E} \bullet d\vec{A} = \frac{\sum q_{inside}}{\varepsilon_0} \quad \text{(where } d\vec{A} = \hat{n} \, dA \text{)}$$

The circle superimposed on the integral sign means "integrate over the entire closed surface."

Results

We used Gauss's law to find the electric field of a plate and a sphere. We also proved some important properties about metals, especially that excess charge is on the surface, and that there is no field in an empty hole in a metal in static equilibrium. We used Ampere's law to find the magnetic field of a thick wire, a solenoid, and a toroid.

Advanced

*Differential form of Gauss's law

$$\vec{\nabla} \bullet \vec{E} = \frac{\partial E_x}{\partial x} + \frac{\partial E_y}{\partial y} + \frac{\partial E_z}{\partial z} = \frac{\rho}{\varepsilon_0}, \text{ or } \nabla^2 V = \frac{\partial^2 V}{\partial x^2} + \frac{\partial^2 V}{\partial y^2} + \frac{\partial^2 V}{\partial z^2} = -\frac{\rho}{\varepsilon_0}$$

*Differential form of Ampere's law

$$\vec{\nabla} \times \vec{B} = \mu_0 \vec{J}$$

21.11 Example problem: Charge at an interface

The diagram shows a portion of a circuit in which a steady-state electron current i is running as shown. A wire made of Metal 1 joins a wire made of Metal 2; both wires have uniform circular cross section of radius r. The electron mobility u_2 in Metal 2 is much smaller than the electron mobility u_1 in Metal 1. There are n_1 mobile electrons per unit volume in Metal 1 and n_2 in Metal 2; $n_1 > n_2$.

Determine the amount of charge on the interface between Metal 1 and Metal 2, including sign. Explain carefully. Explicitly state all assumptions, and show how all quantities were evaluated.

Solution

Start from fundamental principles:

By the current node rule, $i_1 = i_2$, so $n_1 A u_1 E_1 = n_2 A u_2 E_2$

and therefore $E_2 = \dfrac{n_1 u_1}{n_2 u_2} E_1$

This makes sense, because E_2 must be greater than E_1. In order to create a bigger electric field in wire 2, it is likely that there is some excess charge at the interface.

Since we don't want to include charge on the outer surface of the wire, we choose a Gaussian surface inside the wire. We consider a tube-shaped Gaussian surface just inside the surface of the wire, shown by the dotted line in Figure 21.55. Any charge inside the Gaussian surface must be on the interface between the wires, since in section 21.4.7 we showed that in a uniform section of wire there could be no charge inside the wire.

By Gauss's law:

$$\oint \vec{E} \bullet d\vec{A} = \frac{\sum q_{\text{inside}}}{\varepsilon_0}$$

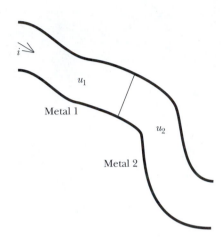

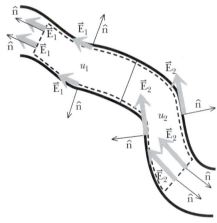

Figure 21.54 Current runs through a part of a circuit containing two different kinds of wire.

Figure 21.55 A Gaussian surface just inside the wire, with electric field and normal vector shown at several locations.

Assumptions: the electric field inside the wire is everywhere parallel to the wire, and is uniform in magnitude inside a wire of uniform composition. (We proved these assertions in Chapter 18.)

Given the direction of the electron current, the electric fields in the wire must be as shown in Figure 21.55. Evaluating the flux on different parts of the surface:

outer surface of wire: flux = 0 because $\vec{E} \perp \hat{n}$.

top (Metal 1) end: flux $= E_1(\pi r^2)\cos 0° = +E_1(\pi r^2)$

bottom (Metal 2) end: flux $= E_2(\pi r^2)\cos 180° = -E_2(\pi r^2)$

$$\oint \vec{E} \bullet d\vec{A} = 0 + E_1(\pi r^2) - E_2(\pi r^2)$$

So the amount of charge on the interface is:

$$\sum q_{\text{inside}} = \varepsilon_0(E_1(\pi r^2) - E_2(\pi r^2))$$

$$\sum q_{\text{inside}} = \varepsilon_0 E_1 \pi r^2 \left(1 - \frac{n_1 u_1}{n_2 u_2}\right)$$

This quantity is negative, which makes sense, because the charge on the interface should be negative in order to increase E_2 and decrease E_1.

21.12 Review questions

Patterns of electric field on closed surfaces

RQ 21.1 Figure 21.56 shows a Gaussian cylinder drawn around some hidden charges. Given this pattern of electric field on the surface of the cylinder, how much net charge is inside? Invent a possible charge distribution inside the cylinder that could give this pattern of electric field.

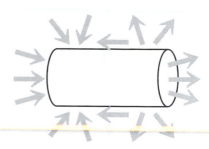

Figure 21.56 Electric field pattern on a cylinder (RQ 21.1).

Definition of electric flux

RQ 21.2 Figure 21.57 shows a disk-shaped region of radius 2 cm, on which there is a uniform electric field of magnitude 300 volts/meter at an angle of 30 degrees to the plane of the disk. Calculate the electric flux on the disk, and include the correct units.

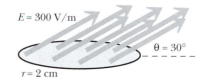

Figure 21.57 Electric flux on a disk-shaped region (RQ 21.2)

Gauss's law

RQ 21.3 Figure 21.58 shows a box on whose surfaces the electric field is measured to be horizontal and to the right. On the left face (3 cm by 2 cm) the magnitude of the electric field is 400 volts/meter, and on the right face the magnitude of the electric field is 1000 volts/meter. On the other faces only the direction is known (horizontal). Calculate the electric flux on every face of the box, the total flux, and the total amount of charge that is inside the box.

RQ 21.4 The electric field on a closed surface is due to all the charges in the universe, including the charges outside the closed surface. Explain why the total flux nevertheless is proportional only to the charges that are inside the surface, with no apparent influence of the charges outside.

RQ 21.5 In Chapter 15 we calculated the electric field at a location on the axis of a ring. Explain why we can't use Gauss's law to determine the electric field at that location without all those calculations.

RQ 21.6 In Figure 21.59 the magnetic field in a region is vertical and was measured to have the values shown on the surface of a cylinder. Why should you suspect something is wrong with these measurements?

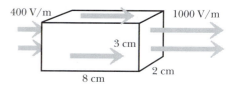

Figure 21.58 What is in the box? (RQ 21.3)

Properties of metals

RQ 21.7 Explain why you are safe and unaffected inside a car that is struck by lightning. Also explain why it might not be safe to step out of the car just after the lightning strike, with one foot in the car and one on the ground. (Note that it is current that kills; a rather small current passing through the region of the heart can be fatal.)

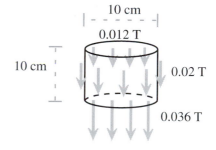

Figure 21.59 What's wrong with these measurements? (RQ 21.6)

Ampere's law

RQ 21.8 The magnetic field has been measured to be horizontal everywhere along a rectangular path 20 cm long and 4 cm high, shown in Figure 21.60. Along the bottom the average magnetic field $B_1 = 1.5 \times 10^{-4}$ tesla, along the sides the average magnetic field $B_2 = 1.0 \times 10^{-4}$ tesla, and along the top the average magnetic field $B_3 = 0.6 \times 10^{-4}$ tesla. What can you conclude about the electric currents in the area that is surrounded by the rectangular path?

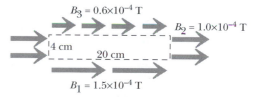

Figure 21.60 Magnetic field measured along a rectangular path (RQ 21.8).

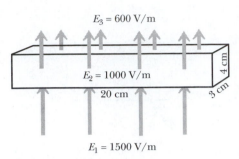

Figure 21.61 The electric field on the surface of a box (Problem 21.1a).

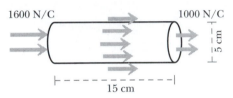

Figure 21.62 Electric field on the surface of a cylinder (Problem 21.1b).

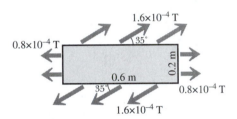

Figure 21.64   How much current flows through the shaded area? (Problem 21.1d)

21.13 Homework problems

Problem 21.1 Simple applications

(a) The electric field has been measured to be vertically upward everywhere on the surface of a box 20 cm long, 4 cm high, and 3 cm deep, shown in Figure 21.61. All over the bottom of the box $E_1 = 1500$ volts/m, all over the sides $E_2 = 1000$ volts/m, and all over the top $E_3 = 600$ volts/m. What can you conclude about the contents of the box? Include a numerical result.

(b) The electric field is horizontal and has the values indicated on the surface of a cylinder shown in Figure 21.62. What can you deduce from this pattern of electric field? Include a numerical result.

(c) The electric field has been measured to be horizontal and to the right everywhere on the closed box shown in Figure 21.63. All over the left side of the box $E_1 = 100$ volts/m, and all over the right, slanting, side of the box $E_2 = 300$ volts/m. On the top the average field is $E_3 = 150$ volts/m, on the front and back the average field is $E_4 = 175$ volts/m, and on the bottom the average field is $E_5 = 220$ volts/m. How much charge is inside the box? Explain briefly.

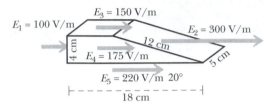

Figure 21.63 The electric field on an oddly shaped closed box (Problem 21.1c).

(d) Figure 21.64 shows a measured pattern of magnetic field in space. How much current I passes through the shaded area? In what direction?

Problem 21.2 A spherical shell with charge inside the shell

A negative point charge $-Q$ is at the center of a hollow insulating spherical shell, which has an inner radius R_1 and an outer radius R_2. There is a total charge of $+3Q$ spread uniformly throughout the volume of the insulating shell, not just on its surface. Determine the electric field for (a) $r < R_1$, (b) $R_1 < r < R_2$, and (c) $R_2 < r$.

Problem 21.3 Charged coaxial cable

You may have seen a "coaxial cable" connected to a television set. A coaxial cable consists of a central copper wire of radius r_1 surrounded by a hollow copper tube (typically made of braided copper wire) of inner radius r_2 and outer radius r_3. Normally the space between the central wire and the outer tube is filled with an insulator, but in this problem assume for simplicity that this space is filled with air. Assume that no current runs in the cable.

Suppose the coaxial cable is straight and has a very long length L, and that the central wire carries a charge $+Q$ uniformly distributed along the wire (so that the charge per unit length is $+Q/L$ everywhere along the wire). Also suppose that the outer tube carries a charge $-Q$ uniformly distributed along its length L. The cylindrical symmetry of the situation indicates that the electric field must point radially outward or radially inward. The electric field cannot have any component parallel to the cable. In this problem, draw mathematical Gaussian cylinders of length d (with d much less than the cable length L) and appropriate radius r, centered on the central wire.

(a) Use a mathematical Gaussian cylinder located inside the central wire ($r < r_1$), and another Gaussian cylinder with a radius in the interior of the outer tube ($r_2 < r < r_3$), to determine the exact amount and location of charge on the inner and outer conductors. (Hint: What do you know about

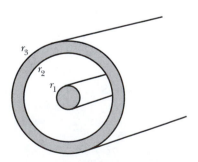

Figure 21.65   A coaxial cable (Problem 21.3).

the electric field in the interior of the two conductors? What do you know about the flux on the ends of your Gaussian cylinders?)

(b) Use a mathematical Gaussian cylinder whose radius is in the air gap ($r_1 < r < r_2$) to determine the electric field in the gap as a function of r. (Don't forget to consider the flux on the ends of your Gaussian cylinder.)

(c) Use a mathematical Gaussian cylinder whose radius is outside the cable ($r > r_2$) to determine the electric field outside the cable. (Don't forget to consider the flux on the ends of your Gaussian cylinder.)

Problem 21.4 Capacitor

Here is a close-up of the central region of a capacitor made of two large metal plates of area A, very close together and charged equally and oppositely. There is $+Q$ and $-Q$ on the inner surfaces of the plates and small amounts of charge $+q$ and $-q$ on the outer surfaces.

(a) Knowing Q, determine E: Consider a Gaussian surface in the shape of a shoe-box, with one end of area A_{box} in the interior of the left plate and the other end in the air gap (surface 1 in the diagram). Using only the fact that the electric field is expected to be horizontal everywhere in this region, use Gauss's law to determine the magnitude of the field in the air gap. Check that your result agrees with our earlier calculations in Chapter 15. (Be sure to consider the flux on all faces of your Gaussian box.)

(b) Proof by contradiction: Consider Gaussian box 2, with both ends in the gap. Use Gauss's law to prove that the electric field is constant in magnitude throughout the gap.

(c) Knowing Q, determine E: Consider Gaussian box 3, with its left end in the gap and its right end in the interior of the right plate. Show that Gauss's law applied to this Gaussian box gives the correct magnitude of the electric field in the gap.

(d) Knowing E, determine Q: Finally, consider Gaussian box 4, with its left end outside the capacitor and its right end in the interior of the left plate. In Chapter 15 we determined that the fringe field is approximately given by

$$E_{fringe} \approx \frac{Q/A}{\varepsilon_0}\frac{s}{L}$$

where s is the gap width and L is the length or width of the plate ($L = 2R$ for a disk). Use this information and Gaussian box 4 to determine the approximate density of charge q/A on the outer surface of the plate, where q is the charge on the outer surface.

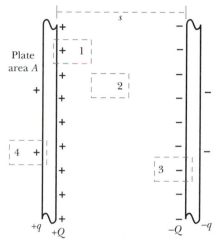

Figure 21.66 Inside and outside a capacitor (Problem 21.4).

Problem 21.5 Force on a current-carrying coil

On page 742 we saw that the non-uniform direction of the magnetic field of a magnet is responsible for the force $F = I(2\pi R)B\sin\theta$ exerted by the magnet on a current loop. However, we didn't know the angle θ, which is a measure of the non-uniformity of the direction of the magnetic field. Gauss's law for magnetism can be used to determine this angle θ.

(a) Apply Gauss's law for magnetism to a thin disk of radius R and thickness Δx located where the current loop will be placed (Figure 21.67). Use Gauss's law to determine the component $B_3\sin\theta$ of the magnetic field that is perpendicular to the axis of the magnet at a radius R from the axis. Assume that R is small enough that the x component of magnetic field is approximately uniform everywhere on the left face of the disk, and also on the right face of the disk (but with a smaller value). The magnet has moment μ.

(b) Now imagine placing the current loop at this location, and show that $F = I(2\pi R)B_3\sin\theta$ can be rewritten as $F = \mu|dB/dx|$, which is the result we obtained from potential-energy arguments (page 742).

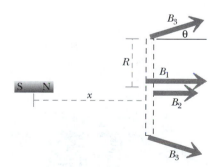

Figure 21.67 A bar magnet exerts a force on a current-carrying loop (Problem 21.5).

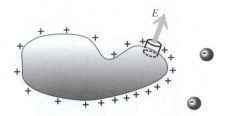

Figure 21.68 What is *E* near the surface of the metal? (Problem 21.6)

Problem 21.6 The electric field near a chunk of metal

(a) Figure 21.68 shows a solid chunk of metal that carries a net positive charge and is also polarized by nearby charges. The electric field very near the surface of a metal in static equilibrium is perpendicular to the surface: any component of the electric field parallel to the surface would drive currents in the metal. Let *S* be the "local" surface charge density—the surface charge per unit area (C/m^2) at a particular location on the surface. Calculate the magnitude *E* of the electric field in the air very near a location on the surface of the metal, in terms of the surface charge density *S* at that location. Use a cylindrical Gaussian surface as shown, with one face just inside the metal and one face just outside. Don't skip steps in your analysis!

(b) Show that nearby charges contribute half of *E*, and distant charges contribute the other half.

Problem 21.7 Electric field in the region of a cube

Consider a cube with one corner at the origin and with sides of length 10 cm positioned along the *xyz* axes. There is an electric field $\vec{E} = \langle 50, 200y, 0 \rangle$ N/m throughout the region which has a constant *x* component and a *y* component which increases linearly with *y*.

(a) Draw a picture of each side showing $\hat{n}$ and $\vec{E}$ for that side.

(b) Determine how much charge is inside the cube.

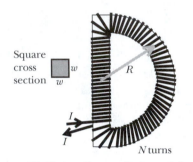

Figure 21.69 A coil on a D-shaped frame (Problem 21.8).

Problem 21.8 A coil on a D-shaped frame

A D-shaped frame is made out of plastic of small square cross section and tightly wrapped uniformly with *N* turns of wire as shown in Figure 21.69, so that the magnetic field has essentially the same magnitude throughout the plastic. (*R* is much bigger than *w*.)

With a current *I* flowing, what is the magnetic field inside the plastic? Show the direction of the magnetic field in the plastic at several locations.

Problem 21.9 Sheets of current-carrying wires

Figure 21.70 shows a large number *N* of closely packed wires, each carrying a current *I* out of the page. The width of this sheet of wires is *L*.

(a) Show (and explain) the direction of the magnetic field at the two indicated locations, a distance *d* from the middle wire, where *d* is much less than *L* (*d* ≪ *L*).

(b) Using the rectangular path shown on the diagram, calculate the magnitude of the magnetic field at the indicated locations, in terms of the given physical quantities *N*, *I*, and *L*.

(c) Using your result from part (b), find $\vec{B}$ in the region between two parallel current sheets with equal currents running in opposite directions.

(d) What is $\vec{B}$ outside both sheets?

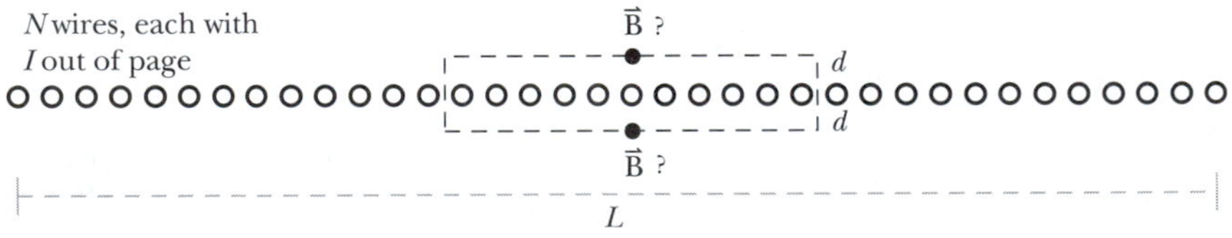

Figure 21.70 A sheet of current-carrying wires (Problem 21.9).

Problem 21.10 A coil on a plastic rod

A long plastic rod with rectangular cross section $h \times w$ is wound with a coil of length *d* consisting of *N* closely packed turns of wire with negligible resistance, and a current *I* runs through the coil as shown in Figure 21.71.

Using the rectangular path of length *b* shown on the diagram, prove that the magnetic field inside the coil is $B = \mu_0 \, NI/d$, and describe the direction of $\vec{B}$ inside the coil.

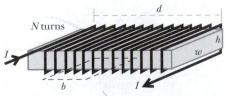

Figure 21.71 A coil on a plastic rod (Problem 21.10).

Problem 21.11 Current-carrying coaxial cable

A coaxial cable, shown in Figure 21.72, consists of an inner metal wire of radius r_1 and an outer metal cylinder of radius r_2. There is current I_1 to the right in the wire and current $I_2 < I_1$ to the left in the cylinder. Consider a circular Amperian path of radius R centered on the wire and perpendicular to the wire. Determine the magnitude and direction of the magnetic field at a location R above the wire, far from the ends of the coaxial cable. Explain in detail, including all assumptions, directions, calculations, etc. Don't skip steps.

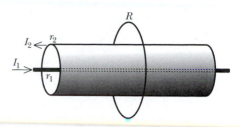

Figure 21.72 A coaxial cable carrying current (Problem 21.11).

Problem 21.12 Magnetic field inside a thick wire

A thick wire of radius R carries a current I, as shown in Figure 21.73. Find the magnetic field inside the wire, a distance r from the center of the wire, where $r < R$.

Figure 21.73 A thick wire with radius R carrying current I (Problem 21.12).

21.14 Answers to exercises

21.1 (page 771)   866 volts/meter; 1.73×10^{-3} volt·m

21.2 (page 771) 1.73×10^{-6} m^2; 1.73×10^{-3} volt·m

21.3 (page 780)   Electric field cannot be blocked (superposition principle). Electric field produced by the polarized metal is equal and opposite to the field due to the outside charges, making a net field that is zero.

21.4 (page 784)   0.2 A, into the page

21.5 (page 790)   N/C/m or volt/m^2

21.6 (page 790) 1.0×10^{36} N/C/m; 1.0×10^{36} N/C/m;

                            1.0×10^{36} N/C/m

21.7 (page 792) $-6\varepsilon_0 a$

21.8 (page 792)   0

21.9 (page 794)   tesla/m

Chapter 22

Faraday's Law

Chapter 22

Faraday's Law

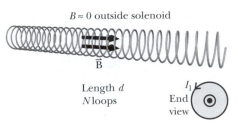

Figure 22.1 The magnetic field inside a long solenoid is $B_1 = \mu_0 NI_1 / d$.

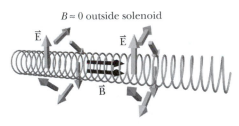

Figure 22.2 A time-varying magnetic field in the solenoid generates a curly electric field!

Charges make electric fields, and electric fields affect charges. Moving charges make magnetic fields, and magnetic fields affect moving charges. In Chapter 20 we glimpsed a deeper relationship among these phenomena in the oddly different electric and magnetic forces observed by a stationary Jack and a moving Jill, and in field transforms between reference frames.

In this chapter we will see another fundamental aspect of the relationship between the electric and magnetic fields. It turns out that a time-varying magnetic field can produce an electric field! So there are two different ways to produce an electric field: by charges according to Coulomb's law, or by time-varying magnetic fields. In the latter case, we say that the time-varying magnetic field "induces" an electric field, and the phenomenon is referred to as "magnetic induction."

No matter how an electric field is produced, it has the same effect on a charge q (that is, $\vec{F} = q\vec{E}$), but the "non-Coulomb" electric field induced by a time-varying magnetic field has a different pattern in space than the "Coulomb" electric field due to charges. In particular, a round-trip path integral of the non-Coulomb electric field is not zero, unlike the situation with the Coulomb electric field. Because of this, time-varying magnetic fields can induce an emf around a circuit loop.

22.1 Changing magnetic fields and curly electric fields

Consider a long solenoid, a long hollow coil of current-carrying wire (Figure 22.1). In Chapter 17 using the Biot-Savart law and Chapter 21 using Ampere's law we found that with a current I_1 in the solenoid, the magnetic field B_1 inside a tightly-wound solenoid (not too near the ends) is $B_1 = \mu_0 NI_1 / d$, where N is the number of turns of wire and d is the length of the solenoid. The magnetic field outside the solenoid (not too near the ends) is very small and for a very long solenoid can be taken to be zero.

If the current is constant, then B_1 is constant in time. In that case a charge that is in motion somewhere outside the solenoid will not experience an electric or magnetic force, because the electric and magnetic fields outside the solenoid are essentially zero. But something remarkable happens if we vary the current, so that the magnetic field B_1 inside the solenoid varies with time (Figure 22.2). There is still almost no magnetic field outside the solenoid, but we observe a curly electric field both inside and outside of the solenoid! This peculiar electric field curls around the axis of the solenoid (end view; Figure 22.3). The electric field is proportional to dB_1 / dt, the rate of change of the magnetic field.

Inside the solenoid the electric field is proportional to r, the distance from the axis (smaller near the axis). Outside the solenoid the curly electric field is proportional to $1/r$; the electric field gets smaller as you go farther away from the solenoid. The curly electric field has the usual effect on charges: a charge q experiences a force $\vec{F} = q\vec{E}$. For example, a proton placed above the solenoid in Figure 22.3 will be initially pushed to the right.

While the curly electric field affects charges in the usual way, it isn't produced by charges according to Coulomb's law. Rather, this electric field is associated with the time-varying magnetic field, and we call such an electric field a "non-Coulomb" field $\vec{E}_{NC}$.

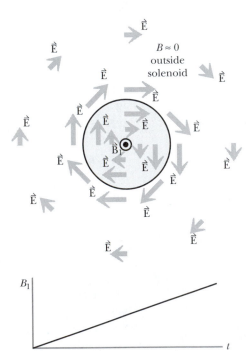

Figure 22.3 There is a curly electric field in the presence of a time-varying magnetic field. In this case B_1 is increasing with time.

? Explain how you know that this pattern of non-Coulomb electric field cannot be produced by an arrangement of stationary charges.

Traveling in a complete loop around the solenoid, $\oint \vec{E}_{NC} \bullet d\vec{l} \neq 0$. Since the round trip integral of the electric field due to stationary charges is always zero, this electric field cannot be produced by stationary charges.

TWO WAYS TO PRODUCE ELECTRIC FIELD

- A Coulomb electric field is produced by charges according to Coulomb's law:

$$\vec{E} = \frac{1}{4\pi\varepsilon_0}\frac{q}{r^2}\hat{r}$$

- A non-Coulomb electric field $\vec{E}_{NC}$ is associated with time-varying magnetic fields $d\vec{B}/dt$. Outside of a long solenoid inside of which the magnetic field is B_1, the induced electric field is proportional to dB_1/dt and decreases with distance like $1/r$.

No matter how an electric field $\vec{E}_1$ is produced, it has the same effect on a charge q_2: $\vec{F}_{21} = q_2\vec{E}_1$.

Figure 22.4 shows what is observed experimentally in four different cases, where the magnetic field points out or into the page, and increases or decreases with time. From these results you can see that it is not the direction of $\vec{B}_1$ that determines the direction of $\vec{E}_{NC}$, but rather the direction of the rate of change of $\vec{B}_1$. Here is a right-hand rule that summarizes the experimental observations, and which you should memorize:

DIRECTION OF THE CURLY ELECTRIC FIELD

With the thumb of your right hand pointing in the direction of $-d\vec{B}_1/dt$, your fingers curl around in the direction of $\vec{E}_{NC}$.

Hints on using this right-hand rule

In order to use this right-rule, you need to be able to determine the direction of the vector quantity $-d\vec{B}_1/dt$. It helps to think about this quantity in the form $-\Delta\vec{B}_1/\Delta t$, where Δt is a small, finite time interval. From this form you can see that the direction of $-d\vec{B}_1/dt$ is the same as the direction of $-\Delta\vec{B}_1$, the negative of the change in direction of the magnetic field during a short time.

So a good way to find the direction of $-d\vec{B}_1/dt$ is to draw the magnetic field $\vec{B}_1(t)$ at a time t, and the magnetic field $\vec{B}_1(t+\Delta t)$ at a slightly later time $t+\Delta t$, and observe the change $\Delta\vec{B}_1$ (Figure 22.5). Then $-\Delta\vec{B}_1$ is the direction of $-d\vec{B}_1/dt$.

Ex. 22.1 A magnetic field near the floor points up and is increasing. Looking down at the floor, does the non-Coulomb electric field curl clockwise or counter-clockwise?

Ex. 22.2 A magnetic field near the ceiling points down and is decreasing. Looking up at the ceiling, does the non-Coulomb electric field curl clockwise or counter-clockwise?

22.1.1 Driving current with a non-Coulomb electric field

Suppose we place a circular metal ring of radius r_2 around a solenoid (Figure 22.6), with the magnetic field in the solenoid increasing with time (Fig-

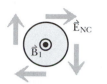

Figure 22.4 Four cases: magnetic field out or in, increasing or decreasing.

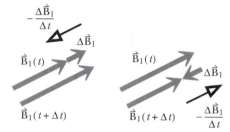

Figure 22.5 Find the change in the magnetic field as a basis for determining the direction of $-d\vec{B}_1/dt$.

Figure 22.6 A metal ring is placed around the solenoid.

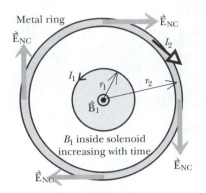

Figure 22.7 End view: The non-Coulomb electric field drives a current I_2 in the ring.

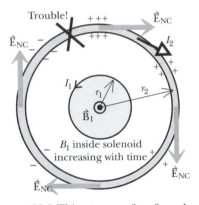

Figure 22.8 This pattern of surface charge is impossible, because it would imply a huge E at the marked location, and in the wrong direction!

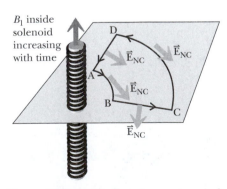

Figure 22.9 A path that does not encircle the solenoid.

ure 22.7). The non-Coulomb electric field inside the metal will drive conventional current clockwise around the ring ($-d\vec{B}_1/dt$ points into the page; point your right thumb into the page and see how your fingers curl clockwise). The technological importance of this effect is that it can make current run in a wire just as though a battery were present.

The current I_2 in the ring is proportional to the electric field E_{NC} inside the metal, as in an ordinary circuit. However, in ordinary circuits there are charges on the surface of the wires that, together with charges on the battery, produce the electric field inside the metal that drives the current.

? Think about a possible pattern of surface charge on this ring. Why is it impossible to draw a plausible gradient of surface charge on the ring in this situation?

The symmetry of the ring makes it impossible to have a surface charge gradient along the ring. We reason by contradiction: If you pick one point and draw positive surface charge there, then gradually decrease the amount of positive charge and increase the amount of negative surface charge, you find that when you get back to the starting point there is a huge change in surface charge (from − to +), which would produce a huge E, in the wrong direction (Figure 22.8). But there is nothing special about the point you picked, so it can't have a different electric field from all other points on the ring. So there cannot be a varying surface charge around the ring. (There will be a small pile-up of electrons on the outside of the ring, which then provides the radially-inward force that turns the electron current.)

The emf is the (non-Coulomb) energy input per unit charge. The (non-Coulomb) force per unit charge is the (non-Coulomb) field E_{NC}.

? Therefore, what is the emf in terms of E_{NC} for this ring, if the ring has a radius r_2?

We have emf $= \oint \vec{E}_{NC} \bullet d\vec{l} = E_{NC}(2\pi r_2)$, since E_{NC} is constant and parallel to the path. The current in the ring of radius r_2 is $I_2 = \text{emf}/R$, where R is the resistance of the ring. It is as though we had inserted a battery into the ring.

? If the metal ring had a radius r_2 twice as large, what would the emf be, since emf $= \oint \vec{E}_{NC} \bullet d\vec{l}$? Remember that the experimental observations show that the non-Coulomb electric field outside the solenoid is proportional to $1/r$.

Double the radius implies half the electric field, so the product $E_{NC}(2\pi r_2)$ stays the same. Apparently the emf in a ring encircling the solenoid is the same for any radius. (In fact, we get the same emf around any circuit surrounding the solenoid, not just a circular ring.)

? Consider a round-trip path that does not encircle the solenoid (Figure 22.9). The electric field is shown all along the path A-B-C-D-A. Use the fact that $E_{NC} = \text{emf}/(2\pi r)$ to explain why we have the result emf $= \oint \vec{E}_{NC} \bullet d\vec{l} = 0$ around this path that does not encircle the solenoid.

As you make a round trip around the path A-B-C-D-A, the contribution to $\oint \vec{E}_{NC} \bullet d\vec{l}$ is positive along A-B, zero along B-C (the parallel component of electric field is zero), negative along C-D, and zero along D-A. Moreover, the magnitude of the contribution along C-D is equal to that along A-B, because the longer path is compensated by the smaller electric field (which is proportional to $1/r$). Therefore we have a zero emf along this path. The non-Coulomb electric field would polarize a wire that followed this path but would not drive current around the loop.

We see that in order to drive current in a wire the wire must encircle a region where the magnetic field is changing. If the wire doesn't encircle the region of changing magnetic field, there is no emf.

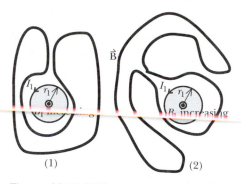

Figure 22.10 Will current run in these wires?

Ex. 22.3 In Figure 22.10, will current run in wire (1)? In wire (2)?

Ex. 22.4 On a circular path of radius 10 cm in air around a solenoid with increasing magnetic field, the emf is 30 volts. What is the magnitude of the non-Coulomb electric field on this path?

Ex. 22.5 A wire with resistance 4 ohms is placed along the path in the previous exercise. What is the current in the wire?

22.2 Faraday's law

So far we have a right-hand rule for determining the direction of the curly electric field, but we don't have a way to determine the magnitude of the non-Coulomb electric field. Faraday's law is a quantitative relationship between the rate of change of the magnetic field and the magnitude of the non-Coulomb electric field. To establish this quantitative relationship, in principle we could vary the magnetic field and measure E_{NC} by observing the effect of the curly electric field on an individual charged particle, but it can be difficult to track the path of an individual charged particle. Alternatively, we can construct a circuit following a path where we expect a non-Coulomb electric field and measure the current in the circuit as a function of dB/dt. We will describe circuit experiments that lead to Faraday's law.

22.2.1   Observing current caused by non-Coulomb electric fields

We can use an ammeter to measure the induced current in a circuit that encircles a solenoid. In Figure 22.11 we omit showing the power supply and connections to the solenoid. Initially the solenoid current I_1 is constant, so B_1 is constant, and no current runs through the resistive wire and the ammeter.

If we vary the magnetic field B_1 by varying I_1, we can infer something about the resulting non-Coulomb electric field E_{NC} from its integral around the circuit, which is the emf. Unfortunately you can't observe the phenomenon with the circuit equipment you have been using, because you need a more sensitive ammeter than is provided by your compass. Perhaps your instructor will demonstrate the effects or arrange for you to experiment with appropriate equipment.

Suppose that we vary the current I_1 in the long solenoid, thus causing the magnetic field B_1 to vary, and we observe the current I_2 in the outer wire. In the solenoid we first increase the current rapidly, then hold the current constant, and then slowly decrease the current, at half the rate we used at first (Figure 22.12). If we know the resistance R of the circuit containing the ammeter we can determine the emf from the ammeter reading, since emf $- RI_2 = 0$; the wire acts as though a battery were inserted.

1) While the solenoid current I_1 is increasing from t_0 to t_1 (Figure 22.12), B_1 is increasing, and the current I_2 runs clockwise in the circuit, out of the "+" terminal of the ammeter. The ammeter is observed to read a negative current. Remember that conventional current flowing into the positive terminal of an ammeter gives a positive reading.

2) While the solenoid current I_1 is held constant from t_1 to t_2 (Figure 22.12), the magnetic field in the solenoid isn't changing, and there is no emf in the circuit. The ammeter reading is zero:

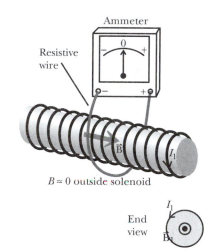

Figure 22.11 An ammeter measures current in a loop surrounding the solenoid. Initially I_1 is constant, so B_1 is constant, and no current runs through the ammeter.

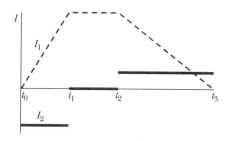

Figure 22.12 Vary the solenoid current I_1 and observe the current I_2 that runs in the outer wire, through the ammeter.

dB/dt makes E_{NC} (and associated emf).
No dB/dt, no E_{NC}, even if there is a large (constant) B.

3) While the solenoid current I_1 is decreasing from t_2 to t_3 (at half the initial rate; see Figure 22.12), B_1 is decreasing, and the current I_2 runs counterclockwise in the circuit, into the "+" terminal of the ammeter. The ammeter is observed to read a positive current. The current I_2 is observed to be half what it was during the first interval.

E_{NC} (and associated emf) outside the solenoid are
proportional to dB/dt inside the solenoid.

4) There is one more crucial experiment. If we use a solenoid with twice the cross-sectional area but the same magnetic field, we find that the current I_2 is twice as big (which means that the emf is twice as big).

E_{NC} (and associated emf) outside the solenoid are
proportional to the cross-sectional area of the solenoid.

22.2.2 Magnetic flux

Putting together these experimental observations, and being quantitative about the emf, we find experimentally that the magnitude of the induced emf is numerically equal to this:

$$|\text{emf}| = \left| \frac{d}{dt}(B_1 \pi r_1^2) \right|$$

The quantity $(B_1 \pi r_1^2)$ is called the "magnetic flux" Φ_{mag} on the area encircled by the circuit (Φ is the capital Greek letter Phi). Magnetic flux is calculated in the same way as the electric flux introduced in Chapter 21 on Gauss's law.

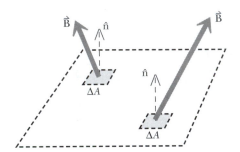

The magnetic flux on a small area ΔA is $\vec{B} \bullet \hat{n} \Delta A = B_\perp \Delta A$, where $\hat{n}$ is a dimensionless unit vector perpendicular to the area ΔA ($\hat{n}$ is called the "normal" to the surface); $B_\perp$ is the perpendicular component of magnetic field. We add up all such contributions over an extended surface to get the magnetic flux over that surface (Figure 22.13):

Figure 22.13 The magnetic flux on an area is the sum of the magnetic flux on each small subarea.

DEFINITION OF MAGNETIC FLUX

$$\Phi_{mag} \equiv \int \vec{B} \bullet \hat{n}\, dA = \int B_\perp\, dA$$

Note that magnetic flux can be positive, negative, or zero, depending on the orientation of the magnetic field $\vec{B}$ relative to the normal $\hat{n}$.

Ex. 22.6 A uniform magnetic field of 3 tesla points 30° away from the perpendicular to the plane of a rectangular loop of wire 0.1 m by 0.2 m (Figure 22.14). What is the magnetic flux on this loop?

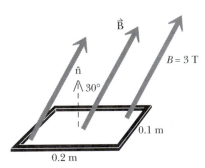

Figure 22.14 Calculate the magnetic flux on the area enclosed by this loop of wire.

22.2.3 Faraday's law—quantitative

The experimental fact that a time-varying magnetic field produces an emf whose magnitude is equal to the rate of change of magnetic flux is called "Faraday's law" (discovered by Michael Faraday in 1831). Faraday's law summarizes a great variety of experimental data, not just the data for the experiments we have discussed. Faraday's law is a major physical law concerning time-varying magnetic fields. Unlike the motional emf we studied in Chapter 20, Faraday's law cannot be derived from any of the other fundamental principles we have studied.

FARADAY'S LAW

$$\text{emf} = -\frac{d\Phi_{mag}}{dt}$$

$$\text{where emf} = \oint \vec{E}_{NC} \bullet d\vec{l} \text{ and } \Phi_{mag} \equiv \int \vec{B} \bullet \hat{n} \, dA$$

In words: The induced emf along a round-trip path is equal to the rate of change of the magnetic flux on the area encircled by the path.

Direction: With the thumb of your right hand pointing in the direction of $-d\vec{B}/dt$, your fingers curl around in the direction of $\vec{E}_{NC}$.

FORMAL VERSION OF FARADAY'S LAW

$$\oint \vec{E}_{NC} \bullet d\vec{l} = -\frac{d}{dt}[\int \vec{B} \bullet \hat{n} \, dA] \quad \text{(sign given by right-hand rule)}$$

Faraday's law summarizes the experiments we've described so far:
- A faster rate of change of magnetic flux induces a bigger emf (if the magnetic flux is constant there is no induced emf).
- If there is a bigger area with the same perpendicular magnetic field there is a bigger emf.

The meaning of the minus sign

If the thumb of your right hand points in the direction of $-d\vec{B}/dt$ (that is, the opposite of the direction in which the magnetic field is increasing), your fingers curl around in the direction along which the path integral of electric field is positive:

$$\oint \vec{E}_{NC} \bullet d\vec{l} = -\frac{d}{dt}[\int \vec{B} \bullet \hat{n} \, dA] \quad \text{(sign given by right-hand rule)}$$

For most of our work it is simplest to calculate the magnitude of the effect, ignoring signs and directions, and then give the appropriate sign or direction based on the right-hand rule. We will include the minus sign in Faraday's law, as a reminder of what we need to do to get directions.

Flux and path

What area exactly do we use when calculating the magnetic flux? Imagine a soap film stretched over the closed path around which we are calculating $\oint \vec{E}_{NC} \bullet d\vec{l}$. We want to compute Φ_{mag} on the area covered by that soap film.

Ex. 22.7 A wire of resistance 10 ohms and length 2.5 m is bent into a circle and is concentric with a solenoid in which the magnetic flux changes from 5 tesla·m^2 to 3 tesla·m^2 in 0.1 seconds. What is the emf in the wire? What is the non-Coulomb electric field in the wire? What is the current in the wire?

22.2.4   The Coulomb electric field can be included in Faraday's law

We will usually write emf $= \oint \vec{E}_{NC} \bullet d\vec{l}$ as a reminder that it is the curly non-Coulomb electric field that has a nonzero round-trip path integral, which we call the emf around that path. However, we could also calculate the emf in terms of the net electric field $\vec{E} = \vec{E}_C + \vec{E}_{NC}$, where $\vec{E}_C$ is the Coulomb electric field due to charges:

$$\text{emf} = \oint \vec{E} \bullet d\vec{l}$$

This is true because we have

$$\text{emf} = \oint \vec{E} \bullet d\vec{l} = \oint (\vec{E}_C + \vec{E}_{NC}) \bullet d\vec{l} = 0 + \oint \vec{E}_{NC} \bullet d\vec{l}$$

since $\oint \vec{E}_C \bullet d\vec{l} = 0$ (round-trip integral of Coulomb electric field is zero).

22.2.5 Application: A circuit surrounding a solenoid

We can apply Faraday's law to analyze quantitatively the situation of a circuit with an ammeter surrounding a solenoid (Figure 22.15; a cross section of the solenoid is shown). There is magnetic field $\vec{B}_1$ only over the circle of radius r_1 (the solenoid), and it points in the same direction as $\hat{n}$, the unit vector perpendicular to the surface.

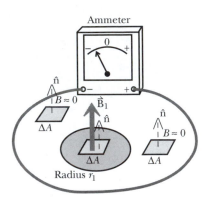

? Calculate the magnetic flux Φ_{mag} on the area enclosed by the solenoid, the flux on the other portions of the area encircled by the circuit, and the total flux on the area encircled by the circuit.

On the area enclosed by the solenoid, $\Phi_{mag} = B_1(\pi r_1^2)$, because $B_\perp$ is constant and equal to B_1 throughout the cross-section of the solenoid. On every small area outside the solenoid, the magnetic field is nearly zero, so $\Phi_{mag} = 0$. Therefore the total flux through the outer wire is $\Phi_{mag} = B_1(\pi r_1^2)$.

Figure 22.15 A cross section through the solenoid; calculate the flux enclosed by the wire.

What counts is the magnetic flux encircled by the circuit, not the magnetic flux on a closed surface such as a box or sphere. Unlike electric flux, the total magnetic flux on a closed surface is always zero, because "magnetic monopoles" seem not to exist.

Experimentally we find that the emf around the circuit is numerically equal to $d\Phi_{mag}/dt$, as predicted by Faraday's law. For example, suppose the magnetic field in the solenoid increases from 0.1 tesla to 0.7 tesla in 0.2 seconds, and the area of the solenoid is 3 cm^2.

? What is the emf around the circuit?

Faraday's law predicts the following average emf:

$$\text{emf} = \frac{\Delta \Phi_{mag}}{\Delta t} = \frac{(0.6\text{T})(3\times10^{-4}\text{ m}^2)}{(0.2\text{ s})} = 9\times10^{-4} \text{ volts}$$

Ammeter reading

? If the resistance of the wire plus ammeter is 0.5 ohms, what current will the ammeter display?

The circuit acts as though a battery were inserted, $\text{emf} - RI = 0$, so

$$I = \frac{\text{emf}}{R} = \frac{(9\times10^{-4}\text{ volts})}{(0.5\text{ohm})} = 1.8\times10^{-3} \text{ ampere}$$

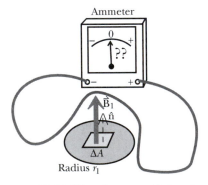

Figure 22.16 What happens if the circuit does not encircle the solenoid?

One more experiment: let's reconnect the wire and ammeter so that they don't encircle the solenoid (Figure 22.16).

? As we increase the current (and magnetic field) in the solenoid, what would you predict from Faraday's law about the ammeter reading? Why?

Within the area bounded by the wire outside the solenoid, there is practically no magnetic field, so no magnetic flux Φ_{mag}, and no rate of change of magnetic flux $d\Phi_{mag}/dt$. When we try the experiment, we do indeed find that the ammeter shows little or no current.

Voltmeter readings

Since a voltmeter is just an ammeter with a large resistance in series, a voltmeter may give a puzzling reading in the presence of time-varying magnetic flux. Consider the two voltmeters shown in Figure 22.17. You normally expect that when voltmeter leads are connected to each other, the voltmeter must read zero volts.

? But do these voltmeters read zero?

The voltmeter leads on the left of Figure 22.17 encircle a region of changing magnetic field, so the voltmeter will read an emf equal to $d\Phi_{mag}/dt$. The leads of the other voltmeter don't encircle a region of changing magnetic field, so the voltmeter reads zero.

Because of the effect of time-varying magnetic fields, you have to be a bit careful in interpreting a voltmeter reading when there are varying magnetic fields around. For example, when there are sinusoidally alternating currents ("AC") there are time-varying magnetic fields due to those time-varying currents. If the leads of an AC voltmeter happen to surround some AC magnetic flux, this will affect the voltmeter reading.

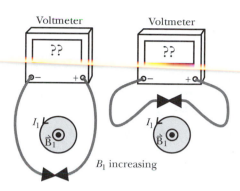

Figure 22.17 What do these voltmeters read?

Ex. 22.8 The magnetic field in a solenoid is $B = \mu_0 NI/d$. A circular wire of radius 10 cm is concentric with a solenoid of radius 2 cm and length $d = 1$ meter, containing 10,000 turns. The current increases at a rate of 50 A/s. What is the emf in the wire? What is the non-Coulomb electric field in the wire?

22.2.6 The emf for a coil with multiple turns

In many devices, instead of just one loop of wire surrounding a time-varying flux there is a coil with many turns, which increases the effect: the emf in a coil containing N turns is approximately N times the emf for one loop of the coil. This point needs some discussion.

In the space around the increasing flux in the coil, there is a curly pattern of induced (non-Coulomb) electric field. If the loops are tightly wound so that they are very close together, E_{NC} is about the same in each loop (Figure 22.18). The emf from one end of the coil to the other is the integral of the non-Coulomb field:

$$\text{emf} = \int_{N\ \text{turns}} \vec{E}_{NC} \bullet d\vec{l} = N(E_{NC}L_{\text{one turn}}) = N(\text{emf}_{\text{one turn}})$$

Because of this, Faraday's law for a coil is written in the following form:

FARADAY'S LAW FOR A COIL

$$\text{emf} = -N\frac{d\Phi_{mag}}{dt} \quad \text{(sign given by right-hand rule)}$$

The induced emf in a coil of N turns is equal to N times the rate of change of the magnetic flux on one loop of the coil.

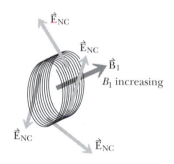

Figure 22.18 The induced emf of a thin coil is approximately N times the emf of one loop.

Ex. 22.9 A thick copper wire connected to a voltmeter surrounds a region of time-varying magnetic flux, and the voltmeter reads 10 volts. If instead of a single wire we use a coil of thick copper wire containing 20 turns, what does the voltmeter read?

Ex. 22.10 A thin Nichrome wire connected to an ammeter surrounds a region of time-varying magnetic flux, and the ammeter reads 10 amperes. If instead of a single wire we use a coil of thin Nichrome wire containing 20 turns, what does the ammeter read?

22.2.7   Faraday's law and moving coils or magnets

Figure 22.19 Move coil 1 toward coil 2, and there is a time-varying magnetic field inside coil 2.

Figure 22.20 Moving a magnet toward coil 2 creates a time-varying magnetic field inside the coil.

Figure 22.21 Rotating a bar magnetic (or coil 1) produces a time-varying magnetic field inside coil 2.

A time-varying magnetic field produces a curly electric field. One way to create a time-varying field is by varying the current in a coil, but this isn't the only way to do produce a time-varying field. With a steady current in one coil in Figure 22.19, you can move that coil closer to a second coil. This increases the magnetic field (and magnetic flux) inside the second coil, and while the magnetic field is increasing there is an emf (and a current in the second coil).

You can also induce an emf by moving a bar magnet toward or away from coil 2, since this creates a time-varying magnetic field (and magnetic flux) inside the coil (Figure 22.20).

You can also get an induced emf by rotating the first coil (or a bar magnet; Figure 22.21), since this changes the flux through the second coil as long as you are rotating and therefore changing the magnetic field (and magnetic flux) in the region of space surrounded by the second coil.

These various experiments give additional confirmation that the emf in one or more loops of wire is due to a time-varying magnetic field, and numerically equal to the rate of change of the magnetic flux enclosed by the loops:

$$\text{emf} = -N\frac{d\Phi_{\text{mag}}}{dt} = -N\frac{d}{dt}\left(\sum \vec{B}_1 \bullet \hat{n}\Delta A_2\right)$$

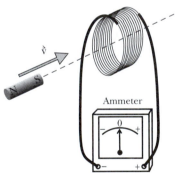

Figure 22.22 Will the ammeter read positive or negative? (Ex 22.11 and 22.12.)

Figure 22.23 Will the current run clockwise or counterclockwise? (Ex. 22.13.)

Ex. 22.11 Suppose you move a bar magnet toward the coil in Figure 22.22, with the "S" end of the bar magnet closest to the coil. Will the ammeter read positive or negative?

Ex. 22.12 Now move the bar magnet away from the coil, with the "S" end still closest to the coil. Will the ammeter read + or –?

Ex. 22.13 A bar magnet is held vertically above a horizontal metal ring, with the south pole of the magnet at the top (Figure 22.23). If the magnet is lifted straight up, will current run clockwise or counterclockwise in the ring, as seen from above?

22.2.8   An interesting complication

The current I_1 in a coil of wire makes a magnetic field B_1. If I_1 varies with time, there is a time-varying magnetic field dB_1/dt in a second coil that can induce an emf and a current I_2 in the second coil. This induced current I_2 *also* makes a magnetic field B_2, an effect that we have ignored until now. If I_2 is constant (due to a constant dB_1/dt), the additional magnetic field B_2 is constant and so does not contribute additional non-Coulomb electric field.

However, if $d^2B_1/dt^2 \neq 0$ (there is a time-varying rate of change), then I_2 isn't constant, and there is a time-varying magnetic field dB_2/dt that contributes a non-Coulomb electric field and emf of its own. In many cases this effect is small compared to the main effect and can be ignored. If the effect is sizable, you can see that a full analysis could be rather difficult, because you have a changing I_1 making a changing I_2 that contributes to the change, which... Whew!

We will mostly avoid trying to analyze this complicated kind of situation in this introductory course, though we will see an example in the discussion of superconductors later in this chapter. A related issue is self-inductance, which we will also study later in this chapter.

22.3 Relationship between Faraday's law and motional emf

In Chapter 20 we studied motional emf: when a wire moves through a magnetic field, magnetic forces drive current along the wire. We calculated the motional emf in a moving bar of length L as $(q|\vec{v} \times \vec{B}|)L/q$, the (non-Coulomb) work per unit charge done by the magnetic force $q\vec{v} \times \vec{B}$. If the velocity of the bar is perpendicular to the magnetic field, the motional emf is simply vBL. There is another way to calculate motional emf that is often mathematically easier, in terms of magnetic flux.

As shown in Figure 22.24, in a short time Δt the bar moved a distance $\Delta x = v\Delta t$, and the area surrounded by the current-carrying pieces of the circuit increased by an amount $\Delta A = L\Delta x = Lv\Delta t$. There is an increased magnetic flux through the circuit of amount

$$\Delta\Phi_{mag} = B_\perp \Delta A = B(Lv\Delta t)$$

Dividing by Δt we find this:

$$\frac{\Delta\Phi_{mag}}{\Delta t} = BLv$$

But BLv is equal to the emf vBL generated by the magnetic force, and in the limit of small Δt we have the following:

MOTIONAL EMF = RATE OF CHANGE OF MAGNETIC FLUX

$$|\text{emf}| = \left|\frac{d\Phi_{mag}}{dt}\right| \quad \text{(sign given by direction of magnetic force)}$$

Although we have proved this result only for a particular special case, it can be shown that this result applies in general to calculating motional emf—that is, an emf associated with the motion of a piece of a circuit in a magnetic field that is not varying in time.

As an example, consider a generator (Figure 22.25). When the normal to the loop is at an angle $\theta = \omega t$ to the magnetic field, the flux through the loop is $Bhw\cos(\omega t)$, so emf $= |d\Phi/dt| = \omega Bhw\sin(\omega t)$, which is the same result we obtained in Chapter 20 when we calculated the emf directly in terms of magnetic forces acting on electrons in the moving wires (see Section 20.9.2 on page 745).

Ex. 22.14 A uniform, non-time-varying magnetic field of 3 tesla points 30° away from the perpendicular to the plane of a rectangular loop of wire 0.1 m by 0.2 m (Figure 22.26). The loop as a whole is moved in such a way that it maintains its shape and its orientation in the uniform magnetic field. What is the emf around the loop during this move?

Ex. 22.15 In 0.1 s the loop is stretched to be 0.12 m by 0.22 m. What is the average emf around the loop during this time?

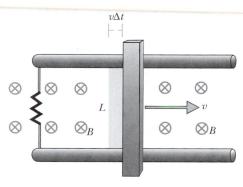

Figure 22.24 There is increased flux through the circuit as the bar moves.

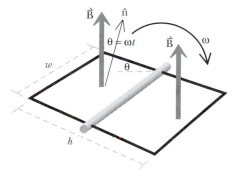

Figure 22.25 We can compute the emf of a generator from the rate of change of flux.

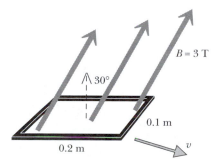

Figure 22.26 Move this rectangular circuit around in a uniform, constant field.

22.3.1 The two pieces of the flux derivative

A time-varying magnetic field is associated with a curly (non-Coulomb) electric field. We can quantitatively predict the behavior of circuits enclosing a time-varying magnetic field by calculating the rate of change of magnetic flux on the surface bounded by the circuit.

However, in the preceding section we saw that there is another way that magnetic flux can change. If the magnetic field is constant but the shape or orientation of the loop changes, so as to enclose more or less flux, the flux changes with time even though the magnetic field doesn't, and this is the phenomenon of motional emf associated with magnetic forces.

Any change in magnetic flux, whether due to a change in magnetic field or a change in shape or orientation, produces an emf equal to the rate of change of flux. Both effects can be present simultaneously. Suppose that as you drag a metal bar along rails, the magnitude of the uniform magnetic field is also increasing with time. The flux is $B_\perp A$, and the rate of change of the flux is due in part to the change in the magnetic field, and in part to the change in the area enclosed by the changing path:

$$\frac{d\Phi_{mag}}{dt} = \frac{d}{dt}(B_\perp A) = \frac{dB_\perp}{dt}A + B_\perp \frac{dA}{dt} = \frac{dB_\perp}{dt}A + B_\perp \frac{Ldx}{dt} = \frac{dB_\perp}{dt}A + B_\perp Lv$$

Change of B,
fixed path

Change of path,
fixed B

In some but not all cases the two effects can be related through a change of reference frame. If you move a magnet toward a stationary coil, the magnetic field in the coil is increasing with time, and there is a curly non-Coulomb electric field that drives current in the coil.

If on the other hand you move a coil toward a stationary magnet, the magnetic field is not changing with time anywhere, but the coil is moving in a magnetic field, so there is a magnetic force on the mobile charges in the wire, and motional emf.

In both situations you can calculate the same emf from the same rate of change of flux, and you observe the same current to run in the coil, though the mechanism looks different in the two cases. You can shift between one situation and the other simply by changing your reference frame (move with the magnet, or move with the coil). The principle of relativity implies that you should predict the same emf and the same current in either frame of reference, and you do.

It is not always possible to relate the two kinds of effects simply by changing reference frame. For example, consider a bar sliding along rails. In the reference frame of the bar, the bar is at rest but the resistor is moving. In neither reference frame is there a time-varying magnetic field.

(A full analysis in the reference frame of the bar is complicated. If the bar is at rest in this frame, why is it polarized? We saw in Chapter 20 that the answer is given by special relativity: when you transform to the frame of the bar in the presence of a uniform magnetic field, a transverse electric field of magnitude vB appears, and in this reference frame it is this new electric field that polarizes the bar.)

22.4 Summary: Non-Coulomb fields and forces

If the magnetic field changes ($d\vec{B}/dt$ nonzero), there is a non-Coulomb electric field $\vec{E}_{NC}$ which curls around the region of changing flux, and emf $= \oint \vec{E}_{NC} \bullet d\vec{l}$ is nonzero. In purely motional emf, however, where the path changes but the magnetic field doesn't change, the non-Coulomb forces are magnetic forces, not electric.

The non-Coulomb electric field $\vec{E}_{NC}$ affects stationary as well as moving charges, but the (non-Coulomb) magnetic forces only affect moving charges, such as electrons in the moving metal bar. It is somewhat odd that the magnitude of the emf in both of these situations is $d\Phi_{mag}/dt$, but that's how it works.

We have identified three different kinds of electric and magnetic effects on the electrons inside a wire:

1) Coulomb electric fields due to surface charges (superposition of contributions of the form $\vec{E} = \dfrac{1}{4\pi\varepsilon_0}\dfrac{q}{r^2}\hat{r}$);

2) non-Coulomb electric fields associated with time-varying magnetic fields ($\dfrac{d\vec{B}}{dt}$); and

3) magnetic forces $q\vec{v} \times \vec{B}$ if the wire is moving (yielding a motional $|emf| = \left|\dfrac{d\Phi_{mag}}{dt}\right|$ due to change of path).

All of these effects can be present simultaneously. For example, if the magnetic field is changing while you pull a bar along rails, there is a Coulomb electric field due to the charges that build up on the surfaces of the circuit elements, there is a non-Coulomb electric field due to the changing magnetic field, and there are non-Coulomb magnetic forces on electrons inside the moving bar. The round-trip integral of the Coulomb electric field is zero. The round-trip integral of the non-Coulomb electric field and the non-Coulomb magnetic force per unit charge together gives the emf, and the current I is given by $I = emf/R$, where $|emf| = |d\Phi_{mag}/dt|$.

22.5 The character of physical laws

We certainly haven't explained *why* there is a curly, non-Coulomb electric field around a solenoid when the magnetic field in the solenoid changes. All we have done is show that Faraday's law correctly predicts the observed electric field (and emf) in this admittedly strange situation. That is why Faraday's law is called a physical "law": it summarizes a wide variety of experiments and predicts the outcomes of experiments yet to be done, but in a deeper sense it doesn't "explain" anything. It doesn't tell us "why" there is a non-Coulomb electric field.

However, the status of Faraday's law is really no different than the status of Coulomb's law. Coulomb's law correctly summarizes a wide variety of experiments and predicts the outcomes of electrical experiments yet to be done, but it explains nothing at all about "why" there are "electrically charged particles" or "why" these "charged particles" attract and repel each other with a $1/r^2$ force.

When we ask you to "explain" electric or magnetic phenomena, we ask you to explain a wide range of phenomena in terms of a small number of fundamental physical laws: Coulomb's law or Gauss's law, the Biot-Savart law or Ampere's law, the Lorentz force, Faraday's law. We don't and can't ask you to go one level deeper and explain these fundamental laws, because these laws are in fact summaries of a wide variety of experiments and may not have an explanation in terms of principles that are even more basic.

It would be satisfying to find deeper physical laws that would explain the laws we have been studying and explain other laws as well, because that would reduce even further the number of principles required to predict an extremely wide range of phenomena. In fact, we have seen that in the framework of Einstein's theory of special relativity, magnetic phenomena appear to be essentially electric phenomena as seen from a different reference frame. Quantum electrodynamics goes even further in unifying electricity and magnetism. And recent work by theoretical and experimental physicists has provided the basis for physical laws that encompass both electromagnetism and the so-called "weak force" responsible for certain kinds of radioactivity.

The search goes on for an all-encompassing "theory of everything" that would include gravitation and the physics of quarks, thought to be the building blocks of protons and neutrons. But many physicists suspect that even if a "theory of everything" explained all observable physical phenomena, it wouldn't explain itself.

We would love to be able to tell you "why" there is a curly electric field near a region of changing magnetic flux, but we can't. That's just how it is!

22.6 Superconductors

As you lower the temperature, the resistivity of an ordinary metal decreases but remains nonzero even at absolute zero (Figure 22.27), while some materials lose all resistance to electric current. Such materials are called "superconductors." It isn't that the resistance merely gets very small. Below a critical temperature T_C that is different for each superconducting material, the resistance vanishes completely, as indicated in Figure 22.28. To put it another way, the mobility is infinite. A nonzero electric field in a superconductor would produce an infinite current! As a result, the electric field inside a superconductor must always be zero, yet a nonzero current can run anyway.

In a ring made of lead that is kept at a temperature below 7.2 degrees Kelvin (that is, only 7.2 degrees above absolute zero), you can observe a current that persists undiminished for years (Figure 22.29)!

? How can you check that the current in the lead ring hasn't changed since last month, without inserting an ammeter into the ring or touching the ring in any way?

Since the current loop produces a magnetic field, you could use a compass or other more sensitive device to measure the magnetic field produced by the current.

? Does this persistent current violate the fundamental principle of conservation of energy? Why or why not?

No energy is being dissipated in the ring. Since its resistance is zero, the power used up is zero (power = RI^2). Energy conservation is not violated.

? Do the persistent atomic currents in your little bar magnet violate the principle of conservation of energy? Why or why not?

The "currents" in the bar magnet are due to the spin and orbital motion of electrons in the atoms of the magnet. The energy of these electrons remains constant—no energy is used up, so no energy input is required. Energy conservation is not violated.

A fruitful way to think about a superconductor is to say that the persistent current is similar to the persistent atomic currents in a magnet, but the diameter of a persistent current loop in a superconductor can be huge compared to an atom. In both the magnet and the superconductor the persistent currents can be explained only by means of quantum mechanics. In both cases the explanation hinges on the fact that these systems have discrete energy levels. Since the energy levels are separated by a gap, there can be no small energy changes, and hence there is no mechanism for energy dissipation.

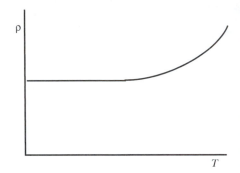

Figure 22.27 Resistivity *vs.* temperature for an ordinary metal.

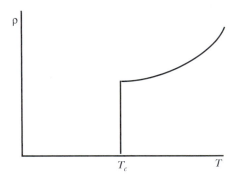

Figure 22.28 Resistivity *vs.* temperature for a superconductor.

Figure 22.29 A current in a superconducting ring persists for years.

22.6.1 Effect of magnetic fields on a superconductor

The lack of resistance in a superconductor offers obvious advantages for power transmission and in electromagnets. But for a long time after 1911, when the Dutch physicist Kamerlingh Onnes discovered superconductivity in mercury below 4.2 degrees Kelvin, all superconductors required very low temperatures. Recently materials have been discovered that are superconducting at temperatures above the boiling point of liquid nitrogen (77 Kelvin, −196 degrees Celsius). Because liquid nitrogen is rather inexpensive to produce, this opens up new opportunities for the use of superconductors. There is even hope that some material might be found that would be a superconductor at room temperature.

Magnetic flux through a superconducting ring

The complete lack of electric resistance has some odd consequences for the effect of magnetic fields on a superconductor. Suppose you try to change the flux through a superconducting metal ring. This would produce a curly electric field (whose round-trip integral is an emf equal to $d\Phi_{mag}/dt$). This curly electric field would drive an infinite current in the ring, because there is no resistance! Because an infinite current is impossible, we conclude that it is not possible to change the flux through a superconducting ring.

What does happen when you try to change the flux through the ring by an amount $\Delta\Phi_{mag}$? Since the flux cannot change, what happens is that a (noninfinite) current runs in the ring in such a way as to produce an amount of flux $-\Delta\Phi_{mag}$, so that the net flux change is zero! We'll consider a specific experimental situation to illustrate the phenomenon.

Start with a ring and a magnet at room temperature, so that there is a certain amount of flux Φ_0 through the ring (Figure 22.30), then cool the ring below the critical temperature at which it becomes superconducting (obviously the ring has to be made of a material that does become superconducting at some low temperature).

Next move the magnet away from the ring, thus decreasing the flux made by the magnet inside the ring (Figure 22.31); there is a change in the flux due to the magnet, of some amount $-\Delta\Phi$.

In a nonsuperconducting material, an emf would be induced in the ring, with a current emf = RI. But since the resistance R of the superconducting ring is zero, the emf must be zero (0 times I is 0). Assuming that Faraday's law applies to superconductors, it must be that just enough current runs in the superconductor, in the appropriate direction, to increase the flux by an amount $+\Delta\Phi$. The net change in the flux is zero, and the flux remains equal to the original flux Φ_0.

Experimentally, that's exactly what happens. Enough current runs in the ring to make exactly the right amount of magnetic flux to make up for the decrease in the flux due to moving the magnet farther away, so that the flux through the ring remains constant (Figure 22.32). Even though a current runs, the electric field inside the superconductor is zero.

Ex. 22.16 When the magnet is moved very far away, how much flux is inside the ring compared to the original flux Φ_0? How much of this flux is due to the current in the ring?

22.6.2 The Meissner effect

For many years it was assumed that a similar effect would happen with a solid disk instead of a hollow ring. It was expected that as you cooled the disk below its critical temperature the magnetic flux through the disk would stay

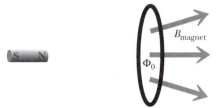

Figure 22.30 Start with a ring with initial flux Φ_0, then cool the ring down to make it become superconducting.

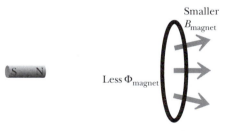

Figure 22.31 At low temperature, move the magnet away from the ring, reducing the flux in the ring contributed by the magnet.

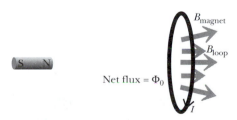

Figure 22.32 Current runs in the superconducting ring in such a way that the net change in the flux in the ring is zero.

the same, and that when you moved the magnet away the induced current in the disk would maintain the original flux. In 1933 physicists were astonished to discover that something else happened, something quite dramatic. As soon as the disk was cooled down below the critical temperature for the onset of superconductivity, the magnetic field throughout the interior of the solid disk suddenly went to zero!

This peculiar phenomenon is called the Meissner effect, and it applies to a particular class of materials called "Type I" superconductors. The configuration of net magnetic field (due to the magnet plus the currents inside the superconductor) looks something like Figure 22.33, with no net magnetic field inside the superconductor. Persistent currents are created in the superconductor which make a magnetic field that exactly cancels the magnetic field due to the magnet, everywhere inside the superconductor. Note that the currents in Figure 22.33 are oriented so that the disk and magnet repel each other.

The Meissner effect was totally unexpected, and it cannot be explained simply in terms of the lack of electric resistance in a superconductor (plus Faraday's law) as we did for the hollow ring. The Meissner effect is a special quantum-mechanical property of superconductors, and its explanation along with the quantum-mechanical explanation of other properties of superconductors was a triumph of the Bardeen-Cooper-Schrieffer (BCS) theory published in 1957.

Given that both the electric field and the magnetic field must be zero inside the superconductor, by using Gauss's law and Ampere's law it can be shown that the superconducting currents can run only on the surface of the superconductor, not in the interior. In particular, current in the interior would produce magnetic field throughout the material.

The electric field in an ordinary conductor in static equilibrium goes to zero due to polarization. However, the fact that electric and magnetic fields must be zero in a superconductor even when not in static equilibrium is quite a different, quantum-mechanical effect.

Magnetic levitation

The Meissner effect is the basis for a dramatic kind of magnetic levitation. A magnet can hover above a superconducting disk, because no matter how the magnet is oriented, currents run in the superconducting disk in such a way as to repel the magnet. One design for maglev trains uses this superconducting effect.

This is in contrast to the situation with ordinary magnets. If you try to balance a magnet above another magnet, with the slightest misalignment the upper magnet flips over and is attracted downward rather than repelled upward.

The "Levitron" toy achieves levitation with ordinary magnets by giving the upper magnet a high spin, so that the magnetic torque makes the spin direction precess rather than flip over. When a superconductor supports the upper magnet, gyroscopic stabilization is not needed.

22.7 Inductance, and energy in a magnetic field

Changing the current in a coil can induce an emf in a second coil. A related effect happens even with a single coil: an attempt to change the current in a coil induces an emf in the same coil, because the coil surrounds a region of time-varying magnetic field (dB/dt), produced by itself. We can show that this self-induced emf acts in a direction to oppose the change in the current. As a result, there is a kind of sluggishness in any coil of wire: it is hard to change the current, either to increase it or to decrease it. It is not difficult to calculate this self-induction effect quantitatively for a solenoid.

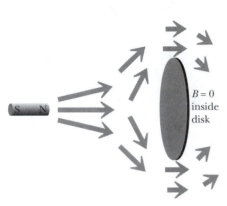

$B = 0$ inside disk

Figure 22.33 The Meissner effect: Currents run in the solid superconducting disk in such a way that the net magnetic field is zero inside the disk.

A very long plastic cylinder with radius R, and length $d \gg R$, is wound with N closely-packed loops of wire with negligible resistance. The solenoid is placed in series with a power supply and a resistor, and current I runs through the solenoid in the direction shown in Figure 22.34. This current makes a magnetic field B inside the solenoid that points to the right.

If the current is steady, the electric field inside the metal of the solenoid coil is nearly zero (very small resistance of this wire). Now suppose that you try to increase the current, by turning up the voltage of the power supply. If the current increases with time, there will be an increasing magnetic flux enclosed by each loop of the solenoid. In Chapter 17 we found that the magnetic field inside a solenoid of length d is $B = \mu_0 NI/d$.

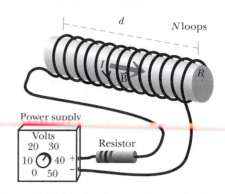

Figure 22.34 A circuit contains a solenoid.

? How much emf is generated in one loop of the coil if the current changes at a known rate dI/dt?

We have this:

$$\text{emf} = \left|\frac{d\Phi_{\text{mag}}}{dt}\right| = \frac{d}{dt}\left[\frac{\mu_0 NI}{d}\pi R^2\right] = \frac{\mu_0 N}{d}\pi R^2 \frac{dI}{dt} \quad \text{(one loop)}$$

? There are N loops, and each loop encloses approximately the same amount of time-varying magnetic flux. Therefore, what is the net emf induced along the full length of the solenoid?

Evidently we add up all the individual one-loop emf's in series to obtain the emf from one end of the coil to the other:

$$\text{emf} = N\left[\frac{\mu_0 N}{d}\pi R^2 \frac{dI}{dt}\right] = \frac{\mu_0 N^2}{d}\pi R^2 \frac{dI}{dt} \quad \text{(entire solenoid)}$$

? If we increase the current I, what is the direction of the induced curly electric field, clockwise or counterclockwise as seen from the right end of the solenoid? Does this induced electric field act to assist or oppose the change in I?

If the current is increasing, the non-Coulomb electric field $\vec{E}_{\text{NC}}$ curls clockwise around the solenoid, opposing the increase in the current and polarizing the solenoid (Figure 22.35). The new surface charges produce a Coulomb electric field $\vec{E}_{\text{C}}$ which follows the wire and points opposite to the non-Coulomb field. If the resistance of the solenoid wire is very small, the magnitude of the two electric fields is nearly equal ($E_{\text{C}} \approx E_{\text{NC}}$).

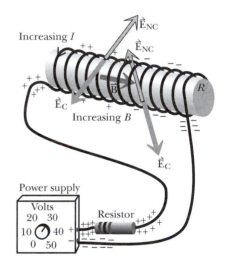

Figure 22.35 The non-Coulomb electric field opposes the increase in the current, and polarizes the solenoid.

Note the similarity to a battery: a non-Coulomb emf_{ind} (the induced emf) maintains a charge separation, and there is a potential drop ΔV_{sol} along the solenoid that is numerically equal to emf_{ind}. If the wire resistance r_{sol} is significant, we have $\Delta V_{\text{sol}} = \text{emf}_{\text{ind}} - r_{\text{sol}}I$, just like a battery with internal resistance. You can think of this self-induced emf as making the solenoid act like a battery that has been put in the circuit "backwards," opposing the change in the current (Figure 22.36).

It is standard practice to lump the many constants together and write that the self-induced emf is proportional to the rate of change of the current, with proportionality constant L:

$$\left|\text{emf}_{\text{ind}}\right| = L\left|\frac{dI}{dt}\right|$$

The proportionality constant L is called the "inductance" or "self-inductance" of the device, which is called an "inductor."

? Use the results above to express the inductance L in terms of the properties of a solenoid.

Evidently the (self-)inductance of a solenoid is this:

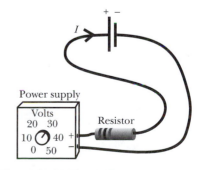

Figure 22.36 The coil acts temporarily like a battery inserted to oppose the change.

$$L = \frac{\mu_0 N^2}{d}\pi R^2$$

The unit of inductance (volt-seconds/ampere) is called the "henry" in honor of the 19th-century American physicist Joseph Henry, who simultaneously with Michael Faraday in 1831 discovered the effects of time-varying magnetic fields.

? What if you decrease the current through the solenoid? What happens to the induced emf? Does it assist or oppose the change?

If you go through the previous analysis but with a decreasing current, you find that the emf goes in the other direction, tending to drive current in the original current direction and therefore opposing the decrease. The solenoid introduces some sluggishness into the circuit: it is more difficult to increase the current or to decrease the current. The sign of the effect should not be surprising, since if the self-induced emf assisted rather than opposing an increase in the current, the current would grow without limit!

Ex. 22.17 To get an idea of the order of magnitude of inductance, calculate the self-inductance in henries for a solenoid with 1000 loops of wire wound on a rod 10 cm long with radius 1 cm.

22.7.1 Energy density

In Chapter 16 we showed that there is an energy density associated with electric field, by expressing the energy in a capacitor in terms of the electric field in the capacitor:

$$\frac{\text{Electric energy}}{\text{Volume}} = \tfrac{1}{2}\varepsilon_0 E^2$$

Now we can show that there is energy density associated with magnetic field, by expressing the energy in an inductor in terms of the magnetic field in the inductor. We obtain an important result about magnetic energy that is quite general, despite the fact that the derivation is for the specific case of an inductor.

It is easy to show that the magnetic energy stored in an inductor is $\tfrac{1}{2}LI^2$. The power going into an inductor is $I\Delta V$ as for any device, and

$$P = I\Delta V = I(\text{emf}) = I\left(L\frac{dI}{dt}\right)$$

Integrating over time, we have

$$\text{energy input} = \int (I\Delta V)\,dt = L\int_{t_i}^{t_f} I\frac{dI}{dt}\,dt = L\int_{I_i}^{I_f} I\,dI = L[\tfrac{1}{2}I^2]_{I_i}^{I_f} = \Delta(\tfrac{1}{2}LI^2)$$

Since $B = \dfrac{\mu_0 NI}{d}$ and $L = \dfrac{\mu_0 N^2}{d}\pi R^2$, we have

$$\text{Magnetic energy} = \tfrac{1}{2}\left(\frac{\mu_0 N^2}{d}\pi R^2\right)\left(\frac{Bd}{\mu_0 N}\right)^2$$

$$\text{Magnetic energy} = \tfrac{1}{2}\frac{1}{\mu_0}(\pi R^2 d)B^2$$

The magnetic field of the solenoid is large in the volume $(\pi R^2 d)$. There-fore we have the following result for the electric and magnetic energy densities:

ELECTRIC AND MAGNETIC ENERGY DENSITIES

$$\frac{\text{Energy}}{\text{Volume}} = \frac{1}{2}\varepsilon_0 E^2 + \frac{1}{2}\frac{1}{\mu_0}B^2 \text{, measured in joules/m}^3$$

Although we have calculated the magnetic energy density for the specific case of a long solenoid, this is a general result. The interpretation is that where there are electric or magnetic fields in space, there is an associated energy density, proportional to the square of the field (E^2 and/or B^2).

22.7.2 *Current in an RL circuit

A circuit containing a resistor R and an inductor L is called an "RL" circuit. Figure 22.37 shows a series RL circuit some time after a switch was closed. The energy-conservation loop rule for this series circuit is this:

$$\Delta V_{\text{battery}} + \Delta V_{\text{resistor}} + \Delta V_{\text{inductor}} = 0$$

$$\text{emf}_{\text{battery}} - RI - L\frac{dI}{dt} = 0$$

The term $-LdI/dt$ in the loop equation has a minus sign because, as we have seen, the emf of the inductor opposes the attempt to increase the current when we close the switch. The properties of this differential equation are similar to those of the differential equation of a resistor-capacitor (RC) circuit analyzed in Chapter 19.

Let the time when we closed the switch be $t = 0$. The current was of course zero just before closing the switch, and it is also zero just after closing the switch, since the sluggishness of the inductor does not permit an instantaneous change from $I = 0$ to $I = $ nonzero. In fact, such an instantaneous change would mean that dI/dt would be infinite, which would require an in-finite battery voltage to overcome the infinite self-induced emf in the inductor.

? Prove that

$$I = \frac{\text{emf}_{\text{battery}}}{R}\left[1 - e^{-\left(\frac{R}{L}\right)t}\right]$$

by substituting I and dI/dt into the energy-conservation equation. What is the final current in the circuit (that is, after a very long time)? Does that make sense?

For large t, the exponential goes to zero, and we have $I = \text{emf}_{\text{battery}}/R$. This makes sense, because after a long time there is a steady state current, and the flux does not change any more. With no change in the flux, no emf is generated in the inductor, and the current has the value that it would have if the inductor were not in the circuit. The current builds up slowly to this value. Figure 22.38 shows a graph of the current vs. time. One measure of the time it takes for the current to build up is the "time constant" L/R; at this time the exponential factor has dropped to $1/e$:

$$e^{-\left(\frac{R}{L}\right)t} = e^{-\left(\frac{R}{L}\right)\left(\frac{L}{R}\right)} = e^{-1} = \frac{1}{e}$$

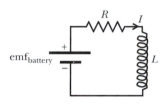

Figure 22.37 An RL circuit containing a resistor and an inductor.

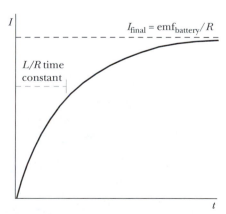

Figure 22.38 A graph of current vs. time in the RL circuit. The switch was closed at $t = 0$.

You see that having an inductor in a circuit makes the circuit somewhat sluggish. It takes a while for the current to reach the value that it would have reached right away if the circuit had consisted just of a battery and a resistor.

? It is interesting to see what happens when you try to open the switch, after the steady current has been attained. If the current were to drop from I to zero in a short time Δt, explain why the emf induced along the solenoid is very large, and can be much larger than the battery emf.

The induced emf depends on dB/dt, which is proportional to dI/dt. If the time interval is very short, then dI/dt is very large, and the induced emf is very large. As a result of this large emf, you may see a spark jump across the opening switch. This phenomenon makes it dangerous to open a switch in an inductive circuit if there are explosive gases around.

22.7.3 *Current in an LC circuit

In this course we have studied many examples of static equilibrium and of steady-state currents. We have also seen examples of a slow approach to static equilibrium (an RC circuit) or to a steady-state current (the RL circuit we just examined).

There is another possibility: a circuit might oscillate, with charge sloshing back and forth forever—the system never settles down to a final equilibrium or a steady state. Of the systems we've analyzed or experimented with, none of them oscillated continuously, because there was always some resistance in the system that would have damped out any such oscillatory tendencies. It is possible that during the first few nanoseconds when surface charge is rearranging itself in a circuit, there may be some sloshing of charge back and forth, but this oscillation dies out due to dissipation in the resistive wires.

A circuit containing an inductor L and a capacitor C is called an LC circuit (Figure 22.39). Such a circuit can oscillate if the resistance is small. In this circuit, the connecting wires are low-resistance thick copper wires. Suppose the capacitor is initially charged, and then you close a switch, connecting the capacitor to the inductor.

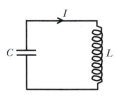

Figure 22.39 An LC circuit consists of an inductor and a capacitor.

At first it is difficult for charge on the capacitor to flow through the inductor, because the inductor opposes attempts to change the current (and the initial current was zero). But the inductor can't completely prevent the current from changing, so little by little there is more and more current, which of course drains the capacitor of its charge.

We'll see that just at the moment when the capacitor runs out of charge, there is a current in the inductor which can't change to zero instantaneously, due to the sluggishness of the inductor. So the system doesn't come to static equilibrium but overshoots the equilibrium condition and pours charge into the capacitor. When the capacitor gets fully charged (opposite in sign to the original condition), the capacitor starts discharging back through the inductor.

This oscillatory cycle repeats forever if there is no resistance in the circuit. If there is some resistance, the oscillations eventually die out, but the system may go through many cycles before static equilibrium is reached.

The energy-conservation loop rule for this circuit is

$$\Delta V_{\text{capacitor}} + \Delta V_{\text{inductor}} = \frac{1}{C}Q - L\frac{dI}{dt} = 0$$

where Q is the charge on the upper plate of the capacitor, and I is the conventional current leaving the upper plate and going through the inductor.

? Can you explain why $I = -dQ/dt$?

dQ/dt is the amount of charge flowing off of the capacitor plate per second. This is the same thing as the current. Because charge is leaving the plate, dQ is negative, so $I = -dQ/dt$. Since $I = -dQ/dt$, we can rewrite the energy-conservation equation as

$$\frac{1}{C}Q + L\frac{d^2Q}{dt^2} = 0$$

? Show that

$$Q = Q_i\cos\left(\frac{1}{\sqrt{LC}}t\right)$$

is a possible solution of the rewritten energy-conservation equation, by substituting Q and its second derivative into the equation.

Moreover, this solution fits the initial conditions at $t = 0$, just after closing the switch, if Q_i is the initial charge on the upper plate, since this expression reduces to $Q = Q_i\cos(0) = Q_i$. (For other initial conditions, the correct solution is a sine, or a combination of sine and cosine.)

? Now that you know Q on the upper plate of the capacitor as a function of time, calculate the current I through the inductor as a function of time.

The current is given by the following:

$$I = -\frac{dQ}{dt} = \frac{Q_i}{\sqrt{LC}}\sin\left(\frac{1}{\sqrt{LC}}t\right)$$

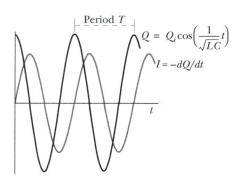

Now that we have expressions for Q and I as a function of time, we can graph both quantities (Figure 22.40). This is very special: charge oscillates back and forth in the circuit forever (if there is negligible resistance). There is no static equilibrium, and there is no steady state. No battery or other source of energy input is needed, because there is no dissipation. The charge just keeps going back and forth on its own. If there is some resistance, the oscillations slowly die out, as shown in Figure 22.41.

Figure 22.40 Capacitor charge and inductor current in an LC circuit with no resistance.

Note that the current I does not become large instantaneously (due to the sluggishness of the inductor). Also note that the current, which is the rate of depletion of the capacitor charge, reaches a maximum just when Q goes to zero. This maximum current starts charging the bottom plate of the capacitor positive, which makes the current decrease. When the current becomes zero, the system is in a state very similar to its original state, but inverted (the bottom plate is positive). Now the system runs the other way and eventually gets back after one complete cycle to a state in which the top plate is again positive, and the current is momentarily zero. Then the process repeats.

After one complete cycle, $(1/\sqrt{LC})t = 2\pi$, so the period of the oscillation is $T = 2\pi\sqrt{LC}$. The frequency $f = 1/T = 1/(2\pi\sqrt{LC})$.

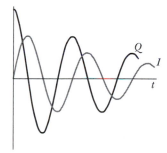

Figure 22.41 An LC circuit with resistance.

Ex. 22.18 What is the oscillation frequency of an LC circuit whose capacitor has a capacitance of 1 microfarad and whose inductor has an inductance of 1 millihenry? (Both of these are fairly typical values for capacitors and inductors in electronic circuits. See page 820 for a numerical example of inductance.)

22.7.4   *Energy in an LC circuit

Another way to talk about an LC circuit is in terms of stored energy. The original energy was the electric energy stored in the capacitor (equal to $Q^2/(2C)$; see Ex. 19.28 on page 685). At the instant during the oscillation when the charge on the capacitor is momentarily zero, there is no energy stored in the capacitor; all the energy at that moment is stored in the inductor, in the form of magnetic energy. As the system oscillates, energy is passed back and forth between the capacitor and the inductor.

Earlier in this chapter we showed that the magnetic energy stored in an inductor is $\frac{1}{2}LI^2$. At $t = 0$, the current is zero, and all the energy in the system is the electric energy of the capacitor. At the instant when $Q = 0$ on the capacitor, the current is a maximum, and all the energy in the system is the magnetic energy of the inductor.

? If Q_i is the initial charge on the capacitor, use an energy argument to find the maximum current I_{max} in the circuit.

Initially, all of the stored energy is electric:

$$U = U_{electric} + U_{magnetic} = \frac{1}{2}\frac{Q_i^2}{C}$$

When $I = I_{max}$, $Q = 0$, so all of the stored energy is magnetic:

$$U = U_{electric} + U_{magnetic} = \frac{1}{2}LI_{max}^2$$

Since the total energy in the circuit does not change, because the circuit has no resistance and no energy is dissipated as heat, the following must be true:

$$\frac{1}{2}LI_{max}^2 = \frac{1}{2}\frac{Q_i^2}{C}, \text{ and } I_{max} = \frac{Q_i}{\sqrt{LC}}$$

? Show that this result is consistent with $Q = Q_i\cos(t/\sqrt{LC})$ by calculating $I = -dQ/dt$ and seeing what the maximum I is.

Starting from the charge as a function of time, we have this:

$$I = -\frac{dQ}{dt} = -\frac{Q_i}{\sqrt{LC}}\sin\left(\frac{t}{\sqrt{LC}}\right), \text{ so that } I_{max} = \frac{Q_i}{\sqrt{LC}}$$

This is the same result we obtained by using an energy argument.

We have been discussing "free" oscillations, in which the oscillations (perpetual or dying out) proceed with no inputs from the outside. If you try to "force" the oscillations by applying an AC voltage from the outside (with an AC power supply), you find that it is difficult to get much current to run unless you nearly match the "free" oscillation frequency $f = 1/(2\pi\sqrt{LC})$. This is an example of what is called a "resonance" phenomenon (see end of Chapter 5), in which a system (the LC circuit in this case) responds significantly only when the forcing of the system is done at a frequency close to the free-oscillation frequency.

22.7.5   *AC circuits

For completeness, it should be mentioned that an important use of inductors is in AC circuits (sinusoidally alternating current). You have just seen that LC circuits are characterized by sinusoidally alternating currents, as are generators with loops rotating in a magnetic field. Whenever there is an AC current passing through an inductor, there is an AC emf generated in the inductor due to the time-varying flux. If the current is a sine function, the emf is a cosine function, and vice versa.

There is an elaborate mathematical formalism for dealing with AC circuits, including accounting for the fact that the current and voltage in an AC circuit need not reach their maxima at the same time (a "phase shift", which was 90° in the LC circuit we just analyzed). We will not go into these complications here, but it is worth noting that an important element in understanding such circuits is to recognize the role of the self-induced emf in an inductor, and the fact that in an inductor a sinusoidal current is associated with a cosinusoidal voltage.

22.8 *Some peculiar circuits

This section offers examples of surprising aspects of circuits when there is a time-varying magnetic field.

22.8.1 *Two bulbs near a solenoid

We can show you a very simple circuit that behaves in a rather surprising way. Consider two light bulbs connected in series around a solenoid with a varying magnetic field (Figure 22.42). The solenoid carries an alternating current (AC) in order to provide a time-varying magnetic flux through the circuit, so that there is an emf to light the bulbs. If the flux varies as $\Phi_{mag} = \Phi_0 \sin(\omega t)$, in the bulb circuit we have a varying emf = $\text{emf}_0 \cos(\omega t)$, since the emf is proportional to the rate of change of the magnetic flux. If each bulb has a resistance R, there is an alternating current through the bulbs of this amount:

$$I = \frac{\text{emf}}{2R} = \frac{\text{emf}_0}{2R}\cos(\omega t)$$

Figure 22.42 Two light bulbs connected around a long solenoid with varying B.

Notice that a uniformly increasing flux ($\Phi_{mag} = \Phi_0 t$) would yield a constant emf, but a sinusoidally varying flux yields a cosinusoidally varying emf. It is very convenient to use AC currents to study the effects of time-varying magnetic fields, because both the inducing current and the induced current vary as sines and cosines, and AC voltmeters and ammeters can be used to measure both parts of the circuit.

Now we alter the circuit by connecting a thick copper wire across the circuit (Figure 22.43). You may find it surprising that the top bulb gets much brighter and the bottom bulb no longer glows. Let's try to understand how this happens.

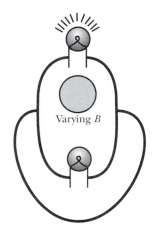

Figure 22.43 Add a thick copper wire to the two-bulb circuit.

As usual, we need to write charge-conservation node equations and energy-conservation loop equations, so we need to label the nodes and loops as in Figure 22.44 (any two of these three loops would be enough, but we'd like to show how to handle all of them).

Here is the key to analyzing this peculiar circuit. The round trip around loop 1 or loop 2 takes in an emf (just as though there were a battery in the loop). But the round trip around loop 3 does not enclose any time-varying flux and so has no emf. Therefore the loop and node rules are the following, where R_1 and R_2 are the resistances of the hot top bulb and cold bottom bulb, and the resistance of the added copper wire is essentially zero:

loop 1: $\text{emf} - R_1 I_1 - R_2 I_2 = 0$ node N: $I_1 = I_2 + I_3$
loop 2: $\text{emf} - R_1 I_1 = 0$
loop 3: $R_2 I_2 = 0$ (no flux enclosed)

From the loop 3 equation we see that $I_2 = 0$, which is why the bottom bulb doesn't glow.

With $I_2 = 0$, the node N rule says that $I_1 = I_3$: all the current in the top bulb goes through the copper wire (and none through the bottom bulb). Both

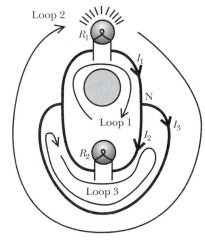

Figure 22.44 Add a thick copper wire to the two-bulb circuit.

the loop 1 and loop 2 equations reduce to $R_1 I_1 = $ emf, so the current through the top bulb is $I_1 = $ emf$/R_1 = ($emf$_0/R_1)\cos(\omega t)$, which is nearly twice what it was before we added the extra copper wire ($R_1 > R$ due to higher temperature).

Ex. 22.19 Here are four different ways to connect the copper wire. Based on the analysis we have just carried out, involving identifying whether or not there is a battery-like emf in a loop, what is the brightness of both bulbs in circuits 1, 2, 3, and 4?

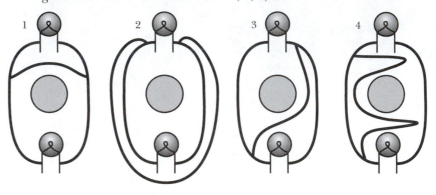

22.8.2   *Coulomb and non-Coulomb electric fields in a nonuniform ring

In a circular ring encircling a solenoid with a time-varying magnetic field, there is current driven by a non-Coulomb electric field. There is no gradient of surface charge along the ring, although there is a small transverse polarization required to turn the electrons in a circle (top of Figure 22.45).

If there is a thin section of the ring, however, charge will pile up at the ends of the thin section, and there will be surface charges all along the ring (bottom of Figure 22.45). These surface charges will produce a Coulomb electric field $\vec{E}_C$ which adds vectorially to the non-Coulomb electric field $\vec{E}_{NC}$ that is due to the changing magnetic flux. The net electric field $\vec{E}_{NC} + \vec{E}_C$ is what drives the electrons around the ring.

Inside the thin section the net electric field is larger than it was for the uniform ring, and in the thick sections the net electric field is smaller than it was for the uniform ring.

?   If both rings are made out of the same material, how do you know that the current must be smaller in the ring with the thin section than it was in the uniform ring?

In the thick section of the nonuniform ring, the only thing that changed is that the net electric field is smaller, so the current in the thick section must be smaller. (Of course this same smaller current runs through the thin section of this ring.)

Note that the Coulomb electric field $\vec{E}_C$ is large and goes clockwise in the thin section, but $\vec{E}_C$ is small and goes counter-clockwise in the thick section. Since $\vec{E}_C$ is due to point charges, the round-trip integral of $\vec{E}_C$ is zero.

On the other hand, the non-Coulomb electric field $\vec{E}_{NC}$ due to the changing magnetic flux goes clockwise all around the ring, and its round-trip integral is emf $= d\Phi_{mag}/dt$, which is not zero.

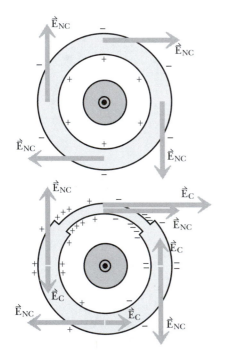

Figure 22.45 A uniform metal ring, and a metal ring with a thin section.

22.8.3 *Two competing effects in a shrinking ring

Suppose we stretch a springy metal ring of radius R and place it in a region of uniform magnetic field B pointing out of the page, which is increasing at a rate of dB/dt (Figure 22.46). When we let go of the ring, it contracts at a rate dR/dt (which is a negative number). What is the emf around the metal ring?

In terms of flux, there are two competing effects here. The increasing magnetic field B tends to increase the flux, but the decreasing area tends to decrease the flux. Whether or not a current will flow depends on the relative magnitudes of these two contributions to the net flux.

Since the magnetic field is the same throughout the ring and is perpendicular to the page, the flux through the ring at any instant is $\Phi_{mag} = B(\pi R^2)$. The emf is the rate of change of the flux:

$$\text{emf} = \frac{d\Phi_{mag}}{dt} = \frac{d}{dt}[B(\pi R^2)] = \frac{dB}{dt}(\pi R^2) + B\frac{d(\pi R^2)}{dt}$$

$$\text{emf} = \frac{dB}{dt}(\pi R^2) + B(2\pi R)\frac{dR}{dt}$$

What is the physical significance of these two contributions to the emf? Because there is a changing magnetic field, there is a clockwise non-Coulomb electric field in the wire. The first term, involving dB/dt, is associated with this non-Coulomb electric field in the wire.

Because the wire is moving inward, there is a magnetic force on the electrons in the wire. In this case, $\vec{v} \times \vec{B}$ (the magnetic force per unit charge) is counter-clockwise in the ring. The second term, involving the negative quantity dR/dt, is associated with the magnetic force on the moving pieces of the ring (motional emf). Note that $|dR/dt|$ is the speed of pieces of ring toward the center, and $2\pi r$ is the circumference, so the contribution to the emf is similar to $\text{emf} = BLv$ for a rod of length L sliding on rails.

Note that if $(dB/dt)(\pi R^2) = B(2\pi R)|(dR/dt)|$, the emf would be zero even though we had both a changing magnetic field and a changing area. In this case the magnetic force on an electron inside the ring is equal and opposite to the non-Coulomb electric force, due to the non-Coulomb electric field.

There is no surface charge gradient around this symmetrical ring, and no Coulomb electric field around the ring. The forces on electrons in the metal ring are due to the non-Coulomb electric field and to the non-Coulomb magnetic force on the moving pieces of the ring.

22.9 *The differential form of Faraday's law

As with Gauss's law and Ampere's law, there is a differential form of Faraday's law, valid at a particular location and time. Using properties of vector calculus one can show that the integral form of Faraday's law is equivalent to the following, where the derivative is a "partial derivative" with respect to time (holding position constant):

DIFFERENTIAL FORM OF FARADAY'S LAW

$$\text{curl}(\vec{E}) = -\frac{\partial \vec{B}}{\partial t} \quad \text{or} \quad \vec{\nabla} \times \vec{E} = -\frac{\partial \vec{B}}{\partial t}$$

22.10 *Lenz's rule

With the thumb of your right hand pointing in the direction of $-d\vec{B}_1/dt$, your fingers curl around in the direction of $\vec{E}_{NC}$. Another rule, called "Lenz's rule," can be used to get the direction of the non-Coulomb electric

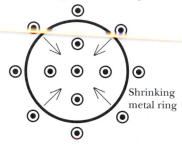

B out of page and increasing

Figure 22.46 A springy metal ring that had been stretched and is now shrinking.

field. We describe Lenz's rule briefly, although it is unnecessary to learn and use Lenz's rule in this introductory course on electricity and magnetism, for reasons that we will give below.

Imagine that you place a wire around the changing flux, so that the non-Coulomb electric field drives current in the wire:

LENZ'S RULE

The induced (non-Coulomb) electric field would drive current
in a direction to make a magnetic field
that attempts to keep the flux constant.

To see how this works, consider the first example in Figure 22.47, where $\vec{B}_1$ points out of the page and is increasing, and the non-Coulomb electric field curls clockwise around the solenoid. If a wire encircles the solenoid, conventional current runs clockwise in the wire, and this induced current produces an additional magnetic field in the region, pointing into the page.

This additional magnetic field is in the opposite direction to the change in $\vec{B}_1$, so we say that the induced magnetic field attempts to keep the magnetic flux constant despite the increase in B_1. The induced current does not succeed at keeping the net flux constant (unless the wire is made of superconducting material). In particular, if B_1 is increasing at a constant rate, the current I_2 in the ring and the induced magnetic field are constant by Faraday's law, so the net flux does increase.

If on the other hand $\vec{B}_1$ points out of the page and is decreasing (the second example in Figure 22.47), there is a counter-clockwise non-Coulomb electric field that would drive conventional current counter-clockwise in a wire, which would make a magnetic field out of the page, in the direction to attempt to keep the magnetic flux constant despite the decrease in B_1. (Again, the net flux will nevertheless decrease, unless the ring is superconducting.)

> **?** Go through this analysis for the third and fourth examples shown on the previous page, and convince yourself that Lenz's rule does correctly summarize the experimental data for the direction of the non-Coulomb electric field.

An interesting property of Lenz's rule is that it also correctly gives the direction of current in the case of motional emf, or in situations where both a time-varying magnetic field and motion contribute to a change in the flux enclosed by a part of a circuit.

We have not emphasized Lenz's rule for two reasons. We have emphasized time-varying magnetic field and non-Coulomb electric field (rather than just time-varying flux and emf), to give a stronger sense of mechanism and to distinguish the effects of time-varying magnetic fields from the effects of motion of a wire in a magnetic field. Also, it is very easy to make serious conceptual mistakes using Lenz's rule when first studying the effect of time-varying magnetic fields, due to a natural tendency to try to use it for predicting the magnitude of the induced current flow, not just its direction. In particular, except for superconductors, it is not true that "enough current runs to cancel the change in the magnetic field."

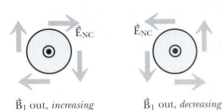

$\vec{B}_1$ out, *increasing* $\vec{B}_1$ out, *decreasing*

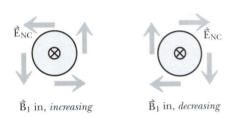

$\vec{B}_1$ in, *increasing* $\vec{B}_1$ in, *decreasing*

Figure 22.47 Using Lenz's rule to determine the direction of the non-Coulomb electric field.

22.11 Summary

Fundamental principles

The new phenomenon introduced in this chapter is that a time-varying magnetic field is accompanied by a non-Coulomb electric field that curls around the region of varying field. Unlike the Coulomb electric field produced by point charges, the non-Coulomb electric field has a nonzero round-trip path integral if the path encircles a time-varying flux.

FARADAY'S LAW

$$\text{emf} = -\frac{d\Phi_{mag}}{dt}$$

$$\text{where emf} = \oint \vec{E}_{NC} \bullet d\vec{l} \text{ and } \Phi_{mag} \equiv \int \vec{B} \bullet \hat{n} \, dA$$

In words: The induced emf along a round-trip path is equal to the rate of change of the magnetic flux on the area encircled by the path.

Direction: With the thumb of your right hand pointing in the direction of $-d\vec{B}/dt$, your fingers curl around in the direction of $\vec{E}_{NC}$.

FORMAL VERSION OF FARADAY'S LAW

$$\oint \vec{E}_{NC} \bullet d\vec{l} = -\frac{d}{dt}[\int \vec{B} \bullet \hat{n} \, dA] \quad \text{(sign given by right-hand rule)}$$

Faraday's law also covers the case where the flux changes due to changes in the path ("motional emf"), and the non-Coulomb force in this case is a magnetic force.

In general, there can be any combination of three effects acting on electrons in a circuit:
    1) Coulomb electric field due to surface charges;
    2) Non-Coulomb electric field due to time-varying magnetic field; and
    3) Magnetic forces on moving portions of the circuit (motional emf).

Results

Superconductors

Superconductors have zero resistance, and a consequence of this is that a superconducting ring maintains a constant flux in the area enclosed by the ring. The magnetic field inside a superconductor is always zero (the Meissner effect), and this important property cannot be explained merely in terms of zero resistance.

Inductance

When you vary the current in a coil, the varying flux enclosed by the coil induces an emf in the *same* coil which makes it difficult to change the current. This sluggishness plays a role in RL and LC circuits, and it is quantified by the "inductance" L of the coil. The emf is proportional to the rate of change of the magnetic flux, which is proportional to the rate of change of the current:

$$\left|\text{emf}_{ind}\right| = L\left|\frac{dI}{dt}\right|$$

We calculated the inductance in one particular case, that of a long solenoid:

$$L_{solenoid} = \frac{\mu_0 N^2}{d}\pi R^2$$

Electric and magnetic energy densities

$$\frac{\text{Energy}}{\text{Volume}} = \tfrac{1}{2}\varepsilon_0 E^2 + \tfrac{1}{2}\frac{1}{\mu_0}B^2 \text{ , measured in joules/m}^3$$

RL (resistor-inductor) circuits

The current in an RL series circuit varies with time:

$$I = \frac{\text{emf}_{\text{battery}}}{R}\left[1 - e^{-\left(\frac{R}{L}\right)t}\right]$$

LC (inductor-capacitor) circuits

An LC circuit oscillates sinusoidally with a frequency $f = 1/(2\pi\sqrt{LC})$:

$$Q = Q_i\cos\left(\frac{1}{\sqrt{LC}}t\right)$$

If there is no resistance, an LC circuit never reaches a final static equilibrium or a final steady state.

*The differential form of Faraday's law

$$\text{curl}(\vec{E}) = -\frac{\partial\vec{B}}{\partial t} \text{ or } \vec{\nabla}\times\vec{E} = -\frac{\partial\vec{B}}{\partial t}$$

22.12 Example problems

22.12.1 Example 1: Wire loop and a solenoid

A length of wire whose total resistance is R is made into a loop with two quarter-circle arcs of radius r_1 and radius r_2, and two straight radial sections as shown in Figure 22.48. A very long solenoid is positioned as shown, going into and out of the page. The solenoid consists of N turns of wire wound tightly onto a cylinder of radius r and length d, and it carries a counter-clockwise current I_s which is increasing at a rate dI_s/dt.

What is the magnitude and direction of the magnetic field at C, at the center of the two arcs? The "fringe" magnetic field of the solenoid (that is, the magnetic field produced by the solenoid outside of the solenoid) is negligible at point C. Show all relevant quantities on a diagram.

Figure 22.48 Upper: A long solenoid passes through a loop of wire. Lower: A perspective view.

Solution

As usual, start from fundamentals. Initially the only sources of field are the moving charges in the solenoid (which make B_s in the solenoid) and dB_s/dt in the solenoid (which makes a curly non-Coulomb electric field around the solenoid). The curly electric field constitutes an emf that drives current I_{loop} in the surrounding wire. The current-carrying wires produce a magnetic field at location C.

Minor point: Because the wire is not circular, pile-up leads to tiny amounts of surface charge on the wire. The net electric field in the wire (the superposition of the non-Coulomb and Coulomb fields) follows the wire and is uniform in magnitude throughout the wire.

1) Inside solenoid, $B_s = \dfrac{\mu_0 N I_s}{d}$ out of page.

2) Flux through *outer* loop $= B_s A_s \cos(0°) + (0) A_{\text{rest of loop}} = \dfrac{\mu_0 N I_s}{d} \pi r^2$

3) $-d\vec{B}/dt$ into page, so $\vec{E}_{NC}$ and induced current I_{loop} clockwise as shown.

$$|\text{emf}| = \frac{d}{dt}\left(\frac{\mu_0 N I_s}{d}\pi r^2\right) = \frac{\mu_0 N}{d}\pi r^2 \frac{dI_s}{dt}, \text{ so } I_{loop} = \frac{\text{emf}}{R} = \frac{\mu_0 N \pi r^2}{d}\frac{1}{R}\frac{dI_s}{dt}$$

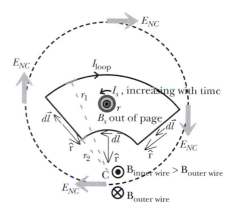

Figure 22.49 Fields and currents.

4) $B_{\text{at C}}$ due to straight segments, inner wire, and outer wire:

$B_{\text{straight segments}} = 0$ because $d\vec{l} \times \hat{r} = 0$ (see diagram)

$$B_{\text{inner wire}} = \left| \int_{\text{quarter loop}} \frac{\mu_0}{4\pi} \frac{I_{loop}\, d\vec{l} \times \hat{r}}{r_1^2} \right| = \frac{\mu_0}{4\pi}\frac{I_{loop}}{r_1^2} \int_{\text{quarter loop}} dl \sin(90°)$$

$$B_{\text{inner wire}} = \frac{\mu_0}{4\pi}\frac{I_{loop}}{r_1^2}\left(\frac{2\pi r_1}{4}\right) = \frac{\mu_0}{4\pi}\frac{\pi I_{loop}}{2 r_1}, \text{ and } B_{\text{outer wire}} = \frac{\mu_0}{4\pi}\frac{\pi I_{loop}}{2 r_2}$$

$B_{\text{inner wire}}$ is out of the page, and $B_{\text{outer wire}}$ is into the page at location C. Since B is proportional to $1/r$, $B_{\text{inner wire}}$ is larger than $B_{\text{outer wire}}$. So the net field is out of the page, with this magnitude:

$$B_{\text{net at C}} = B_{\text{inner wire}} - B_{\text{outer wire}} = \frac{\mu_0}{4\pi}\frac{\pi I_{loop}}{2}\left(\frac{1}{r_1} - \frac{1}{r_2}\right)$$

$$B_{\text{net at C}} = \frac{\mu_0}{4\pi}\left(\frac{\pi}{2}\right)\left(\frac{\mu_0 N \pi r^2}{d}\frac{1}{R}\frac{dI_s}{dt}\right)\left(\frac{1}{r_1} - \frac{1}{r_2}\right), \text{ out of the page}$$

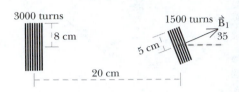

Figure 22.50 Two coils.

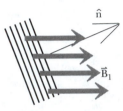

Figure 22.51 Assume the magnetic field made by coil 1 is uniform through the interior of coil 2.

22.12.2 Example 2: Two coils

Two coils of wire are near each other, 20 cm apart, positioned on a common axis, as shown in Figure 22.50. Coil 1 has a radius 8 cm and contains 3000 loops of wire. It is connected to a power supply whose output voltage can be changed, so that the current I_1 in coil 1 can be varied. Coil 2 contains 1500 loops of wire and has radius 5 cm, and is rotated 35° from the axis, as shown in Figure 22.50. The current in Coil 1 changes from 0 to 3 amperes in 4 milliseconds. Calculate the emf in Coil 2.

Solution

Start from fundamental principles. For a given current I_1 in the first coil, we can find B_1, the magnetic field that this current makes at the location of Coil 2. We know (dI_1/dt), so we can find $(d\Phi_{mag}/dt)$ in Coil 2, which gives us the emf in Coil 2.

Approximations: Assume the coils are far enough apart that the magnetic field of the first coil is approximately uniform in direction and magnitude throughout the interior of the second coil (Figure 22.51), and can be approximated as a magnetic dipole field. Also assume that the length of the coils is small compared to the distance between them (essentially treating all loops as if they were located at the center of the coil).

Work out everything symbolically first, plug in numbers at the end.

Magnetic field B_1 at the location of coil 2:

$$B_1 \approx N_1 \frac{\mu_0}{4\pi} \frac{2I_1(\pi r_1^2)}{z^3}$$

Magnetic flux through one loop of coil 2:

$$\Phi_{mag} \approx B_1(\pi r_2^2)\cos\theta$$

$$= N_1 \frac{\mu_0}{4\pi} \frac{2I_1(\pi r_1^2)}{z^3}(\pi r_2^2)\cos\theta$$

I_1 is the only quantity that is changing, so emf around one loop of coil 2 is:

$$\text{emf} = \left|\frac{d\Phi_{mag}}{dt}\right| \approx N_1 \frac{\mu_0}{4\pi} \frac{2(\pi r_1^2)(\pi r_2^2)}{z^3}\cos\theta \frac{dI_1}{dt}$$

total emf for all of coil 2:

$$\text{emf} = N_2\left|\frac{d\Phi_{mag}}{dt}\right| \approx N_1 N_2 \frac{\mu_0}{4\pi} \frac{2(\pi r_1^2)(\pi r_2^2)}{z^3}\cos\theta \frac{dI_1}{dt}$$

$$\frac{(1500)(3000)\left(1\times10^{-7}\,\frac{\text{T}\cdot\text{m}}{\text{A}}\right)(2)(0.08\ \text{m})^2(\pi^2)(0.05\ \text{m})^2\cos 35°(3\ \text{A})}{(0.20\ \text{m})^3(4\times10^{-3}\ \text{s})}$$

$$= 10.9\ \text{V}$$

22.13 Review questions

Faraday's law

RQ 22.1 The north pole of a bar magnet points toward a thin circular coil of wire containing 40 turns (Figure 22.52). The magnet is moved away from the coil, so that the flux through one turn inside the coil decreases by 0.3 tesla·m² in 0.2 seconds. What is the average emf induced in the coil during this time interval? Viewed from the right side (opposite the bar magnet), does the induced current run clockwise or counterclockwise? Explain briefly.

Figure 22.52 Bar magnet moving away from a coil (RQ 22.1).

RQ 22.2 Figure 22.53 shows a circular region of radius r_1 where there is a uniform magnetic field B pointing out of the paper (the magnetic field is essentially zero outside this region). The magnetic field is *decreasing* at this moment at a rate $|dB/dt|$. A wire of radius r_2 and resistance R lies entirely *outside* the magnetic-field region (where there is no magnetic field). In which direction does conventional current I flow in the ring? What is the magnitude of the current I? Show the pattern of (non-Coulomb) electric field in the ring. What is the magnitude E_{NC} of the non-Coulomb electric field?

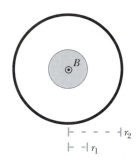

Figure 22.53 A metal ring surrounding a region of decreasing magnetic field (RQ 22.2).

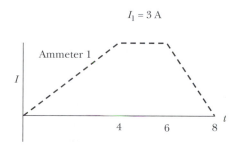

Figure 22.55 The ammeter readings as a function of time (RQ 22.3).

RQ 22.3 Two coils of wire are near each other, positioned on a common axis (Figure 22.54). Coil 1 is connected to a power supply whose output voltage can be adjusted by turning a knob, so that the current I_1 in coil 1 can be varied, and I_1 is measured by ammeter 1. Current I_2 in coil 2 is measured by ammeter 2. The ammeters have needles that deflect positive or negative depending on the direction of current passing through the ammeter, and ammeters read positive if conventional current flows into the "+" terminal. Figure 22.55 shows a graph of I_1 vs. time.

Draw a graph of I_2 vs. time over the same time interval. Explain your reasoning.

Inductance

RQ 22.4 A thin coil has 12 rectangular turns of wire. When a current of 3 amperes runs through the coil, there is a total flux of 10^{-3} tesla·m² enclosed by one turn of the coil (note that $\Phi = kI$, and you can calculate the proportionality constant k). Determine the inductance in henries.

RQ 22.5 Would the inductance of a solenoid be larger or smaller if the solenoid is filled with iron? Explain briefly.

RL and LC circuits

RQ 22.6

(a) When an RL circuit has been connected to a 1.5-volt battery for a very *short* time, what is the potential difference from one end of the resistor to the other?

(b) When an RL circuit has been connected to a 1.5-volt battery for a very *long* time, what is the potential difference from one end of the resistor to the other?

(c) Explain briefly the difference between static equilibrium, steady-state current, and the behavior of an LC circuit.

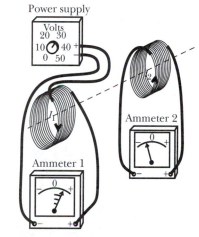

Figure 22.54 Vary the current in coil 1, observe the current in coil 2 (RQ 22.3).

Figure 22.56 Two metal rings on a table (Problem 22.1a).

Figure 22.57 A very long wire passes through a metal ring perpendicular to the wire (Problem 22.1b).

Figure 22.58 A bar magnet falls slowly through a copper tube (Problem 22.1c).

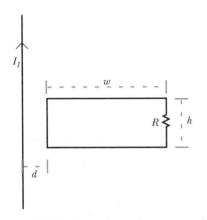

Figure 22.59 A long wire and a rectangular loop (Problem 22.2).

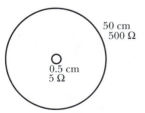

Figure 22.60 Two concentric metal rings (Problem 22.3).

22.14 Homework problems

Problem 22.1 Some small problems on time-varying magnetic fields

(a) Two metal rings lie side-by-side on a table (Figure 22.56). Current in the left ring runs clockwise and is increasing with time. This induces a current to run in the right ring. Does this induced current run clockwise or counterclockwise? Explain, using a diagram. (Hint: Think carefully about the direction of magnetic field in the right ring produced by the left ring, taking into consideration what sections of the left ring are closest.)

(b) A very long straight wire (essentially infinite in length) carries a current of 6 ampere (Figure 22.57). The wire passes through the center of a circular metal ring of radius 4 cm and resistance 2 ohms that is perpendicular to the wire. If the current in the wire increases at a rate of 0.25 ampere/s, what is the current induced in the ring? Explain carefully.

(c) A bar magnet is dropped through a vertical copper tube and is observed to fall very slowly, despite the fact that mechanical friction between the magnet and the tube is negligible (Figure 22.58). Explain carefully, including adequate diagrams.

Problem 22.2 A wire and a rectangular loop

A very long wire carries a current I_1 *upward* as shown, and this current is *decreasing with time* (Figure 22.59). Nearby is a rectangular loop of wire lying in the plane of the long wire, and containing a resistor R; the resistance of the rest of the loop is negligible compared to R. The loop has a width w and height h, and is located a distance d to the right of the long wire.

(a) Does the current I_2 in the loop run clockwise or counterclockwise? Explain, using a diagram.

(b) Show that the magnitude of the current I_2 in the loop is this:

$$I_2 = \frac{\mu_0}{4\pi} \frac{(2h)\ln\left[(d+w)/d\right]}{R}\left|\frac{dI_1}{dt}\right|$$

(Hint: divide the area into narrow vertical strips along which you know the magnetic field.)

Problem 22.3 Two concentric metal rings

A very small circular metal ring of radius $r = 0.5$ cm and resistance $x = 5$ ohms is at the center of a large concentric circular metal ring of radius $R = 50$ cm and resistance $X = 500$ ohms (Figure 22.60). The two rings lie in the same plane. At $t = 3$ seconds, the large ring carries a *clockwise* current of 3 ampere. At $t = 3.2$ seconds, the large ring carries a *counter-clockwise* current of 5 ampere.

(a) What is the average electric field induced in the small ring during this time? Give both the magnitude and direction of the electric field. Draw a diagram, showing all relevant quantities.

(b) What is the magnitude and direction of the average current in the small ring?

Problem 22.4 Connect a battery to a solenoid

A cylindrical solenoid 40 cm long with a radius of 5 mm has 300 tightly-wound turns of wire uniformly distributed along its length (Figure 22.61). Around the middle of the solenoid is a two-turn rectangular loop 3 cm by 2 cm made of resistive wire having a resistance of 150 ohms.

One microsecond after connecting the loose wire to the battery to form a series circuit with the battery and a 20-ohm resistor, what is the magnitude of the current in the rectangular loop and its direction (clockwise or counter-clockwise in the diagram)?

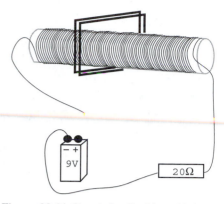

Figure 22.61 Circuit for Problem 22.4.

Problem 22.5 Induced electric field

The magnetic field is uniform and out of the page inside a circle of radius R (Figure 22.62). The magnetic field is essentially zero outside the circular region. The magnitude of the magnetic field as a function of time is $(B_0 + bt^3)$. B_0 and b are positive constants, and t = time.

(a) What is the direction and magnitude of the induced electric field at location P, at a distance r_1 to the left of the center ($r_1 < R$)?

(b) What is the direction and magnitude of the induced electric field at location Q, at a distance r_2 to the right of the center ($r_2 > R$)?

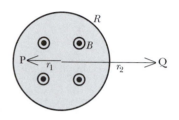

Figure 22.62 Electric field inside and outside a region of changing magnetic field (Problem 22.5).

Problem 22.6 Design a circuit

It is now possible to buy capacitors that have a capacitance of one farad.

(a) Design a solenoid so that when it is connected to a charged one-farad capacitor, the circuit will oscillate with a period of one second. Give all the relevant parameters of the solenoid (length, etc.) so that someone could build the solenoid from your design specifications. Assume that there is wire available with low enough resistance that the resistance of the solenoid is negligible, although this may be difficult to achieve in practice unless the wire is superconducting.

(b) If the one-farad capacitor is initially charged to 3 volts, what is the maximum current that will run through the inductor?

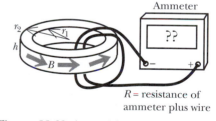

Figure 22.63 A toroid connected to an ammeter (Problem 22.7).

Problem 22.7 Toroid

A toroid has a rectangular cross section with an inner radius r_1 = 9 cm, an outer radius r_2 = 12 cm, and a height h = 5 cm, and it is wrapped around by many densely-packed turns of current-carrying wire (not shown in Figure 22.63). The direction of the magnetic field inside the windings is shown on the diagram. There is essentially no magnetic field outside the windings. A wire is connected to a sensitive ammeter as shown. The resistance of the wire and ammeter is R = 1.4 ohms.

The current in the windings of the toroid is varied so that the magnetic field inside the windings, averaged over the cross section, varies with time as shown in Figure 22.64.

Make a careful graph of the ammeter reading, including sign, as a function of time. Label your graph, and explain the numerical aspects of the graph, including signs.

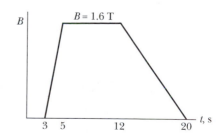

Figure 22.64 Magnetic field in toroid as a function of time (Problem 22.7).

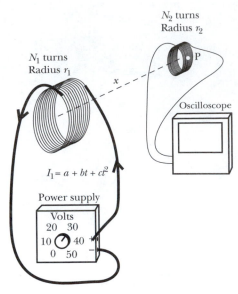

Figure 22.65 Observing with an oscilloscope (Problem 22.8).

Problem 22.8 A coil attached to an oscilloscope

In Figure 22.65, a thin circular coil of radius r_1 with N_1 turns carries a current $I_1 = a + bt + ct^2$, where t is the time and a, b, and c are positive constants. A second thin circular coil of radius r_2 with N_2 turns is located a long distance x along the axis of the first coil. The second coil is connected to an oscilloscope, which has very high resistance.

(a) As a function of time t, calculate the magnitude of the voltage that is displayed on the oscilloscope. Explain your work carefully, but you do not have to worry about signs.

(b) At point P on the drawing (on the right side of the second coil), draw a vector representing the non-Coulomb electric field.

(c) Calculate the magnitude of this non-Coulomb electric field.

Problem 22.9 Throw a bar magnet

You throw a bar magnet downward with its South end pointing down. The bar magnet has a magnetic dipole moment of 1.2 A·m^2. Lying on the table is a nearly flat circular coil of 1000 turns of wire, with radius 5 cm. The coil is connected to an oscilloscope, which has a very large resistance.

(a) On a diagram, show the pattern of non-Coulomb electric field in the coil. Explain briefly.

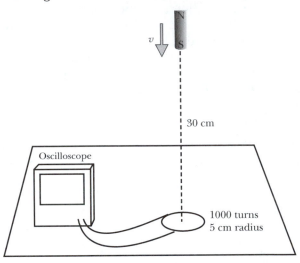

Figure 22.66 Throw a bar magnet (Problem 22.9).

(b) At the instant when the magnet is 30 cm above the table, the oscilloscope indicates a voltage of magnitude 2 millivolts. What is the speed v of the magnet at that instant?

Problem 22.10 Magnetic monopoles

One of the methods physicists have used to search for magnetic monopoles is to monitor the current produced in a loop of wire. Draw graphs of current in the loop *vs.* time for an electrically uncharged magnetic monopole passing through the loop, and for an electrically uncharged magnetic dipole (such as a neutron) passing through the loop with its North end head-first. Don't worry too much about the details of the exact moment when the particle goes through the plane of the loop; concentrate on the times just before and just after this event. Explain the differences in the two graphs.

(A signal corresponding to a magnetic monopole was seen once in such an experiment, but no one has been able to reproduce this result, and most physicists seem to believe that the supposed event was due to extraneous noise in the system or other malfunction of the apparatus.)

Problem 22.11 A proton in a magnetic field

A single circular loop of wire with radius R carries a large clockwise constant current $I_{loop} = I_0$, which constrains a proton of mass M and charge e to travel in a small circle of radius r at constant speed around the center of the loop, in the plane of the loop. The orbit radius r is much smaller than the loop radius R: $r \ll R$.

(a) Draw a diagram of the proton orbit, indicating the direction that the proton travels, clockwise or counter-clockwise. Explain briefly.

(b) What is the speed v of the proton, in terms of the known quantities I_0, R, r, M, and e? Explain your work, including any approximations you must make.

(c) The current was constant for a while, but at a certain time $t = t_0$ it began to decrease slowly, so that after t_0 the current was $I_{loop} = I_0 - k(t - t_0)$. On your diagram of the proton orbit, draw electric field vectors at four locations (one in each quadrant) and explain briefly.

(d) When the current decreases, does the proton speed up or slow down?

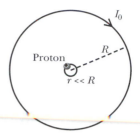

Figure 22.67 A proton in a magnetic field (Problem 22.11).

Problem 22.12 A rectangular loop

A very long, tightly-wound solenoid has a circular cross section of radius 3 cm (only a portion of the very long solenoid is shown in Figure 22.68). The magnetic field outside the solenoid is negligible. Throughout the inside of the solenoid the magnetic field B is uniform, to the left as shown, but varying with time t: $B = [0.07 + 0.03t^2]$ tesla. Surrounding the circular solenoid is a loop of 4 turns of wire in the shape of a rectangle 10 cm by 15 cm. The total resistance of the 4-turn loop is 0.1 ohm.

(a) At $t = 2$ seconds, what is the direction of the current in the 4-turn loop? Explain briefly.

(b) At $t = 2$ seconds, what is the magnitude of the current in the 4-turn loop? Explain briefly.

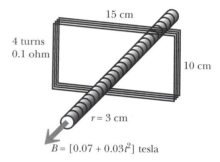

Figure 22.68 A long solenoid inside a rectangular coil (Problem 22.12).

Problem 22.13 Solenoid and ring

A very long solenoid of length d and radius r_1 is tightly wound uniformly with N turns of wire (Figure 22.69). A variable power supply forces a current to run in the solenoid of amount $I_1 = p - kt$, where p and k are positive constants (so that the current is initially equal to p but decreases with time). A circular metal ring of radius r_2 and resistance R is centered on the solenoid and located near the middle of the solenoid, very far from the ends of the solenoid (so that the solenoid contributes essentially no magnetic field outside the solenoid).

(a) What are the magnitude and the direction of the magnetic field at the center of the ring, due to the solenoid and the ring, as a function of the time t? Explain your work; part of the credit will be given for the clarity of your explanation, including clarity of appropriate diagrams.

(b) Qualitatively, how would these results have been affected if an iron rod had been inserted into the solenoid? Very briefly, why?

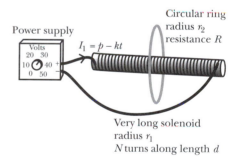

Figure 22.69 A metal ring around a long solenoid (Problem 22.13).

Problem 22.14 A bar magnet and a coil

In Figure 22.70 a coil with 3000 turns and radius 5 cm is connected to an oscilloscope. You move a bar magnet toward the coil along the coil's axis, moving from 40 cm away to 30 cm away in 0.2 seconds. The bar magnet has a magnetic moment of 0.8 ampere·m².

(a) On the diagram, draw the non-Coulomb electric field vectors at locations 1 and 2. Briefly explain your choices graphically.

(b) What is the approximate magnitude of the signal observed on the oscilloscope?

(c) What approximations or simplifying assumptions did you make?

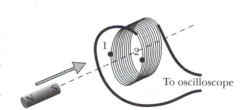

Figure 22.70 A bar magnet is moved toward a coil (Problem 22.14).

Problem 22.15 Oscilloscope and coils

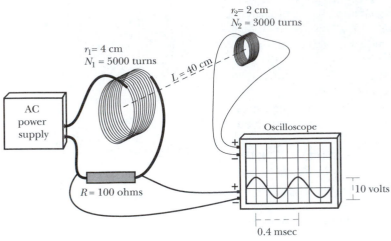

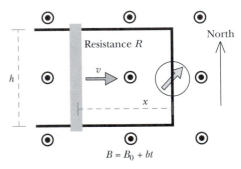

Figure 22.71 Oscilloscope and coils (Problem 22.15).

A thin coil of radius $r_1 = 4$ cm, containing $N_1 = 5000$ turns, is connected through a resistor $R = 100$ ohms to an AC power supply running at a frequency $f = 2500$ hertz, so that the current through the resistor (and coil) is $I_1 \sin(2\pi \times 2500\, t)$. The voltage across the resistor triggers an oscilloscope which also displays this voltage, which is 10 volts peak-to-peak and therefore has an amplitude of 5 volts (and therefore the amplitude of the current is $I_1 = 0.05$ ampere).

A second thin coil of radius $r_2 = 2$ cm, containing $N_2 = 3000$ turns, is a distance $L = 40$ cm from the first coil. The axes of the two coils are along the same line. The second coil is connected to the upper input of the oscilloscope, so that the voltage across the second coil can be displayed along with the voltage across the resistor.

(a) Assume you have adjusted the VOLTS/DIVISION knob on the upper input so that you can easily see the signal from the second coil. Sketch the second coil voltage along with the resistor voltage on the oscilloscope display, paying careful attention to the shape and positioning of the two voltage curves with respect to each other. Explain briefly.

(b) Calculate the amplitude (maximum voltage) of the second coil voltage. If you must make simplifying assumptions, state clearly what they are.

Problem 22.16 A copper bar slides on rails

A copper bar of length h and electric resistance R slides with negligible friction on metal rails which have negligible electric resistance. The rails are connected on the right with a wire of negligible electric resistance, and a magnetic compass is placed under this wire (the diagram is a top view). The compass needle deflects to the right of North, as shown on the diagram.

Throughout this region there is a uniform magnetic field B pointing out of the page, produced by large coils which are not shown. This magnetic field is increasing with time, and the magnitude is $B = B_0 + bt$, where B_0 and b are constants, and t is the time in seconds. You slide the copper bar to the right and at time $t = 0$ you release the bar when it is a distance x from the right end of the apparatus. At that instant the bar is moving to the right with a speed v.

(a) Calculate the magnitude of the initial current I in this circuit.

(b) Calculate the magnitude of the net force on the bar just after you release it.

(c) Will the bar speed up, slow down, or slide at a constant speed? Explain briefly.

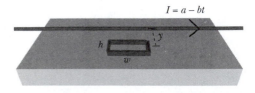

Figure 22.72 A copper bar slides on rails (Problem 22.16).

Problem 22.17 A wire and a rectangular loop

A thin rectangular coil lies flat on a low-friction table. A very long straight wire lying on the table a distance y from the coil carries a conventional current I to the right as shown in Figure 22.73, and this current is decreasing: $I = a - bt$, where t is the time in seconds, and a and b are positive constants. The coil has width w and height h, where $h \ll y$. It has N turns of wire with total resistance R.

What are the initial magnitude and direction of the nonzero net force that is acting on the coil? Explain in detail. If you must make simplifying assumptions, state clearly what they are, but bear in mind that the net force is not zero.

Figure 22.73 A thin rectangular coil lies on a low-friction table a distance y from a very long current-carrying wire (Problem 22.17). Not to scale: $h \ll y$.

Problem 22.18 Power lines

Tall towers support power lines 50 m above the ground and 20 m apart that run from a hydroelectric plant to a large city, carrying 60-hertz alternating current with amplitude 10^4 ampere (Figure 22.74). That is, the current in both of the power lines is $I = (10^4\,A)\sin(2\pi \cdot 60\,\text{hertz} \cdot t)$.

(a) Calculate the amplitude (largest magnitude) and direction of the magnetic field produced by the two power lines at the base of the tower, when a current of 10^4 ampere in the left power line is headed out of the page, and a current of 10^4 ampere in the right power line is headed into the page.

(b) This magnetic field is not large compared to the earth's magnetic field, but it varies in time and so might have different biological effects than the earth's steady field. For a person lying on the ground at the base of the tower, approximately what is the maximum emf produced around the perimeter of the body, which is about 2 meter long by half a meter wide?

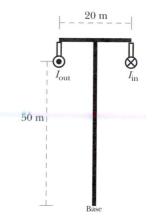

Figure 22.74 Power lines carrying alternating current (Problem 22.18).

Problem 22.19 The betatron

Suppose you have an electron moving with speed comparable to the speed of light in a circular orbit of radius r in a large region of uniform magnetic field B.

(a) What must be the relativistic momentum p of the electron?

(b) Now the uniform magnetic field begins to increase with time: $B = B_0 + bt$, where B_0 and b are positive constants. In one orbit, how much does the energy of the electron increase, assuming that in one orbit the radius doesn't change very much? (This effect was exploited in the "betatron," an electron accelerator invented in the 1940's.)

Note: it turns out that the electron's energy increases by less than the amount you calculated, for reasons that will become clear in Chapter 23.

Problem 22.20 A current-carrying wire moves toward a coil

A long straight wire carrying current I is moving with speed v toward a small circular coil of radius r containing N turns, which is attached to a voltmeter as shown. The long wire is in the plane of the coil. (Only a small portion of the wire is shown in the diagram.)

At the instant when the long wire is a distance x from the center of the coil, what is the voltmeter reading? Include both magnitude and sign. (Remember that a voltmeter reads "+" if the higher potential is connected to the "+" terminal of the voltmeter.) Explain. State what approximations or simplifying assumptions you make.

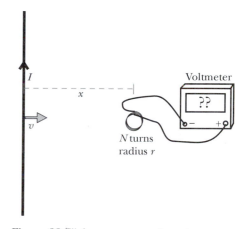

Figure 22.75 A current-carrying wire moves toward a coil (Problem 22.20).

Problem 22.21 A moving superconducting ring

A superconducting ring of radius R has a permanent conventional current I in the direction shown and is oriented with its axis along the x axis. The ring is moving to the right along the x axis with uniform speed v. Calculate the electric field $\vec{E}$ (magnitude and direction) at location $<x, y, 0>$ relative to the center of the ring. If you must make any simplifying assumptions or approximations, state them explicitly.

Figure 22.76 A moving superconducting ring (Problem 22.21).

22.15 Answer to exercises

22.1 (page 805)   Clockwise.

22.2 (page 805)   Counterclockwise.

22.3 (page 807)   (1) no current. (2) current.

22.4 (page 807)   47.7 V/m

22.5 (page 807)   7.5 amperes

22.6 (page 808)   0.052 tesla·m^2

22.7 (page 809)   20 volts; 8 V/m; 2 amperes

22.8 (page 811) 7.9×10^{-4} volts ; 1.3×10^{-3} V/m

22.9 (page 811)   200 volts

22.10 (page 812)   10 amperes; 20 times emf, but 20 times resistance

22.11 (page 812)   Ammeter reading is positive. Remember that an ammeter reads positive if conventional current flows into the "+" terminal of the ammeter.

22.12 (page 812)   Ammeter reading is negative.

22.13 (page 812)   Current in loop runs clockwise, seen from above.

22.14 (page 813)   0

22.15 (page 813)   0.166 volts

22.16 (page 817)   The flux must still be equal to Φ_0. Since the magnet is too far away to contribute, all the flux must be due to the induced current in the ring.

22.17 (page 820) 4×10^{-3} henry

22.18 (page 823)   5000 hertz

22.19 (page 826)   In the first three cases, for the bulb labeled "bright" a loop can be drawn which consists of just that bulb plus a wire which encircles the varying magnetic flux; for the bulb labeled "off" a loop can be drawn which consists of just that bulb plus a wire which does not encircle the varying magnetic flux. 1) lower bulb bright, upper off; 2) upper bulb bright, lower off; 3) upper bulb bright, lower off.

For the fourth case, neither bulb is in a loop that doesn't encircle the varying magnetic flux. The current is smaller than in the other circuits, because the induced emf drives two bulbs in series.

Chapter 23

Electromagnetic Radiation

Chapter 23

Electromagnetic Radiation

We are approaching the climax of our study of electricity and magnetism: the classical model of electromagnetic radiation, or light. In Chapter 6 we studied aspects of the quantum mechanical model of light, which involves particles called photons which have energy and momentum, but zero rest mass. However, there are important properties of light, and aspects of the interaction of light and matter, that we cannot explain with our simple particle model. These phenomena, which have to do with the wavelike nature of light, can be explained by the classical model of electromagnetic radiation, which we will explore in this chapter. Aspects of this classical model are important to a fuller understanding of the quantum mechanical model of light, as well.

23.1 Maxwell's equations

The four Maxwell equations provide a complete description of possible spatial patterns of electric and magnetic field in space. So far we have studied three of these equations, Gauss's law, Gauss's law for magnetism, and Faraday's law. However, we have an incomplete version of the fourth equation, Ampere's law. Before we can go further, we need to complete this equation. Once we have all of Maxwell's equations, we will be able to show that some surprising configurations of electric and magnetic fields in space are not only possible but extremely important.

23.1.1 The Ampere-Maxwell law

As we saw in Chapter 22, Faraday's law expresses a connection between a time-varying magnetic field in a region of space and a curly (non-Coulomb) electric field throughout space. You may have wondered whether a time-varying electric field in a region might be associated with a magnetic field throughout space. In fact, this does turn out to be the case. We will see that Ampere's law needs to be modified to include this relation.

What is missing from Ampere's law?

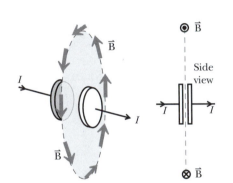

Figure 23.1 The soap film is stretched flat over the circular Amperian path.

If we try to apply Ampere's law to a circuit containing long straight wires carrying current to charge a capacitor, we will find what appears to be a contradiction. We choose a circular "Amperian" path around a charging capacitor, and stretch an imaginary soap film over the chosen path. Now consider two different soap film positions.

In Figure 23.1 the path does not enclose any current because the soap film is flat and passes between the capacitor plates. We must therefore conclude that $\oint \vec{B} \cdot d\vec{l}$ is zero along this circular path.

However, in Figure 23.2 the path does enclose the current I because the soap film bulges out and is pierced by the long straight wire on the left. In this case we conclude that $\oint \vec{B} \cdot d\vec{l}$ is equal to $\mu_0 I$. This contradicts our previous result.

? Which result seems more reasonable? Would you expect the magnetic field along the chosen path to be zero or nonzero?

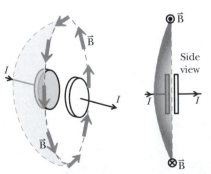

Figure 23.2 The soap film bulges out and is pierced by the current on the left.

Presumably the magnetic field along the path is really nonzero, because the current in the wire contributes to the magnetic field at all locations. We would expect the magnetic field around the capacitor to be nearly the same as the magnetic field around a nearby section of the wire, as indicated in Figure 23.3.

In Chapter 17 we found by applying the Biot-Savart law applied to a long wire that the magnetic field near the wire was:

$$B = \frac{\mu_0}{4\pi}\frac{2I}{r} \text{ (a distance } r \text{ from the center of a long wire)}$$

This should be approximately the magnetic field along the circular path in Figure 23.1, because there is only a very short piece of wire missing from the long straight wire. Therefore around the Amperian path in Figure 23.1 we should have

$$\oint \vec{B} \bullet d\vec{l} \approx \frac{\mu_0}{4\pi}\frac{2I}{r}2\pi r = \mu_0 I$$

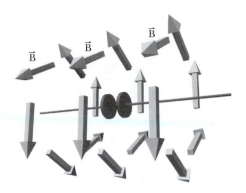

Figure 23.3 Probable magnetic field near the capacitor.

A time-varying electric field

This dilemma was resolved by James Clerk Maxwell, a British physicist. Knowing of Faraday's discovery that a time-varying magnetic field produces an electric field, he guessed that a time-varying electric field might make a magnetic field. Between the plates of the capacitor there is a time-varying electric field, because the electric field between the plates is growing as the capacitor becomes more and more charged. This time-varying electric field could contribute to the magnetic field around the capacitor.

Electric flux

Let's explore this idea quantitatively. By analogy to Faraday's law, we will guess that the time rate of change of the electric flux in a region is related to the integral of the magnetic field along a path surrounding the region.

First we need an expression for the electric flux through the surface shown in Figure 23.4.

$$\Phi_{\text{electric}} = \int \vec{E} \cdot \hat{n}\, dA$$

As shown in Figure 23.4, inside the capacitor the electric field is parallel to $\hat{n}$ and has magnitude $(Q/A)/\varepsilon_0$. Outside the capacitor the electric field is very small if the plates are large and close together. Therefore

$$\Phi_{\text{electric}} = \left(\frac{Q}{A\varepsilon_0}\right)A\cos 0° = \frac{Q}{\varepsilon_0}$$

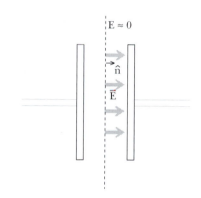

Figure 23.4 Calculating flux through a surface between the plates of the capacitor (side view).

? What is the time derivative of the electric flux in this situation?

In a time Δt the amount of charge ΔQ that flows onto the capacitor is $\Delta Q = I\Delta t$. Thus, the rate of change of Q, the amount of charge on the capacitor, is

$$\frac{dQ}{dt} = I$$

Therefore, the time rate of change of electric flux is

$$\frac{d\Phi_{\text{electric}}}{dt} = \frac{d}{dt}\left(\frac{Q}{\varepsilon_0}\right) = \frac{I}{\varepsilon_0}$$

The quantity $\varepsilon_0\dfrac{d\Phi_{\text{electric}}}{dt}$ has units of Amperes, so we will combine it with I in Ampere's law.

We will call this more general form of Ampere's law the "Ampere-Maxwell law":

THE AMPERE-MAXWELL LAW

$$\oint \vec{B} \bullet d\vec{l} = \mu_0 \left[\sum I_{\text{inside path}} + \varepsilon_0 \frac{d\Phi_{\text{electric}}}{dt} \right]$$

If our imaginary soap film is outside the capacitor, there is just the term involving I but no term involving changing electric flux. If our imaginary soap film goes between the capacitor plates, there is just the term involving changing electric flux but no term involving I. The Ampere-Maxwell law handles both cases.

It also handles the case of a small bulge in the soap film near the center of the capacitor plate (Figure 23.5). In this case the film doesn't enclose all of the electric flux. The soap film is pierced not only by the incoming current I but also by a lesser amount of radial current I_2 flowing outward from the center of the capacitor plate toward the edge of the plate, so that the net current piercing the soap film is $(I - I_2)$. In this case the Ampere-Maxwell equation contains both a current term and a flux term, adding up to $\mu_0 I$.

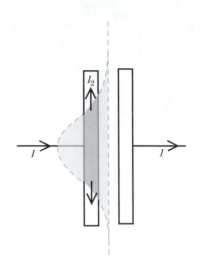

Figure 23.5 The soap film has a small bulge near the center of the plate.

Verification of the Ampere-Maxwell law

We interpret the Ampere-Maxwell law as saying that a time-varying electric field can produce a magnetic field. It is quite difficult to verify Maxwell's ingenious guess by a direct experiment, because in order to get a changing electric field you have to run a current to deposit a growing amount of charge, and this unavoidable current may be the source of the magnetic field you observe. Unless you are very clever in your geometrical arrangement, the real currents may mask the effects of the time-varying electric field.

There has in fact been direct experimental verification of the electric flux term, but the most compelling confirmation of Maxwell's inspired guess is indirect, in the explanation he was able to provide for the nature of light and radio waves, as we will soon see.

23.1.2 Maxwell's equations (integral form)

We can now list the four equations that summarize all electromagnetic effects and are collectively called "Maxwell's equations" (although Maxwell himself did not write them in this modern form). Note that the first three are familiar equations from previous chapters:

MAXWELL'S EQUATIONS

$$\oint \vec{E} \bullet \hat{n} \, dA = \frac{1}{\varepsilon_0} \sum Q_{\text{inside the surface}} \qquad \text{Gauss's law}$$

$$\oint \vec{B} \bullet \hat{n} \, dA = 0 \qquad \text{Gauss's law for magnetism}$$

$$\oint \vec{E} \bullet d\vec{l} = -\frac{d}{dt} \left[\int \vec{B} \bullet \hat{n} \, dA \right] \qquad \text{Faraday's law}$$

$$\oint \vec{B} \bullet d\vec{l} = \mu_0 \left[\sum I_{\text{inside path}} + \varepsilon_0 \frac{d\Phi_{\text{electric}}}{dt} \right] \qquad \text{Ampere-Maxwell law}$$

These are the "integral" versions of Maxwell's equations. The "differential" versions of Maxwell's equations are given at the end of this chapter.

In addition to Maxwell's equations, to complete our list of fundamental relations of electricity and magnetism we need to add the Lorentz force, which defines what we mean by electric and magnetic fields in terms of their effects on charges:

THE LORENTZ FORCE

$$\vec{F} = q\vec{E} + q\vec{v} \times \vec{B} \qquad (d\vec{F} = Id\vec{l} \times \vec{B} \text{ for currents})$$

23.2 Fields travelling through space

A time-varying magnetic field can make an electric field, and a time-varying electric field can make a magnetic field. Might it be possible for time-varying electric and magnetic fields in empty space to create each other, without any charges or currents around?

If we can show that this is possible (that is, if this would not be inconsistent with one or more of Maxwell's equations), we will still need to answer the question of how such a process might get started. Eventually we will show that such a configuration of fields could be produced by an accelerated charge, but first we will concentrate on establishing that such a configuration of electric and magnetic fields is indeed possible.

Solving Maxwell's equations

In principle, a straight-forward way to show how this could work would be to solve the partial differential equations represented by Maxwell's equations in their differential form, but the required mathematics is beyond the scope of this introductory course. However, we can do what we did in finding a solution to the differential equation for an RC circuit in Chapter 19. In that case we guessed that an exponential function would be a solution for the current as a function of time. We then showed that this function was a solution of the differential equation, by taking derivatives of the guessed function and plugging them into the differential equation. This is one of the most common schemes for solving a differential equation, and it can be used to solve integral equations as well.

Solution plan

We will take the following approach to the problem:
- Propose a particular configuration of electric and magnetic fields in space and time.
- Show that this is consistent with all four Maxwell equations, and is therefore a possible solution.
- Show that an accelerated charge produces such fields.
- Identify the effects that such fields would have on matter.
- Analyze a variety of phenomena involving such fields.

23.2.1 A simple configuration of travelling fields

We could consider various configurations of time-varying electric and magnetic fields, in the absence of any charges, and we would find that most of them do not satisfy Maxwell's equations, and hence are not possible. However, if we add the assumption that the fields are "travelling" through space at a velocity $\vec{v}$, we can find a solution that works. This idea—that the region where the fields are nonzero must move through space—is the key to finding time-varying electric and magnetic fields that can create and sustain each other.

You may recognize this phenomenon—a disturbance moving through space—as a wave. The idea of a travelling region of $\vec{E}$ and $\vec{B}$ is not quite as unmotivated as it may appear here, because to someone familiar with differential equations, the differential form of Maxwell's equations suggests a wave as a possible solution. Many waveforms are possible, including the familiar sinusoidal wave. To simplify our analysis, we will first consider the simplest possible wave: a single pulse, as shown in Figure 23.6.

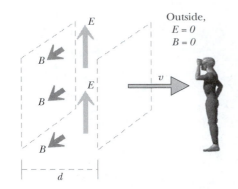

Figure 23.6 A slab contains electric and magnetic fields, with no fields outside the slab.

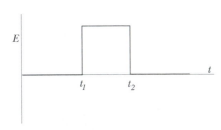

Figure 23.7 Magnitude of electric field observed at a particular location in space, as a function of time, as the slab passes. Note that at times t_1 and t_2, the electric field at this location is changing.

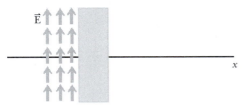

Figure 23.8 At a given instant the electric field is nonzero in the region shown. In the next instant, the region of nonzero field will propagate to the shaded area. For clarity, magnetic field is not shown.

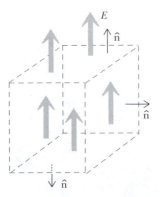

Figure 23.9 Applying Gauss's law, we see that the net electric flux on the box is 0.

A moving slab (or pulse)

Consider a rectangular "slab" of space, of thickness d and very large height and depth, filled with a uniform electric field $\vec{E}$ pointing up and a uniform magnetic field $\vec{B}$ pointing out of the page (Figure 23.6). The electric and magnetic fields are zero outside the slab. We imagine the field-filled slab moving to the right with an unknown speed v, by which we mean that at a later time the slab-shaped region of space that contains fields is somewhere to the right of the present position (and at that time the fields are zero in the original position).

Time-varying fields

If you stand at a location to the right of this region, as shown in Figure 23.6, for a while you don't experience any electric or magnetic fields. Then, as shown in Figure 23.7, the fields at your location change. For a short time you can detect an upward-pointing electric field, and a magnetic field coming out of the page. After the slab passes by, you again experience no electric or magnetic fields. You might describe your experience by saying that a short "pulse" of fields passed by. As the edges of the pulse pass a location, there are time-varying fields at that location. (In a real pulse, the magnitude of the field would not change instantaneously, so dE/dt and dB/dt at the edges of the pulse would not be infinite.)

Propagation

We say that in our example the fields "propagate" to the right, or that the "direction of propagation" is to the right. "Propagation" means the spreading of something (in this case, electric and magnetic fields) into a new region. Strictly speaking, the fields don't "move." Rather, it's just that at one instant in time there are fields in a particular slab-shaped region of space, and at a later time there are what might be called "different" fields in a different slab-shaped region of space, to the right of the original region (Figure 23.8) At a particular instant in time, the fields vary with location. At a particular location in space, the fields vary with time (Figure 23.7).

We will show that this configuration of electric and magnetic fields is consistent with all four of Maxwell's equations. Of course, we have chosen to consider these particular directions for $\vec{E}$, $\vec{B}$, and $\vec{v}$ because we know that they will work. But if we try other sets of relative directions, say with $\vec{E}$ and $\vec{B}$ both pointing down, and $\vec{v}$ up, we find that these directions are not consistent with all four of Maxwell's equations. Only the arrangement we treat here turns out to satisfy Maxwell's equations, with $\vec{E}$, $\vec{B}$, and $\vec{v}$ perpendicular to each other as shown in Figure 23.6. (Of course, $\vec{v}$ can be in any direction; it is the relative orientation of $\vec{E}$, $\vec{B}$, and $\vec{v}$ that is important.)

23.2.2 Gauss's law

First we'll consider Gauss's law,

$$\oint \vec{E} \cdot \hat{n}\, dA = \frac{1}{\varepsilon_0}\sum Q_{\text{inside the surface}}$$

We need to show that our chosen configuration of electric fields is permissible in a region of space where there are no charged particles. We choose an imaginary closed box, fixed in space, through which the field-filled slab is moving (Figure 23.9). The positive flux on the top of the box is exactly canceled by an equal negative flux on the bottom. On the sides $\vec{E}$ is perpendicular to $\hat{n}$, so the flux on the sides is zero. Since the net flux is zero, there must be no charge in the box, as we expected.

23.2.3 Gauss's law for magnetism

Next we consider Gauss's law for magnetism,

$$\oint \vec{B} \bullet \hat{n} \, dA = 0$$

Using the same box-shaped surface (Figure 23.10), we find that the positive magnetic flux on the front of the box is cancelled by an equal amount of negative flux on the back of the box. The other sides contribute no flux, because $\vec{B}$ is perpendicular to $\hat{n}$. So, this configuration of magnetic field can exist in a region where there are no magnetic monopoles, which is a good thing, because as far as we know there are none anywhere!

23.2.4 Faraday's law

Faraday's law relates a changing magnetic flux to a non-Coulomb electric field,

$$\oint \vec{E} \bullet d\vec{l} = -\frac{d}{dt}[\int \vec{B} \bullet \hat{n} \, dA]$$

Because the region of space in which $\vec{E}$ and $\vec{B}$ are nonzero is "moving," the magnetic field at a particular location in space changes with time, so this time-varying magnetic field creates an electric field. To consider Faraday's law, we draw a vertical rectangular path, h high and w wide, partly inside the slab and partly outside as shown in Figure 23.11, which includes a side view. This rectangular path is fixed in space, whereas the slab "moves" to the right. Let's apply Faraday's law to the rectangular path.

? If the slab is moving to the right at speed v, calculate the rate of change of magnetic flux enclosed by the path in terms of B, by calculating how much additional flux there is in a time Δt, then dividing by Δt. Note in the side view in Figure 23.11 the additional region of magnetic flux that appears in the time Δt.

In a time Δt the area of magnetic field inside our chosen path increases by an amount $(v\Delta t)h$, and since the magnetic field B is uniform within the slab, the magnetic flux increases by an amount $B(v\Delta t)h\cos 0 = Bvh\Delta t$. Therefore the rate of change of magnetic flux inside our chosen path is this:

$$\frac{d\Phi_{\text{magnetic}}}{dt} = Bvh$$

By Faraday's law, this rate of change of magnetic flux is equal to the emf around the path, which is a path integral of the electric field.

? Keeping in mind the fact that part of the path is outside the slab, what is the path integral of the electric field in terms of the electric field E?

The only portion of the path along which there is a parallel component of nonzero electric field is along the piece of length h, so we have this:

$$\text{emf} = \left|\oint \vec{E} \bullet d\vec{l}\right| = Eh\cos 0 = Eh$$

? Therefore Faraday's law establishes an algebraic relation between E and B. State this relation.

The emf is equal to the rate of change of magnetic flux, $Eh = Bvh$, so we have the following relationship between B and E in the slab:

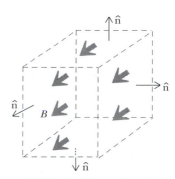

Figure 23.10 Applying Gauss's law for magnetism, we find that the net magnetic flux on the box is 0.

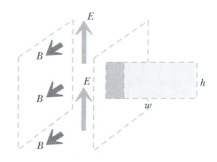

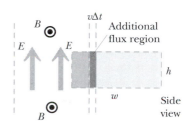

Figure 23.11 Applying Faraday's law to the slab. The lower diagram shows a side view.

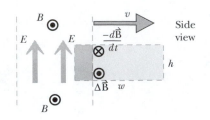

Figure 23.12 The direction of the electric field is consistent with Faraday's law.

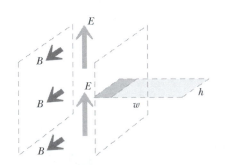

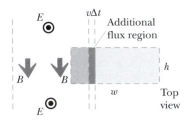

Figure 23.13 Applying the Ampere-Maxwell law to the slab. The lower diagram shows a view from above.

RESULT FROM FARADAY'S LAW

$$E = vB$$

? Explain how the direction of $\vec{E}$ is consistent with Faraday's law.

In the region ahead of the slab, the magnetic field changes from being zero to pointing out of the page, so $\Delta\vec{B}$ points out of the page and $-d\vec{B}/dt$ points into the page (Figure 23.12). With the right thumb pointing in the direction of $-d\vec{B}/dt$, the fingers of the right hand curl clockwise, which is the sense of the electric field around the path.

We have shown that our chosen configuration of moving fields is consistent with Faraday's law if $E = vB$.

23.2.5   The Ampere-Maxwell law

Because the field-containing slab is "moving," the electric field at a location in space changes with time, and by the Ampere-Maxwell law this changing electric field creates a magnetic field. To consider the Ampere-Maxwell law, we draw a rectangular path, fixed in space, that is horizontal (so this rectangle is perpendicular to the electric field). Figure 23.13 includes a view from above, looking down on the upward-pointing electric field.

? Let's apply the Ampere-Maxwell law to this path. If the slab is moving to the right at speed v, calculate the rate of change of electric flux enclosed by the path in terms of E, just as you calculated the rate of change of magnetic flux in the preceding section. Note in the view from above in Figure 23.13 the additional region of electric flux that appears in the time Δt.

In a time Δt the area of electric field inside our chosen path increases by an amount $(v\Delta t)h$, and since the electric field E is uniform within the slab, the electric flux increases by an amount $E(v\Delta t)h\cos 0 = Evh\Delta t$. Therefore the rate of change of electric flux inside our chosen path is this:

$$\frac{d\Phi_{\text{electric}}}{dt} = Evh$$

? Keeping in mind the fact that part of the path is outside the slab, what is the path integral of the magnetic field in terms of B?

The only portion of the path along which there is a parallel component of nonzero magnetic field is along the piece of length h, so we have this:

$$\oint \vec{B} \bullet d\vec{l} = Bh\cos 0 = Bh = \mu_0\left[\sum I_{\text{inside path}} + \varepsilon_0 \frac{d\Phi_{\text{electric}}}{dt}\right]$$

? Therefore the Ampere-Maxwell law establishes another algebraic relation between E and B. State this relation.

There is no current I, so the path integral of the magnetic field is equal to $\mu_0\varepsilon_0$ times the rate of change of electric flux, $Bh = \mu_0\varepsilon_0 Evh$, so we have the following relationship between B and E in the slab:

RESULT FROM AMPERE-MAXWELL LAW

$$B = \mu_0\varepsilon_0 vE$$

23.2.6   Speed of propagation

Since we have shown that our chosen configuration of fields is possible if the region of nonzero field is travelling with speed v, we are very interested in finding out what v must be. We have two different algebraic relations involving E and B, both of them involving the unknown speed v with which the pulse of electric and magnetic fields is traveling through space:

$$E = vB$$

$$B = \mu_0\varepsilon_0 vE$$

? Solve for v and evaluate the result numerically to see with what speed the fields must propagate in order that this pattern of fields be consistent with all four of Maxwell's equations.

By combining $E = vB$ and $B = \mu_0\varepsilon_0 vE$, we find that

$$v^2 = \frac{1}{(\mu_0\varepsilon_0)}$$

Evaluating this numerically we obtain this:

$$v = \sqrt{\frac{1}{\mu_0\varepsilon_0}} = \sqrt{\left(\frac{1}{4\pi\varepsilon_0}\right)\left(\frac{4\pi}{\mu_0}\right)}$$

$$v = \sqrt{(9\times10^9\ \text{N·m}^2/\text{C}^2)(10^7\ \text{A/T·m})} = 3\times10^8\ \text{m/s}$$

This is a spectacular result: the pulse is predicted to advance at a speed that is the same as the known speed of light. At the time that Maxwell went through a similar line of reasoning in the mid-1800's, no one had any idea as to the real nature of light, but the speed of light was known to be 3×10^8 m/s. Maxwell went through a theoretical analysis leading to the possibility of "moving" fields, and he predicted the speed of propagation of such fields by using values for μ_0 and ε_0 that were determined from simple experiments on electricity and magnetism.

When the predicted speed turned out to be the known speed of light, Maxwell concluded that light must be a combination of electric and magnetic fields that feed off each other and can therefore propagate through otherwise empty space, far from any charges or currents. This theoretical analysis was brilliantly confirmed in the period 1885-1889 by Heinrich Hertz, who accelerated charges in a spark and observed the first radio waves, which also propagate at the speed of light.

It is useful to note that the direction of propagation $\hat{v}$ of the pulse is the same as the direction of the cross product $\vec{E}\times\vec{B}$ (Figure 23.14). We can summarize the relation between E and B in a pulse:

RELATION OF *E* TO *B* IN AN ELECTROMAGNETIC WAVE

$E = cB$, where c is the speed of light.

Direction of $\hat{v}$ is direction of $\vec{E}\times\vec{B}$

The fact that $E = cB$ is easy to remember if you note that electric force is $q\vec{E}$ while magnetic force is $q\vec{v}\times\vec{B}$, so cB has the same units as E.

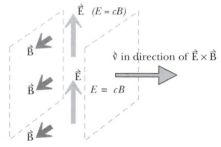

Figure 23.14 The direction of propagation of the electromagnetic pulse.

Ex. 23.1 If the magnetic field in a particular pulse has a magnitude of 10^{-5} tesla (comparable to the Earth's magnetic field), what is the magnitude of the associated electric field?

Ex. 23.2 A pulse of radiation propagates with velocity $\vec{v} = \langle 0, c, 0 \rangle$. The electric field in the pulse is $\vec{E} = \langle 0, 0, 1\times10^6 \rangle$ N/C. What is the magnetic field in the pulse?

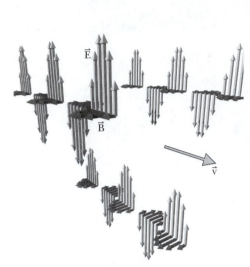

Figure 23.15 A sinusoidal electromagnetic wave.

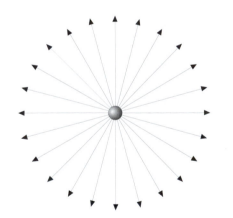

Figure 23.16 Directions of electric field surrounding a stationary positive charge.

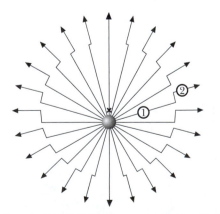

Figure 23.17 Directions of electric field shortly after the charge was kicked. The × marks the original location of the charge.

23.2.7 Electromagnetic radiation

The pulse of propagating electric and magnetic fields is called "electromagnetic radiation"—"electromagnetic" because it involves both electric fields and magnetic fields, and "radiation" because it "radiates" outward from a source. Visible light—electromagnetic radiation that can be detected by the human eye—makes up only part of the spectrum of electromagnetic radiation. When we say "light," we often mean not only visible light, but any kind of electromagnetic radiation—microwaves, radio waves, infrared, ultraviolet, x-rays, or gamma-rays. Later in this chapter we will discuss the electromagnetic spectrum in more detail.

In any case, there is only one arrangement of propagating electric and magnetic fields that satisfies Maxwell's equations in empty space: $\vec{E}$ and $\vec{B}$ at right angles to each other, with $E = cB$, and the region containing non-zero fields propagating in the direction of $\vec{E} \times \vec{B}$ at the speed of light. Now that we have successfully shown that these propagating fields are consistent with Maxwell's equations, our next task is to investigate what effect such fields would have on matter.

Other waveforms

The simple pulse of radiation that we've been considering is not the only possible kind of electromagnetic wave. In fact, any waveform that can be made up from successive short pulses of different magnitudes would also satisfy Maxwell's equations. So, for example, a square wave or a sinusoidal wave (Figure 23.15) would also be possible. We will see later that a sinusoidal electromagnetic wave is particularly easy to produce.

23.3 Accelerated charges produce radiation

We have seen that a pulse of electromagnetic radiation propagating with the speed of light is consistent with Maxwell's equations. The next question we need to answer is how such a configuration of propagating fields could be produced. It turns out that the way to produce electromagnetic radiation is to accelerate a charge.

Consider a positive charge q sitting at rest (Figure 23.16). The direction of the electric field produced by this charge is radially outward in all directions, which we indicate with "electric field lines" showing the direction (but not the magnitude) of the electric field at each point.

We kick the charge downward, so that in a very short time it picks up a little bit of speed and then coasts downward at that slow speed (there is no friction). Some time after the kick, we take a snapshot of the direction of electric field throughout this region, and the snapshot has the surprising and peculiar appearance shown in Figure 23.17. We will discuss what is observed by two people stationed at location 1 and location 2. Observer 1 sees an electric field that points away from the present location of the slowly moving charge.

> **?** Explain why at this instant in time observer 2 sees an electric field that points away from the original position of the charge rather than the present location of the charge.

The key issue is retardation. Information about the change in the electric field has not yet reached observer 2. Since changes in the electric field propagate at the speed of light, it takes a while for the new field to appear at observer 2's position.

Between observer 1 and observer 2 there is a kink in the electric field direction. At some later time this kink will be at location 2 (Figure 23.18).

? If the distance from the original location of the charge to observer 2 is r, how long after we kicked the charge does observer 2 detect a sideways ("transverse") electric field?

Since the transverse electric field propagates at the speed of light, we have $r = ct$, and $t = r/c$. If observer 2 is very far away from the original position of the charge, the radius of curvature r of the propagating radiation is very large, and observer 2 experiences a brief pulse of transverse electric field that looks very much like the transverse electric field associated with the flat moving slab we discussed earlier. An accelerated charge produces electromagnetic radiation. Of course, these flat diagrams show only a cross-sectional slice through a 3-D pattern of radiation, shown in Figure 23.19.

It is clear from our application of Faraday's law and the Ampere-Maxwell law to the moving slab that a pulse of radiation must contain both electric and magnetic fields, so that the pulse can move forever through otherwise empty space, far from any charges. The transverse electric field is accompanied by a transverse magnetic field perpendicular to the electric field, with $\vec{E} \times \vec{B}$ in the direction of propagation, which means that the radiative magnetic field must point out of the page on the right of Figures 23.18 and 23.19, and into the page on the left of these figures.

Outside the expanding sphere of radiation there is no magnetic field because the charge was initially at rest. Inside the expanding sphere there is a charge moving downward with constant velocity, and it makes an ordinary magnetic field (by the Biot-Savart law) that is out of the page on the right and into the page on the left. In this case the Biot-Savart magnetic field is in the same direction as the radiative magnetic field, but this need not be the case. If the charge's acceleration is opposite to its velocity (slowing down rather than speeding up), the two kinds of magnetic fields point in opposite directions.

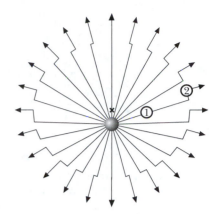

Figure 23.18 Information about the change has reached observer 2.

Figure 23.19 Radiation expands in a spherical pattern from a charge (center) which has been briefly accelerated downwards. The lighter vectors show electric fields, and the shorter, darker vectors show the accompanying magnetic fields.

Ex. 23.3 A charge is initially at rest. Then you move it to another position, leaving it again at rest. Is there any radiation?

23.3.1 Direction of the transverse electric field

If we draw only the transverse electric field at some time after the kick, we see a pattern like that shown in Figure 23.20. Bear in mind that Figure 23.20 is a two-dimensional slice through the three-dimensional phenomenon that is shown above in Figure 23.19, with the pulse headed out of the page and into the page as well as moving in the plane of the diagram.

Ex. 23.4 In what directions is the electromagnetic radiation most intense? In what directions is there no electromagnetic radiation?

23.3.2 Magnitude of the transverse electric field

The qualitative argument we have just made gives the direction of the radiative fields. This argument, due originally to Purcell, can be made quantitative by expert application of Gauss's law, to derive expressions for the magnitude and direction of the radiative electric and magnetic fields. We will not present the details of this derivation, which is mathematically advanced, but if you are interested, we recommend the excellent presentation in Purcell, E.M., *Electricity and Magnetism*, Berkeley Physics Course Vol. II,

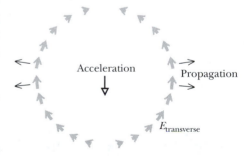

Figure 23.20 Just the transverse electric field, which propagates away from the accelerated charge.

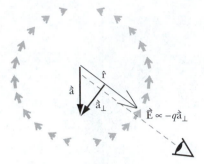

Figure 23.21 The transverse electric field is proportional to the transverse component of acceleration.

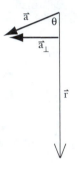

Figure 23.22 The magnitude of $\hat{a}_\perp$ is $a\sin\theta$.

1984, McGraw-Hill. The results of that derivation give the following description of the radiative fields.

The transverse electric field is associated with the kink in the electric field direction associated with accelerated motion of the source charge, and from the diagrams of this process it is plausible that the magnitude of the radiative field is proportional to $-q\hat{a}_\perp$, where $\hat{a}_\perp$ is the "projected" acceleration of the charge (projected onto a plane perpendicular to the line of sight, as in Figure 23.21). Note that the magnitude of $\hat{a}_\perp$ is just $a\sin\theta$ (Figure 23.22)

The direction of propagation of the radiation is outward from the source charge, in the direction of the unit vector $\hat{r}$, drawn as usual from the source charge toward the observation location. This is also the direction of $\vec{E}\times\vec{B}$.

The radiative (transverse) electric field is proportional to the charge q and the projected acceleration $\hat{a}_\perp$, and it turns out to be inversely proportional to the distance r. The full relationship is the following:

PRODUCING A RADIATIVE ELECTRIC FIELD

$$\vec{E}_{\text{radiative}} = \frac{1}{4\pi\varepsilon_0}\frac{-q\hat{a}_\perp}{c^2 r}$$

There are two important points to notice in this formula:
- The direction of the transverse electric field is opposite to $q\hat{a}_\perp$.
- The field falls off with distance at the rate of $1/r$.

This is very different from the $1/r^2$ behavior of the ordinary electric field of a stationary charge or of a charge moving at a constant velocity. It is this much slower fall-off with distance that makes it possible for electromagnetic radiation to affect matter that is very far from the accelerated charges that produced the radiation. We see extremely distant stars because of this slow fall-off with distance.

In earlier chapters we showed that there is energy density associated with electric and magnetic fields, proportional to the squares of the field magnitudes. Similarly, later in this chapter we will show that the radiative power flow in watts per square meter is proportional to the square of the electric field, so the power flow per square meter falls off like $1/r^2$. Energy outflow over larger and larger spherical surfaces is constant, because the surface area $4\pi r^2$ grows proportional to r^2.

Note that if the accelerated object is an electron or other negative charge, all the radiative electric field vectors in Figure 23.21 would point in the opposite direction, because the electric field is proportional to the charge q.

Ex. 23.5 An electron is briefly accelerated in the direction shown in Figure 23.23. The resultant electromagnetic radiation is observed at a location indicated by the vector $\hat{r}$ in the diagram. On a copy of the diagram, draw two vectors: $\hat{a}_\perp$, and $\vec{E}$ at the observation location.

Ex. 23.6 An electric field of 10^6 N/C acts on an electron, resulting in an acceleration of 1.8×10^{17} m/s^2 for a short time. What is the magnitude of the radiative electric field observed at a location a distance of 2 cm away along a line perpendicular to the direction of the acceleration?

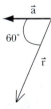

Figure 23.23 An electron is accelerated in the direction shown (Exercise 23.5).

23.3.3 Stability of atoms

Classically (pre-quantum mechanics), any accelerated charge must emit electromagnetic radiation. Since electromagnetic radiation can accelerate other charges, there must be a flow of energy associated with the radiation. This implies that an accelerated charge loses some mechanical energy simply by virtue of emitting radiation.

Before quantum mechanics, this posed a puzzle for the stability of atoms. If we think of an atom as a miniature solar system, with electrons orbiting the nucleus, the electrons are undergoing accelerated motion (radial acceleration = v^2/r) and ought to lose energy by radiation and spiral into the nucleus. When one calculates how long an atom should last, it is a tiny fraction of a second! This paradox also applies to the persistent atomic currents in a magnet or the persistent currents in a superconducting ring.

This paradox was resolved by quantum mechanics. The basic idea is that the energy of a system is quantized, and in some cases (atoms, magnets, superconductors) there is no lower-energy state to which the system can go, and so there is no way to radiate energy and in the process drop to a lower-energy state.

On the other hand, an atom that is in a state with energy above the "ground state" can and does drop to a lower-energy state and emit energy in the form of light.

23.3.4 Sinusoidal electromagnetic radiation

When we give just one kick to a charge to get it moving, during the brief acceleration the charge emits a single brief pulse of electromagnetic radiation. A common way to produce continuous radiation is to move a charge sinusoidally (Figure 23.24), so that the position of the charge is described by $y = y_{max}\sin(\omega t)$, where ω is the "angular frequency" in radians per second (the frequency f in hertz or cycles per second is $f = \omega/2\pi$). The time between maxima at one location is the period T of the oscillation, where $T = 1/f$ (Figure 23.25).

? If $y = y_{max}\sin(\omega t)$, what is the acceleration of the charge as a function of time?

Take the second derivative: $d^2y/dt^2 = -\omega^2 y_{max}\sin(\omega t) = -\omega^2 y$. So the sinusoidal motion produces a sinusoidal acceleration which produces a sinusoidal electromagnetic radiation. If we are a large distance away from the accelerated charge, a snapshot of the electric and magnetic fields at a particular instant looks like Figure 23.24.

? This pattern moves to the right with a speed c. If you stand in one place, how would you describe the electric field that you would observe as this pattern goes by you?

The electric field starts out at zero, then is small, pointing up. It increases to a maximum (still pointing up), decreases to zero, then begins to point down. It increases to a maximum pointing down, decreases back to zero, and the cycle starts over again. At a particular observation location, the electric field varies with time as in Figure 23.25.

Wavelength

Sinusoidal electromagnetic radiation is often described in terms of wavelength λ instead of frequency f, and $c = \lambda f$, because the wave advances a distance λ in one period $T = 1/f$, as illustrated in Figure 23.26. It should also be evident that electromagnetic radiation need not be sinusoidal. The moving slab that we treated earlier represented a single pulse and was not at all like a sinusoid. But sinusoidal radiation is of great practical importance because so many different kinds of systems emit sinusoidal radiation, from atoms to radio stations.

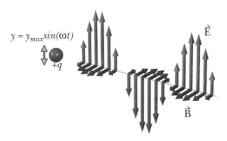

Figure 23.24 A charge that is moved sinusoidally emits continuous radiation.

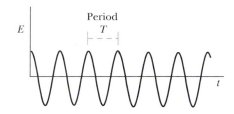

Figure 23.25 A plot of E vs. t at a single location. The period T is the time between two maxima. The frequency $f = 1/T$.

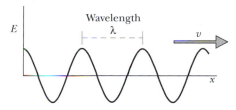

Figure 23.26 A plot of E vs. x at a single instant in time. The wavelength λ is the distance between two maxima.

23.4 Production of electric and magnetic fields: Summary

Here is a summary of the most basic properties of how electric and magnetic fields are produced:

FIELDS MADE BY CHARGES

- A charge at rest makes a $1/r^2$ electric field, but no magnetic field.
- A charge moving with constant velocity makes a $1/r^2$ electric field and a $1/r^2$ magnetic field.
- An accelerated charge in addition makes electromagnetic radiation, with a $1/r$ electric field and an accompanying $1/r$ magnetic field.

Remember that so far, no one has found an isolated "magnetic monopole," but if there were such objects, they would presumably produce fields as follows:

FIELDS MADE BY (HYPOTHETICAL) MAGNETIC MONOPOLES

- A magnetic monopole at rest would make a $1/r^2$ magnetic field, but no electric field.
- A magnetic monopole moving with constant velocity would make a $1/r^2$ magnetic field and a $1/r^2$ electric field.
- An accelerated magnetic monopole would in addition make electromagnetic radiation, with a $1/r$ magnetic field and an accompanying $1/r$ electric field.

23.5 Energy and momentum in electromagnetic radiation

In Chapter 6 we saw that according to the particle model of light, photons have both energy and momentum, which they can transfer to matter. In the classical model that we are now studying, it is also the case that electromagnetic radiation carries both momentum and energy, and can impart both energy and momentum to matter.

23.5.1 Energy in electromagnetic radiation

How would a pulse of electromagnetic radiation interact with ordinary matter? Since electromagnetic radiation is composed of electric and magnetic fields, we can analyze its effects straightforwardly, in terms of the effects of electric and magnetic fields on charged particles. We can see the main effects by examining what would happen to a single charge exposed to the electromagnetic radiation. We'll hang a small ball with a positive charge $+q$ from a spring, so that the ball can move up and down in the direction of the electric field, and we'll watch the ball to see what happens when the radiation goes by (Figure 23.27).

A charged ball on a spring

Of course nothing happens until the pulse reaches the ball, at which time there is a short pulse of electric and magnetic fields that act on the ball. If the slab containing the fields has a thickness d, the pulse lasts for a short time d/c (c is the speed of light).

    For example, if $d = 30$ cm (1 foot), the pulse duration is only one nanosecond. During that brief time, the charged ball experiences an upward electric force $F = qE$, where E is the magnitude of the electric field in the electromagnetic pulse. This brief impulse $F\Delta t$ gives the ball some upward momentum:

$$\Delta p = p - 0 = F\Delta t = (qE)\left(\frac{d}{c}\right)$$

The interaction is over so quickly that the ball hardly has time to move a significant distance upward during the short duration of the pulse, but because

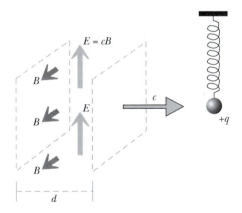

Figure 23.27 A hanging charged ball is affected by electromagnetic radiation.

the ball acquires some upward momentum, it will now oscillate up and down on the spring for a while after the pulse has passed by (until mechanical friction damps out the oscillations).

Energy transfer

Since the kinetic energy of the ball has to come from somewhere, it must be the case that there is energy carried by electromagnetic radiation, and that when the ball is accelerated some of this energy is transferred from the radiation to the charged ball. Since the momentum of the ball is proportional to the magnitude of the electric field in the pulse,

$$\Delta K = K - 0 \approx \frac{p^2}{2m} = \left(qE\frac{d}{c}\right)^2\left(\frac{1}{2m}\right)$$

assuming the speed of the ball remains low compared to the speed of light during this brief interaction. This implies that the amount of energy in a pulse of electromagnetic radiation is proportional to the square of the electric field:

$$\text{Radiative energy} \propto E^2$$

Conservation of energy

Since the ball has gained energy, the radiation should have lost an equal amount of energy. Apparently the electric field E should be smaller after the pulse has accelerated the ball, and later in the chapter we will see how this can come about.

Energy density

We can be quantitative about the amount of energy carried by electromagnetic radiation. Earlier we found that there is an energy density (energy per unit volume) associated with electric and magnetic fields:

ENERGY DENSITY IN ELECTRIC AND MAGNETIC FIELDS

$$\frac{\text{Energy}}{\text{Volume}} = \tfrac{1}{2}\varepsilon_0 E^2 + \tfrac{1}{2}\frac{1}{\mu_0}B^2 \text{, measured in joules}/\text{m}^3$$

For the particular case of electromagnetic radiation, $E = cB$, so

ENERGY DENSITY IN ELECTROMAGNETIC RADIATION

$$\frac{\text{Energy}}{\text{Volume}} = \tfrac{1}{2}\varepsilon_0 E^2 + \tfrac{1}{2}\frac{1}{\mu_0}\left(\frac{E}{c}\right)^2 = \tfrac{1}{2}\varepsilon_0 E^2\left[1 + \frac{1}{\mu_0\varepsilon_0 c^2}\right] = \varepsilon_0 E^2$$

since $\mu_0\varepsilon_0 = 1/c^2$; the electric and magnetic energy densities are equal.

The Poynting vector

Energy density is related to energy "flux" in joules per second per square meter, or watts per square meter (in this case, "flux" denotes a flow of energy through a bounded surface). Consider a cross sectional area A of an advancing slab of electromagnetic radiation (Figure 23.28). In a time Δt, a volume $A(c\Delta t)$ of electric and magnetic fields passes through the area A. The amount of energy passing through area A in the time Δt is $\varepsilon_0 E^2(Ac\Delta t)$, and therefore the energy flux is $\varepsilon_0 E^2 c$ in watts/m^2. Since $E = cB$, and $\mu_0\varepsilon_0 = 1/c^2$, we can write this:

$$\text{energy flux} = \varepsilon_0 EBc^2 = \frac{1}{\mu_0}EB$$

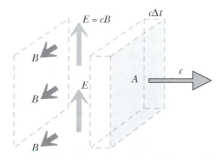

Figure 23.28 The advancing slab of radiation carries an energy flux in W/m^2.

Since the direction of the energy flow is given by $\vec{E} \times \vec{B}$, and since $\vec{E}$ and $\vec{B}$ are perpendicular to each other, the magnitude $|\vec{E} \times \vec{B}| = EB$. We define the "Poynting vector" $\vec{S}$, whose magnitude is the rate of energy flux in watts per square meter and whose direction is the direction of propagation of the electromagnetic radiation:

ENERGY FLUX (THE "POYNTING VECTOR")

$$\vec{S} = \frac{1}{\mu_0} \vec{E} \times \vec{B}, \text{ in W/m}^2$$

Ex. 23.7 In the vicinity of Earth's orbit around the Sun, the energy intensity of sunlight is about 1400 W/m^2. What is the approximate magnitude of the electric field in the sunlight? (What you calculate is actually the "root-mean-square" or "rms" magnitude of the electric field, because in sunlight the magnitude of the electric field at a fixed location varies sinusoidally, and the intensity is proportional to E^2.)

Ex. 23.8 A small laser used as a pointer produces a beam of red light 5 mm in diameter, and has a power output of 5 milliwatts. What is the magnitude of the electric field in the laser beam?

23.5.2 Momentum in electromagnetic radiation

The classical model of electromagnetic radiation can also account for the transfer of momentum from a pulse of radiation to an object.

In the previous section we saw that the electric field in a pulse of radiation exerted a force on a charged object, starting it moving in a direction perpendicular to the direction of propagation of the pulse. While the main effect of electromagnetic radiation is this sideways ("transverse") kick of the electric field, the magnetic field also has an interesting effect, although it is very small and hard to observe. Consider again the charged ball hanging from the spring (Figure 23.29). As soon as the electric field imparts some upward momentum Δp to the ball, the magnetic field will exert a force on the moving ball, which has a speed v upward.

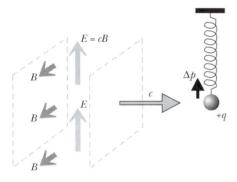

Figure 23.29 What is the direction of the magnetic force on the positively charged ball?

? What is the direction of the magnetic force on the ball, which carries a positive charge $+q$?

The direction of $+q\vec{v} \times \vec{B}$ is to the right.

? Now consider what happens if the ball carries a negative charge $-q$ (Figure 23.30). What is the direction of the magnetic force on the negatively charged ball?

The direction of $-q\vec{v} \times \vec{B}$ is again to the right, because the charge is negative (q is a positive value) and the velocity of the charge is downward. So the direction of the magnetic force is the same no matter whether the ball is positively charged or negatively charged. This effect that electromagnetic radiation has on matter is called "radiation pressure." It even acts on neutral matter, because it affects the protons and electrons in the material in the same way. It is difficult to observe because it is a small effect.

We can estimate the magnitude. A charge q acquires a speed v due to the passage of the electromagnetic pulse. The average speed is $v/2$, and the average magnetic force is $|q\vec{v} \times \vec{B}| = q(v/2)B\sin(90°) = q(v/2)(E/c)$ since $B = E/c$ in the slab of radiation. So the ratio of the average magnetic force to the electric force is this:

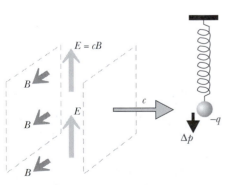

Figure 23.30 What is the direction of the magnetic force on the negatively charged ball?

$$\frac{\overline{F_m}}{F_e} = \frac{q(v/2)(E/c)}{qE} = \frac{1}{2}\frac{v}{c}$$

Normally the speed v acquired due to the passage of an electromagnetic pulse is much less than the speed of light, so the effect of the magnetic field is very small compared to the effect of the electric field.

Electromagnetic radiation carries momentum

A charged particle exposed to sinusoidal electromagnetic radiation (Figure 23.31) would gain no net momentum in a direction perpendicular to the direction of propagation of the radiation, since it would alternately be accelerated in opposite directions by the alternating electric field, and the average electric force would be zero. However, the average magnetic force would not be zero, so the particle's momentum in the direction of $\vec{v}$ would increase.

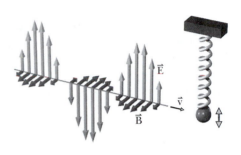

Figure 23.31 Sinusoidal radiation incident upon a charged particle.

Recall that relativistic momentum p and energy E are related to each other for a particle of mass m by the equation $E^2 = (pc)^2 + (mc^2)^2$ (here E is energy, not electric field!). In the quantum view of light as photons, light can be considered to have particle-like properties with zero mass, in which case $E = pc$. This relation between energy E and momentum p is also valid in Maxwell's classical theory of light, and the energy and momentum carried by electromagnetic radiation are both proportional to the square of the electric field. The momentum flux is $1/c$ times the energy flux given by the Poynting vector:

MOMENTUM FLUX

$$\frac{\vec{S}}{c} = \frac{1}{\mu_0 c}\vec{E}\times\vec{B}, \text{ in N/m}^2$$

The units of the Poynting vector are W/m^2, and dividing by c means that the momentum flux is in $W\cdot s/m^3$, which is J/m^3, or $N\cdot m/m^3$, which is N/m^2. Hence the momentum flux has dimensions of pressure.

Conservation of momentum

If electromagnetic radiation imparts momentum to a particle, the radiation should lose an equal amount of momentum. Again, we see that the electric and magnetic fields should be smaller in magnitude after an interaction with matter. In Section 23.5.3 we will see how this can come about.

Radiation pressure

When electromagnetic radiation strikes neutral matter, both the protons and the electrons experience a force in the direction of propagation of the radiation ("radiation pressure"). Despite the fact that radiation pressure is a small effect, there have been serious proposals to build spaceships with huge solar sails, many kilometers in diameter, that could move around the solar system thanks to the radiation pressure of sunlight on the sails. The effect is doubled if the sails are highly reflective, and the plans call for sails made of very thin aluminized plastic.

Ex. 23.9 Since force is dp/dt, the force due to radiation pressure reflected off of a solar sail can be calculated as 2 times the radiative momentum striking the sail per second. In the vicinity of Earth's orbit around the Sun, the energy intensity of sunlight is about 1400

W/m^2. What is the approximate magnitude of the pressure on the sail? (For comparison, atmospheric pressure is about 10^5 N/m^2.)

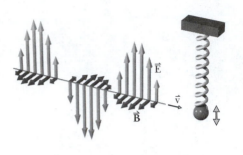

Figure 23.32 Effect of sinusoidal radiation on a hanging charged ball.

23.5.3 Re-radiation or "scattering"

Consider once again the charged ball hanging from the spring, exposed to electromagnetic radiation and therefore oscillating at the radiation frequency (Figure 23.32).

Wait a minute! Won't the ball radiate? After all, the oscillating ball is a sinusoidally moving (and sinusoidally accelerating) charged object, so shouldn't it emit electromagnetic radiation? Yes!

The oscillating ball will re-radiate in (nearly) all directions (see Figure 23.19 on page 851). In Figure 23.33 we show a few of the re-radiation directions. The re-radiation by the charged ball redistributes the energy in new directions. This re-radiation is often called "scattering" because part of the original energy moving to the right has been "scattered" into other directions.

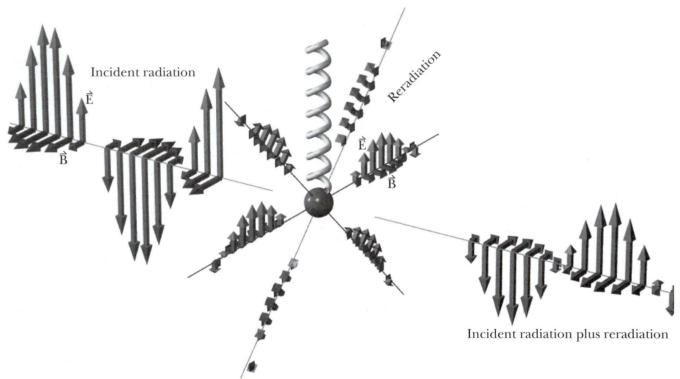

Figure 23.33 Re-radiation by the accelerated charged ball. The accelerated charge reradiates in almost all directions; a sample plane is shown in the picture. The outgoing radiation is the superposition of the incident radiation and the re-radiated fields. *E* outgoing is smaller than *E* incoming; some energy has been scattered in other directions.

? Explain why scattering (reradiation) is never observed in the direction of or directly opposite to the direction of the original transverse electric field (that is, up or down in Figure 23.33).

Straight above or below the accelerated charge, the projected acceleration $\overset{\circ}{a}_\perp$ is zero, so there is no re-radiation in those directions.

We have seen in Section 23.5.1 that the incoming radiation loses energy by giving kinetic energy to the charged ball, proportional to E^2. The re-radiation by the accelerated charge provides a mechanism for decreasing the energy of the original radiation. To the right of the charged ball, the outgoing electromagnetic radiation is the superposition of the original fields plus the fields radiated by the accelerated ball. If an incoming electric field points upward, it accelerates a positively charged ball upwards, and to the right of the ball the re-radiated electric field points downward.

As a result, the net outgoing electric field to the right of the ball in Figure 23.33 is smaller than the original incoming electric field. The energy in the electromagnetic radiation, which is proportional to E^2, is smaller after giving kinetic energy to the ball.

Ex. 23.10 Why doesn't light go through a piece of cardboard (Figure 23.34)? According to the superposition principle, the electric and magnetic fields radiated by the accelerated charges in the light bulb are definitely present to the right of the cardboard, so why isn't there any light there? What must be true of the light reradiated by charges in the cardboard?

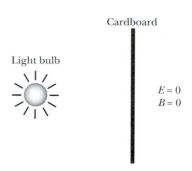

Figure 23.34 Why isn't there light behind the cardboard?

23.6 The effect of electromagnetic radiation on matter

Because radiation pressure is such a small effect, the most important aspect of the interaction between electromagnetic radiation and matter is the effect that the electric field has on the charged particles inside atoms and molecules. For the rest of the chapter, we will concentrate on the electric field in electromagnetic radiation, and neglect the magnetic field. Of course the magnetic field must be present in electromagnetic radiation, because the electric and magnetic fields feed off each other, but usually only the electric field need be considered when we try to understand the main effects of electromagnetic radiation on matter, in terms of our model of electromagnetic radiation.

In the following sections we will consider a wide range of important and interesting phenomena in terms of what we have just learned about how to produce electromagnetic radiation consisting of propagating electric and magnetic fields, and what we already know about the effects electric and magnetic fields have on the charged particles in matter.

23.6.1 The effect of radiation on a neutral atom

The main effect that a pulse of radiation has on matter is that it gives a brief electric kick to charged objects. This kick is sideways or "transverse" to the direction of propagation of the fields. What would a pulse of electromagnetic radiation do to a neutral atom or molecule? A simple model of an atom or molecule is adequate to get the main idea. Consider a positive core (nucleus plus tightly bound inner electrons) and a single, loosely bound outer electron (Figure 23.35).

? What happens to this atom when the pulse of electromagnetic radiation goes by? Which part of the atom picks up more speed, the electron or the positive core?

The electric field polarizes the atom. During this process the electron moves farther than the much more massive ion (nucleus plus inner electrons); the magnitude of the force eE is the same on both objects. For many purposes the interaction of electromagnetic radiation with an atom can be analyzed by modeling an outer electron as being connected to the rest of the atom by a spring; see Problem 15.22 (page 543).

According to our model of electromagnetic radiation, this is the basic mechanism for the interaction of light and matter: the electric field of the light exerts forces which move charged particles inside atoms and molecules. Although electromagnetic radiation is rather unfamiliar compared to topics we studied early in the course, there is nothing unfamiliar about the effect of the associated electric field on charged particles, which is simply $\vec{F} = q\vec{E}$.

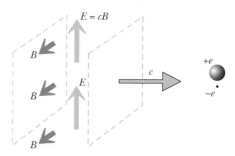

Figure 23.35 A neutral atom is affected by electromagnetic radiation.

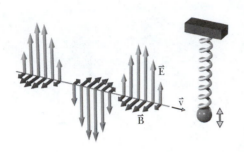

Figure 23.36 Effect of sinusoidal radiation on a hanging charged ball.

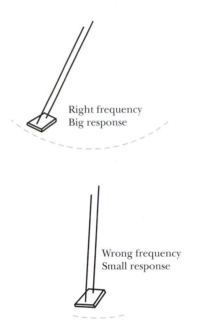

Right frequency
Big response

Wrong frequency
Small response

Figure 23.37 Resonance behavior in pushing a swing.

23.6.2 Effects that depend on frequency

Let us again suspend a charged ball from a spring and watch what happens as this sinusoidally varying electromagnetic radiation goes by (Figure 23.36). In the vertical direction the electric force on the ball varies like $qE_{max}\sin(\omega t)$, and this forces the ball to oscillate up and down at the same angular frequency ω. How large an oscillation we get depends on how close ω is to the angular frequency at which the spring-ball system would freely oscillate in the absence of the radiation. If the radiation frequency nearly matches the free-oscillation frequency, the ball will oscillate up and down with a large motion. If on the other hand the radiation frequency is very different from the free-oscillation frequency, the ball will move up and down very little, even if the radiative electric field is quite large.

Resonance

This phenomenon is called "resonance." It is easy to observe resonance when you push someone in a swing (Figure 23.37). If you give little pushes at exactly the same frequency as the swing's free-oscillation frequency (for example, every time the swing comes all the way back), even these little pushes add up to produce a very large motion, and very small pushes are sufficient to maintain this large motion. However, if you push at a frequency different from the free-oscillation frequency, the swing is forced to oscillate at your frequency but will hardly move at all, because your inputs don't reinforce each other at the right times.

There is an extremely important consequence of the fact that systems respond strongly to sinusoidal radiation only if the frequency of the radiation closely matches the frequency at which these systems would oscillate freely. Different frequencies of sinusoidal electromagnetic radiation affect matter differently. For example:

- Sinusoidal electromagnetic radiation with a very high frequency, $f = \omega/2\pi$ of about 10^{15} hertz, has a big effect on the organic molecules in your retina, and you detect "visible light." However, your radio does not respond to visible light.

- Sinusoidal electromagnetic radiation with a frequency of about 10^6 hertz (1000 kilohertz) has a big effect on the mobile electrons in the metal of a radio antenna, and the radio receiver detects "radio waves." You can tune a radio to be sensitive to electromagnetic radiation at 1020 kilohertz and be insensitive to all other radio frequencies, so that you hear just one station at a time. However, your eye does not detect radio waves.

- The very high frequency of the sinusoidal electromagnetic radiation known as "x-rays" has very little effect on the atoms of your body and so mostly passes right through with no interaction; there is just enough interaction that shadows of the denser parts of your body appear on film sensitive to x-rays.

For many purposes involving the classical interaction of light and matter, a neutral atom can be modeled as though an electron in the atom acts like a mass on a spring, whose free-oscillation frequency determines the resonance response to electromagnetic radiation. See Problem 15.22 on page 543.

The electromagnetic spectrum

Figure 23.38 is a summary of the "electromagnetic spectrum," consisting of the ranges of wavelengths and frequencies that are characteristic of various mechanisms for producing and responding to electromagnetic radiation.

Color vision

In your retina there are three kinds of organic molecules which we might call R, G, and B. Type R responds strongly to electromagnetic radiation near 700 nanometers (0.7×10^{-6} m, or 4.3×10^{14} hertz); type G responds strongly to electromagnetic radiation near 550 nm (0.55×10^{-6} m, or 5.5×10^{14} hertz); and type B responds strongly to electromagnetic radiation near 400 nm (0.4×10^{-6} m, or 7.5×10^{14} hertz). Each of these molecules responds less or not at all to the frequencies favored by the other molecules. Your brain interprets signals from R molecules as indicating "red" light; G molecules indicate "green" light; and B molecules indicate "blue" light.

Combinations of these responses are interpreted as all of the other colors that you can perceive, many of which are not in the rainbow. For example, a mixture of electromagnetic radiation at 700 nm ("red") and at 400 nm ("blue") is interpreted by the brain as "magenta." In a rainbow, red and blue are separated from each other, so magenta doesn't occur. Equal stimulation of the R, G, and B molecules is interpreted by the brain as white light, another color that is not in the rainbow.

There is some overlap in wavelengths in the sensitivity of the R, G, and B receptors. For example, light containing the single wavelength of 588 nm (the yellow light of burning sodium) stimulates both the R and the G receptors, so you also perceive yellow with a suitable mixture of two wavelengths, one red and one green. Color television and computer screens use closely-spaced red and green dots to fool the brain into seeing a yellow dot.

Transmitters

How can we produce electromagnetic radiation of a desired frequency? Radio transmitters have an LC circuit connected to the transmitting antenna, and either the inductance *L* or the capacitance *C* can be adjusted to choose the desired frequency (see Section 22.7.3 on page 822). This circuit accelerates electrons in the antenna, and the accelerated electrons radiate.

Visible light is often produced by making some material so hot that the thermal oscillation of the atoms reaches frequencies in the visible region. This is the situation in the filament of a flashlight bulb. In this case there is a broad mixture of different frequencies which we perceive as white light.

Energy transitions between quantized atomic levels produce light of a single frequency, as in a laser.

23.6.3   Polarized electromagnetic radiation

Figure 23.39 illustrates a demonstration of electromagnetic radiation that is sometimes presented in physics lectures. A radio transmitter consists of a high-frequency AC voltage supply (frequency around 300 megahertz) connected to a transmitting antenna in the form of two metal rods, and a radio receiver is made of two metal rods connected by a flashlight bulb. When positioned as shown, the bulb lights although it is not connected to the circuit!

The high-frequency AC voltage drives the mobile electrons back and forth in the transmitting antenna, and the sinusoidal motion of these electrons implies a sinusoidal acceleration, which produces electromagnetic radiation (Figure 23.40). For the radiation that heads toward the right, the oscillating electric field is horizontal, and it can accelerate the mobile electrons in the receiving antenna, driving them back and forth through the flashlight bulb, and lighting the bulb.

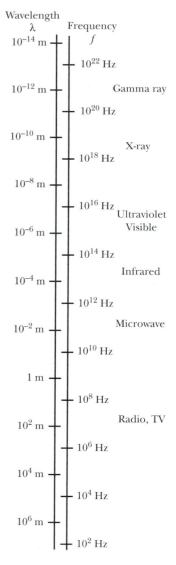

Figure 23.38 The electromagnetic spectrum: wavelength and frequency (logarithmic scale).

Figure 23.39 A high-frequency radio transmitter and receiver.

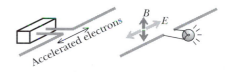

Figure 23.40 The radiative electric field drives current through the bulb.

Figure 23.41 When the distance is greater, the bulb is dimmer.

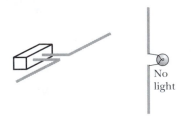

Figure 23.42 With the receiving antenna vertical, the bulb doesn't light.

Figure 23.43 Unpolarized light produced by randomly-oriented emitters.

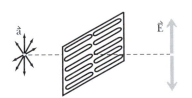

Figure 23.44 Making polarized light from unpolarized light.

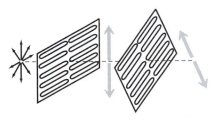

Figure 23.45 A second polarizer rotates the polarization direction.

If you move the receiver and transmitter farther apart (Figure 23.41), the bulb gets dimmer, which is related to the $1/r$ fall-off of the electric field.

If you turn the receiving antenna vertical, the bulb does not light (Figure 23.42).

? Explain why the bulb does not light when the receiving antenna is vertical.

The radiated electric field is in the plane perpendicular to the receiving antenna. The electrons in the antenna are accelerated back and forth across the width of the wire (not its length), so the current does not go through the bulb.

We say that the emitted radiation is "polarized" in the direction of the transmitting antenna, and the direction of polarization clearly affects the response of the receiver. Not all electromagnetic radiation is polarized. The visible light emitted by the Sun or by an incandescent bulb is unpolarized because the atomic emitters are randomly aligned in all directions, giving rise to a mixture of electric field directions transverse to the direction of propagation (Figure 23.43).

You can produce polarized light by passing unpolarized light through a sheet of a special plastic whose long molecules are aligned with each other, such as the material in Polaroid® sunglasses (Figure 23.44). Light that is polarized in the same direction as these long molecules is able to drive electrons along the length of the molecule and lose energy to this relatively large-scale motion. Light that is polarized perpendicular to these long molecules can't move the electrons very far and consequently loses much less energy in the plastic. For light that is polarized at some other angle only the component of electric field that is perpendicular to the long molecules manages to show up after the polarizer. The net effect is that polarized light is produced.

If you place a second polarizer after the first one, its orientation determines whether all, some, or none of the polarized light is seen after the polarizer, and if any component of the light is seen after the second polarizer it is polarized in alignment with the second polarizer (Figure 23.45).

? Polaroid sunglasses are useful at the beach because unpolarized sunlight that glances off the water happens to become polarized in the process, with the electric field horizontal. In order to block this glare but let other (unpolarized) aspects of the scene come through, should the long molecules in the plastic lenses be horizontal or vertical?

Since we want to diminish the horizontally polarized radiation (glare), the molecules should also be horizontal. Radiation with a horizontal electric field will be absorbed and re-radiated, with only a fraction seen in the original direction. Vertically polarized light will be seen unaffected.

You can observe these effects with radio waves, too. For a polarizer, use a piece of cardboard that has a series of parallel metal foils taped to the cardboard. When the foils are aligned with the antennas, the bulb does not glow; but when the foils are aligned perpendicular to the antennas, the bulb does glow (Figure 23.46).

Ex. 23.11 Briefly explain the observations in the previous paragraph (Figure 23.46). (Hint: Think about the re-radiation that may occur from the wires.)

Ex. 23.12 Figure 23.47 shows several different orientations (A, B, and C) of the receiving antenna discussed on page 862. In each case, predict the brightness of the bulb and explain briefly.

23.6.4 Why the sky is blue (and polarized)

Why is the light we see coming from the sky blue (and polarized)? Suppose you look straight up at the blue sky, in the early morning or late afternoon, so that the Sun is near the horizon (Figure 23.48).

? First, why do you see any light at all coming from the sky overhead? After all, if you stand on the Moon and look up, the sky is black (except for stars).

The reason why you see any light at all coming down from straight overhead on the Earth is that electrons in the air molecules all through the atmosphere above your head are accelerated by the sunlight and re-radiate. The Moon has no atmosphere to scatter the sunlight.

Second, the sky light coming from overhead is polarized, despite the fact that light from the Sun is unpolarized due to being emitted by many atoms at random orientations. This polarization is easy to observe for yourself. While wearing Polaroid sunglasses on a cloudless day, look straight overhead in early morning or late afternoon and turn your head to vary the orientation of the polarizer. You will see a marked change in the intensity of the sky light.

? Explain why the sky light is polarized, and state what the direction of polarization is.

The electric field of light from the Sun is perpendicular to the direction of propagation. Air molecules excited by a horizontal field radiate downward toward you, but air molecules excited by a vertical field do not radiate downward toward you (the projected acceleration is zero), so the sky light is polarized with the electric field direction perpendicular to the direction to the Sun.

? Third, why is the sky light blue?

Light from the Sun is a mixture of many different frequencies of sinusoidal electromagnetic radiation. The spectrum of sunlight runs from red to blue, and the mixture is perceived by you as nearly white (but somewhat yellowish). So why does the light from the sky look blue instead of yellow? After all, the electrons are forced to oscillate at the same frequency as the light that hits them.

The electrons in the atom are shaken by the oscillating electric field. If the position of the electron can be represented by $x = A\cos(\omega t)$, then the acceleration is $a = -\omega^2 A\cos(\omega t)$. The magnitude of the reradiated electric field is proportional to the acceleration, and the power (energy/second) of the reradiation is proportional to the square of the acceleration. We expect the reradiation that comes to your eye to be proportional to ω^4. This implies that reradiation of the high-frequency blue component of sunlight will be much greater than the reradiation of the low-frequency red component of sunlight. The ratio of blue-to-red frequencies is nearly a factor of 2, so the ratio of reradiated intensities is about a factor of $2^4 = 16$.

Ex. 23.13 If you look toward the setting Sun, you are looking through a very long column of air: Why does the sky look red in this case? (Take into account what you have just learned about the frequency dependence of the response of air molecules.) Why is the light unpolarized?

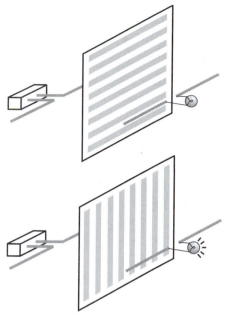

Figure 23.46 A polarizer for radio waves.

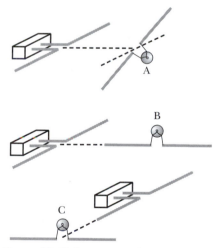

Figure 23.47 Antenna and receiver orientations for Exercise 23.12.

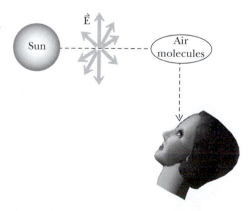

Figure 23.48 Looking up at the sky—why is there any light?

23.6.5 Why distant mountains appear bluish

On a very clear day, with little dust or haze in the air, distant mountains nevertheless look bluish. An indication that there is little haze is the fact that distant mountain ridges appear with crisp and sharp lines, not blurred as they would look in haze or fog. The explanation is essentially the same as the explanation of why the sky is blue! When you look up at the sky and see a brilliant blue color, you are looking through the equivalent of only a few kilometers of sea-level air (the atmosphere extends out to many tens of kilometers, but the air has exponentially decreasing density as you go higher).

When you look at a distant mountain through many kilometers of air, sunlight coming from the side scatters off that column of air into your eyes, adding a bluish light to the image of the mountain. You need a dark mountain as a background in order to be able to notice this relatively low-intensity bluish light.

23.7 *The differential form of Maxwell's equations

Here are the differential forms of Maxwell's equations, where $\vec{J}$ is the current density in amperes per square meter:

MAXWELL'S EQUATIONS—DIFFERENTIAL FORMS

$$\text{div}(\vec{E}) = \vec{\nabla} \bullet \vec{E} = \frac{\rho}{\varepsilon_0} \qquad\qquad \text{Gauss's law}$$

$$\text{div}(\vec{B}) = \vec{\nabla} \bullet \vec{B} = 0 \qquad\qquad \text{Gauss's law for magnetism}$$

$$\text{curl}(\vec{E}) = \vec{\nabla} \times \vec{E} = -\frac{\partial \vec{B}}{\partial t} \qquad\qquad \text{Faraday's law}$$

$$\text{curl}(\vec{B}) = \vec{\nabla} \times \vec{B} = \mu_0\left[\vec{J} + \varepsilon_0\frac{\partial \vec{E}}{\partial t}\right] \qquad \text{Ampere-Maxwell law}$$

THE LORENTZ FORCE

$$\vec{F} = q\vec{E} + q\vec{v} \times \vec{B}$$

23.8 Summary

Fundamental physical principles

Maxwell's equations

$$\oint \vec{E} \bullet \hat{n}\, dA = \frac{1}{\varepsilon_0} \sum Q_{\text{inside surface}} \qquad \text{Gauss's law}$$

$$\oint \vec{B} \bullet \hat{n}\, dA = 0 \qquad \text{Gauss's law for magnetism}$$

$$\oint \vec{E} \bullet d\vec{l} = -\frac{d}{dt}[\int \vec{B} \bullet \hat{n}\, dA] \qquad \text{Faraday's law}$$

$$\oint \vec{B} \bullet d\vec{l} = \mu_0\left[\sum I_{\text{inside path}} + \varepsilon_0 \frac{d\Phi_{\text{electric}}}{dt}\right] \qquad \text{Ampere-Maxwell law}$$

The Lorentz force

$$\vec{F} = q\vec{E} + q\vec{v} \times \vec{B} \qquad (d\vec{F} = Id\vec{l} \times \vec{B} \text{ for currents})$$

Producing a radiative electric field

$$\vec{E}_{\text{radiative}} = \frac{1}{4\pi\varepsilon_0}\frac{-q\vec{a}_\perp}{c^2 r}; E = cB \text{ in electromagnetic radiation}$$

Direction of $\vec{v}$ is direction of $\vec{E} \times \vec{B}$

New concepts

The main effect that electromagnetic radiation has on matter is due to $\vec{F} = q\vec{E}_{\text{radiative}}$, but the magnetic field is responsible for radiation pressure, which is a small effect.

If $\vec{E}_{\text{radiative}}$ makes a charge accelerate, that charge radiates in turn ("re-radiation" or "scattering").

Different systems respond differently to sinusoidal electromagnetic radiation of a particular frequency f ($c = f\lambda$). Systems that freely oscillate sinusoidally with a frequency f respond very strongly to sinusoidal electromagnetic radiation of the same frequency f ("resonance" phenomenon).

Results

Energy flux of electromagnetic radiation

$$\vec{S} = \frac{1}{\mu_0}\vec{E} \times \vec{B} \text{, in W/m}^2 \text{ (the "Poynting vector")}$$

Momentum flux of electromagnetic radiation

$$\frac{\vec{S}}{c} = \frac{1}{\mu_0 c}\vec{E} \times \vec{B} \text{, in N/m}^2$$

The sky is blue and polarized due to reradiation of sunlight, with blue light reradiated more strongly than red light.

23.9 Example problem: Electron enters region of electric field

An electron is traveling at nonrelativistic speed along the z axis in the $+z$ direction in a region of nearly zero fields (Figure 23.49). At $t = 0$ it reaches the origin $<0, 0, 0>$ and enters a region where the electric field is $<0, 0, 2\times10^6 \text{ V/m}>$. You observe fields at location $<5, 3, 0>$ m. In the following, explain your work fully.

(a) When do you first observe radiative fields? Explain.

(b) On the diagram, show the direction of the initial radiative electric field at your location. Explain.

(c) On the diagram, show the direction of the initial radiative magnetic field at your location. Explain.

(d) What is the magnitude of the initial radiative electric field at your location?

(e) What is the magnitude of the initial radiative magnetic field at your location?

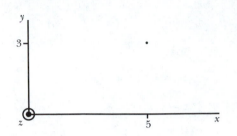

Figure 23.49 An electron enters a region of electric field.

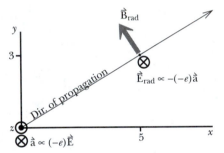

Figure 23.50 Directions of radiative fields.

Solution

(a) It takes $\dfrac{\sqrt{(5 \text{ m})^2 + (3 \text{ m})^2}}{3\times10^8 \text{ m/s}} = 1.9\times10^{-8}$ s for light to propagate from the origin to the observation location.

(b) See Figure 23.50. The electron undergoes an acceleration (deceleration) in the $-z$ direction. This produces a radiative electric field in the $-z$ direction, as shown.

(c) The direction of propagation is given by the direction of $\vec{E} \times \vec{B}$, which requires that $\vec{B}$ point in the direction shown.

It is important to note that the direction of the magnetic field predicted by the Biot-Savart law is in the opposite direction, because it depends on the velocity of the electron (in the $+z$ direction), not the acceleration (in the $-z$ direction). At a far distance the radiative magnetic field $(1/r)$ is much larger than the Biot-Savart magnetic field $(1/r^2)$.

(d) Nonrelativistic, so $a_\perp = a = eE/m$, and $E_{\text{rad}} = \dfrac{1}{4\pi\varepsilon_0}\dfrac{e(eE/m)}{c^2 r}$

$$E_{\text{rad}} = \left(9\times10^9 \frac{\text{N}\cdot\text{m}^2}{\text{C}^2}\right)\frac{(1.6\times10^{-19} \text{ C})^2(2\times10^6 \text{ V/m})}{(3\times10^8 \text{ m/s})^2(\sqrt{(5 \text{ m})^2 + (3 \text{ m})^2})(9\times10^{-31} \text{ kg})}$$

$$E_{\text{rad}} = 9.8\times10^{-10} \text{ V/m}$$

(e) $B_{\text{rad}} = \dfrac{E_{\text{rad}}}{c} = \dfrac{9.8\times10^{-10} \text{ V/m}}{3\times10^8 \text{ m/s}} = 3.2\times10^{-18}$ T

23.10 Review questions

Maxwell's Equations

RQ 23.1 For each one of Maxwell's Equations, match the physical situation with the equation that would be useful in analyzing the situation:

1) Gauss's law

    a) Find the magnetic field inside a current-carrying coaxial cable.

2) Gauss's law for magnetism

    b) Predict what an AC voltmeter reads whose leads encircle a coil.

3) Faraday's law

    c) Find the amount of charge inside a box.

4) Ampere-Maxwell law

    d) Relate the magnetic flux on one face of a cube to the magnetic flux on the other faces.

Electromagnetic radiation

RQ 23.2 Electromagnetic radiation is moving to the right, and at this time and place the electric field is horizontal and points out of the page (Figure 23.51). The magnitude of the electric field is $E = 3000$ N/C. What are the magnitude and direction of the associated magnetic field at this time and place?

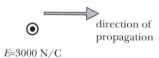

direction of propagation

E=3000 N/C

Figure 23.51 Maxwell's equations (RQ23.2).

Effect of electromagnetic radiation on matter

RQ 23.3 When discussing the effect of electromagnetic radiation on matter informally, physicists often talk about the effects that electromagnetic radiation has on the electrons in the matter. Why would they omit commenting on the effects on the nucleus, which is also charged?

RQ 23.4 Explain why electromagnetic radiation moving to the right applies radiation pressure to the right on a neutral grain of dust in the air, even though the grain has zero net charge.

Producing electromagnetic radiation

RQ 23.5 (a) If the electric field inside a capacitor exceeds about 3×10^{6} V/m, the few free electrons in the air are accelerated enough to trigger an avalanche and make a spark. In the spark shown in Figure 23.52, electrons are accelerated upward and positive ions are accelerated downward. Qualitatively, explain the directions and relative magnitudes of the radiative electric fields at location A, to the left of the capacitor, due to the indicated motions of the electrons and of the ions (actually, in a spark there are also accelerations in the opposite direction when the electrons and ions collide with air molecules).

  (b) Repeat the analysis of part (a) for locations B (to the right) and C (to the right and down a bit).

  (c) If you are at location A, 3 meters to the left of the capacitor, how long after the initiation of the spark could you first detect a magnetic field? What is the direction of the radiative magnetic field? (Before the spark occurs there is no magnetic field anywhere in this region.)

  (d) In addition to producing electromagnetic radiation, the moving charges in the spark produce electric fields according to Coulomb's law and magnetic fields according to the Biot-Savart law. Explain why these fields are much smaller than the radiative fields, if you are far away from the capacitor.

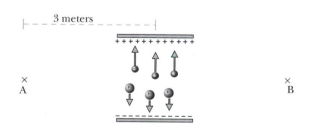

Figure 23.52 Producing electromagnetic radiation (RQ 23.5) (not to scale).

Sinusoidal electromagnetic radiation

RQ 23.6 Calculate the wave length for several examples of sinusoidal electromagnetic radiation:

radio, 1000 kilohertz, $\lambda = ?$
television, 100 megahertz, $\lambda = ?$
red light, 4.3×10^{14} hertz, $\lambda = ?$
blue light, 7.5×10^{14} hertz, $\lambda = ?$

(Note for comparison that an atomic diameter is about 10^{-10} m.)

Polarization

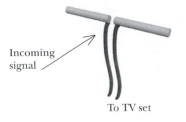

RQ 23.7 You may have noticed that television antennas on housetops have metal rods mounted horizontally (Figure 23.53), at least in the United States (we're not talking about satellite dishes). What does this imply about the construction of the transmitting antennas used by television stations? Why?

Figure 23.53 A simple antenna (RQ 23.7).

RQ 23.8 At dawn, with the Sun just rising in the east, you face the Sun and bend your head back to look straight up, and you examine the blue sky light with a Polaroid filter. Why is the light polarized? What is the direction of the electric field, east-west or north-south? Explain carefully.

Re-radiation

RQ 23.9 On a cloudless day, when you look away from the sun at the rest of the sky, the sky is bright and you cannot see the stars. On the Moon, however, the sky is dark and you *can* see the stars even when the sun is visible. Explain briefly.

23.11 Homework problems

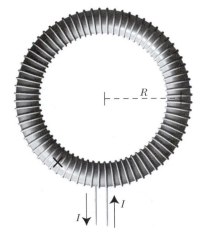

Figure 23.54 A toroid (Problem 23.1). Current runs in the direction indicated.

Problem 23.1 Fields associated with a toroid
N closely spaced turns of wire are wound in the direction indicated on a hollow plastic ring of radius *R*, with circular cross section (Figure 23.54).

(a) If the current in the wire is *I*, determine the magnetic field *B* at the location indicated by the "x," at the center of the cross section of the ring, and indicate the direction by drawing a vector at that location. Give a complete derivation, not just a final result. Use a circular path.

(b) Throughout this region there is a uniform electric field *E* into the paper. This electric field begins to increase at a rate dE/dt, and there continues to be a current *I* in the wire. Now what is the magnitude of the magnetic field at the indicated point?

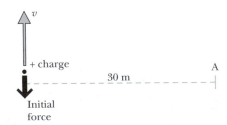

Figure 23.55 A decelerated positive charge (Problem 23.2).

Problem 23.2 A decelerated positive charge
A positive charge coasts upward at a constant velocity for a long time. Then at $t = 0$ a force acts downward on it for 1 ns (10^{-9} s); then it coasts upward at a smaller constant speed for 1 ns; then a force acts upward for 1 ns and it resumes its original speed (Figure 23.55). The new position reached at $t = 3$ ns is much less than a millimeter from the original position.

You stand at location A, 30 meters to the right of the charge, with instruments for measuring electric and magnetic fields. What will you observe due to the motion of the positive charge, at what times? You do not need to calculate the magnitudes of the electric and magnetic fields, but you do need to specify their directions, and the times when these fields are observed.

Problem 23.3 An accelerated electron

An electron is initially at rest. At time $t_1 = 0$ it is accelerated upward with an acceleration of 10^{18} m/s^2 for a very short time (this large acceleration is possible because the electron has a very small mass). We make observations at location A, 15 meters from the electron (Figure 23.56).

(a) At time $t_2 = 1$ ns (10^{-9} s), what is the magnitude and direction of the electric field at location A due to the electron?

(b) At what time t_3 does the electric field at location A change?

(c) What is the direction of the electric field at location A at time t_3?

(d) What is the magnitude of this electric field?

(e) Just after time t_3, what is the direction of the magnetic force on a positive charge that was initially at rest at location A? Explain with a diagram.

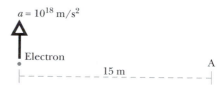

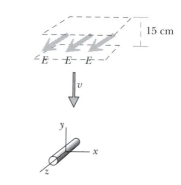

Figure 23.56 An accelerated electron (Problem 23.3).

Problem 23.4 Radiation and a copper wire

A slab (pulse) of electromagnetic radiation that is 15 cm thick is propagating downward (in the $-y$ direction) toward a short horizontal copper wire located at the origin and oriented parallel to the z axis, as shown in Figure 23.57.

(a) The direction of the electric field inside the slab is out of the page (in the $+z$ direction). On a diagram, show and describe clearly the direction of the magnetic field inside the slab.

(b) You stand on the x axis at location $\langle 12, 0, 0 \rangle$ m, at right angles to the direction of propagation of the pulse. Your friend stands on the z axis at location $\langle 0, 0, 12 \rangle$ m. The pulse passes the copper wire at time $t_1 = 0$, and at a later time t_2 you observe new non-zero electric and magnetic fields at your location, but your friend does not. Explain. What is t_2? (give a numerical answer.)

(c) On a diagram, show and describe clearly the directions of these new non-zero fields ($\vec{E}$ and $\vec{B}$) at your location. Explain briefly but carefully.

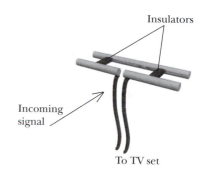

Figure 23.57 Radiation and a copper wire (Problem 23.4).

Problem 23.5 Possible field configurations

Show that a moving region of space in which $\vec{E} = \langle 0, E, 0 \rangle$, $\vec{B} = \langle 0, -B, 0 \rangle$, and $\vec{v} = \langle -c, 0, 0 \rangle$ does not satisfy Maxwell's equations, and therefore is not a possible configuration of travelling fields.

Problem 23.6 An antenna with two bars

Television antennas often have one or more horizontal metal bars mounted behind the receiving antenna and insulated from it (Figure 23.58). Explain qualitatively why adding a second metal bar can make the television signal either stronger or weaker, depending on the distance between the front receiving antenna and the second bar.

Figure 23.58 An antenna with a second metal bar insulated from it (Problem 23.6).

Problem 23.7 Electric field in a spotlight

A 100 watt light bulb is placed in a fixture with a reflector which makes a spot of radius 20 cm. Calculate approximately the amplitude of the radiative electric field in the spot.

Problem 23.8 Effect on a proton

At time $t = 0$, electrons in a vertical copper wire are accelerated downward for a very short time Δt (by a power supply which is not shown). A proton, initially at rest, is located a very large distance L from the wire.

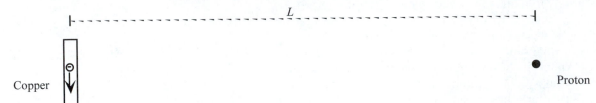

Copper

Proton

Explain in detail what happens to the proton (neglect gravity), and at what time.

Problem 23.9 Magnetic field in empty space

A rectangle 4 cm by 8 cm is drawn in empty space, in a region where there is no matter present, and the magnetic field is measured at all points along the rectangle as shown in Figure 23.59. What can you conclude about the region enclosed by the rectangle? Be as quantitative as you can.

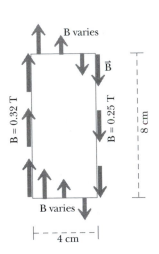

Figure 23.59 Pattern of magnetic field in a region of empty space (Problem 23.9).

23.12 Answers to exercises

23.1 (page 849)   3000 N/C

23.2 (page 849) $\vec{B} = \langle 3.3\times10^{-3}, 0, 0 \rangle$ T

23.3 (page 851)   Yes. Moving the charge involves accelerating from rest, moving at some speed, then accelerating (opposite to the velocity) to rest again. So there are at least two pulses of radiation emitted.

23.4 (page 851)   Most intense to the sides; zero radiation in the direction of the acceleration (and the opposite direction).

23.5 (page 852)   See Figure 23.60.

23.6 (page 852) 1.44×10^{-7} N/C

23.7 (page 856)   725 N/C

23.8 (page 856)   310 N/C

23.9 (page 857) 9.3×10^{-6} N/m^2

23.10 (page 859)
• Since electric fields are not blocked by matter, the radiated fields must be present on the other side of the cardboard.

• Yet, there is clearly no light there, so the net field must be zero there.

• There must be an additional source of electric and magnetic field somewhere, producing a field that adds to the radiative field to produce a zero net field.

• The electrons in the atoms of the cardboard must be accelerated by the electric field of the incoming radiation in such a way that the re-radiated field is equal and opposite to the incident field. The net field on the other side of the cardboard is the superposition of the two fields, and is zero.

23.11 (page 862):
Top: Horizontally polarized radiation accelerates electrons back and forth along the horizontal metal foils. The electrons reradiate, with much of the radiation going off in other directions. The net field reaching the antenna does accelerate electrons along the antenna, but does not produce enough current to light the bulb. Bottom: Horizontally polarized radiation across the narrow width of the foils is not very effective in causing a large response in the foils. There is minimal re-radiation, so most of the original radiation accelerates electrons in the receiver along its length, thus lighting the bulb.

23.12 (page 862):
A: medium bright; $\vec{E}$ has some component along the length of the receiving antenna. B: no light; no component of $\vec{E}$ along the length of the receiving antenna. C: no light; no radiation is emitted in the direction of the acceleration of the charges in the transmitting antenna.

23.13 (page 863)
Since blue light is scattered to the side by molecules in the air more than red light is, the blue portion of the light doesn't get through very well, leaving an excess of red. Looking toward the source, you will see more red light. It is unpolarized because you see re-radiation for all polarizations of the sunlight.

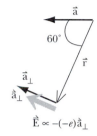

Figure 23.60 An electron is accelerated in the direction shown (Exercise 23.5).

Chapter 24

Waves and Particles

Chapter 24

Waves and Particles

In our study of mechanics (Chapters 6 and 8) we found that a particle model of light was useful in explaining a variety of phenomena. For example, considering light as discrete particles called photons, each carrying a fixed amount of energy and momentum, we could explain both elastic and inelastic collisions of particles and nuclei, and could also explain the emission and absorption of light by quantum systems such as diatomic molecules.

However, a particle model of light is not adequate to explain a number of important aspects of the behavior of light. For example, phenomena such as these seem to imply wavelike behavior:

> interference (how can two beams of light "cancel" each other at some locations?)

> diffraction (why does a beam of light spread out after passing through an opening?)

> reflection (why do we see light reflected from a shiny surface only at one particular angle?)

In this chapter we will apply the wave model of electromagnetic radiation to these and other phenomena.

Two models of light?

Although we will find that the wave model has significant explanatory and predictive power, it isn't very satisfying to have two completely different models of light. In our previous work, we have usually looked for one model adequate to explain all behavior we observe within a given range of parameters. In the second part of this chapter we'll look at phenomena for which wave and particle models of light make different predictions, such as the photoelectric effect and Compton scattering. Insights from analyzing these phenomena will help us construct a unified model of light that includes both wave and particle characteristics.

24.1 Wave phenomena

In the first part of this chapter we will examine phenomena which can be explained with a wave model of light. Phenomena which require a wave model of light include interference, diffraction, and reflection of electromagnetic radiation. Other phenomena which can be understood with a wave model of light include reflection and light scattering.

24.1.1 Description of sinusoidal waves

Before we study interference phenomena in detail, we need to review the basic vocabulary used to describe sinusoidal waves that propagate away from a source, such as a radio transmitting antenna that emits sinusoidal electromagnetic waves or an audio speaker that emits sinusoidal sound waves.

Wavelength, period, speed, and frequency

> Wavelength λ (Greek lambda): The distance between crests of a sinusoidal wave (see Figure 24.1). Units: meters.

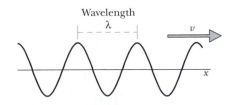

Figure 24.1 A sinusoidal wave propagating at speed v; the wavelength λ is the distance between crests of the wave.

Period T: The time between crests passing by a particular fixed location. Units: seconds.

Speed of propagation v: The distance one crest moves per second. Units: meters/second.

Frequency f: The number of crests passing by a location every second. Units: cycles per second, or "hertz" (Hz).

Angular frequency $\omega = 2\pi f$: Units: radians per second.

SPEED OF PROPAGATION, WAVELENGTH, PERIOD, FREQUENCY

$$v = \frac{\lambda}{T}$$ because one crest travels a distance λ in a time T.

$$f = \frac{1}{T}$$ because T is the time for one cycle.

$$v = f\lambda$$ by combining the two equations above.

You should remember the definitions of these quantities, and all three of the equations above, because we will use them frequently. Note that you can always check your memory by checking units.

Do the following calculations to get a feel for orders of magnitude (remember that $c = 3 \times 10^8$ m/s):

Ex. 24.1 A particular AM radio station broadcasts at a frequency of 1020 kilohertz. What is the wavelength of this electromagnetic radiation? How much time is required for the radiation to propagate from the broadcasting antenna to a radio 4 km away?

Ex. 24.2 The wavelength of violet light is about 400 nm (1 nanometer = 10^{-9} m). What are the frequency and period of the light waves?

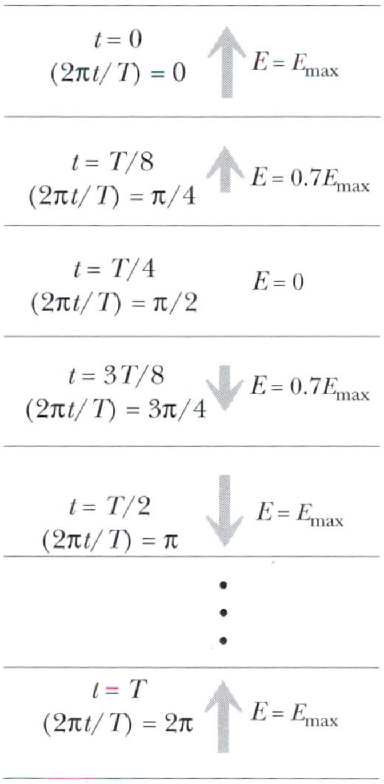

Variation in time

If you stand at one particular location as a sinusoidal or cosinusoidal electromagnetic wave goes by, you observe the electric field varying in time like $E\cos(2\pi t/T)$. The time-varying quantity $(2\pi t/T)$ is called the *phase* of the cosine function. The phase $(2\pi t/T)$, increases by 2π radians (360°) whenever t increases by an amount T. Figure 24.2 shows as a series of snapshots what you would observe at various times, at one location in space, while Figure 24.3 shows the same observations as a graph of E_y *vs.* t.

Of course there is also a cosinusoidally varying magnetic field as well, at right angles to the electric field. We are concentrating on the electric field in electromagnetic radiation because it has a much bigger effect on matter than does the magnetic field.

Variation in space

On the other hand, if you take a snapshot at one particular time, you see a sinusoidal or cosinusoidal pattern of electric field varying in space as shown in Figure 24.4.

This pattern in space can be expressed as $E\cos(2\pi x/\lambda)$, since the total phase of the cosine, $(2\pi x/\lambda)$, increases by 2π radians (360°) whenever x increases by an amount λ. Notice that if we replace x by t, and λ by T, this graph could just as easily represent the time variation of electric field at a particular fixed location in space.

$t = 0$ $(2\pi t/T) = 0$	$E = E_{max}$
$t = T/8$ $(2\pi t/T) = \pi/4$	$E = 0.7E_{max}$
$t = T/4$ $(2\pi t/T) = \pi/2$	$E = 0$
$t = 3T/8$ $(2\pi t/T) = 3\pi/4$	$E = 0.7E_{max}$
$t = T/2$ $(2\pi t/T) = \pi$	$E = E_{max}$
$\vdots$	
$t = T$ $(2\pi t/T) = 2\pi$	$E = E_{max}$

Figure 24.2 Snapshots of the electric field at one particular location, observed at successive times.

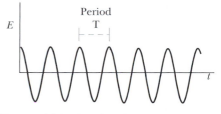

Figure 24.3 Graph of E_y *vs.* time as observed at one particular location.

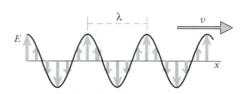

Figure 24.4 Space-varying electric field at one particular time.

Time and space

A cosine wave that moves to the right can be described in terms of x and t as follows:

$$E\cos\left(2\pi\frac{t}{T} - 2\pi\frac{x}{\lambda}\right)$$

This formula captures both the time and space aspects, because as time t increases, you need a larger x (motion to the right) to get the same total phase $(2\pi t/T - 2\pi x/\lambda)$ corresponding to a particular crest of the wave.

? In what direction is a wave traveling that is represented by the formula $E\cos(2\pi t/T + 2\pi x/\lambda)$? (Note the "+" sign.)

In this case a larger t requires a smaller x in order to get the same total phase, so this formula represents a wave moving to the left.

Phase shift

Depending on how we define $t = 0$ or $x = 0$, a wave may not be described exactly in the form $E\cos(2\pi t/T - 2\pi x/\lambda)$ because it may not start at its maximum value. We can add a constant phase "shift" ϕ and write $E\cos(2\pi t/T - 2\pi x/\lambda + \phi)$ to account for this. The phase shift ϕ can have any value between 0 and 2π (360°). When the phase shift between two waves is 0, we say the waves are *in phase*. When the phase shift is not zero, we describe the waves as *out of phase*.

Ex. 24.3 Consider two examples of cosine waves in time, at the location $x = 0$ (Figure 24.5). Study the two curves, and make sure you understand the relationship between them.

Label an x axis in units of half-periods ($t = T/2$, $2T/2$, $3T/2$, etc.). Then sketch a graph of $E\cos(2\pi t/T + \pi)$ to make sure you understand what a phase shift means. Your curve can be considered either as a phase shift by π radians, or alternatively as the negative of the original curve. Both points of view are useful.

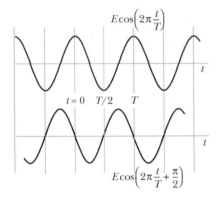

Figure 24.5 A phase shift affects the time at which the maximum occurs. The bottom curve differs from the top curve by a phase shift of $\pi/2$.

Angular frequency

Most of the time in our study of interference we only need the time variation of a wave at a particular location, in which case a cosine wave can be represented by any of these equivalent forms:

$E\cos(2\pi t/T + \phi)$　basic form
$E\cos(2\pi ft + \phi)$　since $f = 1/T$
$E\cos(\omega t + \phi)$　where the "angular frequency" ω is defined as $\omega = 2\pi f = 2\pi/T$

The angular frequency ω (familiar from our study of orbits and vibrations in classical mechanics) is measured in radians per second (2π radians per T seconds). Because of its compactness, we will usually use $E\cos(\omega t + \phi)$.

Amplitude

In an electromagnetic wave the maximum absolute value of the electric field is called the "amplitude" (Figure 24.6). The amplitude is a positive number. In the formula $E\cos(\omega t + \phi)$, E is the amplitude, and the component of electric field along an aligned axis varies between the extremes of $+E$ and $-E$. Note that the amplitude E is a parameter; the time variation is in the cosine function. Note also that the amplitude is half the peak to trough distance.

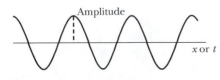

Figure 24.6 The amplitude is the maximum absolute value of the electric field in an electromagnetic wave.

Intensity

The frequency of electromagnetic waves in the visible range is so high that we don't perceive individual maxima and minima. Rather we perceive a kind of time average. This time average is called the "intensity," and it is what we can most easily perceive and measure for many kinds of waves.

In Chapter 23 we saw that the energy flow carried by electromagnetic radiation is proportional to the square of the electric field. At a particular instant, the electric field is $E\cos(\omega t)$ and the instantaneous energy flow is proportional to $[E\cos(\omega t)]^2$. Therefore the time-averaged energy flow is proportional to the square of the amplitude (E^2).

We call the average energy flow the intensity I, measured in watts per square meter (W/m^2; see Figure 24.7). Although the electric field at a particular location varies sinusoidally, and the field is even zero twice during every cycle, in many cases of interest the oscillation is so fast that we just perceive an average energy flow. For example, one cycle of visible light takes only about 10^{-15} seconds (see Exercise 24.2 on page 875), and your eye cannot detect the sinusoidal variation of such high-frequency oscillations.

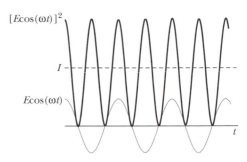

Figure 24.7 The time-averaged energy flow is called the intensity I.

INTENSITY AND AMPLITUDE

Intensity I (watts/m^2) is proportional to (amplitude)^2

$I \propto E^2$ for electromagnetic radiation

Intensity is proportional to the square of the amplitude not only for sinusoidal electromagnetic radiation, but also for other kinds of sinusoidal waves, such as water waves and sound waves. In the case of visible light, the intensity is of course also called the "brightness." For sound waves, intensity corresponds to loudness. It is the intensity I, the time-averaged energy flow, that we typically measure.

Ex. 24.4 If the peak electric field in a sinusoidal electromagnetic wave is tripled, by what factor does the intensity change?

24.1.2 Interference of two electromagnetic waves

According to the superposition principle, the net electric field at a location in space is just the vector sum of the electric fields contributed by *all* sources of electric field. This principle is still valid even if the electric fields change with time, so we can apply the superposition principle to electromagnetic radiation made by several different sources.

In Chapter 14 we saw that a radio antenna driven by sinusoidally alternating voltage (AC) produces propagating electric and magnetic fields whose magnitudes also vary sinusoidally. We begin our study of multiple sources of radiation by considering the simple case of just two such sources emitting sinusoidal radiation. Many situations involve very large numbers of sources, such as when light hits a mirror, but a careful analysis of the case of just two sources brings out the main aspects of the phenomena quite clearly.

Interference pattern

When a laser illuminates two narrow vertical slits in a metal sheet, a pattern of multiple stripes of light can be observed on a distant screen (Figure 24.8). Furthermore, the brightness of the stripes decreases as the distance from the center of the pattern increases (this change in brightness is not shown in Figure 24.8).

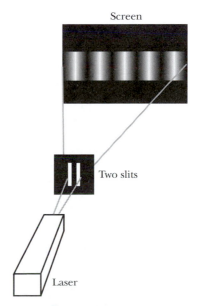

Figure 24.8 Laser light illuminates two narrow slits and makes an interference pattern on a screen.

? Does a particle model of light predict this pattern?

No. A simple particle model of light predicts that we will see only two bright spots, one in front of each slit. However, a wave model of electromagnetic radiation allows us to explain both the number of bright stripes, and their relative brightness. Furthermore, it allows us to make quantitative predictions of the locations of the bright and dark spots on the screen, simply by applying the superposition principle.

Coherent monochromatic radiation

Lasers emit light with rather special properties. The electromagnetic radiation in a laser beam is *coherent* (the phase of the oscillations is the same) and *monochromatic* (all radiation emitted has the same wavelength). Because a laser produces coherent, monochromatic light, it is a good source to use for experiments with light. In contrast, light from an ordinary incandescent light bulb is a mixture of radiation of many wavelengths, and is not coherent (electric fields not oscillating in phase).

We can think of the two slits as two separate sources of electromagnetic radiation, producing two electromagnetic waves which are coherent and monochromatic. A derivation at the end of this chapter shows that for the region between the slits and the screen we can treat the two-slit interference phenomenon as though the two slits were in-phase sources of light.

Constructive interference

Consider two parallel radio antennas placed near each other, both driven by the same transmitter and both emitting electric-field cosine waves of the form $E\cos(\omega t)$, with the same amplitude E and the same angular frequency ω (where $\omega = 2\pi f = 2\pi/T$). For clarity, the accompanying magnetic field is not shown in Figure 24.9, nor do we show the $1/r$ fall-off of the radiative electric field. Since the emitted electromagnetic waves have only a single frequency, they are monochromatic. Because the phase of the waves emitted by the two sources is the same, they are also coherent.

At any location in space the net electric field is of course the superposition (vector sum) of the two radiative electric fields generated by the two antennas. At the location in Figure 24.9 where we show two waves crossing, the two electric fields add up "in phase" with each other:

$$E\cos(\omega t) + E\cos(\omega t) = 2E\cos(\omega t)$$

The maximum electric field (or amplitude) at this particular location is twice as large as the electric field due to a single antenna. We say that at this location the two waves "interfere constructively."

Figure 24.9 The electric field radiated by two radio transmitters. At the particular location shown, the two electric fields are in phase and the net field is twice the field of one transmitter.

Redistribution of energy flow

We have just seen that at a location where two cosine waves add up "in phase," we find an amplitude that is twice as big as the amplitude due to a single antenna. (The electric field at this location still oscillates sinusoidally, with the same frequency as that of the two waves.)

? If the amplitude is twice as big, how much bigger is the intensity (the time-averaged energy flow)?

The intensity at that location (or the brightness in the case of visible light) is four times as large as it would be from one antenna.

Evidently this provides a way to make more intense radiation at that location, but something is peculiar. The two antennas radiate twice as much energy as one antenna, so how can the intensity at this location be four times as large? Evidently there must be some other place where the intensity is smaller than usual.

Destructive interference

Consider a different location where the two cosine waves are "out of phase" by 180°. At this location, one of the waves had to travel $\lambda/2$ farther than the other one did. Since they started in phase, there is now a phase difference of 180° (π radians)

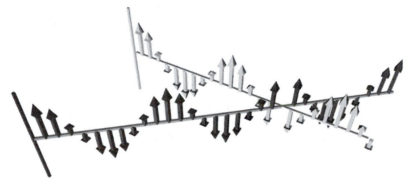

Figure 24.10 At a different location, the two electric fields are 180° out of phase and give a net field of zero at all times.

? What is the amplitude (maximum electric field) at the location in Figure 24.10 where we show two rays crossing?

Evidently the electric field sums to zero at all times at this location, because the two waves are 180° out of phase with each other. We say that these two waves "interfere destructively" at this location, and the general phenomenon is called "interference." This word has taken on a special meaning in physics, where "interference" refers in general to the superposition of waves from multiple sources. We say that at a location where the two waves add to each other, they "interfere constructively," which is rather peculiar language but commonly used in technical discussions.

There is a complex pattern of radiation in space. The intensity of the radiation is larger at some locations than one might expect, and smaller or even zero at other locations. If you look at the intensity in every direction, it can be shown that the total energy radiated away is twice the energy radiated away by one antenna, as you would expect. It's just that the radiated energy is redistributed in a special way.

We can summarize the most important aspects of interference:

INTERFERENCE MAXIMA AND MINIMA

Two cosines can add up to give a *maximum* with twice the amplitude and four times the intensity.

This occurs when the path difference is 0, λ, 2λ, 3λ, etc.

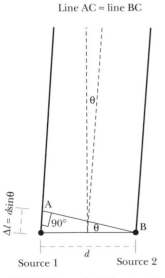

C is very far from the sources

Line AC ≈ line BC

Figure 24.11 The path difference between two sources radiating in phase. Near the sources the paths are nearly parallel.

Two cosines can subtract to give a *minimum* with zero amplitude and zero intensity.

This occurs when the path difference is $\lambda/2$, $3\lambda/2$, $5\lambda/2$, etc.

? Can you observe completely destructive interference (that is, zero intensity) at some location with two sources that have different angular frequencies ω_1 and ω_2? Why or why not?

There is no location at which the two sources cancel each other all the time, because even if the waves canceled each other at some moment, after one cycle of the first wave, the second wave wouldn't have returned to the same value, so they wouldn't cancel at this later time. Not only must the frequencies (or angular frequencies) of the sources be exactly the same, but the sources must also have a constant relative phase (must be coherent).

24.1.3   Predicting the radiation pattern for two sources

We should now be able to predict where two sources will give high-intensity maxima. All we have to do is figure out where the path differences are multiples of one wavelength (if the sources are in phase with each other). Equivalently, we can figure out where the phase difference is a multiple of $360°$ (2π radians). Similarly, we can find a location of zero intensity by figuring out where there is a path difference of half a wavelength or a phase difference of $180°$ (π radians).

Consider two sources that radiate in phase, and consider interference at a location C in a direction θ measured from the perpendicular bisector between the two sources (Figure 24.11). We need to be able to calculate the path difference Δl in terms of the angle θ, in order to determine whether there could be a maximum or a minimum in the intensity at location C.

Approximation for distant observation locations

If the observation location C is far away from the sources (as it usually is in most applications of the theory), we can make the approximation that in the right triangle ABC drawn on the diagram the base AC is almost the same length as the hypotenuse BC. This approximation is valid if the observation location C is far enough from the sources that $\angle ACB$ is a very small angle. In this case, the path from source 1 to C is to an excellent approximation $\Delta l = d\sin\theta$ longer than the path from source 2 (path BC).

PATH DIFFERENCE

$$\Delta l = d\sin\theta$$

This is geometry, not physics. The physics lies in the fact that if the path difference Δl is an integer number of wavelengths, there will be a maximum at the angle θ determined by $\Delta l = d\sin\theta$. If Δl is $\lambda/2$, $3\lambda/2$, $5\lambda 2$, etc., there will be a minimum (zero intensity) at the angle θ determined by $\Delta l = d\sin\theta$.

Note the importance of the spacing d between the sources. In particular, if $d < \lambda$ you can never get completely constructive interference except at $\theta = 0°$ or $\theta = 180°$, because $d\sin\theta$ is always less than λ. If $d < \lambda/2$ you can never get zero intensity (completely destructive interference).

Intensity *vs.* angle

Figure 24.12 is a plot of the intensity pattern *vs.* angle for two sources that are driven in phase with each other and located a distance d apart, where in this particular case $d = 4.5\lambda$. In some directions (Max_0, Max_1, etc.), the difference in path length $\Delta l = d\sin\theta$ for the two cosine waves is zero or an integer number of wavelengths, the amplitudes add up in phase, and the intensity I is four times as large as the intensity of a single source. In some other directions (Min_1, Min_2, etc.), the difference in path length $\Delta l = d\sin\theta$ for the two cosine waves is half a wavelength (or $3\lambda/2$, $5\lambda/2$, etc.), the waves are 180° out of phase with each other, and the intensity I is zero.

? Why is the intensity a maximum at $\theta = 0°$ (location Max_0) and $\theta = 180°$?

At 0° and 180° the path lengths are the same to both sources, so the waves are in phase, giving a maximum intensity.

? In Figure 24.12, $d = 4.5\lambda$. Why is the intensity equal to zero at $\theta = 90°$ (location Min_5) and at $\theta = -90°$?

At 90° and −90° the difference in path lengths to the two sources is 4.5λ, so the two waves are 0.5λ out of phase, which is 180° difference in phase. This gives completely destructive interference.

Starting from $\theta = 0$, as θ increases from zero, the intensity falls rapidly to zero at location Min_1. The exact variation of the intensity with angle can be calculated by adding up two cosine waves with a varying phase difference, and this has been done in the computer program used to create the diagram. However, for many purposes it is sufficient just to figure out where the maxima and minima are. A rough qualitative sketch of the intensity pattern between a maximum and a minimum will be adequate for our purposes in this chapter.

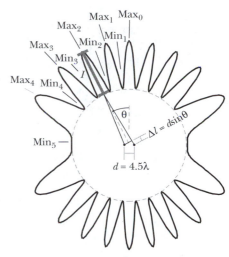

Figure 24.12 The intensity pattern for two sources. The bar marks the intensity I at the location of a maximum (Max_2).

Ex. 24.5 The two sources in Figure 24.12 are separated by a distance $d = 4.5\lambda$. Calculate the angle θ where the first minimum is encountered (location Min_1 in Figure 24.12). (Hint: What must the path difference $d\sin\theta$ be in this direction?)

Ex. 24.6 As θ increases beyond the location of the first minimum, the intensity rises rapidly to a second maximum, then falls to a minimum, rises again, etc. Calculate the angle θ where the fourth maximum is encountered (location Max_4 in Figure 24.12). (Hint: What must $d\sin\theta$ be in this direction?)

Ex. 24.7 Figure 24.13 is an intensity plot for a different situation, where the sources are much closer together, with $d = \lambda/3.5$. We see that the intensity distribution isn't very dramatic when $d < \lambda/2$. Why is the intensity non-zero everywhere? Why isn't the intensity zero somewhere?

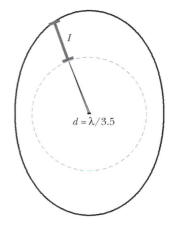

Figure 24.13 Intensity plot for two sources very close together.

Summary: Two source interference

Here is a summary of the major points so far:

LOCATIONS OF MAXIMA AND MINIMA

Maximum intensity ($4I_0$) where $\Delta l = d\sin\theta$ is 0, λ, 2λ 3λ, etc.
Minimum intensity (zero) where $\Delta l = d\sin\theta$ is $\lambda/2$, $3\lambda/2$, $5\lambda/2$, etc.
No zero-intensity minimum is possible if $d < \lambda/2$.

If the sources are coherent but *not* in phase, you need to take that phase difference into consideration in addition to the phase difference due to Δl.

Two-slit interference

Earlier we mentioned the phenomenon of two-slit interference, observable with a laser light source. We are now in a position to relate the slit spacing, angles at which we find maxima and minima, and the wavelength of the laser light.

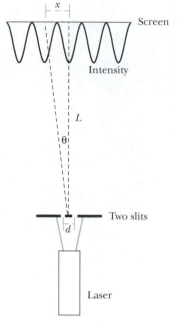

Figure 24.14 Top view of a laser, two slits, and an interference pattern on a screen.

In a top view (Figure 24.14), we show the laser, the two slits, and the intensity pattern on the screen, which can be measured quantitatively with a light meter. Suppose the distance between the slits is $d = 0.5$ mm, the distance from the slits to the screen is $L = 2$ m, and the distance on the screen from the center maximum to the next maximum is measured to be $x = 2.4$ mm.

? Calculate the wavelength of this laser light.

At the second maximum, $d\sin\theta = \lambda$, so $\sin\theta = \lambda/d$. For small angles, $\sin\theta \approx \tan\theta \approx \theta$, so $\lambda/d \approx x/L$. Therefore we have

$$\lambda = \frac{xd}{L} = \frac{(2.4\times10^{-3}\text{ m})(0.5\times10^{-3}\text{ m})}{(2\text{ m})} = 6\times10^{-7}\text{ m} = 600\text{ nm}$$

This wavelength is in the red portion of the visible spectrum. It was experiments of a similar kind that initially established what we now know about the wavelengths of visible light.

Intensity variation of interference maxima

We can predict the locations of interference maxima and minima. However, we have not yet found an explanation for the possible variation in brightness of the maxima. To explain why in some cases the central maximum may be much brighter than the others, we will need to consider diffraction from a single slit, which is discussed in Section 24.4.2.

Wavelength, path length, and distance between sources

The interference effects we have described show up in situations where the difference in path length of two rays is larger than but comparable to the wavelength of the electromagnetic radiation. In the case of radio waves we considered in the previous section, d, the spacing between the antennas, was several times the wavelength, resulting in path differences of up to 4λ. In the case of two slit interference with a laser, the slit spacing (0.5 mm, or 5×10^{-4} m) was nearly 1000 times the wavelength of the light (about 500 nm, or 5×10^{-7} m). However, because the surface on which we saw the interference pattern was quite far from the slits, the difference in path length for the two waves was still only a few times the wavelength of the light.

Interferometry

A modern application of this interference phenomenon is to measure very small distances. A laser beam is split into two beams, and mirrors guide one of the beams to an object of interest, from which the beam is reflected to a screen where there is interference with the other laser beam. If the object moves even a fraction of one wavelength of visible light, there is an easily observable shift in the interference pattern on the screen. This use of interference phenomena to measure small changes in distance is called "interferometry."

24.1.4 Multi-source interference: X-ray diffraction

Later in this chapter we will discuss Compton scattering: an experiment in which light interacts with free electrons. The electromagnetic radiation used in this experiment is in the x-ray region of the spectrum. In order to do this experiment, it is necessary to produce a monochromatic, coherent beam of x-ray radiation, and to find a way to measure precisely the wavelength of this radiation. X-ray diffraction, a technique which can be used to study the structure of crystals or of individual molecules in a crystal, can also be used to prepare a monochromatic x-ray beam and to measure precisely the wavelength of such a beam. Because of its importance, we will discuss x-ray diffraction in some detail. "Diffraction" is the term used to describe interference effects involving a large number of sources; "interference" describes effects involving only a few sources.

In a crystal, atoms or molecules are arranged in a regular three-dimensional array, called a *lattice,* as shown in Figure 24.15. If electromagnetic radiation of a single wavelength hits a crystal, electrons in all the atoms are accelerated and re-radiate. We can expect to find high intensity re-radiation only in those directions where the path difference for adjacent atoms is λ, 2λ, 3λ, etc. Dealing with a three-dimensional array of sources can be geometrically complex, but we will limit this discussion to some simple cases that illustrate the basic principle.

X-ray wavelength: $\lambda \approx d$

The spacing between atoms in a crystalline solid is of the order of magnitude of atomic sizes, or about 10^{-10} meter. We have seen that interference effects require that the wavelength λ must be comparable to the spacing d, or about 10^{-10} meter. Such electromagnetic radiation is called "x-rays." X-rays have very short wavelengths compared to visible light: violet light has a wavelength of 400 nm, or 4000×10^{-10} meter, about 4000 times the wavelength of x-rays.

A common way to produce x-rays is to shoot a beam of very high-energy electrons into a block of metal, such as copper. These high-energy electrons sometimes knock an electron out of an inner electron shell of the copper atom. Very shortly thereafter an electron from an outer electron shell drops into the hole, and the associated large drop in energy of the atom is accompanied by the emission of light of high energy and short wavelength. This is the production mechanism used in medical x-ray equipment.

Conditions for constructive interference

To see the basic idea behind x-ray diffraction, we'll consider a beam of single-wavelength x-rays that hit a simple crystal, one in which the atoms are arranged on a three-dimensional rectangular grid. If the x-rays come in at an angle θ to the surface, the first layer of atoms will re-radiate with constructive interference at an angle θ that looks like simple reflection, because in that direction there is a zero path difference between adjacent sources (atoms in the first layer). Along path 1 in Figure 24.16, there is an extra length s in the outgoing ray, and along path 2, there is the same extra length s in the incoming ray. At other angles the path difference would not be zero, and unless the path difference is an integer number of wavelengths there would not be constructive interference. We will consider just the simple case where we observe a maximum at the simple reflection angle.

The re-radiation from other atomic layers, below the surface layer, need not be in phase with the top layer unless special conditions hold. In Figure 24.17, if the center-to-center distance between layers is d, there is a difference $2d\sin\theta$ in path lengths for electromagnetic radiation accelerating electrons in an atom in the top layer (which re-radiates), and for accelerating electrons in an atom in the next layer (which also re-radiates).

If the difference in path length, $2d\sin\theta$, is not equal to the wavelength λ (or some multiple of λ), the two topmost layers of atoms will not re-radiate in phase at the angle θ. And at some other angle θ, the atoms in the top layer won't radiate in phase with each other. So if you orient the surface of the crystal at an angle θ to the x-ray beam, you may or may not get an intense beam of x-rays re-radiated ("reflected") from the crystal, at the "reflection" angle. You *will* get an intense "reflection" only if the following condition is true:

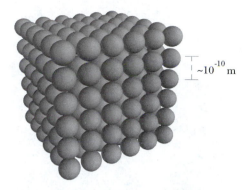

Figure 24.15 Part of a simple cubic crystal lattice. Other more complex geometric arrangements of atoms are also possible.

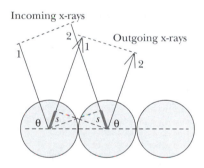

Figure 24.16 Re-radiation from the top layer of atoms in a crystalline solid. Note that θ is measured from the surface, not from the normal to the surface.

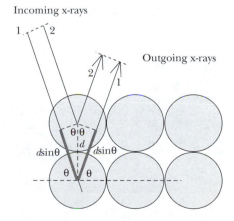

Figure 24.17 Re-radiation from the top two layers involves an additional path difference.

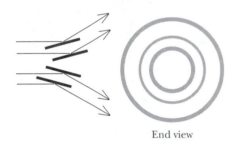

Figure 24.18 X-ray diffraction pattern showing interference maxima from a single crystal of tungsten.

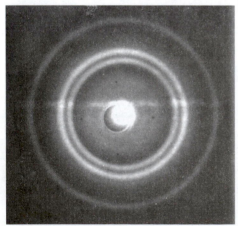

End view

Figure 24.19 X-ray diffraction with a powder of crystals at random orientations.

CONDITION FOR CONSTRUCTIVE X-RAY INTERFERENCE

$2d\sin\theta = n\lambda$, where n is an integer

Note that if the top two layers re-radiate in phase, the re-radiation from all the other layers will also be in phase.

This effect, called "x-ray diffraction" because it involves a huge number of sources, can be used to deduce the spacing d between layers in a crystal. Assume that you have a source of single-wavelength x-rays, whose wavelength λ is known. You direct a beam of these x-rays at an angle θ onto the crystal, and you look for an intense beam at the reflection angle. If there is no intense beam there, you change the angle a bit and look again. (Typically one turns the crystal rather than the large x-ray beam apparatus.) If you find an intense reflection at a particular angle θ, you know that the spacing between layers of atoms is $d = n\lambda/[2\sin\theta]$. You may need to do some additional checking to determine the value of n, which could be 1, 2, 3, etc. Figure 24.18 shows a pattern of constructive interference from a crystal of tungsten.

The x-rays produced by some kinds of x-ray sources have a continuous spectrum of wavelengths. How can one make a beam of single-wavelength x-rays from such sources?

? Suppose you have a crystal with known spacing d between atomic layers. Explain briefly how to use this crystal with a multi-wavelength beam of x-rays to produce a single-wavelength beam, which can be used in experiments on crystals whose structure is not known.

For the given atomic-layer spacing d, and desired wavelength λ, we can get a large intensity for just that wavelength at the angle θ that satisfies $2d\sin\theta = \lambda$. So direct the x-ray beam at the crystal at this angle, and we'll get a strong "reflection" for the desired wavelength, while other wavelengths won't give a strong beam in that direction.

X-ray diffraction is an extremely powerful tool that has been used to determine the structure of a huge variety of crystals. It can also be used with "polycrystalline" materials, materials in the form of a powder of tiny crystals oriented in random directions. In such a powder containing a very large number of tiny crystals, there will be some crystals that happen to be oriented at angles to the x-ray beam that satisfy the condition $2d\sin\theta = n\lambda$, in which case there is intense re-radiation in a ring surrounding the incoming x-ray beam, at the appropriate angle (Figure 24.19). X-ray film exposed to this re-radiation shows rings whose diameters tell a great deal about the crystal structure, even if one doesn't have a single large crystal available. A diffraction pattern from powdered lithium fluoride is shown in Figure 24.20.

A layer of atoms re-radiating in phase need not be parallel to the surface of a crystal. A layer can consist of any geometrical plane drawn through the crystal in such a way that many atoms lie in that plane (so as to give a large re-radiation intensity). In addition, if the crystal is made up of more than one kind of atom (as in NaCl crystals, for example), accurate measurements of intensity can provide information about the arrangement of the different atoms, because different atoms re-radiate different amounts of energy.

Ex. 24.8 It is known that the spacing between neighboring planes of atoms in a particular crystal is 2×10^{-10} m. A monochromatic x-ray beam of wavelength 0.96×10^{-10} m is directed at the crystal. At what angle might one expect to find an interference maximum?

Figure 24.20 X-ray diffraction pattern from polycrystalline lithium fluoride.

24.1.5 Reflection of visible light: λ >> d

Why do we see a reflection of visible light from a smooth surface at only one angle? When visible light instead of x-rays hits a crystal, it accelerates electrons in the material, and there is indeed re-radiation in all directions. However, the x-ray diffraction condition $2d\sin\theta = n\lambda$ for adjacent layers is not relevant for visible light, because the distance between atomic layers is extremely small compared to the wavelength of visible light ($d \approx 10^{-10}$ m, $\lambda \approx 5000 \times 10^{-10}$ m). The difference in path length for re-radiation from adjacent layers is negligible, and adjacent layers are nearly in phase.

Because the interatomic spacing d is much smaller than λ for visible light, there will be constructive interference between two waves only if the difference in path length is zero, so we will see only one high-intensity maximum. As can be seen from Figure 24.21, for interactions with atoms in the surface layer, $\Delta l = 0$ only if the reflection angle equals the incident angle. At any other angle, the difference in path length is nonzero (Figure 24.22), and there is destructive interference. The interaction of the incident radiation with deeper layers of atoms is essentially the same, since the layers are very close together compared to the wavelength of the radiation.

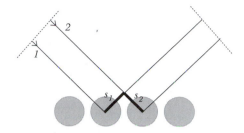

Figure 24.21 Only when incident angle = reflection angle is $s_1 = s_2$, so $\Delta l = 0$.

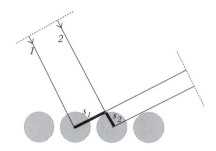

Figure 24.22 At any other angle, $s_1 \neq s_2$, so Δl is nonzero.

Metal surfaces

Reflection of visible light off a metal surface is quite different from the reflection off transparent insulating materials such as water or glass. The mobile electrons in a metal are very easy to accelerate. The mobile electrons near the surface gain a lot of energy, so we conclude that little energy is available to transfer to electrons far from the surface. It must be the case that the intensity of the radiation (and therefore the intensity of the electric field) inside the metal is small. How can this happen? Presumably there is so much re-radiation by accelerated electrons near the surface of the metal that there is very little net field in the interior of the metal. There are no significant interference effects between layers at different depths from the surface. Of course there are the usual interference effects involving electrons in different locations on the surface, so that there is strong intensity only at the "reflection" angle.

Coherence length

There is an effect that limits interference effects in thick films. Most sources of visible light produce sinusoidal waves with a fairly short total length L, which corresponds to the short length of time $\Delta t = L/c$ during which an atom was emitting the light (c is the speed of light). Even if the total length L corresponds to a large number of wavelengths, it might be less than a millimeter. If the total length L is shorter than twice the thickness of the film, by the time the light re-radiated by the last atomic layer emerges from the film, the first atomic layer has stopped emitting, and there is nothing to interfere with. The total length L of the sinusoidal emission is called the "coherence length." One of the important properties of lasers is that they emit light with a very long coherence length. This is particularly important in making holograms, which are based on interference effects.

Thin-film interference

Despite what we just said about the condition $2d\sin\theta = n\lambda$ for adjacent layers in connection with reflection of visible light, there can be interesting interference effects when visible light hits a thin film of transparent material only a few wavelengths thick, such as a soap bubble, or an oil slick floating on a puddle of water. These thin-film interference effects depend critically on the exact thickness of the thin film.

Assume for simplicity that single-frequency incoming light hits the surface of a thin film nearly head-on (rays nearly perpendicular to the surface, as in Figure 24.23). Suppose the thin film has a thickness of only half a wavelength ($\lambda/2$). Though this is a very thin film, it nevertheless contains a large number N of atomic layers: N is about (250×10^{-9} m/10^{-10} m) = 2500 atomic layers.

In Figure 24.23, consider pairing the first layer of atoms with atomic layer number ($N/2 + 1 = 1251$), pairing the second atomic layer with atomic layer number ($N/2 + 2 = 1252$), etc. Atomic layer number ($N/2 = 1250$) is paired with atomic layer number $N = 2500$ (the last layer). The path difference for each of these pairs is $\lambda/2$, so each pair produces zero intensity, and the total re-radiated intensity is zero for a thin film whose thickness is $\lambda/2$.

By similar arguments you can show that there will be zero re-radiation for film thicknesses of $\lambda/2$, $2\lambda/2$, $3\lambda/2$, $4\lambda/2$, etc.—any integer multiple of $\lambda/2$. For film thicknesses half-way between these thicknesses ($\lambda/4$, $3\lambda/4$, $5\lambda/4$, etc.), it can be shown that there is maximum re-radiation, although a formal proof is beyond the scope of this discussion.

? A soap film or oil slick when illuminated by ordinary white light shows a rainbow of colors. Briefly explain why this is.

For a given film thickness, some wavelength might be the right length to give fully constructive interference and be bright, but other wavelengths wouldn't be and would be dim. And if the film thickness varies, or you look at a different angle, the wavelength that gives fully constructive interference would be different.

In a soap bubble that lasts a long time, more and more of the liquid flows to the bottom of the bubble. The top part gets very thin, thinner than a quarter-wavelength of light even for violet light, which has the shortest visible wavelength. This very thin film seems to reradiate almost no light at all (appears "black"), because there are only a small number of atomic layers compared to the several thousand that contribute when the thickness is a quarter-wavelength.

A closely related thin-film interference phenomenon is responsible for the brilliant iridescent colors seen in some butterfly wings and bird feathers, which contain structures with spacing comparable to a wavelength of light. There is constructive interference for some colors and destructive interference for others.

Index of refraction

The relevant wavelength in thin-film interference is the wavelength inside the material. This turns out to be shorter than the wavelength in air! Inside a dense transparent material such as glass or water, individual atoms are significantly affected not only by the incoming electromagnetic radiation but also by the electric fields re-radiated by the other atoms. When you add up all the electric fields, you find that the pattern of the net field has a crest-to-crest distance that is shortened. The factor by which the wavelength is decreased is called the "index of refraction" and is around 1.5 for many kinds of glass; it is 1.33 for water.

In a material with index of refraction n, the wavelength $\lambda' = \lambda/n$, where λ is the wavelength of the radiation in vacuum. The frequency of the wave is not affected, and since $v = f\lambda$, the *apparent* speed of light inside a solid is noticeably slower than 3×10^8 m/s.

This complication does not affect our analysis of x-ray diffraction because the re-radiation by an atom exposed to x-rays is very small compared to the re-radiation by an atom exposed to visible light. In the case of x-rays, we can consider the electric field inside the material to be due almost entirely to

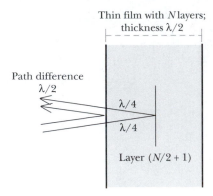

Figure 24.23 Visible light encounters a very thin film of matter. Consider pairs of layers such as layer 1 and layer ($N/2 + 1$), etc.

the incoming radiation, with a negligible contribution from the tiny re-radiation from the other atoms.

Ex. 24.9 A film of oil 200 nm thick floats on water. The index of refraction is 1.6. For what wavelength of light will there be complete destructive interference?

24.2 The wave model *vs.* the particle model of light

We have established that a variety of phenomena can be explained by the wave properties of electromagnetic radiation. Yet, in Chapters 6 and 8 we explained a variety of phenomena by a particle (photon) model of light. As a first step in comparing the wave and particle models of light, let's consider the predictions of the two models of light in two different experimental situations: the photoelectric effect and the collision of a beam of light with free electrons (Compton scattering).

24.2.1 The photoelectric effect

The mobile electrons in a metal, although quite free to move around inside the metal, are nevertheless bound to the metal as a whole: they don't drip out! It takes some work to move an electron from just inside to just outside the surface of the metal. At that point (electron just outside surface), the metal would now be positively charged and would attract the electron, so you would need to do additional work to move the electron far away. The total amount of work required to remove an electron from a metal is often called the "work function" but could just as well have been called the "binding energy." We will refer to it with the symbol W. Typical values for common metals are a few electron volts (eV).

Prediction: the wave model

One way to supply enough energy to remove an electron from a metal is to shine light on the metal (Figure 24.24).

? According to the wave model of light, how could a beam of light have sufficient energy W to overcome the binding energy, and remove an electron from a piece of metal?

Since the energy per square meter (intensity) of an electromagnetic wave is proportional to E^2, we would need a high-intensity electromagnetic wave, in which the amplitude of the electric field is very large, to supply enough energy. Alternatively, if the energy from the electromagnetic wave were not quickly dissipated into thermal energy, we might be able to shine a low-intensity beam on the metal for a long time to supply enough energy.

? According to the wave model of light, would we need to use electromagnetic radiation of any particular wavelength, or would any wavelength be adequate?

According to the wave model, the energy density of an electromagnetic wave depends on E^2, the square of its amplitude, not on its wavelength. All we need is an intense beam of light—it could have any wavelength (or color). We might therefore expect a plot of the kinetic energy of an ejected electron vs. intensity of light hitting the metal to be linear, as in Figure 24.25.

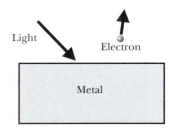

Figure 24.24 The photoelectric effect: light can eject an electron from a metal.

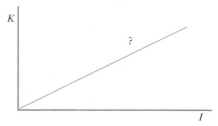

Figure 24.25 The wave model of light predicts that the kinetic energy of an ejected electron should be proportional to the intensity of light hitting the metal.

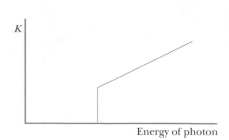

Figure 24.26 The particle model predicts that above a certain threshold, the kinetic energy of an ejected electron should be proportional to the energy of a incoming photon.

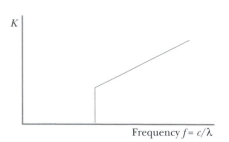

Figure 24.27 What is actually observed is that above a minimum frequency (below a maximum wavelength), the kinetic energy of ejected electrons is proportional to frequency (inversely proportional to wavelength) of incident radiation.

Figure 24.28 A photon can be described as a "wave packet": an electromagnetic wave extending over a very small region of space.

Prediction: the particle model

Recall from Chapter 6 that according to the particle model a beam of light is a collection of particles travelling at speed c, each with zero rest mass, but with a fixed energy and momentum.

? What does the particle model of light predict would be required to eject an electron from a metal?

The particle model suggests that the energy of an individual photon is what is important, since the absorption of a single photon of the appropriate energy would be adequate to free an electron from the metal. This model predicts that intensity of the light is not important; an extremely intense beam (many photons per square meter) will have no effect if the energy of the individual particles is too low (Figure 24.26).

What is observed?

In fact, what is observed experimentally is that the wavelength of the light that falls on the metal surface determines whether or not an electron will be liberated from the metal (Figure 24.27). If the wavelength of the light is too large (the frequency is too low), nothing happens, even if the beam is very intense. However, as the wavelength of the radiation is decreased (frequency increased), at one particular wavelength we begin to see electrons liberated from the metal. As the wavelength of the incident light is decreased even further, the kinetic energy of the electrons liberated from the metal increases.

What happens if we use a very low intensity beam? At low intensity, long-wavelength light does not eject electrons, no matter how long we wait. However, as predicted by the particle model, with a low intensity beam of short wavelength light, we sometimes see an electron ejected almost immediately.

For light of short enough wavelength, we do however observe that the rate at which electrons are ejected from the metal is proportional to the intensity of the light.

Unifying the two models

How do we reconcile our models with the experimental observations? First, it seems clear that this phenomenon is essentially a quantum one. A certain amount of energy must be delivered in one chunk, or packet, to the system, in order to raise the system to an energy state in which the electron is unbound. Thus, the particle model of light seems appropriate here.

The surprising thing, however, is that the wavelength of the incident light is important. Wavelength is certainly a property of a wave—how can this be reconciled with a particle model of light?

Evidently, we need to model light as something with both wavelike and particle-like properties. The wave nature of light explains the interference phenomena we have observed. The particle nature of light explains how a fixed amount of energy can be delivered to a system by a photon. Sometimes a photon is described as a "wave packet": an electromagnetic wave occupying a short region of space (Figure 24.28).

Energy and wavelength

From experimental observations of the photoelectric effect, it is found that the energy of a photon is proportional to the frequency and inversely proportional to the wavelength. Quantitatively, the relation is

$$E = hf = \frac{hc}{\lambda}$$

where $h = 6.6 \times 10^{-34}$ joule-second is Planck's constant, which you may remember from our study of quantized harmonic oscillators in Volume I.

Intensity and number of photons/second

Since the number of electrons per second ejected is proportional to the intensity of light of sufficiently small wavelength, apparently in the particle model the intensity of light is proportional to the number of photons per second striking a surface.

Applying the model

? If the work function (binding energy) $W = 3$ eV for a certain metal, what is the minimum energy a photon must have to eject an electron?

Evidently the photon must have an energy of 3 eV or more. If the photon has less energy, it can't knock an electron out of this metal.

? What is the wavelength of a photon with this much energy?

$$\lambda = \frac{hc}{E} = \frac{(6.6 \times 10^{-34} \text{ J·s})(3 \times 10^8 \text{ m/s})}{(3 \text{ eV})(1.6 \times 10^{-19} \text{ J/eV})} = 4.12 \times 10^{-7} \text{ m} = 412 \text{ nm}$$

This is near the violet end of the visible spectrum, which extends from about 400 nm (3.1 eV) to about 700 nm (1.8 eV). For a metal with $W = 3$ eV, light in most of the visible spectrum or in the infrared cannot eject an electron. UV (ultraviolet) light, with wavelength shorter than 400 nm (photon energy greater than 3.1 eV), can knock electrons out of this metal.

? What if you expose this metal to light whose photon energy is 4 eV (in the UV)? How much kinetic energy would you expect the ejected electron to have when it was far from the metal?

Since it takes 3 eV just to remove the electron, a 4 eV photon will eject the electron with 1 eV to spare, so the electron will have kinetic energy of 1 eV.

Experiments have verified this simple particle model for the photoelectric effect. The main points are these:

- There is a minimum photon energy W required to eject an electron from a metal; photons of lower energy (longer wavelength) do not eject electrons.
- Photons with excess energy eject electrons whose kinetic energy is equal to the photon energy minus W.

Historically, at the beginning of the 20th century the wave model of light was the accepted model. It was Einstein who first proposed the particle interpretation of the photoelectric effect, in 1905. (In the same year Einstein also published the theory of special relativity, and he also correctly analyzed Brownian motion, the random motion of small objects due to molecular collisions, which led to the first accurate determination of atomic sizes. A truly remarkable year!)

Ex. 24.10 What is the energy of a photon whose wavelength is 334 nm?

Ex. 24.11 If a particular metal surface is hit by a photon of energy 4.3 eV, an electron is ejected, and the kinetic energy of the electron is 0.9 eV. What is the work function W (binding energy) of this metal?

Ex. 24.12 If the work function of a metal is 3.9 eV, what wavelength of light would you have to shine on the metal surface in order to eject an electron with negligible kinetic energy?

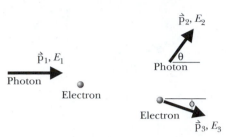

Figure 24.29 Compton scattering: a photon scatters off a stationary electron.

24.2.2 Compton scattering

Figure 24.29 is a diagram of a collision between a photon and a stationary electron. This process is called "Compton scattering" in honor of the physicist who showed experimentally that this process has both particle and wave aspects. In the particle model of the interaction, if the photon changes direction the electron recoils with some kinetic energy, and the photon necessarily has lost energy. But a loss of photon energy corresponds to a longer wavelength in the wave model. Compton showed experimentally that the combined wave-particle model did correctly predict the change of wavelength as a function of the scattering angle of the electron. This was one of the most compelling early pieces of evidence for the combined wave-particle nature of light.

In Chapter 8 you used the momentum principle and the energy principle to study particle collisions, and you treated photons as particles with zero rest mass. The general relationship between energy and momentum for a particle is $E^2 = (pc)^2 + (mc^2)^2$. For zero-mass particles such as photons (and probably neutrinos), this reduces to $E = pc$. We will use this in an analysis of Compton scattering.

Let's use the combined wave-particle model of light to predict the change in wavelength $\Delta\lambda$ for a photon which collides with a free electron. As indicated in Figure 24.29, let the energy of the incoming photon be E_1, the outgoing photon energy be E_2, and the energy of the recoil electron be E_3, where $E_3 = mc^2 + K$, K is the kinetic energy of the recoil electron, and m is the mass of the electron.

$$E_1 + mc^2 = E_2 + E_3 \qquad \text{energy principle}$$

$$E_3 = E_1 - E_2 + mc^2 \qquad \text{solve for recoil electron energy}$$

$$\vec{p}_1 = \vec{p}_2 + \vec{p}_3 \qquad \text{momentum principle}$$

$$\vec{p}_3 = \vec{p}_1 - \vec{p}_2 \qquad \text{solve for recoil electron momentum}$$

To get the square of the magnitude of a vector, we take the dot product of the vector with itself:

$$\vec{p}_3 \bullet \vec{p}_3 = (\vec{p}_1 - \vec{p}_2) \bullet (\vec{p}_1 - \vec{p}_2) \qquad \text{dot product}$$

$$p_3^2 = p_1^2 + p_2^2 - 2\vec{p}_1 \bullet \vec{p}_2 = p_1^2 + p_2^2 - 2p_1 p_2 \cos\theta \qquad \text{expand}$$

$$(p_3 c)^2 = (p_1 c)^2 + (p_2 c)^2 - 2(p_1 c)(p_2 c)\cos\theta \qquad \text{multiply by } c^2$$

$$E_3^2 - (mc^2)^2 = E_1^2 + E_2^2 - 2E_1 E_2 \cos\theta \qquad \text{substitute } (pc)^2 = E^2 - (mc^2)^2$$

$$(E_1 - E_2 + mc^2)^2 - (mc^2)^2 = E_1^2 + E_2^2 - 2E_1 E_2 \cos\theta \qquad \text{substitute } E_3$$

After expanding the squared parenthesis and simplifying, we obtain this:

$$mc^2(E_1 - E_2) = E_1 E_2 (1 - \cos\theta)$$

$$mc^2\left(\frac{1}{E_2} - \frac{1}{E_1}\right) = (1 - \cos\theta) \qquad \text{divide by } E_1 E_2$$

$$mc^2\left(\frac{\lambda_2}{hc} - \frac{\lambda_1}{hc}\right) = (1 - \cos\theta) \qquad \text{use wave-particle relation } E = \frac{hc}{\lambda}$$

$$\lambda_2 - \lambda_1 = \frac{h}{mc}(1 - \cos\theta) \qquad \text{Compton scattering prediction}$$

Here we have a quantitative prediction for the increase in wavelength of the scattered photon, as a function of the scattering angle θ. The quantity

$$\frac{h}{mc} = \frac{(6.6 \times 10^{-34} \text{ J·s})}{(9 \times 10^{-31} \text{ kg})(3 \times 10^{8} \text{ m/s})} = 2.4 \times 10^{-12} \text{ m}$$

is called the Compton wavelength. To make it feasible to measure the wavelength increase, one must use light whose wavelength is comparable to the Compton wavelength. Such light is in the x-ray region of the electromagnetic spectrum.

To study this kind of collision, it is necessary to have a beam of monochromatic (single wavelength) x-rays. However, most x-ray beams start out with a range of wavelengths.

? Given what you know about x-ray diffraction, can you think of a way to produce a single-wavelength beam?

If a beam of x-rays interacts with a crystal, re-radiated x-rays of a given wavelength will be particularly intense at an angle satisfying the x-ray diffraction condition. Such a single-wavelength beam could be used in the experiment.

? How might one measure the wavelength of the x-rays after they interact with an electron?

Again, by using x-ray diffraction from a single crystal. If we know the spacing of atoms in a particular crystal, measuring the angles of diffraction maxima will tell us the wavelength of the x-rays.

Compton did the experiment, shooting x-rays of known, single wavelength at a block of metal (which contains lots of nearly free electrons) and measuring the wavelength of the scattered x-rays at various scattering angles. He found that the wavelength did indeed increase by the amount predicted by the analysis given above.

To summarize, a particle view of light was used to predict the change in energy of the photon, and the wave-particle relation $E = hc/\lambda$ was used to convert the energy change to a change in wavelength. The experiment showed that the combination of the particle model and the wave model did indeed predict what actually happens in this process. Neither a pure wave model of the process nor a pure particle model would make the correct prediction.

Ex. 24.13 In a Compton scattering experiment, if the incoming x-rays have wavelength 5×10^{-11} m , what is the wavelength for x-rays observed to scatter through 90° ?

24.2.3 Two-slit interference revisited

We saw previously that if a beam of monochromatic light passes through two slits, an interference pattern is formed on a distant surface. Using the wave model of light, and applying the superposition principle, we were able to predict accurately the angles at which maxima and minima would occur in the interference pattern.

Suppose we shine a beam light through two slits, as before, but this time we decrease the intensity of the light until the beam is extremely weak. The amount of light hitting a distant screen is too small to be detected by human eyes, so we use a different kind of detector: a photomultiplier (Figure 24.30). The functioning of a photomultiplier tube is based on the photoelectric effect. If a single photon ejects an electron from a metal surface, the

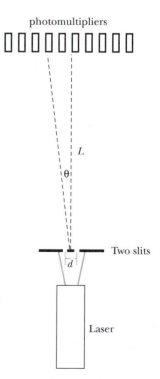

Figure 24.30 A two-slit experiment with a very low-intensity beam, using photomultipliers as photon detectors.

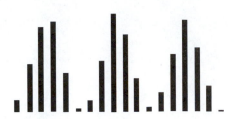

Figure 24.31 Number of photons detected vs. angle for a low-intensity beam passing through two slits.

After 10 electrons

After 50 electrons

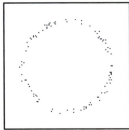

After 100 electrons

After 1000 electrons

Figure 24.32 Longer and longer exposures to electrons passing through a polycrystalline powder.

resulting one-electron "current" is amplified electronically into a signal strong enough to detect. So photomultiplier tubes can detect individual photons. We will put an array of photomultipliers at the location of the screen where we previously saw an interference pattern with a strong beam of light, and make a plot of number of photons detected vs. location.

? What would you expect the appearance of the resulting plot to be?

Although individual photons are detected at particular locations, over time an interference pattern builds up, identical to the one observed with an intense beam (Figure 24.31). This pattern, however, builds up statistically, with photons frequently detected at locations of maxima, and photons never detected at locations of minima. Despite the fact that we are detecting individual particles, we still observe a wave-like phenomenon.

It is clear from this and other experiments that a complete model of light must have both a wave-like and a particle-like character, as we concluded earlier.

24.2.4 Electron diffraction

If we shoot a beam of x-rays at a single atom, re-radiation of x-rays by accelerated electrons in the atom goes in nearly all directions. But if we shoot a beam of x-rays at a crystal made up a huge number of atoms arranged in a three-dimensional array, there is constructive interference of the re-radiation only in certain special directions for which the condition $2d\sin\theta = n\lambda$ is true. This is very much a wave phenomenon.

We also discussed x-ray diffraction by polycrystalline powders (page 884), in which there is strong constructive interference for those tiny crystals that happen to be oriented at the appropriate angle for the condition $2d\sin\theta = n\lambda$ to be true. The result is rings of re-radiated x-rays surrounding the incoming x-ray beam. The larger the wavelength λ, the larger the angle θ, and the larger the rings seen on the x-ray film.

We would expect very different behavior if we used particles instead of waves. Suppose that instead of using x-rays we shoot electrons at a single atom. There is a complicated electric interaction between an incoming electron and the atom, including polarization of the atom by the electron. The electron is deflected by the collision with the atom and can go in almost any direction, depending on how close to center it hits the atom.

? If electrons were shot at a crystal made up of a huge number of atoms arranged in a three-dimensional array, what would you expect to see?

We certainly don't expect to see any interference effects, since we assume that an electron is a particle, not a wave. Nevertheless, if we shoot a beam of electrons into a polycrystalline powder, we get rings of electrons, just as with x-ray diffraction, with the angles of the rings described by the usual x-ray diffraction equation! Experimentally, it appears that electrons also can act like waves.

Moreover, an electron not only exhibits wave-like interference phenomena but even appears to interfere with itself! If you send in electrons one at a time (a very low-current beam of electrons), and detect the electrons one at a time with an array of detectors (or a photographic film), you get an interference pattern, but built up in a statistical way. Figure 24.32 shows what you observe with longer and longer exposures of film with a low-current beam of electrons hitting a polycrystalline powder. Each dot represents a location where an electron hit the film.

This is intriguing! You detect individual electrons, like particles. Yet if you wait long enough, you find that the pattern of detection looks like what you get with x-rays, which are waves. We say that electrons have properties that are both particle-like and wave-like.

In these electron experiments the condition for (statistical) constructive interference is $2d\sin\theta = n\lambda$, just the same as for x-ray diffraction. What is the wavelength λ of the electrons? It was predicted theoretically by de Broglie and verified experimentally by Davisson and Germer that

$$\lambda = \frac{h}{p}$$

where p is the momentum of the electron, and h is Planck's constant, an extremely small quantity measured to be $h = 6.6 \times 10^{-34}$ joule·second. Note that since for a photon $E = pc$, this relation is valid for photons too:

$$\lambda = \frac{hc}{E} = \frac{h}{p}$$

This relationship is easily demonstrated with an electron-diffraction apparatus in which you can control the momentum of the incoming electrons by varying the accelerating voltage ΔV. We calculate the momentum corresponding to a given kinetic energy K:

$$K = \frac{p^2}{2m}, \text{ so } p = \sqrt{2mK}, \text{ where } K = e\Delta V \text{ if the electrons start from rest.}$$

Therefore we predict this wavelength for the electron:

$$\lambda = \frac{h}{p} = \frac{h}{\sqrt{2me\Delta V}}$$

? Based on this relation, and on the diffraction relation $n\lambda = 2d\sin\theta$, how would you expect the spacing of electron diffraction rings to change if you increase the accelerating voltage?

Experimentally we find that if the accelerating voltage ΔV is quadrupled, the electron wavelength is decreased by a factor of 2, and the diffraction rings shrink due to $\sin\theta$ being reduced by a factor of 2.

Electron diffraction is used extensively to study the structure of materials. It has an advantage compared with x-ray diffraction in that electron beams can be focused easily to a very small spot using electric and magnetic fields, whereas this is difficult to do with x-rays. For this reason, electron diffraction is particularly useful in studying small regions of a sample of material. In actual practice, rather high-energy electrons are used in order to achieve good penetration of the material, with typical accelerating potentials of several hundred thousand volts. That means that the wavelength λ is much smaller than a typical interatomic spacing d. Consequently θ is quite small, and the electrons are sent in nearly parallel to the atomic planes of interest.

Neutron diffraction

Perhaps electrons and photons are very special in having both particle and wave aspects? Not really. Similar diffraction patterns have been observed when entire helium atoms are shot at a crystal, and the wavelength of the helium atom is again $\lambda = h/p$ (but with a much larger mass m).

A particularly important case is neutron diffraction. Neutrons have no charge, and only a very small magnetic moment, so neutrons have no electric interactions with electrons and a small magnetic interaction. But neutrons do have a strong nuclear interaction with protons and other neutrons, as is evident by the presence of neutrons in the nuclei of atoms. If you shoot a single neutron at a single atom, the neutron may be hardly affected by the electrons but can be deflected by the nucleus through just about any angle.

When you shoot neutrons at a crystal, the usual formula $2d\sin\theta = n\lambda$ for constructive interference applies.

What is it that is "waving"?

Electromagnetic radiation consists of waves of electric field. Interference of electromagnetic radiation is due to superposition of electric fields, and the intensity is proportional to the square of the field amplitude.

In the case of electron or neutron diffraction, what is it that is "waving"? It turns out that particles can be described by abstract "wave functions" whose amplitude squared at a location in space is the probability of finding the particle there. The same formula $2d\sin\theta = n\lambda$ that predicts the locations of rings of intensity for x-ray diffraction also predicts the locations of rings of probability for electron or neutron diffraction. Classical interference of light is therefore a good foundation for understanding quantum interference of particles. But the interpretation changes from classical intensity to quantum probability.

Ex. 24.14 Roughly, what is the minimum accelerating potential ΔV needed in order that electrons exhibit diffraction effects in a crystal?

Ex. 24.15 Roughly, what is the minimum speed that neutrons should be moving in order to exhibit diffraction effects in a crystal?

24.3 Further applications of the wave model

Having explored the combined wave-particle model of light, we return in this section to the wave model of light. There are a large number of important phenomena that can be most easily analyzed in terms of the wave aspects of light, and we examine some of these in the following sections.

24.3.1 Scattering of light

In Section 23.6.4 we discussed why the sky is blue, which involves frequency-dependent re-radiation by air molecules of light from the Sun, with blue light re-radiated more strongly than red light. This process, in which light passing through a nearly transparent medium is partially re-radiated to the side, is called "scattering" (the light is "scattered" into directions different from the original direction).

We will describe an experiment that brings out some interference aspects that affect scattering phenomena. Perhaps your instructor will demonstrate this for you. A bright white light is directed through a long aquarium tank full of water and onto a screen, where it shows up as a white spot (Figure 24.33). If the water is quite clean, very little light is scattered to the side.

A puzzle is why there is almost no light scattered off to the side. After all, there are huge numbers of molecules in the water, and light striking those molecules accelerates electrons, leading to re-radiation. We'll consider path differences for different molecules and see where we should expect low-intensity and high-intensity light.

For any molecule in the water, find a second molecule a short distance $\lambda/2$ downstream (Figure 24.34). Off to the side, the path difference for the re-radiation from these two molecules is half a wavelength, so the intensity is zero. If we can pair up all the molecules in this way, we can expect to see very little intensity off to the side.

But can we always find a suitable partner for every molecule? The answer for water (or glass) is yes, because in a liquid or solid the molecules are right

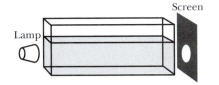

Figure 24.33 Light passes through a tank of water and appears on a screen.

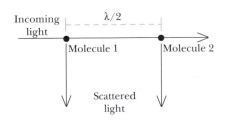

Figure 24.34 Viewed from above: re-radiation (scattering) from water molecules a distance $\lambda/2$ apart.

next to each other, filling the entire space. Go downstream a distance $\lambda/2$, a distance of about 250 nm = 2500 atomic diameters, and you can be sure that you will find a molecule at that location (unless you are within $\lambda/2$ of the end of the water tank, but that short region contains very little water). It is true that molecules in a liquid have some limited freedom to move about, unlike molecules in a solid, but they do fill the space at all times. Note that in the downstream direction, straight through the water toward the screen, the re-radiation from these two molecules is in phase with each other, so we can expect to see light coming out the end of the water tank, and we do.

(It is reasonable to ask what would happen if we paired up molecules that were a distance λ apart: Wouldn't that give constructive interference? Yes, but we would also have to calculate how that re-radiated light would interfere with the light from all the other pairs that were one wavelength apart. This calculation is beyond the scope of this discussion, but the result is that we get zero intensity when we add up all the contributions. The simplicity of the argument in terms of pairs of molecules that are $\lambda/2$ apart comes from the fact that it is easy to add up all the zero intensities from all such pairs and get zero.)

Add some soap to the water

Next we drop some material such as soap into the water to provide a low concentration of specks of material. Now you see some light scattered off to the side, and it is distinctly bluish. At the same time, the spot on the screen becomes somewhat reddish, because the blue portion of the white light has been scattered more than the red portion, leaving light that has been de-pleted of the blue component. As explained in Section 23.6.4, the scattered bluish light is strongly polarized, as you can see by turning a piece of polar-izing material in front of your eye as you look toward the tank from the side. (You also get a big effect by turning a piece of polaroid between the light source and the water tank.)

The visible scattering is due to there being a relatively small number of specks of material in the water that re-radiate a different amount than the water molecules they displace. Moreover, the specks are far away from each other in random locations and do not destructively interfere with each oth-er. That is, if you go a distance $\lambda/2$ downstream from a soap particle, you won't necessarily find another, similar soap particle at that location.

Scattering from a gas

There is a subtle point concerning why the sky is blue. Why is there signifi-cant scattering off to the side of clear air, when we see little scattering from water? A gas is different from a liquid or solid in that the molecules are not in contact with each other but are roaming around quite freely, with lots of space between molecules. Since the air molecules are not in relatively fixed positions, the argument for destructive interference that we used for water is not valid.

Consider a cube of air whose edge is only 0.1λ long, at standard temper-ature and pressure. Such a cube contains a surprisingly large number of molecules, though many fewer than in a comparable volume of a liquid or solid:

$$(0.1 \times 500 \times 10^{-9} \text{ m}) \frac{6 \times 10^{23} \text{ molecules}}{22.4 \text{ liter}} \frac{1 \text{ liter}}{10^3 \text{ cm}^3} \frac{10^6 \text{ cm}^3}{\text{m}^3} \approx 3000 \text{ molecules}$$

If there were exactly 3000 molecules in each tiny cube of air, there would be little scattering, due to destructive interference by radiation from pairs of cubes located a distance of $\lambda/2$ apart. But the number of molecules in each cube fluctuates randomly as air molecules roam in and out of a cube. The scattering from the sky that we see and enjoy is largely due to the fluctuations in the number of molecules per unit volume. This effect is much less pronounced for liquids. Although the molecules do slide past each other in a liquid, the number of molecules per unit volume hardly changes, since the molecules remain in contact with each other. And of course there is no fluctuation at all in a solid, and if there were no impurities or spatial imperfections in a block of glass it would not scatter light passing through it.

The typical statistical fluctuation of N molecules in a given volume of air can be shown to be equal to the square root of N. The electric fields re-radiated by these $\sqrt{N}$ molecules are nearly in phase with each other (because they are in a volume that is small compared to λ). So the intensity is proportional to $(\sqrt{N})^2 = N$. The scattering is what one would naively expect—N times the scattering from one molecule!

24.3.2   Diffraction gratings

A compact disc (CD) looks pretty much like a mirror, and you can see a lamp reflected in the shiny surface. But if you hold the disc in a fixed position and move your eye away from the direct reflection you see a rainbow of colors. This is an interference effect. The compact disc has concentric rings of tiny pits etched into the plastic, and these rings of pits separate undisturbed rings of plastic from each other. You observe interference between reflections from adjacent undisturbed sections of the plastic. The distance d between adjacent rings is comparable to the wavelength of visible light (otherwise there would be no effect). There are a very large number of rings on a compact disc, and the effect is called "diffraction," the term applied to interference phenomena when there are a large number of sources whose waves interfere with each other.

A similar device is the "diffraction grating," a piece of glass or plastic that has been accurately scratched with a large number of closely spaced straight parallel lines a very small distance d apart (Figure 24.35). As with a compact disc, you can see an ordinary reflection in a diffraction grating, but off at an angle you see a rainbow of colors. "Transmission" diffraction gratings are transparent between the scratches and act like a large number of parallel slits through which light passes. As with two-slit interference, each slit can be treated as though it were a source of light (see derivation at the end of this chapter).

The diffraction grating is an important scientific tool because it can be used to measure very accurately the wavelengths (and therefore the frequencies) of light emitted by atoms. Different atoms emit light of different frequencies, and diffraction gratings have not only played an important role in the study of atomic structure but have also provided a convenient way to identify what atoms are present in a sample of material.

A diffraction grating may have as many as 1000 parallel scratches per millimeter! In this case $d = 10^{-3}$ mm $= 10^{-6}$ m; compare with a typical wavelength of light of about 500 nm $= 0.5 \times 10^{-6}$ meter.

? Before we treat this phenomenon quantitatively, see if you can explain very briefly and qualitatively why there are bands of colors off at an angle to a compact disc (or reflective diffraction grating), and why the ordinary direct reflection does not show a rainbow.

The angle θ where you get another maximum depends on the wavelength (because the phase difference between adjacent strips depends on the wavelength). So different wavelengths in the white light show up at different angles. The ordinary reflection is normal, because in this case the path lengths are all equal, so there is no dependence on wavelength.

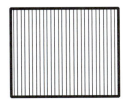

Figure 24.35 A diffraction grating (not to scale) has many fine parallel scratches on its surface.

To analyze a diffraction grating quantitatively, we'll consider a particularly simple geometry but one which illustrates the most important aspects of the device. Suppose a beam of red light strikes a transmission diffraction grating nearly perpendicular to the grating (Figure 24.36). There will be a central peak at $\theta = 0$ since all the slits are in phase with each other, but there can be one or more additional peaks in other directions, where the path difference between adjacent slits is one wavelength.

? If the spacing between adjacent slits is d, and the wavelength of the red light is λ, what is the first non-zero angle θ where you see high intensity? (Hint: Think through what θ is required to get a path difference of one wavelength for adjacent slits of the grating; draw a diagram for adjacent slits.)

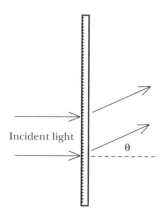

Figure 24.36 A transmission diffraction grating.

This maximum occurs at an angle where $d\sin\theta = \lambda$, so the angle where you will see high intensity for red light ($\lambda \approx$ about 700 nm) is calculable from $\sin\theta = \lambda / d$.

? If you replace the source of red light with a source of violet light (wavelength of about 400 nm), will the high intensity be seen at a larger or smaller angle θ? Why?

Since $\sin\theta = \lambda / d$, a shorter wavelength λ will give a maximum at a smaller angle θ. These results imply that white light from the Sun or from an incandescent lamp must be a mixture of many colors, from red to yellow to green to violet, since you see these various colors coming at different angles from the diffraction grating (or compact disc).

Diffraction gratings are used heavily in science and technology to determine what wavelengths are present in various kinds of light, which is an indicator of what atoms are producing that light. If you look through a diffraction grating at a slit illuminated by a neon lamp, you can see individual lines of color that are characteristic of the neon atoms inside the lamp that are producing the light. This "line spectrum" is different for different atoms, and quite different from the continuous spectrum produced by hot glowing objects such as the Sun or an incandescent lamp.

What's important about having lots of slits?

In the next section we will show formally that the beams of light at the maxima of a diffraction grating are extremely narrow compared to those made by just two slits. It is easy to see from an energy argument that this must be true. With just two slits, the maximum intensity is $2^2 = 4$ times the intensity due to one slit. But in a typical diffraction grating 2 cm wide, with spacing between slits $d = 10^{-3}$ mm, there are 2×10^4 slits, and the maximum intensity is $(2\times10^4)^2 = 4\times10^8$ times the intensity of a single slit! For there to be such a large intensity in a few special directions, there must be wide regions where the intensity is extremely small. The beams of light from a diffraction grating are extremely narrow and sharply defined, which is why diffraction gratings can be used to make very precise measurements of wavelengths.

Ex. 24.16 A certain diffraction grating is known to have a line spacing of $d = 10^{-3}$ mm. A source of single-wavelength light is observed to produce a high-intensity beam at an angle $\theta = 36°$ to the perpendicular. If this is the first maximum at a nonzero angle, what is the wavelength of the light?

This wavelength is emitted by excited sodium, as in sodium vapor lamps. Detecting this wavelength in some light indicates the presence of excited sodium.

24.4 Angular resolution

In this section we will examine a set of phenomena which at first may seem unrelated, but which can be explained by the same basic aspects of the wave model of light. We are interested in the following questions:

- The diameter of a telescope mirror (or lens) determines whether you see two distant stars as two separate bright spots, or one large bright spot.
- In two-slit interference, not all the maxima have the same brightness.
- Light passing through one single slit shows maxima and minima on a screen placed in front of the slit.
- The width of a diffraction grating determines the angular separation of the maxima observed through the grating.

All of these questions involve what is called "angular resolution," and can be explained in terms of the formula

$$\Delta\theta = \frac{\lambda}{W}$$

where W is the width of the device.

24.4.1   Angular width of a maximum

An important property of a diffraction grating is that in the direction of a maximum, the re-radiated beam is extremely narrow. Just how wide (in angle) is one of the maxima observed when monochromatic light passes through a particular diffraction grating? The narrower this maximum, the easier it is to distinguish light of two slightly different wavelengths.

Width of a maximum

How do we define the "width" of a maximum? We could say that the width of the maximum is the angular distance from the brightest area of the maximum to the darkest area next to it—the adjacent minimum. So our task is to calculate the location of the adjacent minimum.

Suppose that θ is the angle of the first maximum, with the path difference for adjacent slits being one wavelength, so that

$$d\sin\theta = \lambda$$

Since the maximum is at angle θ, we'll call the location of the adjacent minimum $(\theta+\Delta\theta)$. Our task is to find the value of $\Delta\theta$.

Conditions for a minimum

We note that there will be a minimum if light passing through the first slit and the $(N/2)$th slit (the middle slit) slit interferes destructively (Figure 24.37). This is reminiscent of the argument we made when considering sideways scattering of light from a liquid. Assume that there is an even number N of slits (or if N is odd, neglect the small amount of light from the last slit).

For a maximum, the path difference between adjacent slits is one wavelength, so the path difference between the first slit and the $(N/2)$th slit is $(N/2)\lambda$. For a minimum, this path difference has an additional $\lambda/2$:

$$\left(\frac{N}{2}\right)\lambda + \frac{1}{2}\lambda$$

Now consider the next pair of slits, slit number 2 and slit number $(N/2 + 1)$. They also have a path difference of $\lambda/2$, so they too interfere destructively. We have succeeded in pairing up all the slits in such a way that they each contribute zero intensity, so the total intensity in the direction $\theta+\Delta\theta$ is zero.

Write the path difference between adjacent slits like this:

$$d\sin(\theta + \Delta\theta) = \lambda + \varepsilon \text{ where } \varepsilon \text{ is a small fraction of } \lambda$$

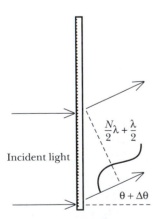

Figure 24.37 The first slit and the middle slit of the diffraction grating interfere destructively if the path difference for adjacent slits is $\lambda+\lambda/N$.

The path difference between the first slit and the $(N/2)$th slit is this:

$$\frac{N}{2}(\lambda + \varepsilon) = \left(\frac{N}{2}\right)\lambda + \frac{1}{2}\lambda$$

Solving for ε we find

$$\varepsilon = \frac{\lambda}{N}$$

Finding $\Delta\theta$

Now, to find $\Delta\theta$ we use trig identities to expand this relation for two adjacent slits:

$$d\sin(\theta + \Delta\theta) = \lambda + \varepsilon = \lambda + \frac{\lambda}{N}$$

$$d\sin\theta\cos\Delta\theta + d\cos\theta\sin\Delta\theta = \lambda + \frac{\lambda}{N}$$

Since $\Delta\theta$ is a small angle, we can make the approximations that

$$\cos\Delta\theta \approx 1 \text{ and } \sin\Delta\theta \approx \Delta\theta, \text{ so}$$

$$d\sin\theta + d\cos\theta\Delta\theta \approx \lambda + \frac{\lambda}{N}$$

Because $d\sin\theta = \lambda$

$$d\cos\theta\Delta\theta \approx \frac{\lambda}{N} \text{ and } \Delta\theta \approx \frac{\lambda}{Nd\cos\theta}$$

The total width of the diffraction grating W is N times the width of one slit:

$$W = Nd$$

Unless θ is close to $90°$, $\cos\theta$ is on the order of 1, so we find that $\Delta\theta$, the "half width" of the maximum, is approximately given by this:

ANGULAR HALF-WIDTH OF A MAXIMUM

$$\Delta\theta \approx \frac{\lambda}{W}$$

where W is the total width of the (illuminated portion) of the device

Although we've proved this only for a diffraction grating, this result is quite general. Whenever a device contains a very large number of sources which interfere, the angular width of a maximum-intensity beam is approximately equal to the wavelength divided by the total width of the device. For visible light hitting a diffraction grating that is 2 cm wide, the angular spreading may be very small: $\Delta\theta \approx \lambda/W \approx (600\times10^{-9} \text{ m})/(2\times10^{-2} \text{ m}) = 3\times10^{-5}$ radians, which is less than 2 millidegrees.

An alternative geometric argument

Instead of using trig identities as we did above, we can find $\Delta\theta$ by a geometric argument involving two adjacent slits (Figure 24.38). In the dashed triangle, the side opposite the very small angle $\Delta\theta$ is $\varepsilon = \lambda/N$, because this is the additional path length that yields a minimum. The hypotenuse is approximately $d\cos\theta$. The angle $\Delta\theta$ in radians is approximately equal to the ratio of the opposite side to the hypotenuse:

$$\Delta\theta \approx \frac{\lambda/N}{d\cos\theta} = \frac{\lambda}{Nd\cos\theta} = \frac{\lambda}{W\cos\theta}$$

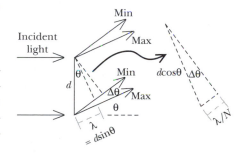

Figure 24.38 Two adjacent slits in a transmission diffraction grating, with a maximum at θ and a minimum at $\theta+\Delta\theta$.

where W is the total width of the grating ($W = Nd$). This is an extremely important result: The wider the grating, the narrower the beam. Unless θ is close to 90°, $\cos\theta$ is of order of 1, so we get, as before:

$$\Delta\theta \approx \frac{\lambda}{W}$$

What about other pairings of slits?

You might wonder whether we would not have gotten complete cancellation if we had chosen other ways of pairing up the slits, rather than pairing the first slit with the middle slit, etc. For example, the first and last slits have a path difference of $N\lambda + \lambda$, so they interfere constructively. However, this is the only pair that does this, and if we were to add up the effects of all the other slits we would find that their combined effects would cancel the re-radiation of the outer slits. Once we have shown that there is some pairing that gives a net intensity of zero, we don't actually have to consider other pairings.

24.4.2   Single-slit diffraction

What does a diffraction grating have to do with maxima and minima created when light passes through a single slit? Just as two narrow slits through which light passes are equivalent to two narrow sources of light, so we can model a single wide slit, of width W, to be equivalent to a very large number N of very narrow sources located a distance $d = W/N$ apart. (Actually, the sum over N finite sources turns into an integral over an infinite number of infinitesimal slits.) There is of course a maximum straight ahead, but there is a minimum at a small angle $\Delta\theta$ given approximately by λ/W (Figure 24.39). The overall pattern of intensity looks like Figure 24.40.

This also explains the variation in intensity of the maxima in two-slit interference. Each slit makes a pattern like Figure 24.40, and these two patterns interfere. The intensity pattern is the combined result of both single-slit diffraction and two-slit interference.

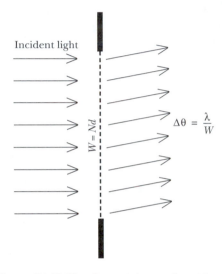

Figure 24.39 The first minimum for a single slit is at $\Delta\theta \approx \lambda/W$.

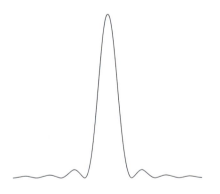

Figure 24.40 The pattern of intensity *vs.* angle for diffraction by a single slit.

Ex. 24.17 To get a feel for the size of these "diffraction" effects, consider a single slit $W = 0.1$ mm wide, illuminated by violet light. The wavelength λ of violet light is about 400 nm (400×10^{-9} m). Light from the slit hits a screen placed a distance $L = 1$ m $= 1000$ mm away from the slit, and the image of the slit is larger than 1 mm due to significant amounts of light heading at angles as large as plus or minus λ/W (for a total angular spread of $2\lambda/W$). Calculate this width in millimeters.

24.4.3   Telescope resolution limited by diffraction

When a telescope makes an image of a distant star on a photographic plate or array of electronic detectors, the image is broadened by diffraction, which spreads the light rays out into a cone with an approximate half-angle $\Delta\theta \approx \lambda/W$. The images of two stars that are close together in the sky may be broadened so much that they overlap each other and cannot be "resolved" into two separate images (Figure 24.41). The "resolving power" or "resolution" of a telescope or camera is limited by diffraction. The wider the lens, the sharper the image (bigger W, smaller $\Delta\theta$). Alternatively, the shorter the wavelength, the sharper the image (smaller λ, smaller $\Delta\theta$). One rule of thumb for resolution states that if ϕ, the angle between the stars, is greater than $2\Delta\theta$, the images will be resolvable.

Radio telescopes used to make astronomical observations are sometimes linked electronically to other radio telescopes located thousands of miles away, in order to make a radio-frequency "lens" that is so large that the diffraction broadening is extremely small. This has made it possible to obtain exquisitely detailed pictures of extremely distant structures.

It is much more difficult to do this with ordinary "optical" telescopes that capture visible light, because you have to bring the light from two different telescopes to the same place, where there can be interference. There are plans to link pairs of large optical telescopes with mirrors to bring light from the two telescopes to an observation location between the telescopes. This will make it possible to obtain pictures with much higher resolution than can be obtained with existing optical telescopes.

Only one large diffraction peak

With a few sources we often find several large completely constructive peaks, corresponding to the path difference between neighboring sources being equal to 0, λ, 2λ, etc. Even a diffraction grating with many sources may produce several maximum-intensity beams. The situation is different with single-slit diffraction.

? Explain briefly why a single slit (or lens) produces just *one* large peak (there are smaller peaks, but only one large peak).

With a single slit, the path difference "d" between adjacent "sources" is infinitesimal in size, so the path difference is an infinitesimal quantity and can never get as big as λ. So there is no direction (except for $\theta = 0$) where all the amplitudes are in phase.

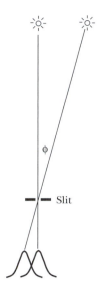

Figure 24.41 Light from two distant stars passes through the aperture of a telescope. The half width of the diffraction-broadened image for each star determines whether or not it will be possible to distinguish the two images.

Ex. 24.18 Satellites used to map the earth photographically are typically about 100 miles (160 kilometers) above the surface of the earth, where there is almost no atmosphere to affect the orbit. If the lens in a satellite camera has a diameter of 15 cm, roughly how far apart do two objects on the earth's surface have to be in order that they can be resolved into two objects, if the resolution is limited mainly by diffraction? The key point is that the angle subtended by the two objects should be comparable to or bigger than the spread in angle $2\Delta\theta$ introduced by diffraction of light going through the lens. Visible light has wavelengths from 400 to 700 nm; consider a wavelength somewhere in the middle of the spectrum. You should find that the numbers on a car license plate would be too blurred to read.

24.5 Standing waves

In this chapter and the previous chapter we have been studying "traveling waves"—waves whose crests travel through space at a speed that is characteristic of the type of wave (the speed of light for electromagnetic radiation, the speed of sound for sound waves, etc.). Next we will study waves that travel back and forth within a confined space. Examples include elastic waves traveling back and forth along a length of string, or sound waves traveling back and forth inside an organ pipe. The behavior of such waves in these classical systems provides hints for understanding some quantum aspects of atoms.

Consider two sinusoidal traveling waves with exactly the same frequency that were initiated some time ago far to the left and far to the right, and which are heading toward each other:

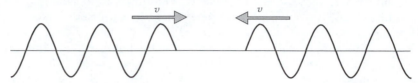

These waves might be electromagnetic waves in empty space, elastic waves propagating from two ends of a very long taut string, sound waves propagating from two ends of a long organ pipe, or water waves propagating from two ends of a lake. The description and analysis is practically the same in each of these very different physical situations. The wide applicability of the analysis we will develop is one of the reasons why we study waves.

When the waves start to overlap, they of course interfere with each other constructively or destructively, depending on the relative phases of the two waves in space. This interference leads to a remarkable effect. Look at the sequence of snapshots in Figure 24.42, taken one-eighth of a period apart. In the overlap region, the "wave" doesn't propagate! It just stays in one place, with the crests waving up and down but not moving to the left or to the right. This is something new, called a "standing wave."

There are even instants when the "wave" is completely zero throughout the entire region! At the instant that this occurs for a string, pieces of the string do have velocity (rate of change of displacement) and kinetic energy, so the waving does continue, even though there is momentarily no transverse displacement of the string anywhere in the region. In the case of electromagnetic radiation, the electric and magnetic fields are changing as they go through zero, and as we saw in the previous chapter it is the rate of change of electric and magnetic fields that generates magnetic and electric fields.

One of the important properties of a standing wave is the existence of what are called "nodes"—locations where the wave is zero *at all times*. Positions of nodes of the standing wave are circled on the bottom snapshot in Figure 24.42. Halfway between the nodes the standing wave periodically gets to full amplitude. These locations are called "anti-nodes."

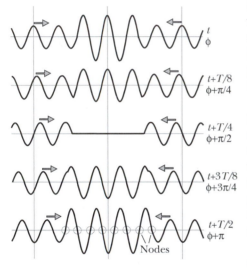

Figure 24.42 Two traveling waves of the same frequency produce a standing wave.

Standing waves vs. traveling waves

A single observer observing the wave as a function of time at one specific location that is not a node can't tell the difference between a traveling wave and a standing wave.

? Briefly explain why this is.

The single observer sees the wave going up and down at this location, which is just what it does when a traveling wave goes by. On the other hand, if the single observer is located at a node, the observer may conclude that nothing interesting is happening at all, or that this is a location where two traveling waves are destructively interfering with each other, as in two-slit interference, without forming a standing wave in space.

Only if we look at the complete pattern of waves in space and time can we see that a standing wave is really very different from a traveling wave. Yet there are some properties which are the same for both kinds of wave. Look carefully at the time sequence of snapshots in Figure 24.42, and then answer the following important questions:

? If the wavelength of the two traveling waves is λ, what is the wavelength (crest-to-crest distance) of the standing wave? If the frequency of the two traveling waves is f, what is the frequency (number of complete cycles per second) of the standing wave?

The wavelength and frequency of the standing wave are the same as for the traveling waves. This means that although the standing wave isn't going anywhere (left or right), we can still use the important relationship $v = f\lambda$ to relate frequency and wavelength for a standing wave through the speed v of a traveling wave.

24.5.1 Confined waves: Standing waves on a string

One of the ways to create a standing wave is to initiate a single traveling sinusoidal wave in a confined space, so that it continually reflects off the boundaries of this confined space, and the reflected wave interferes with the original wave. The frequency doesn't change in reflecting off a stationary boundary, so the original wave and its reflection are guaranteed to have the same frequency, which is necessary in order to be able to create a standing wave.

Examples of such confined waves include elastic waves on a string stretched between two supports, sound waves inside a pipe closed at both ends, and light waves reflecting back and forth between mirrors at each end of a laser. For concreteness we'll emphasize elastic waves on a taut string, but the analysis applies to many other quite different physical situations.

Figure 24.43 shows two possible standing waves on a taut string of length L that is tied to two fixed supports. The wavelength in the upper case is $\lambda = 2L$, as can be verified by noting that the length L contains half a wavelength. The wavelength in the lower case is $\lambda = L$.

? Why can't you have a standing wave on this string with wavelength $3L$, or $1.1L$, or $0.95L$?

These wavelengths don't fit because the waves aren't zero at the ends of the string, where the string is tied to fixed supports. You have discovered that the possible standing waves on a string are "quantized." Not just any old wavelength will do. The possible wavelengths of a standing wave on a taut string are completely determined by the distance between the supports. The same important principle applies to possible wavelengths of standing sound waves in an organ pipe, or possible wavelengths of standing light waves in a laser.

STANDING WAVES ARE QUANTIZED

Only certain wavelengths (and frequencies) are possible.

24.5.2 Enumerating the possible standing waves

Let's generate a more complete catalog of the possible wavelengths for standing waves on a taut string.

? In terms of the length L, write down the six longest possible wavelengths. You already know the first two: $\lambda_1 = 2L$, $\lambda_2 = L$.

Evidently we have $\lambda_3 = 2L/3$, $\lambda_4 = 2L/4$, $\lambda_5 = 2L/5$, $\lambda_6 = 2L/6$. The waves are shown in Figure 24.44. A quick way to determine the possible wavelengths is to note that in the length L there must be an integer number n of half-wavelengths, so $L = n(\lambda_n/2)$, which means that $\lambda_n = 2L/n$.

As long as the amplitude isn't too large, or the wavelength too small, the speed of propagation v of traveling waves on a taut string turns out to be independent of the wavelength.

? In terms of this speed v, what are the frequencies of the six lowest possible standing-wave frequencies?

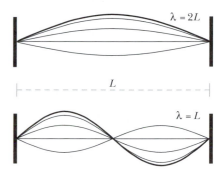

Figure 24.43 Two of the possible standing waves on a taut string of length L.

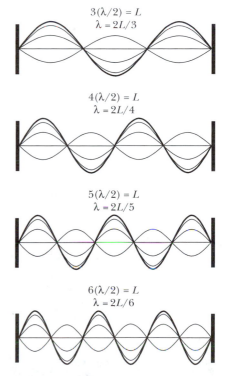

Figure 24.44 The next four possible standing waves.

Solving for $f_n = v/\lambda_n = nv/(2L)$, we have $f_1 = v/(2L)$, $f_2 = 2v/(2L)$, $f_3 = 3v/(2L)$, $f_4 = 4v/(2L)$, $f_5 = 5v/(2L)$, $f_6 = 6v/(2L)$. These different standing waves are called "modes." The technical use of the term is that a "mode" of oscillation of a system is one characterized by a standing wave that is sinusoidal in time with the same frequency at every location, and which necessarily satisfies the boundary conditions (in our case, that the wave is zero at the ends of the string). A mode may also be sinusoidal in space at a particular instant of time, as is true for a string, but it need not be spatially sinusoidal in more complicated systems, such as the two-dimensional surface of a drum.

We've seen that there is a longest possible wavelength $2L$, and a lowest possible frequency $v/(2L)$. Is there a shortest possible wavelength, corresponding to a highest possible frequency? Yes. If the distance between atoms in the string is D, the shortest theoretically possible wavelength for a standing elastic wave is $2D$, with adjacent atoms oscillating out of phase with each other (Figure 24.45). Any shorter wavelength has no physical significance.

As is the case for many other systems, a vibrating string has a number of modes (number of different possible standing-wave patterns) that is not infinite but finite.

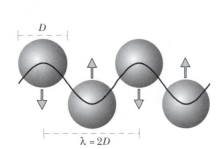

Figure 24.45 The shortest possible wavelength (highest possible frequency).

24.5.3   Building up to the steady state

You may have had the experience of building up a standing wave in a string or rope. You tie one end of the rope to something, pull it taut, then shake your end of the rope up and down. If you shake at one of the mode frequencies, a large oscillation quickly builds up, because the traveling wave you send down the rope reflects off the other end in such a way as to produce a standing wave through constructive interference. But if you try to shake the rope at a frequency that is not right for any mode, you don't build up a large oscillation, because you don't get constructive interference.

As you continue to put energy into the rope by shaking it at an appropriate frequency, the standing wave grows and grows. What limits its growth? What determines the final amplitude of the standing wave?

At first, almost all of your energy input goes into making the standing wave grow. But as the amplitude increases, so does the rate of energy losses due to air resistance and "internal friction" in the rope itself associated with bending the material. The air resistance grows because air resistance is larger at higher speeds, and the rope has higher transverse speeds when the amplitude of the standing wave is larger. Internal friction inside the rope also grows as the bending angle of the rope gets larger with larger amplitude.

At some particular amplitude the losses to air resistance and internal friction in each cycle have grown to be exactly equal to the energy input you make each cycle. During each cycle you simply make up for the losses, but no further growth of the wave occurs. You have reached a steady state.

If the losses are relatively small, you don't have to do much to keep the wave going. In fact, you may have experienced keeping a rope oscillating with a sizable amplitude while your hand hardly moves. In a low-loss situation, your hand is very nearly a node, in that your hand moves a very tiny amount compared with the large motion at one of the anti-nodes (Figure 24.46).

Figure 24.46 With low losses, very little motion of the hand is sufficient to maintain the steady state.

Ex. 24.19 Given the loss mechanisms we have outlined, which should be easier to build up and sustain, a long-wavelength (low-frequency) mode, or a short-wavelength (high-frequency) mode? If you have the opportunity, try this yourself with a rope.

24.5.4 Superposition of modes

We've established that standing waves (modes) are quantized for oscillating strings. Is that all a string can do—be in one of these modes? No! According to the superposition principle, it ought to be possible for a string to be oscillating in a manner described by the superposition of two or more modes. What would that look like? Would it be a standing wave?

Figure 24.47 shows you a time sequence of the superposition of the two lowest-frequency modes (wavelengths $2L$ and L). Both modes have been given the same amplitude. The individual modes are shown with light lines, and their sum is shown with a dark line.

? Is this motion a standing wave? A traveling wave? Why or why not?

Neither. It is not a standing wave, because there are no nodes (places where the amplitude is always zero). It is not a simple traveling wave, because crests move to the left and to the right. Also, there isn't a "wavelength," which is something that both standing and traveling waves have. Despite the complexity of this motion, it is at least periodic—the motion repeats after a time. To understand how the motion repeats, in Figure 24.47 it is instructive to follow mode 1 from top to bottom, then follow mode 2 from top to bottom.

? If the period of the longest-wavelength mode is T ($= 2L/v$), what is the period of this two-mode oscillation?

After T, the low–frequency mode has gone through one complete cycle, and the second mode has gone through two complete cycles, and we're back at the original configuration. So the period is T. Another way to see this is that in Figure 24.47, mode 1 goes through a quarter cycle (from maximum positive to zero), while mode 2 goes through two quarter cycles (from maximum positive on the left to maximum negative on the left). Extending the set of diagrams to four times as much time would bring both modes back to the starting configuration.

Combinations of large numbers of modes can represent pretty complicated motions. Figure 24.48 is a time sequence of the superposition of particular amplitudes of ten different modes, chosen in such a way as to approximate a "square wave" initially.

How complicated can this get? It can be mathematically proven that any function $f(x)$ that is zero at the ends of the string, $f(0) = 0$ and $f(L) = 0$, can be created by adding up carefully chosen amounts of a (possibly infinite) number of modes. In a section on "Fourier analysis" at the end of this chapter we show you how we determined the amplitudes of each of the ten modes we used to approximate the wave shown in Figure 24.48.

This provides an extremely powerful way of analyzing the behavior of confined waves. First figure out what the modes are (the possible sinusoidal standing waves). Then figure out what combination of modes can make a particular shape $f(x)$ at time $t = 0$. What will happen from then on can be calculated simply by following the sinusoidal behavior of the individual modes, each with their own characteristic frequencies, and adding up the individual contributions of the various modes.

24.5.5 Musical tones

When a violin string is bowed, it oscillates with a superposition of various modes, and these oscillations drive the body of the violin to oscillate, which alternately compresses and rarefies the air to produce the sound waves that you hear. Usually the largest-amplitude mode is the longest-wavelength (lowest-frequency) mode, which corresponds to what is called the "pitch" or "fundamental frequency." A pure single-frequency tone sounds very boring. The rich musical tone you hear from a violin is due not just to the funda-

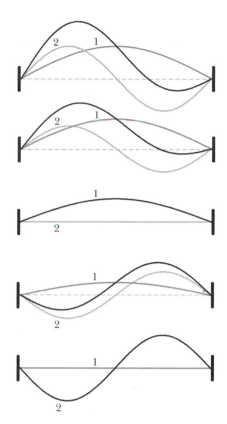

Figure 24.47 A confined wave that is the superposition of the two lowest-frequency modes.

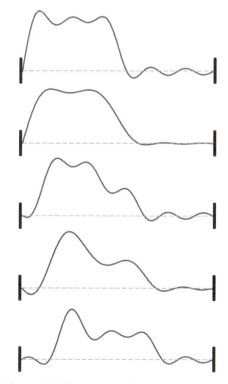

Figure 24.48 An approximation to an initial square wave, using ten modes.

mental frequency but to the superposition of many higher-frequency modes that are also present. Mode frequencies that are simple integer multiples of the fundamental frequency are called "harmonics." It is a peculiarity of the human ear and brain that we find the superposition of harmonics very pleasing. The superposition of frequencies that are not simple integer multiples of a fundamental frequency is considered to be much less pleasing.

24.5.6 More complicated systems

The example of a string is highly representative of a very broad range of standing-wave phenomena. However, we should alert you to some issues that are not illustrated by the relatively simple behavior of a string.

Nonharmonic oscillations

First, while the geometry of the system determines the possible wavelengths (in order to satisfy the boundary conditions), the different wavelengths need not be simply $1/n$ times the longest wavelength. A good example is a drum. The wavelengths of the two-dimensional waves on a drumhead are determined by the requirement that the wave be zero at the drumhead boundary. The relationship among the various mode wavelengths is rather complicated, not simply $1/n$ times the longest wavelength. The fact that the wavelengths of drum modes are not simply related to each other (are not "harmonics") corresponds to the fact that a drum does not produce a musical tone such as is produced by a violin.

Speed dependent on frequency

Second, the speed of propagation in many systems is not constant but depends on the frequency (or wavelength). Since $f = v/\lambda$, in such systems the frequency of a mode need not be simply inversely proportional to the wavelength of that mode, as it is for a string. Such systems are called "dispersive" because traveling waves of different frequencies spread out (disperse) from each other due to differences in their speeds. The analysis in terms of modes for dispersive systems is still valid. It's just that when you calculate $f = v/\lambda$ for a particular λ, you have to take into account the fact that v isn't a constant.

Nonlinear systems

Third, we relied on the superposition principle to combine modes simply by adding them up. More fundamentally, we implicitly assumed that if we double the amplitude of a mode, the frequency won't change. Such systems are called "linear" systems (because the amplitude of the superposition of two waves of the same mode is just the same mode, with an amplitude that is the sum of the amplitudes of the two waves). But many real systems are only approximately linear in this sense. In "non-linear" systems, the superposition of two waves of the same frequency may lead to a wave with a different frequency.

This can even happen with a string. The speed of propagation v for a string turns out to be proportional to the square root of the tension in the string. For small amplitudes, the string is not stretched very much more than it was at rest, and the tension is not changed very much. But for large amplitudes, the string may be stretched quite a bit more than it is when at rest, and that means a change in v, which means a change in frequency for the same wavelength. Hence two medium-sized waves might add up to a large wave with a different frequency.

24.5.7 Confined waves and quantum mechanics

We have been considering quantization in classical (non quantum) systems. We found that standing waves in confined regions couldn't have just any old wavelength, because the possible wavelengths are quantized by the constraints of the boundaries.

There is an analogous situation in the quantum world of atoms, due to the fact that particles such as electrons can have wave-like properties. For example, we found that electron diffraction demonstrates the startling fact that electrons have wave properties as well as particle properties. This implies that there could be three-dimensional "standing waves" of an electron in a hydrogen atom, because the electron is confined to a region near the proton, due to the electric attraction. A hydrogen atom does indeed have quantized states analogous to standing-wave modes. Each of these states has a specific energy—the energy is quantized. A hydrogen atom can drop from a higher-energy to a lower-energy state with the emission of energy in the form of light, and a study of the light emitted by atoms lets us determine the quantized energies of atomic states.

24.5.8 *Fourier analysis

How did we figure out what amount of each of the ten modes to superimpose to get the complicated behavior shown in Section 24.5.4? In principle we could have done it by trial and error—just try various combinations of modes until we get the shape we want. But you can easily imagine that it would be very tedious to try to get this right merely by trial and error.

We'll outline the general scheme for doing this, then apply it to our specific case. We want to express the initial shape of the string, $f(x)$, as a sum of appropriate amounts of many modes, which are sine functions with wavelengths $2L$, $2L/2$, $2L/3$, etc.:

$$f(x) = A_1 \sin\left(2\pi \frac{x}{2L}\right) + A_2 \sin\left(2\pi \frac{2x}{2L}\right) + A_3 \sin\left(2\pi \frac{3x}{2L}\right) + \dots$$

We need to adjust the coefficients A_1, A_2, A_3, etc. in order to choose the appropriate superposition of modes that is equal to $f(x)$. In order to figure out what the nth coefficient should be, consider the following integral:

$$\int_0^L f(x) \sin\left(2\pi \frac{nx}{2L}\right) dx =$$

$$\int_0^L \left[A_1 \sin\left(2\pi \frac{x}{2L}\right) + A_2 \sin\left(2\pi \frac{2x}{2L}\right) + A_3 \sin\left(2\pi \frac{3x}{2L}\right) + \dots \right] \sin\left(2\pi \frac{nx}{2L}\right) dx$$

It can be shown by using trig identities that the following is true:

$$\int_0^L \sin\left(2\pi \frac{kx}{2L}\right) \sin\left(2\pi \frac{nx}{2L}\right) dx = \frac{L}{2} \text{ if } k = n$$

$$\int_0^L \sin\left(2\pi \frac{kx}{2L}\right) \sin\left(2\pi \frac{nx}{2L}\right) dx = 0 \text{ if } k \neq n$$

Therefore the integral picks out just the A_n term:

$$\int_0^L f(x) \sin\left(2\pi \frac{nx}{2L}\right) dx = 0 + 0 + \dots + 0 + A_n \frac{L}{2} + 0 + 0 + \dots$$

Finally, this means that we can calculate the *n*th mode coefficient by the following formula:

$$A_n = \frac{2}{L}\int_0^L f(x)\sin\left(2\pi\frac{nx}{2L}\right)dx$$

This scheme is called "Fourier analysis." It plays an enormously important role in many branches of science and engineering.

A specific example

In section 24.5.4 we wanted to analyze a "square wave" as a sum of modes (Figure 24.49). This function $f(x)$ has the value +1 for $0 < x < L/2$, and 0 for $L/2 < x < L$, so we have

$$A_n = \frac{2}{L}\int_0^L f(x)\sin\left(2\pi\frac{nx}{2L}\right)dx = \frac{2}{L}\int_0^{L/2} f(x)\sin\left(2\pi\frac{nx}{2L}\right)dx + 0$$

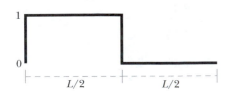

$$A_n = \frac{2}{L}\left[-\frac{L}{\pi n}\cos\left(\pi\frac{nx}{L}\right)\right]_0^{L/2} = \frac{2}{\pi}\frac{1}{n}\left[1 - \cos\left(n\frac{\pi}{2}\right)\right]$$

Figure 24.49 We will express this square wave in terms of a sum of sinusoids.

Dropping the overall factor of $2/\pi$ (which just sets the overall scale of how big the oscillation is), we get the following set of mode coefficients: $A_1 = 1$, $A_2 = 1$, $A_3 = 1/3$, $A_4 = 0$, $A_5 = 1/5$, $A_6 = 1/3$, $A_7 = 1/7$, $A_8 = 0$, $A_9 = 1/9$, $A_{10} = 1/5$, etc.

In section 24.5.4 we used just these first ten modes, so we didn't mimic the desired square-wave function $f(x)$ exactly right, but we were close. If we had used more modes, we would have come even closer to approximating the square wave. If we use an arbitrarily large number of modes we can come arbitrarily close to the desired function.

This is actually a pretty extreme example, because this square wave has discontinuities which require very high-frequency modes in order to approximate the infinite slopes of the function. Functions that have less extreme behavior can often be approximated quite well with only a few modes.

24.6 *Derivation: Two slits are like two sources

The proof that two illuminated slits act like two sources is an application of the superposition principle. Suppose a laser illuminates an opaque sheet (Figure 24.50). According to the superposition principle, electric and magnetic fields produced by the laser must be present behind the sheet, despite the lack of light there. This means that the light from the laser accelerates electrons in the sheet in such a way that the accelerated electrons produce electric and magnetic fields behind the sheet that exactly cancel the fields produced by the laser.

We can consider the sheet to consist of the material that will eventually be removed to make the slits (sections S_1 and S_2 of the sheet), plus the rest of the sheet (section R). The (zero) net electric and magnetic fields beyond the sheet can be considered to be produced by four sources–the laser and the accelerated electrons in sections S_1, S_2, and R of the sheet. The net electric field is

Here, $\vec{E} = \vec{E}_{laser} + \vec{E}_{S_1} + \vec{E}_{S_2} + \vec{E}_R = 0$

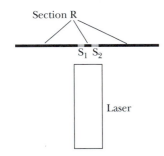

Figure 24.50 Behind the opaque sheet, the net electric field is zero.

Now, $\vec{E} = \vec{E}_{laser} + \vec{E}_R$

$$\vec{E}_{laser} + \vec{E}_{S_1} + \vec{E}_{S_2} + \vec{E}_R = 0$$

When we cut out the slits, the net electric and magnetic fields beyond the sheet are due just to the laser and to the accelerated electrons in section R of the sheet, $\vec{E}_{laser} + \vec{E}_R$ (Figure 24.51).

If we assume that the electron accelerations in section R are about the same in the two situations, whether or not the slits have been cut out, then the electric and magnetic fields produced by section R are about the same

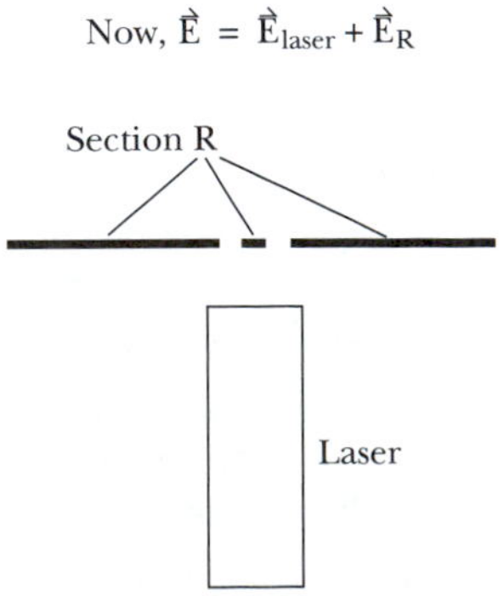

Figure 24.51 With the slits cut out, the net field is due just to the laser and section R.

in both situations. Since $\vec{E}_{laser} + \vec{E}_{S_1} + \vec{E}_{S_2} + \vec{E}_R = 0$, the electric field with the slits cut out is

$$\vec{E}_{laser} + \vec{E}_R = -(\vec{E}_{S_1} + \vec{E}_{S_2})$$

In calculating intensities we don't care about an overall sign, so instead of considering the difficult situation of a laser and a sheet with two slits cut out of it, we can consider the much simpler and already familiar situation of two sources, located where the slits actually are, and producing a net electric field $\vec{E}_{S_1} + \vec{E}_{S_2}$, with no laser and no section R (Figure 24.52).

How good is the assumption that the electron accelerations in section R are about the same whether or not the slits have been cut out? This assumption may not work well very close to the slits, because the presence or absence of the material will affect the edge regions of section R, but we are interested in explaining the interference pattern on a distant screen, not close to the slits.

The assumption also may not work very well if the slits are too small, because in that case their contributions are so small as to be comparable to the edge effects, but we typically deal with slits that are quite wide in the sense of having a width much larger than the wavelength of light. Predictions based on treating the two slits as though they were sources do give results in excellent agreement with observations.

This derivation is due to Richard Feynman (*The Feynman Lectures on Physics*, by R. P. Feynman, R. B. Leighton, and M. Sands, Addison-Wesley 1964, page 31-10.

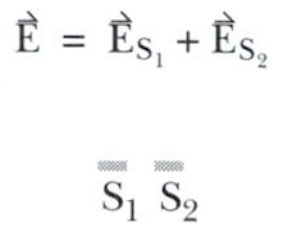

$$\vec{E} = \vec{E}_{S_1} + \vec{E}_{S_2}$$

$$S_1 \quad S_2$$

Figure 24.52 Beyond the slits, we can pretend that the field is due to just these two sources.

24.7 Summary

Fundamental principles

Particles have wave-like properties. The momentum of a particle is related to its wavelength:

$$\lambda = \frac{h}{p}$$

where $h = 6.6 \times 10^{-34}$ joule-second is Planck's constant.

Light has both wave-like and particle-like properties. A photon is a particle of light with zero rest mass, with energy and momentum related to its wavelength:

$$\lambda = \frac{h}{p} \text{ and } E = \frac{hc}{\lambda}$$

Both photons and particles with nonzero rest mass such as electrons and neutrons can exhibit wave-like properties, such as interference.

New concepts

Properties of sinusoidal waves

A wave propagating to the right: $E\cos\left(2\pi\frac{t}{T} - 2\pi\frac{x}{\lambda} + \phi\right)$
    E is the amplitude
    T is the period
    $f = 1/T$ is the frequency
    $\omega = 2\pi f$ is the angular frequency
    λ is the wavelength
    ϕ is a phase corresponding to choices of $t = 0$ and $x = 0$
    $v = f\lambda$ (v is the speed of propagation of the wave)
    Intensity I is proportional to (amplitude)^2

Interference

Waves of the same frequency can interfere with each other. As a result of the superposition principle, some locations have unusually large amplitudes and intensities, while other locations have small or zero amplitudes and intensities.

    Two-slit interference may be observed even if photons go through the slits one at a time.

The photoelectric effect

A photon with energy equal to or greater than the binding energy of an electron and a metal (the "work function") can eject an electron from the metal if it falls on a metal surface.

Compton scattering

A collision between a photon and a free electron results in a change of the wavelength of the photon, corresponding to the amount of energy gained by the electron in the collision.

Standing waves

Standing waves result from the interference of two traveling waves of the same frequency which are traveling in opposite directions. The standing waves have the same frequency and wavelength as the traveling waves.

In a confined space standing waves are quantized—only certain standing-wave wavelengths are allowed. In the simple case of a taut string, the allowed wavelengths are $2L$, $2L/2$, $2L/3$, etc.

The standing-wave motions are called "modes." Arbitrarily complex confined waves can be built up in terms of superpositions of these simple modes.

The steady state is reached when the energy losses are just equal to the input energy.

Results

Path difference and two-source interference

Interference of two in-phase sources that are a distance d apart:

Maximum intensity ($4I_0$) where $\Delta l = d\sin\theta$ is 0, λ, 2λ 3λ, etc.
Minimum intensity (zero) where $\Delta l = d\sin\theta$ is $\lambda/2$, $3\lambda/2$, $5\lambda/2$, etc.
No zero-intensity minimum is possible if $d < \lambda/2$.

If the sources are not in phase, you need to take that phase difference into consideration in addition to the phase difference due to Δl.

X-ray diffraction

Condition for "reflection" is $2d\sin\theta = n\lambda$, where n is an integer, d is the distance between adjacent crystal planes, and θ is the angle of the incoming x-rays to the surface. Polycrystalline powders produce rings around the beam.

Thin films show interference effects. Examples are oil slicks and soap bubbles.

Diffraction gratings

Maxima at angles satisfying $d\sin\theta = n\lambda$, where d is the slit spacing.

Angular resolution of devices

$$\Delta\theta \approx \frac{\lambda}{W}$$

where W is the total width of the (illuminated portion) of the device

Single source of width W: angle to first diffraction minimum is $\theta \approx \lambda/W$.

Scattering of light is affected by interference effects.

From allowed wavelengths in a standing wave one can calculate the corresponding frequencies. In the simple case of a taut string, the allowed frequencies can be calculated from $f = v/\lambda$, and they are $v/(2L)$, $2v/(2L)$, $3v/(2L)$, etc.

Fourier analysis

A confined wave can be expressed as a sum of sinusoids:

$$f(x) = A_1\sin\left(2\pi\frac{x}{2L}\right) + A_2\sin\left(2\pi\frac{2x}{2L}\right) + A_3\sin\left(2\pi\frac{3x}{2L}\right) + \ldots$$

where

$$A_n = \frac{2}{L}\int_0^L f(x)\sin\left(2\pi\frac{nx}{2L}\right)dx$$

Derivation

Two slits act like two sources.

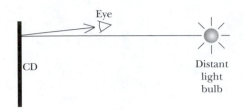

Figure 24.53 Looking straight at a CD, you see an ordinary reflection.

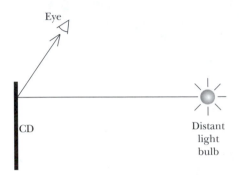

Figure 24.54 Looking at an angle to a CD, you see colors.

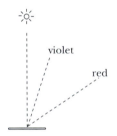

Figure 24.55 Violet and blue appear at the smallest angle. Yellow and red appear at larger angles.

24.8 Example problem: Experimenting with a CD

Hold a compact disc (CD) nearly perpendicular to a line from the CD to a distant light bulb. If you look nearly perpendicular into a region of the CD near the edge, you see an ordinary reflection of the light bulb (Figure 24.53).

While holding the CD stationary in this position, move your head slowly away from the perpendicular (or equivalently, tip the CD) while continuing to look at this region of the CD, until you see the first color of a rainbow (Figure 24.54).

(a) Is the first color you see red or violet? (Try it!) Why? (Violet ≈ 400 nm; red ≈ 700 nm.)

(b) If you keep moving your head further away from the perpendicular (or tipping the CD) you see a complete spectrum. If you continue to even larger angles you may see additional spectra. How many complete spectra do you see?

(c) Use your observations in parts (a) and (b) to estimate the distance between circular tracks on the CD. Report the data on which you base your estimate.

(d) The CD is read with a relatively inexpensive infrared laser (wavelength longer than red). Why can't the tracks be placed closer together, to get more information on the CD? (Newer devices such as DVD use shorter-wavelength lasers.)

Solution

First, try the experiment for yourself!

(a) Moving your head away from the perpendicular, the first color you see is violet, which has the shortest wavelength of the visible spectrum (Figure 24.55). This is what you would expect, because for the light re-radiated by all the grooves to be in phase, $\sin\theta = \lambda/d$, so the smallest wavelength will have a maximum at the smallest angle (d is the spacing between grooves on the CD).

(b) Most people see three spectra, but this is a bit tricky, since you have to hold your head just right.

(c) We found the first maximum for red light to be at an angle θ_1 of about 30 degrees, and we can obtain a value for d from this:

$$d\sin\theta = \lambda \text{ (first maximum)}$$

$$d = \frac{\lambda}{\sin\theta_1} \approx \frac{(700\times10^{-9} \text{ m})}{\sin 30°} = 1400\times10^{-9} \text{ m} = 1400 \text{ nm}$$

Alternatively, if we see three complete spectra, the red of the 3rd spectra is an at angle $\theta_2 \approx 90°$, so we can carry out an alternative, approximate calculation of the spacing d.

$$d\sin\theta = 3\lambda \text{ (third maximum)}$$

$$d = \frac{3\lambda}{\sin\theta_1} \approx \frac{3(700\times10^{-9} \text{ m})}{\sin 90°} = 2100\times10^{-9} \text{ m} = 2100 \text{ nm}$$

(c) If the spacing d between the tracks is shorter than the wavelength of the incident light, it is not possible to resolve the images of the two tracks (they overlap). If the tracks were closer together, the wavelength of light used to read the information would need to be shorter.

24.9 Review questions

Properties of sinusoidal waves—definitions

RQ 24.1 The height of a particular water wave is described by the function

$$y = (1.5 \text{ m})\cos\left[(25.1/\text{s})t - (2.51/\text{m})x + \left(\frac{\pi}{2} \text{ radians}\right)\right]$$

Calculate the angular frequency ω, the frequency f, the period T, the wavelength λ, the speed and direction of propagation (the $+x$ or the $-x$ direction), and the amplitude of the wave. At $x = 0$ and $t = 0$, what is the height of the wave?

RQ 24.2 The relations between wavelength, frequency, period, and speed of propagation apply to all wave phenomena, not just electromagnetic radiation. Middle C on a piano has a fundamental frequency of about 256 hertz. What is the corresponding wavelength of the sound waves? (The speed of sound in air varies with temperature but is about 340 m/s at room temperature.)

> We urge you to try to analyze the following review questions by thinking them through for yourself from fundamental principles, rather than immediately looking up some formulas.

Two-source interference

RQ 24.3 Two audio speakers are side by side, 1 meter apart. They are connected to the same amplifier, which is producing a sine wave of 440 Hertz ("concert A"). Calculate a direction in which you won't hear anything, and make a diagram showing the speakers and this direction. (The speed of sound in air is about 340 m/s.)

Diffraction gratings

RQ 24.4 A ray of violet light (wavelength 400 nm) hits perpendicular to a transmission diffraction grating which has 10,000 lines per centimeter. At what angles to the perpendicular are there bright violet rays, in addition to zero degrees?

RQ 24.5 At night you look through a transmission diffraction grating at a sodium-vapor lamp used for outdoor lighting, which emits nearly monochromatic light of wavelength 588 nanometers. The manufacturer of the grating states that it was ruled with 10,000 lines per centimeter. In addition to zero degrees, at what other angles will you observe bright light?

X-ray diffraction

RQ 24.6 When x-rays with wavelength $0.4{\times}10^{-10}$ m hit a crystal at an angle of 30° to the surface, a strong beam of x-rays is observed in the direction of the "reflection" angle. Assume you know from other evidence that the spacing of the atomic layers parallel to this surface of the crystal is greater than $1.0{\times}10^{-10}$ m and less than $1.8{\times}10^{-10}$ m. What are possible values of the spacing between the layers?

Reflection

RQ 24.7 It is natural to say that "light bounces off mirrors," but is there a physics principle that would account for such behavior? What really happens? And why isn't there light going in all directions, not just in the direction of the "reflection" angle?

Waves and particles

RQ 24.8 Summarize the different predictions of the wave and particle models of light regarding the photoelectric effect. What experimental observations support the particle model?

RQ 24.9 According to the wave model of light, what is the relationship between energy and intensity? According to the particle model of light?

RQ 24.10 In a collision between a photon and a stationary electron, why does the wavelength of the photon change as a result of the collision? Does it increase or decrease?

RQ 24.11 What is the energy of a photon whose wavelength is 690 nm?

RQ 24.12 If the work function of a metal is 3.4 eV, what would be the maximum wavelength of light required to eject an electron from the metal?

RQ 24.13 In electron diffraction, diffraction rings are produced when electrons go through polycrystalline material after being accelerated through an accelerating potential difference. What happens to the size of the diffraction rings when the accelerating potential difference is increased?

RQ 24.14 Electrons are accelerated through a potential difference of 1000 volts and strike a polycrystalline powder whose atomic layers are 10^{-10} m apart. Predict an angle at which you will see an electron-diffraction ring.

Scattering of light

RQ 24.15 Explain why a beam of light can go straight through a rather long tank of clear water or a long rod of clear glass, with hardly any light emitted to the side despite the huge number of atoms whose electrons are accelerated by the incoming light.

Angular resolution

RQ 24.16 A satellite in Earth orbit carries a camera. Explain why the diameter of the camera's lens determines how small an object can be resolved in a photo taken by the camera.

RQ 24.17 Red laser light with wavelength 630 nm goes through a single slit whose width is 0.05 mm. What is the width of the image of the slit seen on a screen 5 m from the slit?

Standing waves

RQ 24.18 How do standing waves differ from traveling waves? How are they similar?

Confined waves

RQ 24.19 The lowest-frequency mode of the lowest (C) string on a cello is 64 hertz. The length of the string between its supports is 70 cm.

(a) What is the speed of propagation of traveling waves on this string?

(b) Calculate the frequencies and wavelengths of the next 4 modes (not including the 64 hertz), and make a sketch of the standing-wave shape for each mode.

Superposition of modes

RQ 24.20 When a cello string is bowed, many different modes are excited, not just the lowest-frequency mode, which contributes to the richness of the sound.

(a) When the bow is taken away the string continues to vibrate. After some time, only the lowest-frequency mode remains in motion. Why?

(b) When a cellist bows while lightly touching the midpoint of the string, without pressing the finger down on the string, the lowest-frequency mode and all modes with odd multiples of that frequency are absent. This changes the tone markedly and is used to achieve a special musical effect. Explain why the odd harmonics are missing.

24.10 Homework problems

We urge you to try to analyze these homework problems by thinking them through for yourself from fundamental principles, rather than immediately looking up some formulas.

Problem 24.1 Two-slit interference

Coherent green light with a wavelength of 500 nm illuminates two narrow vertical slits a distance 0.12 mm apart. Bright green stripes are seen on a screen 2 meters away.

(a) How far apart are these stripes, center-to-center?

(b) Do they get farther apart or closer together if you move the slits closer together?

(c) Do they get farther apart or closer together if you use violet light (wavelength = 400 nm)?

Problem 24.2 One-slit diffraction

A single vertical slit 0.01 mm wide is illuminated by red light of wavelength 700 nm.

(a) About how wide is the bright stripe on a screen 2 meters away?

(b) Does the stripe get wider or narrower if you make the slit narrower?

(c) Does the stripe get wider or narrower if you use light?

Problem 24.3 Communications satellite

Consider a communications satellite that orbits the earth at a height of about 40,000 kilometers. At this height the orbit period is 24 hours, so that it seems to hang motionless above the earth which turns underneath the satellite once every 24 hours. This is called a "synchronous" satellite. The advantage of such an arrangement is that receiving antennas on the ground can point at a seemingly fixed location of the satellite in the sky and not have to be steered. You have probably seen such fixed receiving antennas pointed at the sky.

If the broadcasting antenna on the satellite is 2 meters in diameter and broadcasts at a frequency of 10 gigahertz ($f = 10 \times 10^9$ hertz), approximately what is the diameter of the region on the ground where the satellite transmission can be picked up?

Problem 24.4 X-ray measurements

Suppose you have an x-ray source that produces a continuous spectrum of wavelengths, with the shortest wavelength being 0.3×10^{-10} m. You have a crystal whose atoms are known to be arranged in a cubic array with the distance between nearest neighbors equal to 1.2×10^{-10} m.

(a) Design an arrangement of x-ray beam and crystal orientation that will give you a monochromatic beam of x-rays whose wavelength is 0.5×10^{-10} m. Explain your arrangement carefully and fully in a diagram.

(b) If the shortest wavelength in the x-ray spectrum were 0.2×10^{-10} m instead of 0.3×10^{-10} m, explain why your beam would not be a pure single-wavelength beam.

Problem 24.5 Magenta in a soap bubble

The color magenta consists of a mixture of red (about 700 nm) and blue light (about 450 nm). If you see a magenta-colored section of a soap bubble, about how thick is this section of the soap bubble? Assume that wavelengths in the material are shortened by a factor of 1.3.

Problem 24.6 Light in a swimming pool

At night you look down into an outdoor swimming pool that has a powerful underwater light whose horizontal beam heads to your right. The water is not completely clear, and there is significant scattered light. You have a polarizing film with a line drawn on it showing the direction of electric field that is passed through with little loss. To minimize the brightness of the scattered light, should you hold the film with the line in the direction of the beam (left/right) or perpendicular to the beam (up/down)? Explain briefly, including a diagram.

Problem 24.7 Radiation and copper wires

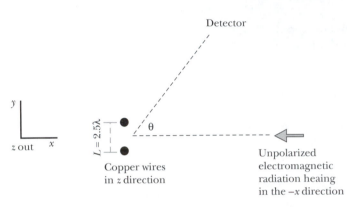

Figure 24.56 Radiation and copper wires (Problem 24.7).

An unpolarized sinusoidal electromagnetic wave with wavelength λ travels along the $-x$ direction in a region where there are two short copper wires oriented along the z direction, a distance $L = 2.5\lambda$ apart.

(a) You place a detector a long ways away, at an angle θ to the x axis. Despite being far outside the region of the unpolarized wave, you detect electromagnetic radiation, and it is polarized. Explain this phenomenon in detail, including the directions of the electric and magnetic fields that you detect.

(b) Calculate an angle θ_1 other than 0° where you will see maximum intensity, and calculate an angle θ_2 where you will see zero intensity. Explain briefly.

Problem 24.8 A 100 watt light bulb

A 100 watt light bulb is placed in a fixture with a reflector which makes a spot of radius 20 cm. Calculate approximately the amplitude of the radiative electric field in the spot, and the number of photons per second hitting the spot.

Problem 24.9 Computer model of an electromagnetic wave

(a) Write a computer program that displays a sinusoidal electromagnetic wave propagating through space. Make the wavelength 600 m (corresponding to a frequency of about 500 kHz, which is in the AM radio frequency band) and the amplitude of the electric field 1.0×10^4 V/m. (This is roughly the amplitude that would be measured a few meters from a 50,000 W radio transmitter. On average, the amplitude of the radiative field from the sun at earth's orbital radius is about 700 V/m.)

Animate the wave as a function of time, making it propagate in the positive x direction, with the electric field polarized in the y direction. Display both the electric and magnetic field vectors, and make sure they have the correct directions relative to each other. Display at least 3 full wavelengths, with enough observation locations per wavelength that you can clearly see the sinusoidal character of the wave. Think carefully about scaling. The length of the arrow objects must be scaled such that both the arrows and the wavelength can be seen. Also, to see smooth wavelike motion, the time step must be a small fraction of the period of the wave.

(b) Place a positron initially at rest in the presence of the electromagnetic wave from part (a). (Using a positron instead of an electron makes it easier to think about signs and directions.) Modify your program to model the motion of the positron due to its interaction with the electromagnetic wave. Leave a trail. This is a relativistic situation since the instantaneous speed of the particle can get quite high. To accurately model the positron's motion, you'll need to use the relativistic relationship between momentum and velocity:

$$\vec{p} = \frac{m\vec{v}}{\sqrt{1 - (v^2/c^2)}} \text{ and therefore } \vec{v} = \frac{\vec{p}}{m}\left(\frac{1}{\sqrt{1 + (p/(mc))^2}}\right)$$

24.11 Answers to exercises

24.1 (page 875)   294 m

24.2 (page 875) $f = 7.5 \times 10^{14}$ hertz (a very high frequency)

$T = 1.3 \times 10^{-15}$ s (a very short time)

24.3 (page 876)   When t = 0, we have $E\cos(\pi) = -E$. which is one point on the curve. The curve is shifted π radians (a half cycle) relative to the first curve.

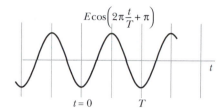

24.4 (page 877)   9

24.5 (page 881)   6.4° (0.11 radian)

24.6 (page 881)   63° (1.1 radian)

24.7 (page 881)   The biggest path difference is at 90° ($\pi/2$ radian) and 270° ($3\pi/2$ radian), where it is only $d = \lambda/3.5 = 0.29\lambda$. This is less than half a wavelength, so the two waves don't cancel each other.

24.8 (page 884)   14°

24.9 (page 887)   640 nm

24.10 (page 889)   3.7 eV

24.11 (page 889)   3.4 eV

24.12 (page 889)   317 nm (3×10^{-7} m)

24.13 (page 891) 5.24×10^{-11} m

24.14 (page 894)   150 volts for $\lambda \approx 10^{-10}$ m

24.15 (page 894)   4000 m/s for $\lambda \approx 10^{-10}$ m

24.16 (page 897)   588 nm

24.17 (page 900)   8 mm

24.18 (page 901)   1.2 m

24.19 (page 904)   The long-wavelength (low-frequency) mode should be easier, because lower speeds mean lower air resistance, and longer wavelength means smaller bending angles (lower internal friction).

Chapter 25

Semiconductor Devices

Chapter 25

Semiconductor Devices

Computers and other modern electronic apparatus are built with semiconductor devices such as diodes and transistors. What is special about semiconductors such as silicon is that we can control the conductivity dynamically, by manipulating the number density of charge carriers.

The simplest semiconductor device is a diode, which conducts current easily in one direction but conducts almost no current in the other direction. A diode can be made by joining two pieces of semiconducting material such as silicon, where one of the pieces contains a small amount of boron and the other contains a small amount of phosphorus. We will study the properties of such a semiconducting junction, which is a fundamental building block of modern electronic circuits. A particularly striking aspect of a diode is that there is an electric field "locked into" the interior of the material, which is something we've never dealt with before.

Understanding the behavior of a semiconducting junction is the foundation for understanding transistors, which are the most common element in digital circuits. Transistors can be constructed from two semiconducting junctions.

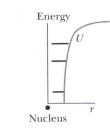

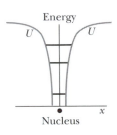

Figure 25.1 Quantized electronic energies in an isolated atom whose potential energy is U, shown as a function of r and as a function of x.

25.1 Energy bands in solids

The possible electronic energy states of an isolated atom are quantized. Figure 25.1 shows the electric potential energy U for an isolated atoms in a row, and the narrow energy levels. We have typically plotted U as a function of r, the separation of nucleus and electron, but in Figure 25.1 we also plot U as a function of x, with the nucleus at the origin, and electrons can be to the left or right of the origin.

If we assemble many such atoms to form a crystalline solid, the electric potential energy of the solid is periodic in x (and y and z) with a period corresponding to the distance between nuclei, as shown in Figure 25.2. In the case of such a periodic potential energy, quantum mechanics predicts and many kinds of experiments confirm that the energy-level scheme is quite different from that of a single atom. The higher energy states are thick bands of very closely spaced levels, often with large separations between these bands of allowed energies. One of these separations is labeled E_{gap} in Figure 25.2. The large number of closely spaced energy levels corresponds to the large number of multielectron states in a solid.

Some electrons in the solid are bound to atomic cores, and electrons in these core states are not mobile. Electrons in an energy band are bound to the solid as a whole rather than to an individual atom, and such electrons can move relatively freely around in the solid, if the energy band is not full, as we will see shortly.

It is an important property of electrons that no two electrons can be in exactly the same quantum state, a property which is referred to as the "Pauli exclusion principle." This means that even at very low temperature, the electrons in an energy band will not all be crowded into the lowest energy state in that band. Rather, electrons will fill the band, with no more than two electrons in any one electronic state (the two electrons must differ in their spin directions in order to obey the Pauli exclusion principle).

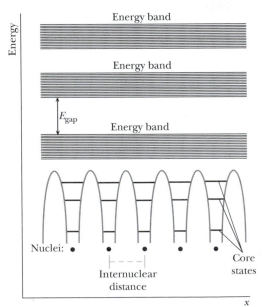

Figure 25.2 In a crystalline solid, the potential energy is periodic, and there are bands of closely spaced allowed energies.

If an energy band is full, other electrons must go into the next higher energy band. The highest energy band that contains electrons is either partially filled, which is the case for a conductor (a metal), or completely filled, which is the case for an insulator (Figure 25.3).

? For the solid to be a conductor, it must be possible to add a small amount of energy to an electron that is at rest, so that it acquires some drift speed. Why will a solid be an insulator if its highest occupied energy band is completely filled?

The highest occupied energy band is separated from the next higher (unoccupied) energy band by the band gap energy E_{gap}. We can't add a small amount of energy to change an electron's energy within the energy band, because all the energy states are filled with two electrons each, and the Pauli exclusion principle forbids putting a third electron into the same energy state (Figure 25.4). In order to increase the energy of an electron in a completely filled energy band, we would need to provide a lot of energy, E_{gap} or more, to move the electron to an unoccupied level near the bottom of the higher, unoccupied energy band.

In some insulators with filled bands such as diamond, the band-gap energy is typically 5 electron volts or more. It is not possible to move an electron to a higher energy state without supplying a very large amount of energy. At ordinary temperatures such a material is a very good insulator and a very poor conductor of electric current.

? In contrast, why is the material a conductor if the energy band is only partially full?

In a partially filled band we can add a small amount of energy to electrons near the top of the occupied levels, moving them to nearby unoccupied levels (Figure 25.5). As a result, the material is a conductor (a metal). The newly acquired extra energy can be lost in a collision, with the electron falling back down to a lower energy. This is an echo of the simple start-stop model of electron conduction in a metal.

Having moved an electron from a lower energy to a higher energy, there is a kind of "hole" in the electron sea of the conductor. Not only can the electron move (opposite to the applied electric field), but the hole can move in the direction of the electric field, like a mobile positive particle. The effect is a bit like the motion of a bubble in water: the water moves down, but the bubble moves up. Unfortunately, a full description of the behavior of electrons and holes in a metal requires advanced solid state and quantum physics. The experimental facts are that most metals behave as though the mobile charges are negative electrons. A few metals, including aluminum and zinc, behave as though the mobile charges are positive holes.

In our simplest model of a metal whose mobile charges are electrons, there is a sea of mobile electrons with a background of positive atomic cores (ions). The drift of mobile electrons due to an applied electric field constitutes an electric current (Figure 25.6). In later discussions it will be important to distinguish between mobile charges and fixed (nonmobile) charges, so we will indicate mobile charges with circles and nonmobile charges without circles in diagrams such as Figure 25.6.

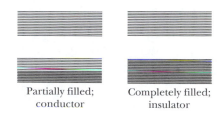

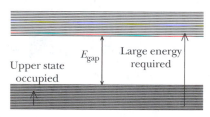

Partially filled;   Completely filled;
conductor   insulator

Figure 25.3 The highest occupied energy band is partially filled or completely filled.

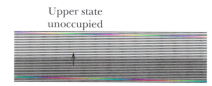

E_{gap} Large energy required

Upper state occupied

Figure 25.4 Cannot promote an electron from one energy state to another within the completely filled energy band of an insulator, and it takes a lot of energy to move to an empty higher energy band.

Upper state unoccupied

Figure 25.5 Can promote an electron from one energy state to another within the partially filled energy band of a conductor.

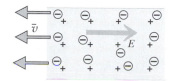

Figure 25.6 Electron conduction in a metal. There are mobile electrons and fixed positive atomic cores (ions). We will indicate mobile charges with circles; nonmobile charges have no circles.

25.2 Semiconductors

In a few materials with completely filled energy bands such as pure silicon or pure germanium, the band-gap energy is relatively small. Such materials are called semiconductors. The band-gap energy E_{gap} in silicon is 1.1 eV, and in germanium it is 0.7 eV. At room temperature these materials do conduct to some extent due to thermal effects. Although their intrinsic conductivity is low compared to a metal, we will see that it is possible to control the conductivity of a semiconductor dynamically, by manipulating the number density of mobile charges.

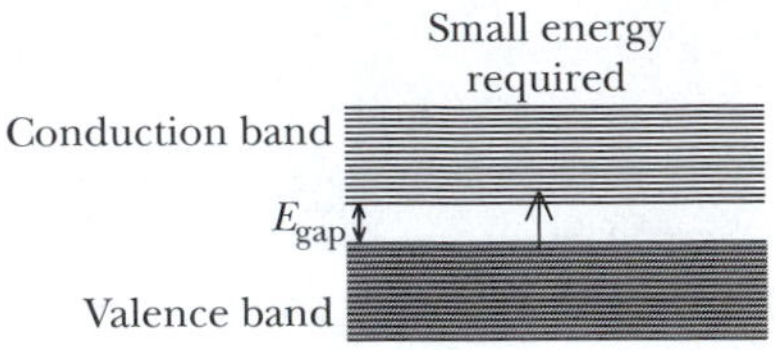

In a semiconductor, thermal energy can occasionally promote an electron into the next higher (unoccupied) energy band (Figure 25.7), and these electrons can pick up energy (and drift speed) from an applied electric field, because other, higher-energy states are nearby. The filled energy band is called the "valence" band, and the nearly unoccupied higher energy band is called the "conduction" band.

Roughly, the number of electrons thermally excited into the conduction band is proportional to the Boltzmann factor $e^{-E_{\text{gap}}/(kT)}$, where k is the Boltzmann constant and kT at room temperature is about $1/40$ eV:

$$kT \approx (1.4 \times 10^{-23}\,\text{J/K})(293\,\text{K})\left(\frac{1\,\text{eV}}{1.6 \times 10^{-19}\,\text{J}}\right) \approx 0.026\,\text{eV} \approx \frac{1}{40}\,\text{eV}$$

Even in germanium with its relatively small band-gap energy ($E_{\text{gap}} = 0.7$ eV) the Boltzmann factor is not very favorable: $e^{-(0.7\,\text{eV})/((1/40)\,\text{eV})} \approx e^{-28} \approx 10^{-12}$. Therefore we might expect the conductivity of pure germanium at room temperature to be about 10^{-12} times smaller than the conductivity of an ordinary metal such as copper. A full analysis shows that the relevant factor is $e^{-E_{\text{gap}}/(2kT)}$, which is about 10^{-6}, which still yields a very low conductivity compared to a metal.

? In germanium or silicon, what will happen to the (low) electric conductivity if you raise the temperature?

When you raise the temperature, thermal energy promotes more electrons to the conduction band. Since the electron current $nA\bar{v} = nAuE$ is proportional to n, the number density of mobile electrons (number per unit volume), the electric conductivity of pure silicon or germanium increases with increasing temperature. In fact, the conductivity increases extremely rapidly with increasing temperature (Figure 25.8). This dramatic effect can be exploited in various ways, including using the current as an indicator of the temperature.

It is true that at higher temperatures the electron mobility decreases, associated with increased thermal motion of the atoms in the material. But this modest decrease in electron mobility is swamped by the huge increase in the number density. The conductivity $\sigma = |q|nu$ depends on both the mobility u and the number density n, and σ actually increases dramatically.

The strongly temperature-dependent conductivity of pure silicon is called *intrinsic* conductivity. In a later section we will discuss another, different mechanism for conductivity in silicon that contains impurities (that is, some silicon atoms are replaced by other kinds of atoms).

25.2.1 Conduction by holes

When thermal energy happens to promote an electron to the conduction band in a semiconductor, the newly mobile electron drifts away, and an

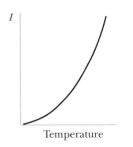

Figure 25.7 A semiconductor is a material with a small band gap.

Figure 25.8 Current in a pure semiconductor grows extremely rapidly with temperature.

atom is now missing one electron. This atom has a net charge of +*e*, and we say that there is a "hole": there is a defective bond with a neighboring silicon atom. An electron in a neighboring atom associated with the bond could fill the hole. As a result, the location of the hole (the positive charge) could move from atom to atom, in the direction of the applied electric field (Figure 25.9). The effect is as though there are actual positive charges, mobile holes, that can move through the material.

If there are any mobile electrons, they drift through the material in a direction opposite to the electric field. Both the mobile electrons and the mobile holes contribute to the conventional current, in the direction of the electric field (Figure 25.10).

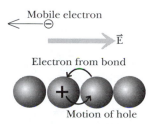

Figure 25.9 A mobile hole moving from atom to atom, to the right, in the direction of the applied electric field. Mobile electrons move to the left.

In a pure semiconductor, where thermal energy creates mobile electrons and mobile hole in pairs, the number density n_n of mobile (negative) electrons is equal to the number density n_p of mobile (positive) holes. In impure semiconductors these number densities need not be equal. Also, the mobilities of the electrons and holes are usually different.

In general, the total conventional current in a semiconductor is this:

$$I = en_n A u_n E + en_p A u_p E$$

Since $J = I/A = \sigma E$, the conductivity is $\sigma = en_n u_n + en_p u_p$.

Differences between metals and semiconductors

? Summarize: What are the major differences between intrinsic conductivity in a semiconductor and conductivity in a metal?

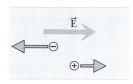

Figure 25.10 Intrinsic semiconductor conduction involves both holes and electrons.

In a metal, there are always lots of mobile charges, because the energy band is only partially filled, and changing the temperature hardly changes the number density of mobile charges.

In most metals there are essentially no mobile holes, and only the mobile electrons contribute to the current when an electric field is applied. Every atomic core is a positive ion, so there is no way for a neighboring atom to give up (another) electron and thereby create a mobile hole.

In contrast, in a pure semiconductor the (low) densities of mobile electrons and holes depend strongly on the temperature, and both electrons and holes contribute to the conductivity. The fraction of atomic cores which are positive ions is very small.

Other pair-creation mechanisms

For completeness, we note that thermal excitation is not the only mechanism for creating mobile electron-hole pairs in a semiconductor. Absorption of a photon can create a pair, and this is the basis for photovoltaic devices ("solar cells").

In another application, the passage of a high-energy charged particle through a semiconductor creates a number of electron-hole pairs that is proportional to the energy loss of the particle, so a semiconductor can be used as a detector of charged particles. Collecting the mobile charges gives a measure of the energy loss of the particle. If the semiconductor is large enough that the particle comes to a stop inside the material, the measured energy loss represents the original kinetic energy of the particle.

25.2.2 Recombination

Sometimes one of the mobile electrons in a semiconductor happens to come near a hole (that is, an atom that is missing an electron and therefore

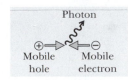

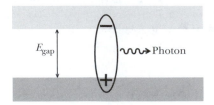

Figure 25.11 In an LED, the recombination of electrons and holes produces photons.

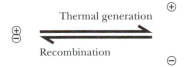

Figure 25.12 Recombination of electrons and holes competes with the thermal generation of electron-hole pairs. In the steady state the two rates are equal.

Figure 25.13 Silicon atoms bond to four neighboring silicon atoms in a tetrahedral arrangement.

has a charge of $+e$). The mobile electron is of course attracted to the atom, and can fill the hole. This reduces both the number of mobile electrons and the number of mobile holes. We call this process recombination.

This process is similar to the recombination of ions and electrons in a spark or plasma, where recombination creates a bound-state atom accompanied by the emission of light, whose energy corresponds to the binding energy of the atom. Because electron-hole recombination takes place inside a solid, the electron-hole binding energy need not be released as light but may be absorbed by neighboring atoms. However, it is possible to make a semiconductor device in which recombination results in the emission of light, with energy close to the band-gap energy (Figure 25.11). Such a device is called a light-emitting diode, or LED.

Recombination, which destroys electron-hole pairs, competes with pair creation by thermal excitation (Figure 25.12). In the steady state, the recombination rate becomes equal to the pair creation rate.

? What happens to the number densities if the temperature is increased?

If the temperature is increased, the pair creation rate goes up, and for a while there is a growth in the number densities of mobile electrons and mobile holes. However, with higher number densities there is a higher probability of an electron encountering a hole and recombining, so eventually a new steady state is established at the higher temperature, with higher but constant densities of mobile electrons and mobile holes.

25.2.3 Summary of types of material

- Metals have a fixed high density of mobile electrons, and high conductivity. There are no mobile holes.
- Insulators have hardly any mobile electrons or mobile holes, and very low conductivity.
- In a pure semiconductor the number density of mobile electrons and mobile holes depends strongly on the temperature. The intrinsic conductivity of pure semiconductors grows extremely rapidly with increasing temperature, but is only about 10^{-6} that of a metal at room temperature. Both mobile electrons and mobile holes contribute to the conductivity.

25.3 Doped semiconductors

In a metal the number density of mobile charges is fixed. We can't vary this density. But in a semiconductor there is a possibility of changing the number densities of mobile electrons or holes, which opens up new possibilities for designing devices with special capabilities. As we have seen, one way to vary the mobile charge densities is to change the temperature in a pure semiconductor. Another way is to deliberately introduce impurities that can contribute mobile charges. This is called "doping."

25.3.1 An n-type semiconductor

In pure silicon, each silicon atom has four outer electrons that form chemical bonds with outer electrons in four neighboring silicon atoms in a tetrahedral arrangement (Figure 25.13). Suppose we replace one silicon atom with a phosphorus atom (Figure 25.14). A phosphorus atom is about the same size as a silicon atom, but it has five outer electrons instead of four (see the periodic table on the inside front cover of this textbook). If we replace a silicon atom with a phosphorus atom, four of the outer electrons of the phosphorus atom participate in bonds with the four neighboring silicon atoms. The fifth electron is only very weakly bound to the phosphorus atom and is very easily promoted into the conduction band by thermal excitation, leaving behind a positively charged phosphorus ion(Figure 25.15). The phosphorus atom is called a "donor" impurity (because it "donates" a mobile electron).

? Does the presence of a phosphorus atom generate a mobile hole?

The positive charge of the phosphorus ion does not constitute a mobile hole. The four outer electrons of each of the neighboring silicon atoms are participating fully in chemical bonds and are not available to fill the hole in the phosphorus atom. This is unlike the case in pure silicon, where a hole in a silicon atom means there is a defective bond with a neighbor, and an electron from a neighboring atom can fill the hole to repair the bond, with the effect of moving the hole.

Doping pure silicon by replacing even a very small fraction of the atoms with phosphorus atoms makes a material in which there are many more mobile electrons than in pure silicon. Such a material is called an "n-type semiconductor" ("n" stands for "negative"). With a typical level of doping of one phosphorus atom in tens of thousands of silicon atoms, at room temperature the mobile electrons donated by the phosphorus atoms are far more numerous than the electrons resulting from thermal creation of electron-hole pairs.

Majority carriers in n-type semiconductors are electrons

Doping with phosphorus greatly reduces the number of mobile holes. Flooding the silicon with mobile electrons means that mobile holes much more frequently encounter mobile electrons and recombine. We say that in an n-type semiconductor the "majority" mobile charges are electrons, and there are few "minority" mobile charges (holes).

It is important to keep in mind that the n-type semiconductor is electrically neutral overall. There are just as many positively charged phosphorus ions as there are mobile electrons. This is similar to an ordinary metal, where the positive charge of the atomic cores balances the negative charge of the mobile electrons (typically one for each atom).

A dilute metal

You can think of an n-type semiconductor as a dilute metal. In both materials there are fixed positive ions and mobile electrons. The mobile electrons have little net force acting on them, and they move around quite freely in the material, which is neutral when averaged over volumes containing many positive atomic cores. The main difference is that in an ordinary metal there are one or more mobile electrons per atom, whereas in an n-type semiconductor only a small fraction of atoms give up an electron. The donor atoms are rather far apart compared to the situation in a true metal where every atom can be an electron donor.

25.3.2 A p-type semiconductor

It may seem that an n-type semiconductor isn't very interesting or useful, since it is basically just like a metal but with a much smaller density of mobile

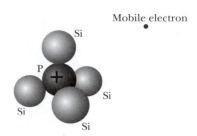

Figure 25.14 A silicon atom is replaced by a phosphorus atom, which contributes a mobile electron.

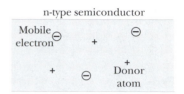

Figure 25.15 An n-type semiconductor, doped with a small number of fixed donor atoms which donate mobile electrons.

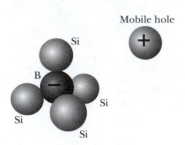

Figure 25.16 A silicon atom is replaced by a boron atom, which grabs an extra electron and creates a mobile hole.

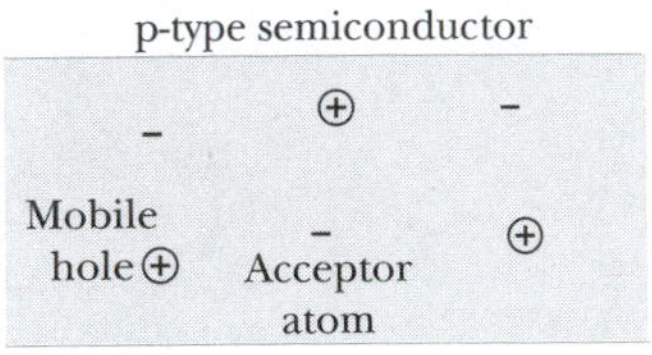

Figure 25.17 A p-type semiconductor, doped with a small number of fixed acceptor atoms which accept electrons, thereby creating mobile holes.

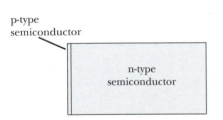

Figure 25.18 An n-type semiconductor with a p-type region introduced by ion implantation.

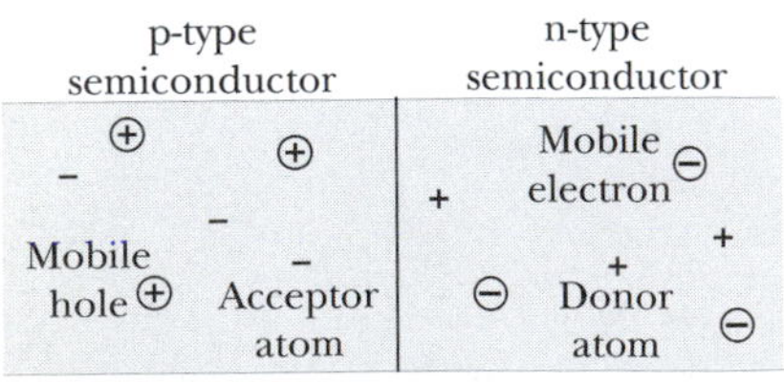

Figure 25.19 Close-up of a p-n junction. The majority mobile charges are holes in the p-type region, and electrons in the n-type region.

electrons. After we discuss a somewhat more unusual material, a p-type semiconductor in which the majority mobile charges are holes rather than electrons, we will be able to discuss a "p-n junction," consisting of p-type and n-type semiconductors right next to each other. P-n junctions have interesting properties and are a basic building block of transistors and integrated circuits.

The analysis of a p-type semiconductor is quite parallel to the analysis of an n-type semiconductor, and the next few paragraphs should sound familiar! A boron atom is about the same size as a silicon atom, but it has only three outer electrons instead of four (see the periodic table on the inside front cover of this textbook). If we replace a silicon atom with a boron atom, the boron atom grabs an additional electron in order to participate in bonds with the four neighboring silicon atoms (Figure 25.16). This leaves a hole in a neighboring silicon atom, and this hole is mobile because an electron from another silicon atom can fill this hole.

The effect is to create a mobile hole and to leave behind a negatively charged boron ion (Figure 25.17). The boron atom is called an "acceptor" impurity (because it "accepts" an electron, thereby contributing a mobile hole). The negative charge of the boron ion obviously does not constitute a mobile electron.

Doping pure silicon by replacing even a very small fraction of the atoms with boron atoms makes a material in which there are many more mobile holes than in pure silicon. Such a material is called a "p-type semiconductor" ("p" stands for "positive"). With a typical level of doping of one boron atom in tens of thousands of silicon atoms, at room temperature the mobile holes contributed by the boron atoms are far more numerous than the holes resulting from thermal creation of electron-hole pairs.

Majority carriers in p-type semiconductors are holes

Doping with boron greatly reduces the number of mobile electrons. Flooding the silicon with mobile holes means that mobile electrons much more frequently encounter mobile holes and recombine. We say that in a p-type semiconductor the "majority" mobile charges are holes, and there are few "minority" mobile charges (electrons).

It is important to keep in mind that the p-type semiconductor is electrically neutral overall. There are just as many negatively charged boron ions as there are mobile holes.

How doping is done

To prepare n-type or p-type silicon, a crystal can be grown from molten silicon that contains precise amounts of donor or acceptor atoms. Another way to dope silicon is to accelerate donor or acceptor ions by electric fields and shoot them into the silicon at high speed. This technique is called "ion implantation" and is used to produce shallow regions of doped material.

Figure 25.18 shows n-type material into which a p-type region has been introduced by ion implantation. This is called a "p-n junction," which we will discuss next.

25.4 A p-n junction

A rectifying diode is an extremely useful device that allows current to run through it in one direction but not the other. Figure 25.19 shows a close-up of a p-n junction, which acts like a diode. We will show that conventional current can run easily from the p-type section of the diode toward the n-type section, but very little current can flow in the opposite direction. The following diagrams represent the essential elements of a p-n junction, but the actual geometry may look rather different in real devices.

Keep in mind that in the p-type material there are lots of mobile holes and few mobile electrons. In the n-type material there are lots of mobile electrons and few mobile holes.

Consider what happens when a mobile electron happens to drift across the junction into the p-type material (Figure 25.20). It has now entered a region that is full of mobile holes, so it may recombine soon after crossing the junction. This removes a hole from circulation in the p-type material, leaving some nearby negatively charged acceptor atom unpaired with a mobile hole. It also leaves an unpaired donor atom in the n-type region.

Similarly, a mobile hole that happens to drift into the n-type material finds itself in a region that is full of mobile electrons, so it may recombine soon after crossing the junction (Figure 25.21). This removes an electron from circulation in the n-type material, leaving some nearby positively charged donor atom unpaired with a mobile electron. It also leaves an unpaired acceptor atom in the p-type region.

The result is that a thin region near the junction becomes depleted of mobile electrons and holes. This is called the "depletion layer" (Figure 25.22). The p-type part of the depletion layer has a negative charge due to the presence of the unpaired acceptor atoms (and lack of compensating mobile holes). The n-type part of the depletion layer has a positive charge due to the presence of the unpaired donor atoms (and lack of compensating mobile electrons). As we will see later, the depletion layer is very thin, typically only a few hundred atoms in width.

This gives rise to a strong electric field in the depletion layer as shown in Figure 25.23. Outside the depletion layer, the charged regions of the depletion layer produce a tiny "fringe field" rather like that outside the plates of a capacitor. This fringe field is very small because, as we will see in section 25.7, the width of the depletion layer is only a few hundred atoms, and the fringe field of a capacitor is proportional to the spacing between plates.

The tiny fringe field drives holes to the right in the p-type material and drives electrons to the left in the n-type material. This polarizes the diode, leading to a tiny amount of charge on distant outer surfaces as shown in Figure 25.24. These charges are bare, positive donor atoms on the surface of the n-type material, and bare, negative acceptor atoms on the surface of the p-type material.

? In equilibrium, why must the net electric field in the material outside the depletion layer be zero?

Outside the depletion layer there are abundant mobile electrons (in the n-type material) or mobile holes (in the p-type material). As long as there is a nonzero electric field, these mobile charges will move, and they move in a direction so as to polarize the material in such a way as to lead to a zero net field. This is another example of the fact that inside a conductor in equilibrium the net electric field must become zero through polarization.

? How can there be a nonzero electric field in the depletion layer?

The situation in the depletion layer is different. There are almost no mobile charges in this region, and in equilibrium there can be a strong electric field, which we will calculate in the next section.

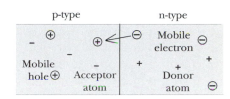

Figure 25.20 If an electron drifts into the p-type region, it almost immediately recombines with a hole.

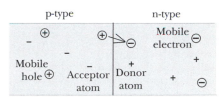

Figure 25.21 If a hole drifts into the n-type region, it almost immediately recombines with an electron.

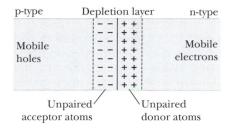

Figure 25.22 In the depletion layer there are unpaired acceptor and donor atoms, and few mobile holes or electrons.

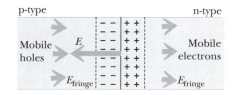

Figure 25.23 Pattern of electric field produced by the depletion layer.

Figure 25.24 The diode polarizes, making $E = 0$ in the p-region and n-region.

The tiny amount of charge on the outer surfaces plays no significant role in the behavior of the p-n junction, so we will omit showing these charges in later diagrams.

25.5 The electric field in the depletion layer

We can use Gauss's law to calculate the nonzero electric field in the depletion layer of the p-n junction. We draw an (imaginary, mathematical) box with one end in the p-type material and the other in the depletion layer (Figure 25.25).

? On which portions of the Gaussian box is there any nonzero electric flux? How much?

The electric field on the left end of the box is zero, because it is in the region where we know the electric field is zero in equilibrium. The electric field on the top, bottom, and sides is either zero (because it is in a zero-field region) or is horizontal (due to the symmetry of the situation), so the flux is zero on all of these sides. Only on the right end of the Gaussian box is there any flux, and this flux is $-EA$, where E is the (unknown) electric field, A is the area of the right end of the box, and the flux is negative because the electric field points into the box.

Choose a coordinate system with the origin at the left end of the depletion layer, whose width is d (Figure 25.25). We'll consider the simple case that the doping density is the same in the p-type and n-type materials, so that the doping charge density is $-\rho$ (in coulombs per m^3) in the left portion and $+\rho$ in the right portion

? Use Gauss's law to calculate the electric field as a function of x, the location of the right end of the Gaussian box (measured from the left end of the depletion layer). It may be easiest to consider the left half of the depletion layer separately from the right half.

When $x < d/2$ (x is in the left portion of the depletion layer), we have the following, because the volume containing charge is Ax:

$$-EA = \frac{Q_{\text{inside}}}{\varepsilon_0} = \frac{-\rho(Ax)}{\varepsilon_0}$$

$$E = \frac{\rho x}{\varepsilon_0}$$

When $x > d/2$ (x is in the right portion of the depletion layer), we have the following:

$$-EA = \frac{Q_{\text{inside}}}{\varepsilon_0} = \frac{-\rho(Ad/2) + \rho[A(x - d/2)]}{\varepsilon_0}$$

$$E = \frac{\rho(d - x)}{\varepsilon_0}$$

Dielectric constant

There is an additional factor that we have omitted from the analysis. Inside this dense material, the electric field contributed by a single bare charge (a donor or acceptor atom) is reduced by a factor K, the dielectric constant (see the end of Chapter 16 for a related discussion). This effect is related to the polarization of the material due to these embedded bare charges. As a

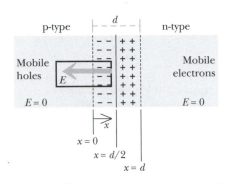

Figure 25.25 A Gaussian box with its right end (with area A) in the depletion layer. Measure x from the left edge of the depletion layer.

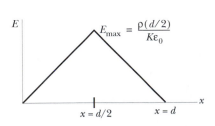

Figure 25.26 The magnitude of the electric field as a function of x, measured from the left edge of the depletion layer.

result, the effective electric field inside the material is $1/K$ as large as the field we have calculated. In silicon this is a sizable effect; K is approximately 12. Figure 25.26 on the previous page is a graph of E vs. x through the depletion layer.

25.6 The potential difference across the depletion layer

Since the electric potential difference is given by $\Delta V = -\int_i^f \vec{E} \cdot d\vec{l}$, the potential rises in going left-to-right through the depletion layer. Figure 25.27 shows the potential as a function of x obtained by integrating the triangle graph of electric field E in Figure 25.26. The result is two parabolas. The total potential difference across the depletion layer is the area of the triangle in Figure 25.26. Taking account of the dielectric constant K, we have this:

$$\Delta V = \frac{d}{2}E_{max} = \frac{\rho d^2}{4K\varepsilon_0}$$

Could we use this locked-in potential difference to drive a perpetual current in a circuit? Not surprisingly, the answer is no. For example, suppose you extend the p-type material into a long wire that comes around and attaches to the end of the n-type material, to form a complete circuit. Now in addition to the p-n junction there is an n-p junction, with a locked-in potential difference in the opposite direction. The round-trip path integral of the electric field is zero, and there is no emf to drive current.

If you connect a metal wire across the p-n junction, a similar situation arises. Although we have not discussed the issue, perhaps you can see that at the junction between a metal and a semiconductor there can be a depletion layer with an associated potential difference.

The band-gap energy

If ΔV is the potential difference across the depletion layer, we can show that $e\Delta V$ is equal to the band-gap energy E_{gap}. To create a mobile electron in a semiconductor one must supply the band-gap energy to promote the electron into an energy level corresponding to a mobile electron.

In Figure 25.28 we plot the electric potential energy $U_{el} = -eV$ of an electron in a silicon diode. The sign of the curve is opposite to that in Figure 25.27, because we are plotting the potential energy U_{el} for a negative electron instead of the potential V. Figure 25.28 refers to a time shortly after the depletion layer begins to form. It takes 1.1 eV of thermal excitation to create a mobile electron in the p-region of the silicon. Once mobile, this electron can "fall" through the potential energy difference into the n-region.

In contrast, at this time an already mobile electron in the n-region in Figure 25.29 needs to obtain very little kinetic energy K from thermal excitation to be able to cross into the p-region, where it can drop 1.1 eV to bind with a silicon atom that is missing an electron (a hole). The smaller energy requirement means that initially it is much more likely for electrons to move from the n-region to the p-region than from the p-region to the n-region.

The net flow of electrons from the n-region to the p-region increases the potential energy difference, as shown in Figure 25.30. The increased energy barrier causes a decrease in the rate of electrons going from the n-region to the p-region, making the net flow smaller. The depletion layer will grow until the potential energy difference is equal to 1.1 eV, at which point there is a dynamic equilibrium and no more growth of the depletion layer.

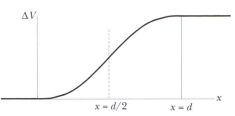

Figure 25.27 A graph of the increasing potential starting from the left edge of the depletion layer.

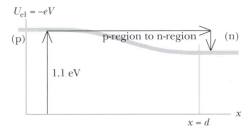

Figure 25.28 A freed-up electron in the p-region falls into the n-region.

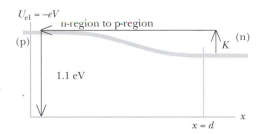

Figure 25.29 An already free electron in the n-region gains enough kinetic energy K to move into the p-region.

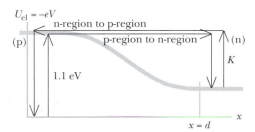

Figure 25.30 Transfer of electrons from the n-region to the p-region increases the potential energy barrier.

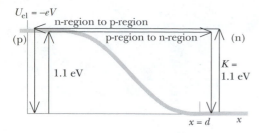

Figure 25.31 A dynamic equilibrium is achieved when the kinetic energy K required to cross the barrier is equal to the band-gap energy.

In Figure 25.31 we plot the electric potential energy in the dynamic equilibrium situation. It takes 1.1 eV of thermal excitation to create a mobile electron in the p-region. Once free, this electron can "fall" through the potential energy difference into the n-region. Similarly, an already-mobile electron in the n-region must obtain kinetic energy K of 1.1 eV of thermal excitation to be able to cross into the p-region, where it can drop 1.1 eV to bind with a silicon atom that is missing an electron (a hole). There is no net advantage to moving in one direction or the other. In dynamic equilibrium the two processes have equal rates.

In terms of the Boltzmann factor, the rate for creating a mobile electron is proportional to $e^{-E_{\text{gap}}/(kT)}$, where ΔE is 1.1 eV for silicon. The rate for electrons in the n-region to obtain kinetic energy K from thermal excitation is proportional to $e^{-K/(kT)}$. When $K = \Delta E$ the two rates are equal.

25.7 Numerical values

We can use these results to show that the depletion layer has a width of just a few hundred atoms. We can also calculate the magnitude of the electric field in the depletion layer.

Using the measured value for the band-gap energy we can write the following for silicon:

$$E_{\text{gap}} = e\Delta V = \frac{e\rho\, d^2}{4K\varepsilon_0} = (1.1 \text{ eV})\left(\frac{1.6\times10^{-19}\text{J}}{\text{eV}}\right)$$

Since $e = 1.6\times10^{-19}$ C we have

$$\Delta V = \frac{\rho\, d^2}{4K\varepsilon_0} = 1.1 \text{ J/C} = 1.1 \text{ volts}$$

We can use this result to determine the width d of the depletion layer if we calculate the charge density ρ. A typical doping level in silicon is 10^{18} atoms per cm^3. The density of silicon is 2.4 grams per cm^3, and there are 28 grams per mole, so

$$\left(\frac{6\times10^{23}\text{ atoms}}{28 \text{ grams}}\right)\left(\frac{2.4 \text{ grams}}{\text{cm}^3}\right) = 5\times10^{22} \text{ silicon atoms/cm}^3$$

In other words, a typical doping is about one impurity atom per 50000 silicon atoms. The charge density of the donor or acceptor atoms is

$$\rho = (1.6\times10^{-19}\text{ C})\left(\frac{10^{18}\text{ atoms}}{10^{-6}\text{ m}^3}\right) = 1.6\times10^5 \text{ C/m}^3$$

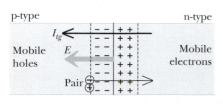

Figure 25.32 Electrons thermally generated on the left are swept through the depletion layer to the right, contributing to the thermal-generation conventional current I_{tg} to the left. $K_i \approx 0$ is sufficient.

$$d = \sqrt{\frac{4K\varepsilon_0\Delta V}{\rho}} = \sqrt{\frac{(4)(12)(9\times10^{-12}\text{ C}^2/\text{N·m}^2)(1.1 \text{ J/C})}{(1.6\times10^5 \text{ C/m}^3)}} = 550\times10^{-10} \text{ m}$$

The depletion layer is very narrow, roughly 500 atoms wide. Now that we have a value for d, we can find the maximum electric field:

$$\Delta V = \frac{d}{2}E_{\text{max}}, \text{ so } E_{\text{max}} = 2\frac{\Delta V}{d} = \frac{2(1.1 \text{ V})}{550\times10^{-10} \text{ m}} = 4\times10^7 \text{ V/m}$$

This is a very large electric field. For example, it is over 10 times the field that causes air to ionize (about 3×10^6 V/m). On the other hand, it is very small compared to the field needed to ionize an isolated atom, which is about 10^{11} V/m.

25.8 Diode action

Now we can explain how a sizable conventional current can run from the p-region toward the n-region, but only a very small current can run through the junction in the other direction. A device with this behavior is called a diode.

In equilibrium, there is no current running from the p-region to the n-region, or vice versa. At the atomic level this is a dynamic equilibrium, not a static one. Occasionally thermal energy creates an electron-hole pair in the p-region (Figure 25.32 on the previous page). If the electron survives long enough to wander into the depletion layer (or is created in the depletion layer), it is quickly swept across the junction by the electric field, and it joins the many mobile electrons in the n-region. This contributes to a "thermal-generation" conventional current I_{tg} to the left.

Occasionally one of the abundant mobile electrons in the n-region happens through random thermal collisions to have enough kinetic energy to move through the depletion layer, against the electric force $(-e)\vec{E}$ (Figure 25.33). When the electron reaches the p-region it recombines with one of the abundant mobile holes found there. This contributes to a "recombination" conventional current I_r to the right.

? Can you think through these two processes with respect to holes?

Similarly, a hole that is thermally generated in the n-region may survive long enough to wander into the depletion layer (or it may be created in the depletion layer), where it is quickly swept across the junction by the electric field, and it joins the many mobile holes in the p-region (Figure 25.34). This too contributes to the thermal-generation conventional current I_{tg} to the left.

Occasionally one of the abundant mobile holes in the p-region happens through random thermal collisions to have enough kinetic energy to move through the depletion layer, against the electric force $(+e)\vec{E}$ (Figure 25.35). When the hole reaches the n-region it recombines with one of the abundant mobile electrons found there. This contributes to the recombination conventional current I_r to the right.

In (dynamic) equilibrium the two processes are balanced: $I_{tg} = I_r$, and the net conventional current to the right $I = I_r - I_{tg}$ is zero. We can create an imbalance in the two types of current and generate a nonzero net current by altering the potential difference across the diode with a battery.

"Reverse" voltage and current

Suppose we connect a battery to the diode as shown in Figure 25.36. This arrangement applies what is called a "reverse" voltage to the diode. We will see that this leads to a nonzero but small current through the diode.

With the "+" end of the battery attached to the n-region, the potential difference ΔV across the diode is increased. The depletion layer becomes somewhat thicker. One way to visualize what happens is to see that the battery will pull some electrons out of the n-region, leaving a thicker positive portion of the depletion layer. Similarly, the battery pulls holes out of the p-region, leaving a thicker negative region of the depletion layer. A thicker depletion layer produces a larger average electric field in the layer, which creates a larger potential difference across the layer.

? Even if a net current runs through the diode, most of this potential difference will be across the depletion layer. Do you see why?

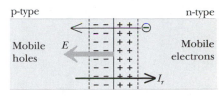

Figure 25.33 Occasionally an electron has enough energy to move from right to left against the electric force, contributing to the recombination conventional current I_r to the right. Need $K_i \approx E_{gap}$.

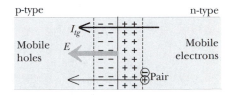

Figure 25.34 Holes thermally generated on the right are swept through the depletion layer to the left, contributing to the thermal-generation conventional current I_{tg} to the left. $K_i \approx 0$ is sufficient.

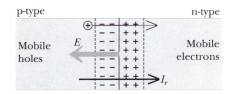

Figure 25.35 Occasionally a hole has enough energy to move from left to right against the electric force, contributing to the recombination conventional current I_r to the right. Need $K_i \approx E_{gap}$.

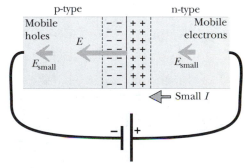

Figure 25.36 Applying a "reverse" voltage to the diode produces a small conventional current to the left.

Remember that the bulk p- and n-regions have lots of mobile charges (holes in the p-region, electrons in the n-region), whereas there are few mobile charges in the depletion layer. In the steady state the electric field in the bulk material will be very small compared to the electric field in the depletion layer, since $I = neAuE$ must be the same throughout the diode. Large number density n means small electric field E. Therefore most of the potential difference appears across the depletion layer.

Now that the potential difference is larger across the depletion layer, it is significantly less likely that an electron in the n-region or a hole in the p-region will happen to have enough kinetic energy to move through the depletion layer against the potential difference.

? As a result of the increased potential difference, what will happen to I_{tg}? To I_r?

The thermal-generation current I_{tg} won't change. It is made up of electrons and holes that are swept quickly through the depletion layer by the electric field anyway, and they will continue to be swept through quickly. A small initial kinetic energy $K_i \approx 0$ is sufficient.

The recombination current I_r however is made up of electrons and holes that had to travel against the electric force, and now it is even harder to go against the larger forces corresponding to the larger potential difference. A large initial kinetic energy $K_i \approx E_{gap}$ is needed. Therefore the recombination current will decrease rapidly with small increases in potential difference across the depletion layer and may be very small compared to the thermal-generation current. (In terms of the Boltzmann factor, the recombination current decreases by an amount $e^{-e|\Delta V_{batt}|/(kT)}$, where ΔV_{batt} is the additional potential difference due to the battery.)

? What is the direction of the net current, which is approximately I_{tg}, since I_r is now very small?

The thermal-generation conventional current flows in the direction of the electric field that is in the depletion layer, so it flows to the left, from the n-type region to the p-type region, in the expected direction for the way the battery is attached.

It is important to understand that this nonzero current is small in an absolute sense because the thermal-generation current is essentially what you would obtain in pure silicon. The mobile electrons and holes contributed by the donor and acceptor atoms play little role. Incidentally, the magnitude of this small current is sensitive to light hitting the junction, since the absorption of light is another way to generate electron-hole pairs. This is the basis for photocells.

After increasing the potential difference enough to make the recombination current negligible, further increases in the potential difference lead to essentially no change in the small current.

"Forward" voltage and current

Next we connect a battery to the diode as shown in Figure 25.37. This arrangement applies what is called a "forward" voltage to the diode. We will see that this battery connection leads to a large current through the diode.

With the "+" end of the battery attached to the p-region, the potential difference ΔV across the diode is decreased. Most of this decreased potential difference appears across the depletion layer, since $neAuE$ must be the same throughout the diode. One way to visualize what happens is to see that

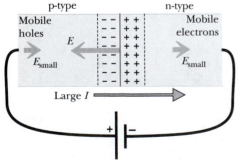

Figure 25.37 Applying a "forward" voltage to the diode produces a large conventional current to the right.

the battery will push some extra electrons into the n-region, leading to a thinner positive portion of the depletion layer. Similarly, the battery pushes holes into the p-region, leading to a thinner negative region of the depletion layer. A thinner depletion layer produces a smaller electric field in the layer, and a smaller potential difference across the layer.

Now that the potential difference is smaller across the depletion layer, it is significantly more likely that an electron in the n-region or a hole in the p-region will happen to have enough kinetic energy to move through the depletion layer against this (smaller) potential difference.

? As a result of the decreased potential difference, what will happen to I_{tg}? To I_r?

The thermal-generation current I_{tg} won't change. It is made up of electrons and holes that are swept quickly through the depletion layer by the electric field anyway, and they will continue to be swept through quickly. A small initial kinetic energy $K_i \approx 0$ is sufficient.

The recombination current I_r however is made up of electrons and holes that had to travel against the electric force, and now it is easier to go against the smaller forces corresponding to the smaller potential difference. A large initial kinetic energy $K_i \approx E_{gap}$ is needed. Therefore the recombination current will increase rapidly with small decreases in potential difference across the depletion layer and may be very large compared to the thermal-generation current.

? What is the direction of the net current, which is approximately I_r, since it is now much larger than I_{tg}?

The recombination conventional current flows in the direction opposite to the electric field that is in the depletion layer, so it flows to the right, from the p-type region to the n-type region, in the expected direction for the way the battery is attached.

It is important to understand that this current is comparatively large because there are so many more mobile electrons and holes than in pure silicon. This large current increases rapidly with decreasing potential difference across the p-n junction (corresponding to increasing emf of the attached battery or power supply). In fact, the current increases exponentially rapidly. (In terms of the Boltzmann factor, the recombination current increases by an amount $e^{+e|\Delta V_{batt}|/(kT)}$, where ΔV_{batt} is the potential difference due to the battery.)

Figure 25.38 shows the net current through the diode for reverse and forward voltages. You see the small, nearly constant current for reverse voltages and the rapidly rising current for forward voltages. (Note that you have to switch the direction of the battery to make measurements corresponding to the left and right sides of this graph.)

A comment is in order about the detailed mechanism for the current in the forward direction in Figure 25.37 on the previous page. On the right, electrons flow into the diode from the battery and stream to the left across the p-n junction thanks to the lowered potential difference there. When they reach the bulk p-region, they recombine with the abundant holes. Electrons in the n-region are continually replenished by electrons coming from the battery.

Holes in the p-region stream to the right across the p-n junction, and when they reach the bulk n-region they recombine with the abundant electrons. It is difficult without a full quantum-mechanical analysis to explain in

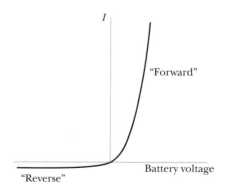

Figure 25.38 Current through a diode as a function of applied battery voltage. The battery connections are switched between the left and right sides of this graph.

a satisfying way what replenishes the supply of holes. Briefly, at the junction of the metal wire with the p-region there is another depletion region, and with the electric field driving holes to the right, mobile electrons are created in the p-region near the junction and move into the metal wire, thereby replenishing the supply of holes near the junction.

25.9 The bipolar transistor

One important kind of transistor, the bipolar transistor, consists of two back-to-back p-n junctions as shown in Figure 25.39. Both pnp and npn configurations are used. These devices have the property that a small current running into or out of the side of the thin middle region (called the base) has a big effect on how much current runs through the transistor between "emitter" and "collector." This makes possible current amplification, where a large collector current is proportional to a small base current. In another use the transistor can act as a kind of gate: depending on the base current, the collector current is on (large) or off (very small).

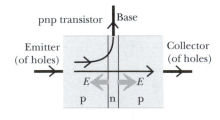

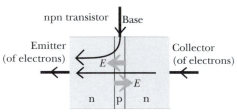

Figure 25.39 Bipolar transistors, pnp and npn. Direction of conventional current is shown. Also shown are the large electric fields in the depletion layers.

We will very briefly describe the basic operation of a pnp transistor. Voltages are applied to the pnp transistor at the top of Figure 25.39 so that hole current runs from the emitter in the "forward" or large-current direction. Small increases in emitter-base voltage lead to large increases in emitter current. The n-type region is made very thin, so that many holes manage to cross the thin n-type region. The large electric field in the depletion layer between the n-type region and the collector points to the right, so it sweeps to the right any holes that reach that point. The base current is quite small compared to the collector current.

With this arrangement, circuit elements attached to the collector are pretty much decoupled from the emitter-base part of the circuit. The large collector current can run through a resistor connected between collector and base and produce a large voltage. The net effect is that a small emitter-base voltage change can make a big change in the collector-base voltage, and the transistor acts as an amplifier. The analysis of an npn transistor, shown at the bottom of Figure 25.39, is similar but in terms of electrons rather than holes.

This description of bipolar transistors is far from complete, but we hope it will give you a basic idea of the functioning of a transistor. A good understanding of p-n junctions provides a useful foundation for further study.

25.10 An alternative calculation method

We mention another calculational technique that is often used in semiconductor physics. Consider again our calculation using Gauss's law for the electric field in the depletion layer of a p-n junction, but instead of a long Gaussian box let the box have a very short length dx (Figure 25.40). The electric field at the left edge of the box is $E_x(x)$, and the electric field at the right edge is $E_x(x + dx)$. The area of the end of the box is again A as before, so the volume is Adx. Applying Gauss's law to this box in a region where the charge density is ρ gives this (the dielectric constant is K):

$$E_x(x + dx)A - E_x(x)A = \frac{Q}{K\varepsilon_0} = \frac{\rho A dx}{K\varepsilon_0}$$

$$\frac{E_x(x + dx) - E_x(x)}{dx} = \frac{\rho}{K\varepsilon_0}$$

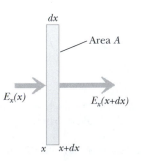

Figure 25.40 Applying Gauss's law to a box whose length is only *dx*.

But $[E_x(x + dx) - E_x(x)]/dx$ in the limit of small dx is dE_x/dx, so

$$\frac{dE_x}{dx} = \frac{\rho}{K\varepsilon_0}$$

We could have started with this equation ("Poisson's equation") instead of Gauss's law to determine the magnitude of the electric field in the depletion layer. For example, in the left half of the depletion region where the charge density is $-\rho$, we have

$$\frac{dE_x}{dx} = \frac{-\rho}{K\varepsilon_0}$$

Solving for E_x, we have the following, where the minus sign expresses the fact that the electric field points to the left in the depletion layer.

$$E_x = \frac{-\rho}{K\varepsilon_0}x$$

This is the same result we obtained earlier by a different use of Gauss's law.

Problem 25.1 Germanium diode
The band-gap energy in germanium has been measured to be about 0.7 eV. A typical doping level in germanium is 10^{18} atoms per cm^3. The density of germanium is 5.4 grams per cm^3, and there are 73 grams per mole. The dielectric constant is 16. Calculate the width of the depletion layer in a germanium p-n junction. Also calculate the maximum electric field in the junction.

Greek alphabet, and its uses in this volume

Alpha	A		α	**alpha particle (helium-4 nucleus)**
Beta	B		β	
Gamma	Γ		γ	**photon**
Delta	Δ	**change of; small quantity of**	δ	
Epsilon	E		ε	
Zeta	Z		ζ	
Eta	H		η	
Theta	Θ		θ	**angle**
Iota	I		ι	
Kappa	K		κ	
Lambda	Λ		λ	**wavelength**
Mu	M	**mega (10^6)**	μ	**micro (10^{-6}); muon**
Nu	N		ν	**neutrino**
Xi	Ξ		ξ	
Omicron	O		o	
Pi	Π		π	**circle: circumference/diameter**
Rho	P		ρ	**density, resistivity**
Sigma	Σ	**sum**	σ	**conductivity**
Tau	T		τ	
Upsilon	Y		υ	
Phi	Φ	**flux**	ϕ	**phase angle**
Chi	X		χ	
Psi	Ψ		ψ	
Omega	Ω	**ohm**	ω	**angular frequency**